Benchmark Papers
in Geology

Series Editor: Rhodes W. Fairbridge
Columbia University

Benchmark Papers in Geology/54

A BENCHMARK® Books Series

MAGNETIC STRATIGRAPHY OF SEDIMENTS

Edited by

JAMES P. KENNETT

University of Rhode Island

A Memorial to Norman D. Watkins

Dowden, Hutchinson & Ross, Inc.

STROUDSBURG, PENNSYLVANIA

211694

LIBRARY OF CONGRESS CATALOGING IN PUBLICATION DATA
Main entry under title:
Magnetic stratigraphy of sediments.
   (Benchmark papers in geology; 54)
   "Bibliography of Norman D. Watkins": p.
   Includes indexes.
   1. Geology, Stratigraphic—Addresses, essays, lectures.   2. Magnetism,
Terrestrial—Addresses, essays, lectures.   3. Paleomagnetism—Addresses,
essays, lectures.   4. Watkins, Norman D., 1934–1977.   I. Kennett,
James P.   II. Watkins, Norman D., 1934–1977.
QE651.M32   551.7'01   79-13662
ISBN 0-87933-354-5

Distributed world wide by Academic Press,
a subsidiary of Harcourt Brace Jovanovich,
Publishers.

MEMORIAL FOR
NORMAN D. WATKINS (1934–1977)

Norman D. Watkins, to whom this volume is dedicated, was recognized as one of the lively spirits in national and international earth sciences during the 1970s. Who was Norman Watkins and what type of person was he? What did he accomplish during his brief 43 years on this planet?

Norman was a highly successful scholar, a strong and complex individual, a dedicated and prolific researcher, and a fine and respected teacher. During his comparatively brief lifetime, through his zeal, scientific idealism, and dedication in all activities in which he participated, he created a major impact far beyond that of the average person.

He strove for excellence in every thing he became involved with: his research, teaching, administration, and care of his family and the education of his sons. He demanded a great deal from himself, perhaps too much at times; and he always aimed high. This was true even in athletics in which, as a young man, he won the bronze medal for the javelin throw in 1958 at the British Empire games in Cardiff, Wales. Foremost to me, however, was his example of shear energy and zest with which he conducted his activities from day to day. He clearly recognized the subtle balance between a competitive and a quiet atmosphere necessary in the maintenance of continuously high creativity at a university.

Norman was a good example of the dedicated researcher also being the successful teacher from freshman through graduate levels. His philosophy of teaching included much about the nature of the scientific method. He attempted to teach his students to think, to synthesize, to differentiate the important from the unimportant, and to examine data from different perspectives. This stimulated his students greatly; so much so that he received a teaching excellence award in 1968 for freshman teaching at Florida State University. He produced about a dozen Master's students and about half that many Ph.D. students during his teaching career.

Norman's high visibility in earth science and oceanography circles around the world resulted largely from the diversity of the problems he tackled, which were related to almost all oceans and to several continents and which included rocks of wide-ranging type and age. The earth was his laboratory. Norman was interested in large problems, and it was easy

Norman D. Watkins, 1934–1977; Professor of Oceanography, University of Rhode Island 1970–1977

to observe that as his experience and scientific maturity developed over the last few years, he was becoming quite skilled at integrating large amounts of data for some big picture. However, this phase of his research career was snipped at the bud.

His personality traits were particularly well known and often discussed. He was flamboyant and talkative, and he held strong views on almost everything, which even if wrong were generally constructive. He was also incredibly blunt and outspoken and, as far as I could establish, would rarely compromise on this even though at times this bluntness led him into rough waters and cost him the friendship of certain colleagues. He was particularly well known at meetings around the world for his commanding presence and for his outspoken criticism of inaccurate approaches and data selectivity. He was often purposely controversial in order to simulate the thinking of those around him. He applied this technique both in his own laboratory and at national and international scientific meetings.

The late 1960s and 1970s have been an incredibly exciting period for earth scientists, with the formulation of plate tectonics and the genuine revolution that has resulted in an understanding of the processes that have molded the ocean basins and the oceanic ridges and that have produced the terrestrial landscapes including the mountain chains and volcanoes. A major component that led to the rapid acceptance of these theories by geologists and geophysicists has been encompassed by paleomagnetism. Norman's research centered on paleomagnetism, and he contributed much to its understanding. At the Graduate School of Oceanography, he created one of the world's best paleomagnetic laboratories. Norman possessed a keen intuition about the most important developing scientific areas and was quick to accept plate tectonics. He later contributed much to the understanding of earth processes within the framework of this theory. Using paleomagnetism as a central theme, he was to branch out in many directions. For instance, he did important work related to the development of oceanic islands; the processes involved in the accumulation of sediments that mantle the deep ocean floor. He was very intrigued with possible relations between magnetism and organisms. More recently he became increasingly interested in the problems related to explosive volcanism, the history of the Antarctic ice cap, and the distribution of manganese nodules in the deep sea—those vast valuable deposits that have become central to the thorny problems related to the formulation of international sea law. One of his last published research contributions considered the nature of the explosive volcanic event in the Mediterranean 1470 B.C., which is believed to have led to the destruction of the Minoan Civilization on Crete. Norman's field work took him to many other unusual corners of the world, especially remote oceanic islands such as Crozet, Amsterdam Is., Kerguelen, Cape Verde, and Madeira.

His professional legacy remains mostly in the form of an unusually large number of published scientific contributions. He authored and coauthored nearly 140 papers. Over a 17-year period of scientific productivity, this represents an output of one paper every 1½ months! These papers were generally of high standard, nearly always useful, and often the result of large amounts of data gained at much effort on his part. In addition to these papers, he published nearly 200 abstracts associated with scientific meetings. He was proud of his scientific output and always attempted to inspire others to emmulate this.

Norman was born an Englishman at Sunbury-on-Thames on 15 February 1934 and in 1968 became a naturalized American citizen. He was passionately interested in politics at all levels; was a devout believer in the democratic process and became a U.S. citizen mostly so that he could actively participate in this process. His political interests transcended to those related to the health of science in North America and to relationships between the politician and the scientist; indeed, it was this interest, I believe, that largely led him to accept a temporary position in 1976 as director of earth science at the U.S. National Science Foundation. While he was in Washington, he was stricken with cancer and cut short, but he was recognized as having started to make an important impact on the future funding of earth and ocean science in North America.

His early geological education was at the University of London and at the University of Birmingham. He received his Ph.D. from the University of London in 1964. More recently he was honored by a Doctor of Science degree from the University of Birmingham in recognition of his research contributions. Before coming to the University of Rhode Island, he had held several postions including geologist with Shell Oil Company of Canada; visiting senior fellow at Chadwick Physics Laboratory at the University of Liverpool, England; research associate and instructor at Stanford University with Allan Cox as his advisor; and as associate professor at Florida State University. Norman was a member of the American Association for the Advancement of Science, American Geophysical Union, Society of Exploration Geophysicists, European Association of Exploration Geophysicists, Seismological Society of America, Royal Astronomical Society, the Geochemical Society, and the Cosmos Club.

The hiatus formed by Norman's passing is nevertheless represented by a solid pavement of knowledge that he created over a brief span of time, rich in information for use in our future studies of earth history.

A colleague summed up his life recently:

In the oceans and on the land we all come and go,
Lucky the few who leave their tracks,
His are wide and solid.

SERIES EDITOR'S FOREWORD

The philosophy behind the Benchmark Papers in Geology is one of collection, sifting, and rediffusion. Scientific literature today is so vast, so dispersed, and, in the case of old papers, so inaccessible for readers not in the immediate neighborhood of major libraries that much valuable information has been ignored by default. It has become just so difficult, or so time consuming, to search out the key papers in any basic area of research that one can hardly blame a busy person for skimping on some of his or her "homework."

This series of volumes has been devised, therefore, as a practical solution to this critical problem. The geologist, perhaps even more than any other scientist, often suffers from twin difficulties—isolation from central library resources and immensely diffused sources of material. New colleges and industrial libraries simply cannot afford to purchase complete runs of all the world's earth science literature. Specialists simply cannot locate reprints or copies of all their principal reference materials. So it is that we are now making a concerted effort to gather into single volumes the critical materials needed to reconstruct the background of any and every major topic of our discipline.

We are interpreting "geology" in its broadest sense: the fundamental science of the planet Earth, its materials, its history, and its dynamics. Because of training in "earthy" materials, we also take in astrogeology, the corresponding aspect of the planetary sciences. Besides the classical core disciplines such as mineralogy, petrology, structure, geomorphology, paleontology, and stratigraphy, we embrace the newer fields of geophysics and geochemistry, applied also to oceanography, geochronology, and paleoecology. We recognize the work of the mining geologists, the petroleum geologists, the hydrologists, and the engineering and environmental geologists. Each specialist needs a working library. We are endeavoring to make the task of compiling such a library a little easier.

Each volume in the series contains an introduction prepared by a specialist (the volume editor)—a "state of the art" opening or a summary of the object and content of the volume. The articles, usually some twenty to fifty reproduced either in their entirety or in significant extracts, are selected in an attempt to cover the field, from the key papers of the last century to fairly recent work. Where the original works are in foreign languages, we have endeavored to locate or commission translations. Geologists, because of their global subject, are often acutely aware of the oneness of our world. The selections cannot therefore be restricted to any one country, and whenever possible an attempt is made to scan the world literature.

To each article, or group of kindred articles, some sort of "highlight commentary" is usually supplied by the volume editor. This commentary should serve to bring that article into historical perspective and to emphasize its particular role in the growth of the field. References, or citations, wherever possible, will be reproduced in their entirety so that the observant reader can assess the background material available to that particular author, or, if desired, he or she too can double-check the earlier sources.

A "benchmark," in surveyor's terminology, is an established point on the ground, recorded on our maps. It is usually anything that is a vantage point, from a modest hill to a mountain peak. From the historical viewpoint, these benchmarks are the bricks of our scientific edifice.

RHODES W. FAIRBRIDGE

PREFACE

The study of magnetic stratigraphy (magnetostratigraphy) of sediments has been a highly active endeavor within the scientific arena in the West for only about 12 years, having been conceived about 3 years earlier than plate tectonism and since forming an integral part of the related revolution in earth sciences. This book attempts to summarize the development of this area in geology and geophysics during the last decade in part by reproducing a series of key papers that have been partly instrumental in the successful development of this field. No previous attempt has been made to compile such papers on magnetostratigraphy, a task made easier because of the youth of this discipline. The greatest difficulty encountered in the production of this book has been in the selection of the papers that best represent the rapid expansion in knowledge that has occurred in the magnetic stratigraphy of sediments. Space limitations have required the deletion of several good papers; even some that added much to the critical foundation of knowledge. These additional contributions are noted in the discussions and in the reading lists provided for each of the eight sections. Most areas within the field of investigation of magnetostratigraphy have been included; but space limitations have not allowed inclusion of papers concerning two important interrelated topics: (1) the origin of magnetism of sediments and (2) the history of secular variation of short-term paleomagnetic excursions as recorded in marine or fresh-water sedimentary sequences. In compiling this set of readings, I have discussed in chronological order the sequence of ideas developed by those involved. Such a compilation provides an opportunity to include accounts of the nature of scientific discovery not normally published in literature and the sense of competition that exists between different laboratories and between different individuals. Such discussions are appropriately included in a volume dedicated to the memory of Norman D. Watkins because he was always very interested in the personalities involved in research, clashes in concepts and approaches, and the scientific method itself. He enjoyed including anecdotes about the activities of earth scientists he knew in lectures to both undergraduates and graduate students who seemed to be stimulated and educated by such information about the character of scientific research.

Preface

The idea of this book was developed as a result of discussions with Henry Spall. I greatly appreciate this. Many others have been helpful in their advice and criticism, especially Drs. C. Amerigan, C. Harrison, N. Opdyke, I. McDougall, and M. Ledbetter. I also thank M. Emery for her valuable editorial assistance during the preparation of this volume and N. Meader for typing the manuscript.

JAMES P. KENNETT

CONTENTS

PART I: THE DEVELOPMENT OF THE TIME SCALE

PART II: EXTENSION OF THE POLARITY SCALE TO DEEP-SEA SEDIMENTS: MULTICHANNEL RECORDERS

Contents

Contents

PART VII: SPECULATIONS: ARE POLARITY CHANGES LINKED TO OTHER PHENOMENA?

PART VIII: THE NOMENCLATURE OF MAGNETOSTRATIGRAPHY

CONTENTS BY AUTHOR

MAGNETIC STRATIGRAPHY
OF SEDIMENTS

INTRODUCTION

The sediments that mantle the earth's surface and ocean floor record vast amounts of information about the history of the earth's environment and life. Sedimentary sequences have provided almost all our knowledge about the history of life on earth and of its changing environments including polar and desert paleoclimatic history; the history of oceanic circulation; lateral and vertical changes in the position of the lithosphere; oceanic geochemical changes and numerous other parameters, including the history of the earth's magnetic field.

Since World War II, great advances have been made in stratigraphy and global stratigraphic correlations of Cenozoic and Cretaceous rocks. This has, to a large extent, been linked with the postwar rapid advancement in the study of oceanography. Ships from the leading oceanographic institutions in the United States, initially funded largely by the Office of Naval Research and later by the National Science Foundation, enabled stratigraphic work to be expanded to include for the first time the world's oceanic areas. Previously nearly all knowledge had been obtained from uplifted marine sections on land. The initial stratigraphic studies were made possible by the collection of large numbers of piston cores from the ocean basins. The collection of these cores ironically was not led by the desire of stratigraphers to obtain such material but by geophysicists who wanted to obtain ages of the sediment near the ocean floor as a supplement for their geophysical interpretations, which were based largely on seismic data and also on magnetic and gravity information. Dr. Maurice Ewing of the Lamont Geological Observatory was the leader in this very active data-accumulation period. The cores that were obtained during these expeditions were largely of Quaternary to Pliocene age, although older Tertiary material was also being collected. These cores required dating and correlating.

Stratigraphers have always been receptive to any new development in technology that might assist in the stratigraphic correlation of sedimentary sequences over both small and large distances and approaches that assist in the dating of the sequences. Predictably there has been an acute interest in magnetostratigraphy, one of the latest stratigraphic approaches to be developed to aid in historical geology and stratigraphy. During deposition of sediments or cooling of molten rock, magnetic iron oxide minerals align themselves with any magnetic field that then exists. This preferred alignment of grains results in a bulk property of the rock referred to as *remanent magnetism,* which in essence acts as a geomagnetic field recorder. Initially it was observed in sequences of volcanic rocks on land that the north and south poles of the earth's magnetic field have reversed fairly frequently. For instance, our present polarity began 7×10^5 years ago, and before that for about one million years, the poles were reversed except for several short episodes of normal polarity. Magnetic stratigraphy, or magnetostratigraphy, is based on the reversals in the earth's magnetic field recorded in sediments, and the utilization of these changes for stratigraphic and chronological purposes. The approach is new, having been accepted as a valid tool only slightly more than a decade ago and yet since that time has been widely used especially in studies of deep-sea sedimentary sequences obtained by piston coring. Typically, however, the first use of paleomagnetic stratigraphy occurred many years before the middle 1960s. A Soviet scientist, Khramov (1957), had recognized the value of paleomagnetism in the subdivision and correlation of sedimentary sequences in the Soviet Union. No immediate impact resulted in the scientific community from the publication of the paper; certainly in the West. For this to happen, the approach had to await its extension to deep-sea sedimentary sequences many years later.

Stratigraphers have used various geochronological approaches, each marked by severe limitations and restrictions. Before the development of magnetostratigraphy, young sediments were dated using the C^{14} method, but this approach was restricted to material younger than about 4×10^4 years old. Likewise, the Thorium 230 method has provided important datums younger than about 3×10^5 years but only in limited numbers. The K/Ar method has been critical in providing the necessary older dates of sedimentary sequences. However, the only sedimentary rocks that can be dated are glauconites, interbedded in volcanic sequences or containing pyroclastic deposits that provide the necessary material for the K/Ar analyses. This approach has been critical in intercalibration of the biostratigraphic zonation with the chronological framework but is limited to only a relatively small number of stratigraphic sections containing volcanic rocks suitable for K/Ar dating (Berggren, Paper 11). It had been hoped that the dating of submarine volcanic rocks immediately beneath oceanic sediments obtained by deep-sea drilling would assist in dating the various biostratigraphic zonations. Unfortunately, degassing of Ar has been too extensive as a result of submarine weathering, and few realistic ages have been obtained. Even in the case of the few available fresh volcanic rocks from these depths, inaccuracies also result from the presence of excess radiogenic ^{40}Ar trapped in rapidly cooled rocks at the time of their formation (Seidemann, 1977). In these cases radiometric ages are too old. Another, more recent approach in correlation has been by the use of oxygen isotopic stratigraphy; correlation using changes in the oxygen isotopic ratio (O^{16}/O^{18}) preserved in fossil foraminiferal tests. Until the early 1970s, however, this method had not developed much beyond the correlation of Late and Middle Quaternary deep-sea sequences (Emiliani, 1966). This has now changed with the exciting prospect of intra- and interoceanic stratigraphic correlations for all the Cenozoic, although the level of resolution of the correlations is as yet still largely untested (Shackleton and Kennett, 1975a, 1975b; Kennett and Shackleton, 1976; Savin et al., 1975). As magnetostratigraphy was beginning to expand in importance as a tool for stratigraphy and correlation, oxygen isotopic stratigraphy was itself only beginning to flourish.

Due to the limitations of other methods of dating sediments and to the urgent need of a practical approach for the extension of chronology to sedimentary sequences older than the late Quaternary, magnetostratigraphy was received with a great deal of excitement in the middle 1960s. Indeed, in some circles it was initially regarded as a panacea for most Late Cenozoic stratigraphic problems. That magnetostratigraphy was not to be a panacea because its own characteristic limitations soon became apparent and will be discussed later in this book. Nevertheless, magnetostratigraphy has provided the following important advances in correlation and chronology:

1. Magnetic reversals are synchronous, worldwide phenomena and thus differ in character from most other stratigraphic criteria employed for correlation that are characteristically diachronous. Other synchronous events such as tephra layers also represent powerful tools for correlation but are too geographically restricted for wide-scale correlations.

2. Magnetostratigraphy for the last 5 million years, established on the basis of radiometric dating of terrestrial lava sequences, has provided stratigraphers with an apparently accurate chronological framework within which to work. Thus a recognizable magnetostratigraphy within the last 5 m.y. provides a source of dates to establish age relationships of sediment and fossil sequences. The chronology in turn has provided the required basis to determine rates of geologic change with good accuracy. This approach is currently being extended to the entire Cenozoic.

3. The generation of magnetostratigraphic data is often carried out on a routine basis requiring little of the principal investigator's own time. This contrasts with biostratigraphic research, which demands a great deal of time toward data production by the principal investigator rather than by assistants.

4. A magnetostratigraphic sequence by itself normally does not provide unequivocal dates for geological events preserved in sediments. This is because magnetic reversals are repetitive and do not possess singular properties. For example, a sequence exhibiting normal polarity may represent the Brunhes Normal Paleomagnetic Epoch or it may even be of Cretaceous age. Independent criteria are required to verify the age of the sequence. This is usually accomplished by independent biostratigraphic analysis carried out in conjunction with the paleomagnetic studies. A great benefit of this approach is that it normally requires little effort by the biostratigrapher to establish the age of a magnetic polarity sequence; for example, whether a sedimentary sequence of normal polarity is Brunhes ($t = 0.69$ m.y. to P.D.); Gauss ($t = 2.43$ to 3.32 m.y.); or Cretaceous in age. This requires examination of the microfossils at only a few stratigraphic levels. Hence large numbers of sequences can be rapidly examined.

Development of the polarity time scale resulted from joint paleomagnetic and radiometric studies of volcanic sequences on land, especially on oceanic islands. The history of this development has already been thoroughly summarized with considerable insight in the book *Plate Tectonics and Geomagnetic Reversals* by Allan Cox (1973), and in papers by Norman D. Watkins (Paper 2) and by Ian McDougall et al. (Paper 3). The establishment of the polarity time scale for about the last 4 m.y. actually owes very little to research on deep-sea sediment cores. Nevertheless, the study of magnetostratigraphy of deep-sea sediments contributed important information about the *duration* of paleomagnetic events. It also helped to stimulate further paleomagnetic studies of volcanic sequences on land by providing those workers with confidence in the polarity time scale they were developing. The study of magnetic stratigraphy of deep-sea sediments and of shallow marine and terrestrial sequences on land is largely one of the applications of the approach to stratigraphic and historical geological problems: i.e., the application of this method as a tool. This approach is now well established and is expected to continue to be an important stratigraphic method in the future. Much of the current work in paleomagnetism of marine sediments is related to several problems:

1. The extension of the polarity scale to sequences older than the late Cenozoic. Few suitable volcanic sequences of this age are available on land, and hence it is expected that the marine sedimentary sequences should play a significant role in the development of the older parts of the Cenozoic polarity time scale.

2. Studies of short-term polarity events and of secular paleomagnetic variation. The most successful of these studies have concentrated on lake deposits that lack bioturbation to disturb the remanent paleomagnetism.

3. Mineralogical and sedimentological studies of sediments, to provide a better understanding of the fundamental nature of the paleomagnetic record; for example, the differentiation of various paleomagnetic signals such as postdepositional chemical overprinting and of the effects of depositional characteristics on remanent magnetism.

4. Studies on possible interrelationships between magnetic polarity change and other phenomena such as faunal extinctions.

Some of these areas are touched on in the book, but its aim is primarily to summarize those discoveries and approaches that have led to the development of magnetostratigraphy as an integral part of the marine sedimentologist's and paleontologist's research.

REFERENCES

Cox, A. 1973. *Plate Tectonics and Geomagnetic Reversals*. San Francisco: W. H. Freeman, 702p.

Emiliani, C. 1966. Palaeotemperature analysis of Caribbean cores P6304–8 and P6304–9 and a generalized temperature curve for the last 425,000 years. *Jour. Geology* **74**:109–126.

Kennett, J. P., and N. J. Shackleton. 1976. Development of the Psychrosphere 38 m.y. ago: Oxygen isotopic evidence from Subantarctic Paleogene sediments. *Nature* **260**:513–515.

Khramov, A. N. 1957. Paleomagnetism: The basis of a new method of correlation and subdivision of sedimentary strata. *Acad. Sci. USSR Doklady, Earth Sci. Sec. Proc.* **112**:129–132.

Savin, S. M., R. G. Douglas, and F. G. Stehli. 1975. Tertiary marine paleotemperatures. *Geol. Soc. America Bull.* **86**:1499–1510.

Seidemann, D. E. 1977. Effects of submarine alteration on K-Ar dating of deep-sea igneous basalts. *Geol. Soc. America Bull.* **88**:1660–1666.

Shackleton, N. J., and J. P. Kennett. 1975a. Paleotemperature history of the Cenozoic and the initiation of Antarctic glaciation: Oxygen and carbon isotope analyses in DSDP Sites 277, 279 and 281. In *Initial Reports of the Deep Sea Drilling Project,* Vol. XXIX, edited by J. P. Kennett et al. Washington, D.C.: U.S. Government Printing Office, pp. 743–755.

Shackleton, N. J., and J. P. Kennett. 1975b. Late Cenozoic oxygen-carbon isotopic changes at D.S.D.P. Site 284: Implications for Glacial History of the Northern Hemisphere and Antarctica. In *Initial Reports of the Deep Sea Drilling Project,* Vol. XXIX, edited by J. P. Kennett et al. Washington, D.C.: U.S. Government Printing Office, pp. 801–807.

Part I

THE DEVELOPMENT OF THE
TIME SCALE

Editor's Comments
on Papers 1, 2, and 3

1   COX
Geomagnetic Reversals

2   WATKINS
Review of the Development of the Geomagnetic Polarity Time Scale and Discussion of Prospects for Its Finer Definition

3   McDOUGALL et al.
Extension of the Geomagnetic Polarity Time Scale to 6.5 m.y.: K-Ar Dating, Geological and Paleomagnetic Study of a 3,500-m Lava Succession in Western Iceland

As is often the case in basic scientific research, the early discoverers of magnetic polarity reversals recorded in volcanic rocks—for example, Matuyama (1929)—had little idea that the later extension of their initial observations into a coherent polarity time scale would lead to a major revolution in earth sciences, involving fundamental understandings of the nature and evolution of the earth's crust, oceans, and land masses. As McDougall (1977) has pointed out, the polarity time scale was developed out of research designed to determine whether the occurrences of reversed remanent magnetization in rocks resulted from the geomagnetic field having opposite polarity to the present at times in the past or whether the rocks themselves had self-reversing properties so that they became magnetized antiparallel to the prevailing magnetic field.

The potential importance and implications of a geomagnetic polarity scale became rapidly apparent to a small group of scientists in the early 1960s working in northern California and in Australia. As a result of their combined and competitive research on the paleomagnetism and chronology of young sequences of volcanic rocks, this small group of workers, in only five years, developed a magnetic polarity sequence. The development of the polarity time scale has been summarized by Cox (1973), Watkins (Paper 2), and McDougall (1977). The role of a paleomagnetism of sedimentary sequences was not particularly important in the development of the polarity time scale but has been of great importance in the development of our understanding of historical geology and the evolution of life.

The initial paleomagnetic work in the early 1960s concentrated on the Quaternary, as Matuyama had done 30 years previously, to determine the nature and age of the youngest paleomagnetic reversals. The dating of the reversals, as recorded in sequences of volcanic rocks, had awaited further developments in mass spectrometry allowing greater dating precision using the K/Ar radiometric approach. This occurred in the late 1950s at the University of California at Berkeley and enabled especially young rocks (<5 m.y.) to be dated accurately. Thus not only could a polarity time scale be established within single sections because of precise dating but paleomagnetic changes exhibited in widely separated locations could now be placed within a single chronostratigraphic framework.

The development of the polarity time scale during the early 1960s was carried out by Cox, Doell, and Dalrymple, all previously students at the University of California at Berkeley and later scientists at the U.S. Geological Survey, and simultaneously by McDougall and Tarling, and later with Chamalaun at the Australian National University, Canberra. The importance of the scientific problem and stimulation from the friendly competition between these two groups led to the rapid publication of a succession of exciting papers on the earth's polarity history during the last 4.5 m.y. in volcanic rocks of numerous oceanic islands such as Hawaii, the Galapagos, and Reunion. Each paper represented an additional step in understanding of the polarity time scale. These developments are summarized in some detail by Watkins (Paper 2) and require no amplification here. The 1969 version of Cox (Paper 1) is still widely used. An understanding of the nature of the polarity time scale has continued to evolve, although more slowly, and the latest versions are not necessarily widely accepted nor widely used (McDougall, 1977), although changes in the ^{40}K decay constants have since required recalculation of the age of the boundaries, and thus later versions of the polarity scale are *required* to be adopted (Mankinen and Dalrymple, 1979).

The development of the polarity time scale during the early 1960s had major scientific repercussions beyond that of dating volcanic and sedimentary sequences. In the middle 1960s when the polarity time scale was becoming well established, Vine and Wilson (1965) and Vine (1966) began to correlate linear magnetic

9

marine anomaly patterns with the polarity time scale. Tuzo Wilson, for instance, had been stimulated by a paper presented by Neil Opdyke on the magnetic stratigraphy of sediments at the International Gondwana Conference in 1965 at Montevedeo that assisted in the formulation of his developing concepts on sea-floor spreading. It was largely the excellent correlations between linear marine magnetic anomaly patterns and the polarity time scale that led to the convincing demonstration of the relationships between marine magnetic anomaly patterns and the sea-floor-spreading hypothesis of Hess (1962) and of Vine and Matthews (1963). This discovery in turn led to the remarkably rapid revolution in earth sciences involving plate tectonism, continental drift, and sea-floor spreading. A polarity time scale was developed by Heirtzler et al. (1968) for the period of time back to 79 m.y. ago based on extrapolation of marine magnetic anomalies from mid-oceanic ridges, assuming constant spreading rates and integration with a chronology independently developed for terrestrial sequences. Later versions (Tarling and Mitchell, 1976; LaBrecque et al., 1977) have introduced additional small corrections in the Heirtzler et al. (1968) time scale as new marine magnetic anomaly data and more accurate data of critical boundaries have become available. Comparison with biostratigraphic ages of Deep Sea Drilling Project (DSDP) drill holes suggests that the original assumption of constant sea-floor spreading is in error and that new versions of the extrapolated time scale need not satisfy this assumption. At about this time magnetostratigraphy was extended to deep-sea sedimentary sequences and began to play an important role in debates related to the finer details of the polarity time scale and of the nature of the shorter term paleomagnetic changes. This topic will be discussed in Part II.

Until the middle 1970s the polarity time scale was confined to rocks younger than about 4.5 m.y. old. An obvious evolutionary step has been its extension to encompass older volcanic and sedimentary sequences. A major difficulty encountered in this extension has resulted from decreasing resolution of the K/Ar dating method as rocks become older than about 4.5 m.y. Nevertheless, there has been some success in the extension of the polarity time scale, which is summarized in McDougall (1977). Watkins and Walker (1977) and McDougall in collaboration with other colleagues (Paper 3) recognized that the best possibility of extending the polarity time scale back beyond about 5 m.y. ago was to work on thick sequences of lavas representing significant amounts of time and where stratigraphic relationships between successive lavas were clear. One of the few areas that offered such a possibility is Iceland, and Watkins, McDougall, and their colleagues proceeded with the field program. Fortunately, they succeeded in this somewhat precarious venture and were able to extend the scale back to 13.6 m.y. as documented in their 1976 and 1977 papers (Watkins and Walker, 1977; McDougall et al. 1976 and Paper 3) and by Harrison et al. (1979). The pattern of the polarity time scale that they generated from the Icelandic volcanic sequences is similar to the extrapolations based on marine magnetic anomaly data. Harrison et al. (1979) found that the assumption of constant spreading is largely sound, although relatively small adjustments are necessary.

REFERENCES

Cox, A. 1973. *Plate Tectonics and Geomagnetic Reversals*. San Francisco: W. H. Freeman, 702p.

Harrison, C. G. A., I. McDougall, and N. D. Watkins. 1979. A geomagnetic field reversal time scale back to 13.0 million years before present. *Earth and Planetary Sci. Letters* **42**:143–152.

Heirtzler, J. R., G. O. Dickson, E. M. Herron, W. C. Pitman, III, and X. Le Pichon. 1968. Marine magnetic anomalies, geomagnetic field reversals and motions of the ocean floor and continents. *Jour. Geophys. Research* **73**:2119–2136.

Hess, H. H. 1962. History of ocean basins. In *Petrologic Studies: A Volume in Honor of A. F. Buddington*, edited by A. E. J. Engel et al. Boulder, Colo.: Geological Society of America, pp. 599–620.

LaBrecque, J. L., D. V. Kent, and S. C. Cande. 1977. Revised magnetic polarity time scale for Late Cretaceous time. *Geology* **5**:330–335.

McDougall, I. 1977. The present status of the geomagnetic polarity time scale. In *The Earth: Its Origin, Structure and Evolution (A volume in honor of J. C. Jaeger and A. L. Hales)*, edited by M. W. McElhinny. London: Academic Press, 34p.

McDougall, I., N. D. Watkins, and L. Kristjansson. 1976. Geochronology and paleomagnetism of a Miocene–Pliocene lava sequence at Bessastadaa, eastern Iceland. *Am. Jour. Sci.* **276**:1078–1095.

Mankinen, E. A., and G. B. Dalrymple. 1979. Revised geomagnetic polarity time scale for the interval 0–5 m.y. B. P. *Jour. Geophys. Research* **84**:615–626.

Matuyama, M. 1929. On the direction of magnetization of basalt in Japan, Tyosen, and Manchuria. *Japan Acad. Proc.* **5**:203–205.

Tarling, D. H., and J. G. Mitchell. 1976. Revised Cenozoic polarity time scale. *Geology* **4**:133–136.

Vine, F. J. 1966. Spreading of the ocean floor: New evidence. *Science* **154**:1405–1415.

Vine, F. J., and D. H. Matthews. 1963. Magnetic anomalies over oceanic ridges. *Nature* **199**:947–949.

Vine, F. J., and J. T. Wilson. 1965. Magnetic anomalies over a young oceanic ridge off Vancouver Island. *Science* **150**:485–489.

Watkins, N. D., and G. P. L. Walker. 1977. Magnetostratigraphy of eastern Iceland. *Am. Jour. Sci.* **277**:513–584.

ADDITIONAL READINGS

Berggren, W. A., M. C. McKenna, J. Hardenbol, and J. D. Obradovich. 1978. Revised paleogene polarity time scale. *Jour. Geology* **86**:67–81.

Berggren, W. A., D. P. McKenzie, J. G. Sclater, and J. D. van Hinte. 1975. World-wide correlation of Mesozoic magnetic anomalies and its implications: Discussion and reply. *Geol. Soc. America Bull.* **86**: 267–272.

Creer, K. M. 1970. A review of palaeomagnetism. *Earth-Sci. Rev.* **6**:369–466.

Hinte, J. E. van. 1972. The Cretaceous time scale and planktonic-foraminiferal zones. *Koninkl. Nederlandse Akad. Wetensch. Ser. B,* **75**:1–8.

Hinte, J. E. van. 1976. A Cretaceous time scale. *Am. Assoc. Petroleum Geologists Bull.* **60**:498–516.

Hinte, J. E. van. 1976. A Jurassic time scale. *Am. Assoc. Petroleum Geologists Bull.* **60**:489–497.

Irving, E. 1964. *Paleomagnetism and Its Application to Geological and Geophysical Problems.* New York: Wiley, 399p.

Irving, E., and R. W. Couillard. 1973. Cretaceous normal polarity interval. *Nature Phys. Sci.* **224**:10–11.

Irving, E., and G. Pullaiah. 1976. Reversals of the geomagnetic field, magnetostratigraphy and relative magnitude of paleosecular variation in the Phanerozoic. *Earth-Sci. Rev.* **12**:35–64.

Johnson, A. H., and A. E. M. Nairn. 1972. Jurassic palaeomagnetism. *Nature* **240**:551–552.

Johnson, A. H., A. E. M. Nairn, and D. N. Peterson. 1972. Mesozoic reversal stratigraphy. *Nature Phys. Sci.* **237**:9–10.

Khramov, A. N. 1957. Paleomagnetism: The basis of a new method of correlation and subdivision of sedimentary strata. *Acad. Sci. USSR Doklady, Earth Sci. Sec. Proc.* **112**:129–132.

Larson, R. L., and C. E. Helsley. 1975. Mesozoic reversal sequence. *Rev. Geophysics and Space Physics* **13**:174–176.

Larson, R. L., and T. W. C. Hilde. 1975. A revised time scale of magnetic reversals for the Early Cretaceous and Late Jurassic. *Jour. Geophys. Research* **80**:2586–2594.

Larson, R. L., and W. C. Pitman, III. 1972. World-wide correlation of Mesozoic magnetic anomalies and its implications. *Geol. Soc. America Bull.* **83**:3645–3662.

Smith, P. J. 1968. The earth's fluctuating dynamo. *New Scientist* **4**:15–17.

Vine, F. J. 1970. Sea-floor spreading and continental drift. *Jour. Geol. Education* **18**:87–90.

Watkins, N. D. 1971. Geomagnetic polarity events and the problem of "the reinforcement syndrome." *Comments on Earth Sci.: Geophys.* **2**: 36–43.

1

Geomagnetic Reversals

Although decreasing rapidly, the earth's
magnetic field is probably not now reversing.

Allan Cox

The periods of the earth's magnetic field that are most crucial to an understanding of the geomagnetic dynamo occur at the ultralow-frequency end of the spectrum. Yet, until recently, this part of the spectrum received relatively little attention, mainly because the spectrum extends to periods longer than the productive lifetimes of individual scientists. The longest periods are, indeed, longer than the entire history of scientific observation. The first hint of the existence of periods greater than 100 years came from observations of changes in field intensity. In 1835—the first year for which C. F. Gauss was able to assemble enough worldwide data to analyze the field through the use of spherical harmonic functions — the earth's dipole moment was 8.5×10^{25} gauss cm³. By 1965, the moment had decreased to 8.0×10^{25} (*1*). The dipole field decreased during this time at a remarkably uniform rate, and it will, if the rate remains constant, pass through a zero point about 2000 years from now and then reverse its polarity (*2*). However, other interpretations are equally consistent with the intensity data—for example, the field might oscillate without changing polarity. Therefore, although there is little question that the geomagnetic spectrum extends to periods considerably greater

The author is professor of geophysics at Stanford University, Stanford, California, and does part of his research with the U.S. Geological Survey. This article is an expanded version of a paper presented 8 April 1968 at a meeting of the American Geophysical Union in Washington, D.C.

than the few centuries spanned by the records of magnetic observatories, the nature of the spectrum has long remained uncertain.

The length of the available magnetic record has now been increased by more than six orders of magnitude through the study of the natural magnetism of rocks and baked clay which retain a magnetic memory of the earth's field in the past. This paleomagnetic research indicates that the earth's field has undergone numerous fluctuations of intensity and that it has also undergone many (although less numerous) reversals in polarity. Recent work suggests that these two phenomena may occupy adjacent parts of the geomagnetic spectrum, the division between them being at periods of about 10^4 years. This part of the spectrum, which is very difficult to resolve experimentally, may hold the key to understanding how polarity reversals are related to intensity fluctuations.

Changes in the Earth's Dipole Moment

The technique for measuring ancient geomagnetic field intensity is based on the observation that, when volcanic rocks and pieces of pottery are cooled in weak magnetic fields, they acquire thermal remanent magnetization which is parallel in direction and proportional in intensity to the applied field. If the natural thermal remanence of rocks is magnetically stable, the ancient field

acting on a sample when it originally cooled may be found by reheating the sample and cooling it in a known field. The ancient field intensity F_o is then given by

$$F_o = F_a \frac{J_o}{J_a} \qquad (1)$$

where J_o is the natural remanence and J_a is the remanence acquired in the known applied field F_a (*3*).

This technique, although simple in concept, is difficult to carry out experimentally because, on being reheated in the laboratory, the ferromagnetic minerals contained in rocks and baked clay commonly undergo chemical changes to form new ferromagnetic minerals. Clearly, the magnetization acquired when the altered samples are cooled in a known field does not provide a measure of the ancient field. However it has proved possible experimentally to identify samples which are chemically stable on heating. These yield values for ancient field intensity which are internally very consistent and appear to be reliable.

In all, 127 paleomagnetic intensity determinations have now been made on samples with ages $0 < t \leqslant 10^4$ years. This work, which was carried out in different laboratories on samples from many parts of the world, was summarized recently by P. J. Smith (*4*), who reduced each ancient field intensity to a virtual dipole moment, defined as the moment of the dipole needed to produce the observed ancient intensity if the earth's field had been entirely dipolar. Virtual dipole moments for samples of the same age from different parts of the world are scattered, with a standard deviation of about 20 percent, because the earth's field consists in part of an irregular nondipole component. This scatter has been reduced in Fig. 1 by averaging virtual dipole moments for different parts of the world. The decrease in dipole moment since 1885, as shown by the slanting bar at left in Fig. 1, is seen to be but the most recent part of a well-defined half cycle with a maximum of 12×10^{25} gauss cm³, which began about 4000 years ago.

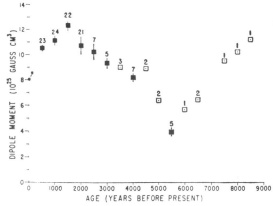

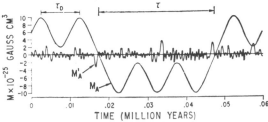

Fig. 1 (left). Variations in geomagnetic dipole moment (12). Changes during the past 130 years, as determined from observatory measurements, are shown by the slanting bar at left. Other values were determined paleomagnetically (4). The number of data that were averaged is shown above each point, and the standard error of the mean is indicated by the vertical lines (12) except for points (open squares) with too few data to provide meaningful statistics.   Fig. 2 (above). Model for reversals used to derive the distribution function for reversals. T_D is the period of the dipole field and T is the length of a polarity interval. A reversal occurs whenever the quantity M'_A, which is a measure of the nondipole field, becomes sufficiently large relative to the dipole moment M_A (12).

An earlier half cycle is weakly suggested by the few older intensity measurements. The important question of whether changes in field intensity are periodic cannot be answered from the present data, but there can be no question that dipole fluctuations occur, with durations of the order of 10^4 years.

Polarity Reversals

Extension of the known geomagnetic spectrum to even longer periods has come about with the discovery that the earth's field undergoes reversals in polarity. Through paleomagnetic research it has been found that the earth's dipole moment alternates between two antiparallel polarity states, a *normal* state, in which the field at the earth's surface is directed northward, and a *reversed* state in which the field has the opposite direction. In either state the dipole undergoes an irregular wobble about the earth's axis of rotation, with an angular standard deviation of 12 degrees (5). The time required to complete a transition between polarity states is estimated to be from 10^3 to 10^4 years (6–8); during this time the field does not go to zero but undergoes an intensity decrease of from 60 to 80 percent (4–9). Averaged over a long sequence of reversals, the total amount of time spent by the dipole in the reversed state is equal to that spent in the normal state (10–12). Moreover, the average intensity of the field in the two states is the same (4), suggesting that the two states have equal energy levels. In all of these characteristics the earth's field

operates like a remarkably symmetrical oscillator or, more precisely, like a bistable flip-flop circuit.

The greatest element of irregularity in reversals is the length of time between successive changes in polarity. The longest known polarity interval lasted for 5×10^7 years (13); intervals with lengths of the order of 10^6 years are common, and short polarity intervals occur, with durations of less than 10^5 years. The latter have proved to be the most difficult to measure. This article deals mainly with recent experimental work on short polarity intervals.

Distribution Function for Polarity Intervals

The results from paleomagnetic studies of reversals may be presented either in the form of a time scale for reversals or, more compactly, in the form of a histogram showing the frequency of occurrence of polarity intervals as a function of their length. The question of what distribution function should be used to describe the observed frequency distribution depends on the more fundamental questions of why reversals occur and what controls the immense variation in the length of time between them.

It is now a generally accepted theory that the earth's field is generated by hydromagnetic processes in the earth's fluid core and that the energy which maintains the field against ohmic dissipation is provided either by thermal convection or by turbulence generated by the earth's precession (14). Unfortu-

nately, however, the theory of the homogeneous fluid dynamo is not sufficiently developed to account quantitatively either for fluctuations in dipole intensity or for polarity reversals. Particularly difficult is the problem of reconciling the long times between polarity reversals, which may exceed 10^6 years, with the much shorter time constants of dipole fluctuation (10^4 years) and fluctuations of the nondipole field (3 years to 10^3 years).

In the absence of a complete theory of geomagnetism, some interesting and possibly relevant analogies are provided by the behavior of nonfluid self-excited dynamos. The simplest one that provides a model for geomagnetic intensity fluctuations and polarity reversals consists of two mutually coupled Faraday disk dynamos (15). In some solutions for this dynamo the current oscillates about a mean current flowing first in one and then in the opposite direction. The distribution function for the lengths of time between successive reversals in current direction depends on (i) the period of the oscillations and (ii) the number of cycles between successive reversals. Both are sensitive to changes in the physical parameters of the model, some solutions for the dynamo having sharply peaked distributions and others having monotonic decreasing functions, the shortest polarity intervals being most frequent.

The earth's field is produced by a much more complex fluid dynamo, so that detailed comparisons with particular solutions for the disk dynamos are of little value. Unlike the rigid disks of the dynamo, the earth's core resembles the atmosphere in possessing a large random component of motion due to turbulence. A direct measure of this component is provided by the rapidly varying nondipole field, the average intensity of which is 20 percent of the dipole field. The sensitivity of the timing of reversals in a solid disk dynamo to small changes in the physical conditions of the model suggests that in fluid dynamos the timing of reversals would also be sensitive to a large random element in the pattern of fluid motions and magnetic fields.

The foregoing considerations suggest the following model, from which a distribution function for polarity intervals can be derived (*12*) (Fig. 2). The main geomagnetic dynamo is assumed to be a steady dipole oscillator which undergoes a polarity reversal only when this is triggered by random fluctuations of the much more rapidly varying nondipole field. The probability *P* that a reversal will occur during one cycle of dipole oscillation is assumed to be the same for all cycles and is determined by the spectrum of nondipole fluctuations and the amplitude of dipole oscillations. The resulting distribution function for variations in the length T of polarity intervals depends only on the probability *P* and the period T_D of dipole oscillation:

$$f[T_c < T \leqslant T_c + T_D] = P(1 - P)^{T_c/T_D} \quad (2)$$

where T_c is an integral number of dipole periods. For small values of *P*, this may be approximated by a simple exponential distribution

$$f'(T) = \lambda \exp(-\lambda T) \, dT \quad (3)$$

where $\lambda = P/T_D$, and $f'(T)$ is the probability for the interval dT.

Before comparing this function with experimental data, one should note the extreme sensitivity of the distribution function to the proportion of short events. This sensitivity is much greater than that of the spectrum of a continuously varying signal to a high-frequency component. Because the earth's field has only two polarity states, inserting a short polarity interval near the middle of a long interval not only adds the short interval to the distribution but also removes a long interval and adds two intervals of intermediate length. This effect may be seen (Fig. 3) in the marked change that occurred in the apparent distribution of polarity intervals as several short polarity intervals were discovered during the course of research between the years 1963 and 1966.

Radiometric Time Scale for Reversals

The first quantitative time scale for reversals was achieved in 1963 by measuring the ages (using the potassium-argon technique) and the magnetic polarities of young volcanic rocks (*16*). This work appeared to confirm earlier paleomagnetic results from undated sedimentary rocks which indicated that all polarity intervals were of nearly equal length (Fig. 3, 1963 scale). These polarity intervals were termed "polarity epochs" and given the names of early workers in the field of geomagnetism (*17*).

As more data were obtained (*18*), some of the ages and polarities were found to be inconsistent with the simple pattern of epochs. This inconsistency led in 1964 (*19*) to the discovery of intra-epoch "polarity events" with unexpectedly short durations of about 10^5 years (Fig. 3). The first polarity events to be recognized were the Olduvai event ($t = 1.9 \times 10^6$ years ago) and the Mammoth event ($t = 3.1 \times 10^6$ years ago), named after the sites of their discovery in Tanzania and California. A paleomagnetic record of polarity events has been found in rocks with similar ages from many parts of the world, demonstrating that the events, like the epochs, are the result of rapid switching of the main dipole field.

With the discovery of polarity events it became apparent that complete resolution of the fine structure of reversals would require an immense number of paleomagnetic and radiometric data. Ideally, such data should be obtained from formations whose ages are uniformly spaced at intervals no greater than 10^4 years; this would require at

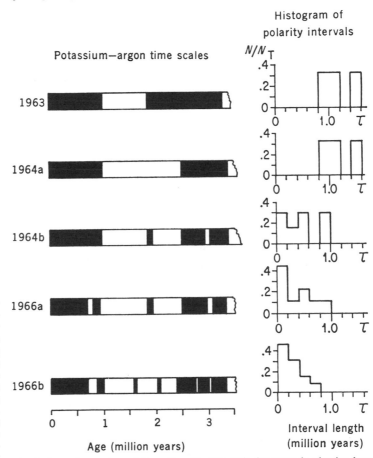

Fig. 3. Successive versions of the radiometric time scale for reversals, showing how the discovery of polarity events changed the apparent distribution of polarity intervals (*16, 18, 19*). In the corresponding histograms, N_T is the total number of polarity intervals and N is the number in each class interval of the histogram.

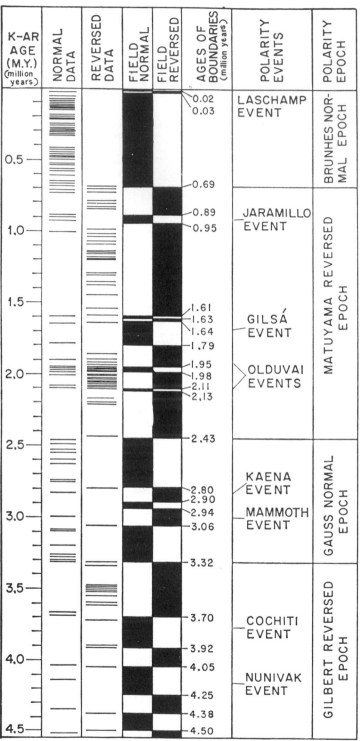

K-AR AGE (M.Y.) (million years)	NORMAL DATA	REVERSED DATA	FIELD NORMAL	FIELD REVERSED	AGES OF BOUNDARIES (million years)	POLARITY EVENTS	POLARITY EPOCH
					0.02 0.03	LASCHAMP EVENT	BRUNHES NORMAL EPOCH
0.5							
					0.69		MATUYAMA REVERSED EPOCH
					0.89	JARAMILLO EVENT	
1.0					0.95		
1.5					1.61 1.63 1.64 1.79	GILSÁ EVENT	
2.0					1.95 1.98 2.11 2.13	OLDUVAI EVENTS	
2.5					2.43		GAUSS NORMAL EPOCH
					2.80 2.90	KAENA EVENT	
3.0					2.94 3.06	MAMMOTH EVENT	
					3.32		GILBERT REVERSED EPOCH
3.5					3.70	COCHITI EVENT	
4.0					3.92 4.05		
					4.25	NUNIVAK EVENT	
4.5					4.38 4.50		

least 450 precisely determined radio-metric dates for the interval $0 < t \leqslant 4.5$ million years. The task of making this many age determinations is formid-able, and that of locating volcanic for-mations with the required ages is even more formidable. The age of a volcanic formation is rarely known to within a factor of 2 prior to determination of the radiometric age. Thus the best sampling scheme that can be hoped for is a rela-tively inefficient one in which the ages are randomly rather than uniformly dis-tributed. Even this is difficult to achieve because of the episodic character of volcanic activity.

For the interval $0 < t \leqslant 4.5$ million years there are now 150 radiometric ages and polarity determinations which meet reasonable standards of reliability and precision (20). The main contribu-tions from the data acquired since 1964 have been more accurate determination of the ages of polarity changes (8) and the identification of several hitherto unknown events, summarized in Fig. 4. Attempts to extend the radiometric time scale for reversals back beyond 4.5 mil-lion years have been unsuccessful be-cause the errors in the radiometric ages of the older rocks are too large. A dat-ing error of 5 percent in a 5-million-year-old sample is 2.5×10^5 years, which is larger than many polarity in-tervals (8, 11, 21).

Have all the polarity events more re-cent than 4.5 million years ago been dis-covered? All of the longer ones appear to have been, to judge both from the density of the present age determina-tions and from the convergence of the more recent versions of the reversal time scale. However, there are gaps of 10^5 years or more in the present data, and the recent discovery by Bonhommet and Babkine (22) of a previously un-suspected reversed event near the end of the Brunhes normal-polarity epoch demonstrates that additional events may exist in gaps even shorter than 10^5 years.

Fig. 4. Time scale for geomagnetic re-versals. Each short horizontal line shows the age as determined by potassium-argon dating and the magnetic polarity (normal or reversed) of one volcanic cooling unit. Included are all published data (37) which meet reasonable standards of reliability and precision (20). Normal-polarity inter-vals are shown by the solid portions of the "field normal" column, and reversed-po-larity intervals, by the solid portions of the "field reversed" column. The duration of events is based in part on paleomagnetic data from sediments (7, 28) and magnetic profiles (11, 23, 24, 27).

Midoceanic Magnetic Anomalies

Additional information about reversals is provided by the magnetic anomalies over the midoceanic ridges (*11, 23, 24*). These anomalies are produced by igneous rocks which become magnetized as they solidify and cool in a narrow zone along the ridge axis. As new material forms, the previously magnetized material spreads to either side. If the rate of spreading is the same on both sides of the ridge, the result is a bilaterally symmetrical pattern of normally and reversely magnetized strips with widths proportional to the lengths of the corresponding polarity intervals. The magnetic anomalies do not in themselves determine an independent reversal time scale because the ages of the igneous rocks are usually not known. However, after being calibrated against known points on the radiometric time scale for reversals, the profiles provide a nearly continuous record of polarity intervals. Of special interest are events that may exist within the gaps in the radiometric data.

The value for the minimum duration of a detectable event depends on the width of the strip of crust that was formed during the time of the event, the distance of the strip from the magnetometers at the sea surface (usually about 3 kilometers), and the level of background noise due to irregularities in the formation and magnetization of the crust. The strong dependence of the minimum value for duration on the rate of crustal spreading, v, may be seen from variations in the size of the anomaly due to the Jaramillo event ($T = 5 \times 10^4$ years). This anomaly is quite large on the profiles across the East Pacific Rise ($v = 4$ to 5 centimeters per year) and the Juan de Fuca Ridge ($v = 3$ centimeters per year) but is not visible on profiles across the Reykjanes Ridge ($v = 1$ centimeter per year) and is at about the limit of resolution on the profile across the Indian Ocean ($v = 2$ centimeters per year) (*11, 23, 24*). The minimum strip width ($v \times T$) detectable at the sea surface is therefore about 1 kilometer, and the shortest detectable event is about 2×10^4, 5×10^4, or 10^5 years long, depending on whether the rate of lateral spreading is 5, 2, or 1 centimeter per year (*25*).

On the profiles one can distinguish small peaks due to polarity events from those due to magnetic noise only by determining which peaks are consistent from profile to profile. A difficulty in

doing this arises from the fact that variations in the rate of lateral spreading along one profile may, over long distances, produce cumulative displacements equal to half the wavelength of anomaly peaks, producing a large loss of signal on cross-correlation. In looking for short events it is desirable to tie the magnetic profiles as closely as possible to well-determined points on the radiometric time scale. Displacements due to variable rates of spreading may then be minimized by interpolating between closely spaced points.

Figure 5 shows the total magnetic field anomaly along the Eltanin 19 profile, which crosses the East Pacific Rise at latitude 52°S and longitude 118°W. As is typical of midoceanic ridges, the positive anomaly over the central zone (M-B to M-B in Fig. 5) is complex, for reasons not yet understood. Elsewhere, many of the polarity transitions can be correlated unambiguously with those of the radiometrically determined time scale. For example, the transition from the Matuyama reversed epoch to the Brunhes normal epoch can be easily recognized (M-B in Fig. 5), as can the transition from the Gauss to the Matuyama epochs (G-M) and the

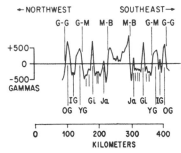

Fig. 5. Magnetic profiles across the East Pacific Rise at latitude 52°S, longitude 118°W (*11, 24*). (G-G, G-M, M-B) Boundaries between the Gilbert, Gauss, Matuyama, and Brunhes polarity epochs. (OG, IG, YG, Gi, Ja) The larger and more consistent positive anomaly peaks. Other small positive peaks are shown with unlabeled lines.

transition from the Gilbert reversed to the Gauss normal epoch (G-G). At high latitudes the steepest gradients in the magnetic profiles occur almost exactly above the boundaries between the normally and the reversely magnetized strips, so that distances between polarity transitions and events may be read directly from the magnetic profiles. It may be seen that the anomalies to the northwest of the rise are more widely spaced; this indicates that the rate of spreading was greater in the direction of the Pacific than toward Antarctica. Therefore the interpolations must be made separately for the two sides of the rise.

The age of the boundary between the Gauss and Matuyama epochs was first found, as a check on the method, by interpolating between the Gilbert-to-Gauss boundary (3.32 million years ago) and the Matuyama-to-Brunhes boundary (0.69 million years ago), both of which have well-determined radiometric ages. Fifteen half profiles were used, all from high latitudes in the South Pacific, North Pacific, and Indian oceans. The mean interpolated age is 2.41 ± 0.03 million years (standard error). This age is consistent with the two radiometrically determined ages of 2.43 million years (reversed magnetization) and 2.45 million years (normal magnetization) which bracket the boundary. When both radiometric and interpolated results are taken into account, the best estimate of the age of the boundary is 2.43 million years.

Three positive anomaly peaks corresponding to the three normal polarity intervals (YG, IG, and OG in Fig. 5) of the Gauss normal epoch appear on those profiles for which spreading rates are rapid. The two reversed events, the Mammoth and the Kaena, which separate the normal intervals have been recognized by radiometric dating (*18, 19*) and from magnetic profiles (*11, 23, 24*). The ages of the midpoints of the two larger anomalies, found by interpolating 13 profiles (Table 1) between the boundaries of the Gauss

Table 1. Ages of the three normal intervals (YG, IG, and OG) in the Gauss normal epoch, as found by interpolation between the Gilbert-to-Gauss boundary (3.32 million years ago) and the Gauss-to-Matuyama boundary (2.43 million years ago) (*36*).

Interval	Number of profiles	Age found by interpolation (million years)	Standard deviation	Standard error	Age of interval midpoint, from K-Ar dating (million years)
YG	13	2.64	0.03	0.01	2.62
IG	8	2.94	.10	.04	2.9
OG	13	3.19	.03	.01	3.19

17

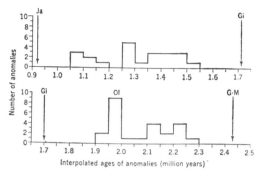

Fig. 6. Histogram of ages of small peaks on magnetic profiles found by interpolating between the Jaramillo event (0.92 million years ago), the Gilsá event (1.71 million years ago), and the Gauss-Matuyama boundary (2.43 million years ago). The most consistent anomaly peak corresponds to the Olduvai event (Ol).

epoch, are in good agreement with the radiometric time scale. The short central anomaly peak has not been dated radiometrically but is bracketed by ages of 3.0 and 2.8 million years obtained for samples from the adjacent reversed events, in agreement with the interpolated age of the normal event of 2.94 ± 0.04 million years.

Two positive anomalies (Ja and Gi in Fig. 5) appear in the Matuyama epoch on almost all magnetic profiles. The smaller and younger anomaly (Ja) has usually been identified with the Jaramillo normal event; the older and larger anomaly, with the Olduvai event (11, 23, 24). The age of 0.93 million years (Table 2) found for the younger anomaly by interpolating between the boundaries of the Matuyama reversed epoch agrees well with the radiometric age of 0.92 million years. However the interpolated age of 1.70 million years for the larger anomaly does not agree with the ages obtained for the Olduvai normal event by potassium-argon dating, most of which are from 1.95 to 2.00 million years. It agrees more nearly with the two ages of 1.60 and 1.65 million years for normally magnetized lava flows from Iceland and Alaska (26), which constitute the evidence for the existence of a normal event termed the Gilsá (26). Vine (27) concluded independently, from detailed study of the Eltanin 19 profile, that the main anomaly in the Matuyama reversed epoch lasted from 1.80 to 1.64 million years ago. Previously Ninkovich et al. (7) had concluded from their paleomagnetic study of marine sediments that the main normal event in the Matuyama reversed epoch began 1.79 million years ago and ended 1.65 million years ago, but they correlated this event with the Olduvai event, as it had been defined previously from radiometric dating (19). It now appears that the main normal-polarity

event in the Matuyama epoch is not the Olduvai but rather the Gilsá event. Its younger boundary is determined from radiometric dating to be at 1.61 million years ago. Its older boundary, although poorly determined radiometrically, appears from the magnetic anomalies and from the paleomagnetism of sediment cores (7) to be at about 1.79 million years ago.

To determine whether the profiles contain evidence for shorter events, ages were found by interpolation for all the small positive peaks in the Matuyama reversed epoch (Fig. 5) on 17 half profiles from high latitudes. The most consistent anomaly appears on 11 half profiles at positions having interpolated ages between 1.94 and 2.00 million years (Fig. 6). The radiometric ages associated with the Olduvai event are almost all in this same range, a fact which indicates that the two are correlative. From the size of the anomaly the duration of the event is estimated to be 3×10^4 years, an estimate which agrees with Vine's (27) interpretation of the Eltanin 19 profile. On the other hand, in paleomagnetic studies of deep-sea sediments no consistent evidence has been found for the existence of both the Gilsá and the Olduvai events (28). Apparently the processes by which sediments become magnetized are sufficiently irregular and integrative in nature as to make it difficult to consistently resolve events as short as 3×10^4 years.

Why have so many normally magnetized rocks been found from the short Olduvai event and so few from the much longer Gilsá event? The answer appears to lie in the uneven distribution of the ages of both the normally and the reversely magnetized samples. Few samples of either polarity have ages in the range between 1.6 and 1.8 million years, whereas many samples of both polarities have ages between 1.9 and 2.1 million years. The overlap in the ages of normally and reversely magnetized samples with ages between 1.9 and 2.1 million years is due to the fact that the duration of the Olduvai event is shorter than many of the discrepancies due to dating errors. The high ratio of reversely to normally magnetized samples is in accord with the conclusion that the Olduvai normal event was short.

Rocks with ages of from 2.2 to 1.9 million years used for these studies are from volcanic formations from Africa, Alaska, the western United States, Cocos Island, Australia, Reunion Island in the Indian Ocean, and Iceland—a range which suggests that an interval of unusually intense volcanic activity occurred 2 million years ago. It was also at about this time that a marked evolutionary change occurred in marine microorganisms, including Radiolaria (29) and Foraminifera (30), marking the beginning of the Pleistocene. Such faunal changes have been noted near several polarity transitions and, in particular, at the time of or slightly before the Olduvai event (29, 30), lending support to earlier suggestions (31) that, when the earth's field decayed during a reversal, the increase in radiation would have been large enough to produce a sudden increase in

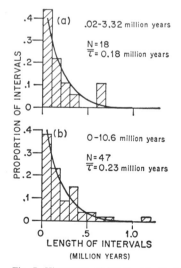

Fig. 7. Histograms of lengths of polarity intervals, (a) from the reversal time scale of Fig. 4, based on ages obtained by potassium-argon dating, and (b) from the reversal time scale of Vine (27) based on the Eltanin 19 magnetic profile. (Solid curves) Distribution function of Eq. 2 for P = 0.05. N is number of intervals.

the rate of evolution. However, recent quantitative studies indicate that the shielding provided by the earth's atmosphere is so great that the increase in radiation on complete collapse of the field would be no more than 12 percent (*32*). Such an increase is comparable with the variations that occur during a normal sunspot cycle, or with the normal variation between equatorial and polar regions. The accompanying increase in mortality rate and in the rate of spontaneous mutation is too small to account for the extinction of a species. An alternative explanation is that the faunal extinctions at the beginning of the Pleistocene are related to volcanic activity which occurred at that time, a possible causal link being increased absorption and reflection of sunlight by volcanic dust in the atmosphere.

An anomaly slightly older than the Olduvai [the X anomaly of Heirtzler *et al.* (*11*)] appears on some profiles. The interpolated ages, although less well grouped than those of the Olduvai anomaly, provide weak evidence that a short event may exist. Support for this is provided by two radiometric ages slightly greater than those of the main group of normally magnetized Olduvai samples (Fig. 4), but again the evidence is rather weak. There are weak suggestions of possible events at 1.26 and 2.24 million years ago, but whether they are real cannot be determined radiometrically because both occur where there are gaps in the data.

In summary, all of the events with durations $T \geqslant 3 \times 10^4$ years that have been found from radiometric dating have also been identified on the magnetic profiles. The radiometric ages of the events and the ages derived through interpolation from the profiles are in excellent agreement. The profiles further establish that there are no additional, undetected events longer than 3×10^4 years, even where there are gaps longer than this between known radiometric ages. Events shorter than 3×10^4 years are below the level of experimental noise and have not been resolved convincingly from the magnetic profiles.

Sedimentary Record of Short Events

The paleomagnetism of marine sediments provides a third possible source of information about short events (*28–30*). In general, the record of polarity epochs in sediments agrees with the

Table 2. Age of the Ja and Gi events, as found by interpolation between the Gauss-to-Matuyama boundary (2.43 million years ago) and the Matuyama-to-Brunhes boundary (0.69 million years ago) (*36*).

Interval	Number of profiles	Age found by interpolation (million years)	Standard deviation	Standard error	Age from K-Ar dating*
Ja	17	0.93	0.05	0.01	0.92
Gi	17	1.70	.10	.02	1.60–1.86

* See Fig. 4 for range of uncertainty.

radiometric time scale. However, the record of short events is obscured by noise due to stratigraphic gaps, variations in the rate of deposition, and delays in the time between deposition and magnetization of the sediments. The amount of delay varies, depending on variations in the amount of reworking by organisms, on the rate of compaction, and on authigenic chemical changes in the ferromagnetic minerals. In addition, there is a loss of information about short polarity intervals because the magnetizing processes in sediments are integrative over time intervals comparable with the durations of the shorter events. As a result, the research on sediments has not produced consistent evidence for the existence of events shorter than those detectable by other methods ($T = 3 \times 10^4$ years)—with one notable exception:

Ninkovich *et al.* (*7*) report rather convincing evidence for two short polarity intervals ($T \sim 10^4$ years) at the end of the Gilsá event. Thus, while the paleomagnetic research on sediments has provided valuable information about the duration of the longer events and has helped extend the time scale back into the Gilbert reversed epoch, it has not resolved the question of the frequency of very short events.

Discussion

Over the interval $0.02 < t \leqslant 3.32$ million years ago (the part of the reversal time scale for which the number of short events is most completely known), a good fit to the observed frequency of polarity intervals (Fig. 7) is obtained from Eq. 2 when P is set

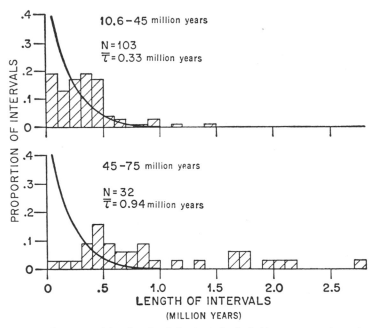

Fig. 8. Histograms of lengths of polarity intervals. Both histograms are from the time scale for intervals of Heirtzler *et al.* (*11*) based on magnetic profiles. Solid curves are as in Fig. 7.

equal to 0.055. The observed number of very small events is consistent with the number predicted by the model. Reversal times scales going back several tens of millions of years have been obtained from magnetic profiles on the assumption that the rate of sea-floor spreading was constant (11, 23, 27). The best fit of Eq. 2 to the data of Vine (27) for the past 10.6 million years is obtained with $P = 0.043$. The observed proportion of very short events (T $<$ 5 $\times$ 10^4 years) is slightly smaller than the proportion found for the interval $0.02 < t \leqslant 3.32$ million years; this suggests that, as had been suspected, some of the shortest events were not resolved on the magnetic profiles.

It is concluded from the analysis discussed here that, on the average, 20 fluctuations in intensity occur between successive polarity reversals. This conclusion, although based on a model in which dipole intensity fluctuations are periodic with a period of 10^4 years, would also be valid if the intensity changes were nonperiodic, with an average time between minima of 10^4 years. The corresponding probability that a geomagnetic reversal will result from the decrease in dipole moment currently in progress is 5 percent.

If sea-floor spreading has occurred at a constant rate, the marine magnetic profiles may be interpreted to yield a reversal time scale going back 75 million years (11). The apparent average duration of the polarity intervals was greater during the time $10.6 < t < 45$ million years than during the past 10.6 million years, and during the time $45 < t \leqslant 75$ million years the average length was still greater (Fig. 8). Part of the difference may be due to variations in spreading rate (11, 33). However, this would account only for an apparent change in the probability P and not for the observed change in the shape of the distribution. The latter change may be more apparent than real, however, if, as is quite possible, a small number of additional short events occurred which have not yet been detected. Numerical experiments indicate that about 30 randomly distributed short events would change the observed distribution to one fitted by Eq. 2 when $P = .02$.

As one goes farther back in time, an average length of 10^7 years has been reported for polarity intervals in the early Paleozoic (34), and the Kiaman reversed-polarity interval during the late Paleozoic was 5 $\times$ 10^7 years long (13). Clearly, the average length of polarity intervals and hence the value of P have changed during the earth's history, reflecting changes in the physical conditions which control the geomagnetic dynamo. The possible importance to magnetohydrodynamic processes of changes in the properties of the core-mantle interface has been pointed out by Hide (35) and by Irving (13). These changes may result from large-scale tectonic processes, such as polar wandering accompanied by mass transport in the lower mantle. Hence it is conceivable that individual changes in polarity are produced by individual tectonic events. The present analysis suggests the alternative possibility that the timing of reversals is controlled on two levels. The *average* length of polarity intervals is controlled by conditions at the core-mantle interface and hence may be related to tectonic processes in the mantle, whereas the length of individual polarity intervals is determined by random processes in the fluid core.

Observationally, the main difference between the two interpretations is in the cutoff anticipated for short periods. In the present model, the cutoff is 10^4 years, the assumed period of fluctuations in the intensity of the dipole field. In Hide's model the cutoff is estimated to be from 10^5 to 10^6 years, depending on the strain rate of the lower mantle and the minimum displacement of the core-mantle interface needed to produce a change in polarity. With the discovery of polarity events at least as short as 3 $\times$ 10^4 years, it appears increasingly likely that the timing of individual reversals is controlled by processes occurring in the fluid core.

References and Notes

1. Recent analyses of the field include those of K. L. McDonald and R. H. Gunst [*ESSA* (*Environ. Sci. Serv. Admin.*) *Tech. Rep. IER 46-IES 1* (1967)], L. Hurwitz, D. G. Knapp, J. H. Nelson, and D. E. Watson [*J. Geophys. Res.* **71**, 653 (1966)], and J. C. Cain, W. E. Daniels, S. J. Hendricks, and D. C. Jensen [ibid. **70**, 3647 (1965)]. Cain and Hendricks [*NASA* (*Nat. Aeron. Space Admin.*) *Tech. Note TN D-4527* (1968)] report that the rate of decrease has dropped since 1900. A summary of the older analyses is given by E. H. Vestine [*J. Geophys. Res.* **58**, 127 (1953)].
2. B. R. Leaton and S. R. C. Malin, *Nature* **213**, 1110 (1967); K. L. McDonald and R. H. Gunst, *ESSA* (*Environ. Sci. Serv. Admin.*) *Tech. Rep. IER 46-IES 1* (1967).
3. A complete account of this technique is given by E. Thellier and O. Thellier, *Ann. Geophys.* **15**, 285 (1959); R. Coe [*J. Geomagnetism Geoelec.* **19**, 157 (1967)] and P. J. Smith [*Earth Planetary Sci. Letters* **2**, 99 (1967)] discuss experimental difficulties.
4. P. J. Smith, *Roy. Astron. Soc. Geophys.* **12**, 321 (1967); ibid. **13**, 417 (1967).
5. A. Cox and R. R. Doell, *Bull. Seismol. Soc. Amer.* **54B**, 2243 (1964).
6. C. G. A. Harrison and L. K. Somayajulu, *Nature* **212**, 1193 (1966).
7. D. Ninkovich, N. Opdyke, B. C. Heezen, J. H. Foster, *Earth Planetary Sci. Letters* **1**, 476 (1966).
8. A. Cox and G. B. Dalrymple, *J. Geophys. Res.* **72**, 2603 (1967).
9. R. Coe, ibid., p. 3247.
10. R. Doell and A. Cox, *Advan. Geophys.* **8**, 221 (1961).
11. J. R. Heirtzler, G. O. Dickson, E. M. Herron. W. C. Pitman, III, X. LePichon, *J. Geophys. Res.* **73**, 2101 (1968).
12. A. Cox, ibid., p. 3247.
13. E. Irving, ibid. **71**, 6025 (1966); B. E. McMahon and D. W. Strangway, *Science* **155**, 1012 (1967).
14. W. V. R. Malkus, ibid. **160**, 259 (1968).
15. T. Rikitake, *Proc. Cambridge Phil. Soc.* **54**, 89 (1958); D. W. Allan, ibid. **38**, 671 (1962); J. H. Mathews and W. K. Gardner, *Naval Res. Lab. Rep. 5886* (1963), p. 1.
16. A. Cox, R. R. Doell, G. B. Dalrymple, *Nature* **198**, 1049 (1963); *Science* **142**, 382 (1963).
17. The alternatives of using the radiometric ages of the boundaries to identify the epochs or of numbering the epochs in sequence were considered. However it was felt that each modification of the time scale would require a confusing change in the nomenclature if numbers were used, especially if (as has happened) additional reversals were discovered.
18. I. McDougall and D. H. Tarling, *Nature* **202**, 171 (1964); R. R. Doell and G. B. Dalrymple. *Science* **152**, 1060 (1966); I. McDougall and S. H. Chamalaun, *Nature* **212**, 1415 (1966).
19. A. Cox, R. R. Doell, G. B. Dalrymple, *Science* **144**, 1537 (1964).
20. The criteria used are as follows. (i) The paleomagnetic study includes laboratory measurements of magnetic stability. (ii) The precision of the potassium-argon age determination is < 0.1 million years for ages $0 < t \leqslant 2.0$ million years and < 5 percent for ages > 2.0 million years. (iii) The magnetic and age measurements were made on rocks and minerals of types known to yield reliable results, and the samples for both measurements were collected from the same volcanic cooling unit (see 8).
21. G. B. Dalrymple, A. Cox, R. R. Doell, S. Grommé, *Earth Planetary Sci. Letters* **2**, 163 (1967).
22. N. Bonhommet and J. Babkine, *Compt. Rend.* **264**, 92 (1967); their results were based on reversed magnetization of the Laschamp volcanic cone in France. The available radiometric data do not precisely determine the age and duration of this event but suggest that it was late Quaternary or Recent and that it was very short. This event is less well documented than the Jaramillo, Gilsá, and Olduvai events, each of which is supported by independent results from two or more volcanic formations from different parts of the world.
23. W. C. Pitman, III, and J. R. Heirtzler, *Science* **154**, 1164 (1966).
24. V. J. Vine, ibid., p. 1405.
25. A boundary zone with low average intensity of magnetization and containing rocks of mixed polarity probably exists between the magnetized strips as the result of irregularities in the process by which the crust is magnetized [F. J. Vine and J. T. Wilson, *Science* **150**, 485 (1965); C. G. A. Harrison, *J. Geophys. Res.* **73**, 2137 (1968)]. The width of the strip of magnetized crust which produces the anomaly is therefore smaller than the quantity $(v \times T)$, but the estimate for the minimum length of an event is not changed if the thickness of the boundary zone is independent of spreading rate.
26. I. McDougall and H. Wensink, *Earth Planetary Sci. Letters* **1**, 232 (1966); A. Cox and G. B. Dalrymple, ibid. **3**, 173 (1967).
27. F. J. Vine, in *History of the Earth's Crust*, R. A. Phinney, Ed. (Princeton Univ. Press, Princeton, N.J., in press).
28. N. D. Watkins and H. G. Goodell [*Earth Planetary Sci. Letters* **2**, 123 (1967)] report that both the Gilsá and Olduvai events are recorded in marine sediments from the Pacific-Antarctic Rise, whereas J. D. Hays and N. D. Opdyke *Science* **158**, 1001 (1967)] dispute this interpretation. However, in the three cores

studied by Hays and Opdyke, the two reversed events in the Gauss normal epoch are resolved on only one of the cores (Eltanin 13, core 3), and this same core also shows a clearly defined normal event with an interpolated age between the Gilsá event and the Gauss-Matuyama boundary of 1.99 million years ago, in agreement with the radiometric age of the Olduvai event.

29. J. D. Hays and N. D. Opdyke, *Science* **158**, 1001 (1967).
30. W. A. Berggren, J. D. Phillips, A. Berteis, D. Wall, *Nature* **216**, 253 (1967).
31. R. J. Uffen, *ibid.* **198**, 143 (1963); J. F. Simpson, *Bull. Geol. Soc. Amer.* **77**, 197 (1966).
32. C. Sagan, *Nature* **206**, 448 (1965); C. J. Waddington, *Science* **158**, 913 (1967); C. G. A. Harrison, *Nature* **217**, 46 (1968); D. I. Black, *Earth Planetary Sci. Letters* **3**, 225 (1968).
33. J. Ewing and M. Ewing, *Science* **156**, 1590 (1967); E. D. Schneider and P. R. Vogt, *Nature* **217**, 1212 (1968).
34. A. N. Khramov, V. P. Rodionov, R. A. Komissarova. Present and past of the geomagnetic field," *Can. Defence Res. Board Pub. T 460 R* (1960), p. 1.
35. R. Hide, *Science* **157**, 55 (1967); E. Irving [*J. Geophys. Res.* **71**, 6025 (1966)] has also suggested that reversals may be due to shifts in the earth's axis of rotation.
36. The profiles used are from the South Pacific (EL20, EL19, S16, S18, MN5, S15) (see *11*), the Indian Ocean (V16) (see *11*), and the Juan de Fuca Ridge (see *20*).
37. A. Cox, R. R. Doell, G. B. Dalrymple, *Geol. Soc. London Quart. J.*, in press.
38. I am indebted to R. R. Doell and George Thompson for reviewing the manuscript of this article, and to F. J. Vine for supplying his interpretation of the Eltanin 19 profile prior to publication.

2

Reprinted from *Geol. Soc. America Bull.* **83**:551–574 (1972)

Review of the Development of the Geomagnetic Polarity Time Scale and Discussion of Prospects for Its Finer Definition

NORMAN D. WATKINS *Graduate School of Oceanography, Narragansett Marine Laboratory, University of Rhode Island, Kingston, Rhode Island 02881*

ABSTRACT

Until recently, studies of the geomagnetic polarity time scale were confined for the most part to the problem of defining the broad structure of the polarity history for the period t = 0 to 6 m.y. Now that this is well advanced, two interrelated directions of research are evolving: the problem of the fine detail of the geomagnetic history is beginning to attract active interest, and the time scale is poised for widespread utilization as the time framework for research on the various phenomena represented in deep-sea sedimentary cores and in outcrops of diverse sediments on land. These latter applied studies are in turn capable of contributing to definition of the polarity scale, but are also very easily capable of distorting the known polarity scale unless rigorous criteria are applied to distinguish polarity changes from other geomagnetic behavior and spurious data. It is therefore appropriate at this time to review the history of the development of the polarity time scale and some of its initial applications to ocean-crust dating, deep-sea sedimentary core studies, and work on continental sediments, in order to illustrate the great vulnerability of the time scale to circular reasoning and the difficulty of reliable interpretation of limited amounts of data. Special attention is paid to the problem of defining short polarity events. This is most efficiently demonstrated by focusing discussion on the polarity history between t = 0 and 2.6 m.y.

The nomenclature conventions and some associated ambiguities are also reviewed. It is concluded that the establishment of an international commission or panel for guiding the acceptance of new and revised details of the polarity history and its nomenclature will be of substantial value in maintaining the future coherence of the polarity scale. This would provide the essential control for resolution with minimum ambiguity of the various geological and geophysical phenomena examined within the framework of the geomagnetic polarity time scale.

INTRODUCTION

The geomagnetic field has periodicities of the order of milliseconds to tens of millions of years. The former are of external origin, but the longer periods are of internal origin. Spherical harmonic analysis of the present field reveals two dominant terms: a centered dipole which is now inclined at 11° to the spin axis, and the non-dipole field. Both originate below the mantle. These mathematically separable components are observed to move at different rates. The non-dipole field, which (depending on geographic position) constitutes from zero to about one-third of the total field at the Earth's surface, changes at a much faster rate than does the dipole field. Within the limits of the available observatory data, the main dipole field has remained virtually stationary for at least the past two hundred years, whereas the present non-dipole field drifts over-all westward at about 0.2° of longitude per year. What is known about the geomagnetic field origins has been summarized by Jacobs (1963).

The major contribution of paleomagnetism to knowledge of geomagnetic field behavior is the discovery and proof that the field has frequently reversed its polarity in the geological past. The reversal spectrum ranges from periods of fixed polarity of 10^4 to 10^7 yrs. For unknown reasons, the Tertiary era has been a time of high reversal frequency, with polarity changes probably occurring on the average at least every 400,000 yrs (Heirtzler and others, 1968). The polarity reversal spectrum for the past

600 m.y. has been examined by McMahon and Strangway (1967), McElhinny (1971), and McElhinny and Burek (1971). These summaries show that the reversal frequency may have been of the order of 10^7 yrs (or more) per polarity change during the Cretaceous and Jurassic periods, and during the Paleozoic era.

By combining paleomagnetism and K : Ar dating methods, the polarity reversal history for the past 5 m.y. (Fig. 1) is becoming well defined. One of the convenient properties of the geomagnetic field is that it can be recorded geologically in different ways. Igneous rocks record the field as thermal or chemical remanent magnetism, and fine sediments as detrital remanent magnetism. Such materials can readily preserve magnetic polarities despite the great possible range of emplacement characteristics involved. The geomagnetic polarity history can also be inferred from oceanic magnetic anomaly patterns since sequential dike injection into mobile oceanic crust is very probably associated with crustal spreading (*summary by* Watkins and Richardson, 1971).

In the eight years since Cox and others (1963a) published the first relevant compilation, the geomagnetic polarity time scale has assumed a perhaps unparalleled importance in the geological sciences. Cenozoic polarity changes, originally detected in basaltic lavas, have now been recognized throughout the world in magnetic anomaly patterns associated with ocean-floor genesis (Vine, 1966, 1968; Heirtzler and others, 1968) and as magnetic polarity boundaries in deep-sea sedimentary cores (Opdyke and others, 1966; Watkins and Goodell, 1966, 1967a) and outcropping sediments of both continental (Bucha, 1970) and oceanic (Kennett and others, 1971) origin. Apart from the fundamental relevance toward an understanding of geomagnetic field mechanisms (Cox, 1968), definition of the isochrons of global extent which constitute the polarity time scale would appear to provide a tool which promises resolution of many stratigraphic, paleontological, paleoclimatological, and other problems (examples *by* Hays and others, 1969; Hays, 1971; Kennett and others, 1971), extension of continental zonation to submarine sequences, and even fine correlation of the timing of oceanic crustal mobility with land-based events. In common with the development of any stratigraphic system, however, definition of the fine structure of the polarity time scale is likely to present several problems (particularly when the nature and limitations

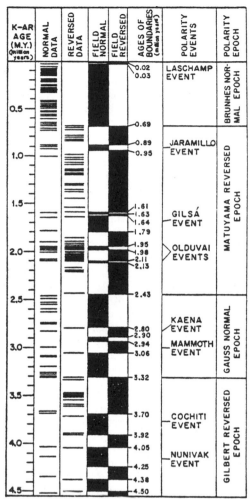

Figure 1. Geomagnetic polarity time scale, as presently known for the period t = 0 to 4.5 m.y. (*from* Cox, 1969). Diagram includes ages of constituent data points, and nomenclature of epochs and events. For further discussion, see text. Published with permission of the author. Copyright 1969 by the American Association for the Advancement of Science.

of the experimental methods are considered), conceivably leading to ambiguities and misconceptions involving not only the polarity history, but also associated diverse applications of the time scale. The inherent potential problem of nomenclature of any stratigraphic method is magnified in the polarity time scale, because of its global relevance, probable incomplete state, and also the ambiguous character of one polarity boundary, which is not necessarily distinguishable from another of a different age.

It is therefore considered appropriate and instructive at this time to review the history of the development of the polarity time scale and its nomenclature as now known (Fig. 1). Of major importance is the need for development of standards or criteria designed to prevent addition of possible spurious and misleading details to the polarity time scale. The period from t = 0.6 to 2.6 m.y. is particularly interesting in this context, since, as will be explained, ambiguities and some confusion have occurred during the evolution of this part of the scale. An understanding of the problems involved in this process is therefore considered as likely to assist in refining other segments of the scale. The relevance of the time scale development to ambiguities in interpretation of paleomagnetic data from deep-sea sedimentary cores and the shorter period polarity behavior will be stressed. It is highly probable that during the next five years, the greatest use of the polarity scale will take place in studies of upper Cenozoic sediments of various types. Micropaleontologists, sedimentologists, and others who will use the scale, should be aware of its evolution to date, in order to assist in evaluation of changes which will undoubtedly be proposed in the future.

POLARITY SCALE EVOLUTION

Irving (1964) has summarized the observations of magnetic polarities in volcanic rocks, bricks, and pottery by nineteenth- and early twentieth-century scientists. Principal among these are results of Melloni (1853), who showed that some historical lavas from Mt. Vesuvius acquired their directions of magnetism roughly parallel to the Earth's field; Brunhes (1906), who showed that several lava flows in the Massif Central of France were magnetized antiparallel to the present magnetic field direction; and Mercanton (1926) who successfully extended observations to both hemispheres, in pursuing the speculation that reversals of geomagnetic fields, if correct, were likely to be worldwide.

Matuyama (1929) obtained the first magnetic stratigraphy: early Quaternary lavas contrasted in polarity with younger lavas in Japan and Manchuria. Roche (1951, 1956) obtained similar results: he noted that in France, upper Pleistocene lavas are magnetized parallel to the present magnetic field, while lavas of lower Pleistocene are magnetized in the opposite direction. Opdyke and Runcorn (1956) made a similar observation using rocks from the southwestern United States. Nagata and others (1957) noted that 28 normal polarity lavas

overlie nine lavas of reversed polarity, in their paleomagnetic study of Quaternary Volcanics near Tokyo. On purely stratigraphic grounds, they suggested that these results represented a change from reversed to normal polarity at about t = 0.9 m.y. Rutten (1959) initiated use of K:Ar data into geomagnetic polarity history studies; he concluded that some Italian volcanos recorded the present geomagnetic polarity ("N_1") for at least the past 470,000 yrs, and that an earlier period of normal polarity ("N_2") existed 2.4 m.y. ago. Rutten and Wensink (1960) applied this concept to correlation between lavas of France and Iceland, and introduced the idea of magnetic definition of a global stratigraphic horizon: the enigmatic Pliocene-Pleistocene boundary was proposed to be equivalent in time to the base of the proposed second period of normal polarity (the "N_2/R_1" boundary).

It became clear that any extension of this proposed magnetic stratigraphy or development of a polarity time scale would require systematic application of K:Ar and paleomagnetic methods.

The critical studies leading to definition of the polarity scale between t = 0.6 and 2.6 m.y. are summarized in chronological order, below, corresponding to Figures 2 and 3. Unless otherwise indicated, the data involved result only from K:Ar and paleomagnetic measurements on basaltic rocks.

1963

1. In June, Cox and others (1963a) published data from six volcanic flows and plugs from California (Fig. 2, i). The K:Ar results were published or made available earlier by the University of California at Berkeley isotope laboratory (Evernden and others, 1957, 1964; Dalrymple, 1963). These ages were all determined by Dalrymple. The favored interpretation was that evidence existed to show a possible one million year periodicity in the polarity history. At that time, the convention of calling the most recent period of normal polarity (in the same sense as the present field) the N_1 epoch, the most recent period of reversed polarity (in the opposite sense of the present field), the R_1 epoch, and earlier corresponding periods N_2, R_2, and so on, was adopted, after Rutten (1959). In retrospect, it must be considered as quite remarkable that these initial six data points led to such early broad delineation of the major characteristics of the polarity scale for the period involved.

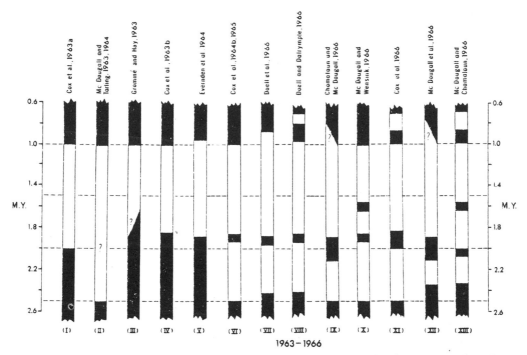

Figure 2. Geomagnetic polarity time-scale evolution between 1963 and 1966, for the period t = 0.6 to 2.6 m.y. Black = normal polarity; clear = reversed polarity. Inclined polarity boundaries indicated range of suspected precision errors. M.Y. = millions of years before present. Horizontal dashed lines are at 0.5-m.y. intervals. References are to those publications suggesting the particular polarity time scale illustrated. All scales are due to K:Ar and paleomagnetic data of igneous rocks. See text for discussion of details.

2. In October, McDougall and Tarling (1963), using results from Hawaii proposed revision of the R_1/N_2 boundary from 2.0 to 2.5 ± 0.1 m.y. (Fig. 2, ii). Even at this early stage, complications were suspected; detection of a reversely magnetized lava of 2.95 m.y. age between normally magnetized lavas led to the conclusion that "behavior of the magnetic field is more complex than has been suggested previously" (*from* McDougall and Tarling, 1963, p. 5). This similarity in results from Hawaii and California provided an early emphasis of the stratigraphic potential of the time scale.

3. In November, Grommé and Hay (1963) published the paleomagnetic results from the basalt member of Bed I (Hay, 1963) of Olduvai Gorge, Tanganyika. The K:Ar age of approximately 1.8 m.y. had been previously made available, prior to publication, by Evernden and Curtis (1965). The measured normal polarity led Grommé and Hay (1963) to suggest that the R_1/N_2 boundary was between t = 1.6 and 1.9 m.y. (Fig. 2, iii).

4. Cox and others (1963b) presented ten new data points from the Sierra Nevada in October, and quoted the Grommé and Hay (1963) result (which was in press at that time) to suggest that N_1 and R_1 were of the order of 1 m.y. duration, and N_2 appeared to be 25 percent longer (Fig. 2, iv).

1964 and 1965

1. In February, Evernden and others (1964) published a table of polarities and K:Ar ages for 28 volcanic units, as part of an evaluation of vertebrate paleontological zonation of North America. Of these dates, 17 had been published earlier by Cox and others (1963a, 1963b). No polarity log was presented, but that given as Figure 2, v in this paper can be deduced from the data.

2. The difference between the polarity scales of Cox and others (1963a, 1963b; and Figs. 2, i and iv) and McDougall and Tarling (1963; and Fig. 2, ii) was reconciled by the recognition of an error in the sampling of what was assumed to be a dated lava, by Cox and others (1964a), as summarized by McDougall and Tarling (1964). The considerable importance of the

25

Olduvai Gorge result (Grommé and Hay, 1963) which at that time was the only known normal polarity lava between t = 1.0 and 2.4 m.y., was stressed by McDougall and Tarling. For reasons which will be discussed in detail later, this result is still of considerable importance. At this stage, the known polarity scale was composed of only 35 data points for the whole of the period t = 0 to 3.5 m.y.

3. Additional results from the Pribiloff Islands, Alaska, led Cox and others (1964b, 1965) to the conclusion that a relatively short period of normal polarity occurred within the R_1 epoch at about t = 1.9 m.y. (Fig. 2, vi). The convention of calling short periods of fixed polarity "events" was established with the event being named for the geographic site where the discovery was made. Thus the Olduvai event was recognized. In addition, realization of the problem involved in the assumption that the N_1, R_1, N_2 assignments were complete on initial definition, led to the convention of calling the epochs after deceased scientists who had contributed to an understanding of Earth magnetism: the Brunhes, Matuyama, Gauss, and Gilbert epochs (Fig. 1) were named for the periods t = 0 to 1.0 m.y., 1.0 to 2.5 m.y., 2.5 to 3.4 m.y., and pre-3.4 m.y., respectively (Cox and others, 1964b). A second event, of reversed polarity within the Gauss epoch at about t = 3.0 m.y., recognized as the result of measurements in a basaltic flow from Mammoth Lake, California, was established as the Mammoth event, although McDougall and Tarling (1963) appear to have first recognized it as a single reversed polarity lava at t = 2.96 ± 0.06 m.y., within a sequence of normal polarity lavas at Waianae on Oahu, Hawaiian Islands. McDougall (1964) noted, however, the possibility that this date may be stratigraphically inconsistent with other K:Ar ages in the Waianae volcano and is therefore questionable. This polarity scale (Fig. 2, vi) was later adopted by Vine and Wilson (1965) and Harrison (1966) for the first attempts at fitting the polarity scale to crustal spreading analyses of marine magnetic anomalies, and for comparison with the polarity sequence in deep-sea sedimentary cores, respectively: it was not realized that the poor results were due to the imperfectly defined state of the polarity scale.

In a note added in proof, Cox and others (1965) noted that new data from the normal polarity Bishop Tuff, which was originally dated by Evernden and others (1964) at t = 0.98 m.y., and incorporated into the original 1963 publication of Cox and others (1963a), suggest an age of about 0.7 m.y. This single modification would place the Brunhes/Matuyama boundary within the period 0.85 ± 0.15 m.y.

1966

1. In January, Doell and others (1966) added 12 new data points from California and Nevada, and proposed the polarity scale for the Matuyama epoch shown in Figure 2, vii. Within the period shown (Fig. 2, vii), 20 data points were available.

The authors also introduced the concept of data acceptability criteria for polarity age methods. Data for the polarity time scale would now require laboratory removal of unstable magnetic components, and K:Ar ages with analytical precision resulting in estimated standard deviations not exceeding 0.1 m.y. for the t = 0 to 3.5 m.y. time range.

2. In May, Doell and Dalrymple (1966) recognized, in results from a series of rhyolite domes, the second normal polarity event within the Matuyama epoch, at about t = 0.9 m.y. This was named the Jaramillo event after the New Mexico discovery site. Full details of the discovery were published later (Doell and others, 1968). The modified scale (Fig. 2, viii) marked a highly significant stage in polarity history definition: it was this scale which was adopted for the early dating of long deep-sea sedimentary cores by Opdyke and others (1966) and the first successful crustal spreading analyses of Vine (1966), Pitman and Heirtzler (1966), and others. Implicit in these spectacular and immensely important applications of the polarity scale was the confirmed existence of a maximum of two events (other than very short events) in the Matuyama epoch. It appeared at this time then, most certainly to the nonspecialist, that the polarity scale between t = 0.6 and 2.6 m.y. was nearing its final true form, since the three separate polarity recording systems were in agreement. The possibility of circular reasoning (by which evidence for additional unknown events was attributed to "noise" or was simply missed) was apparently overlooked.

3. In June, Chamalaun and McDougall (1966), in presenting new results from the island of Réunion recognized inconsistencies in results around t = 1.0 m.y. (Fig. 2, ix). They suggested that a hitherto undiscovered event might exist at about this part of the scale. The

Jaramillo event, discovered almost simultaneously by Doell and others (1966), was involved. A practice of presenting only those data resulting from paleomagnetic and K:Ar analyses of the same body was initiated.

4. In July, McDougall and Wensink (1966) reported, in a manuscript submitted for publication in June, the discovery of another event within the Matuyama epoch (Fig. 2, x). Within a series of reversely magnetized Icelandic lavas, two normal polarity lavas occur, separated by a single, *possibly* reversed polarity, lava. The youngest normal polarity lava yielded a K:Ar age of 1.60 ± 0.05 m.y. The authors considered the result "to be a very reliable date as the rock was quite free of alteration," and so the Gilsa event was proposed. They speculated that the older normal polarity lava might represent the Olduvai event. In one of the early publications on the paleomagnetism of deep-sea sedimentary cores, Watkins and Goodell (1967a), while stressing the dangers of circular reasoning in identification of polarity event in cores, confirmed the existence of three events within the Matuyama epoch, including the proposed Gilsa. The data in slightly earlier studies of other deep-sea sedimentary cores by Ninkovitch and others (1966) and Dickson and Foster (1966) were reinterpreted to support the reality of the Gilsa event, and in the case of the latter, the existence of three discrete normal polarity events within the Matuyama. Many subsequent publications (Berggren and others, 1967; Glass and others, 1967; Hays and Opdyke, 1967; Ninkovitch, 1968; Ericson and Wollin, 1968; Phillips and others, 1968; Hays and others, 1969; Nakagawa and others, 1969; Hays, 1970; Hays and Ninkovitch, 1970; Foster and Opdyke, 1970; Talwani and others, 1971) were to nevertheless employ only two events in the Matuyama, rejecting the possibility that the Gilsa event existed. In the author's opinion, this was to a large extent because of the "reinforcement process" which resulted from crustal-spreading analyses, an incompletely defined polarity scale, and initial deep-sea sedimentary core results combining almost simultaneously to block belief in the existence of the third event.

This ambiguity was unusually significant, since the Pliocene-Pleistocene boundary was later identified on micropaleontological grounds (Glass and others, 1967; Berggren and others, 1967; Hays and Opdyke, 1967; Ericson and Wollin, 1968; Phillips and others, 1968; and Hays and others, 1969) to occur within one of these polarity events. If the Olduvai was accepted as the event detected in cores, then the problem of the age of the Pliocene-Pleistocene boundary would be, by definition, resolved at 1.9 to 2.0 m.y., as stressed by Ericson and Wollin (1968); but if the Gilsa event was identified as the actual event involved, then the age would be 0.3 to 0.4 m.y. younger (Kennett and others, 1971). The importance of unambiguous identification of the events and the dangers of circular reasoning were therefore becoming obvious.

5. Cox and others (1966) reported results from the Pribiloff Islands, Alaska, which included additional data confirming the existence of the Olduvai event (Fig. 2, xi). Four normally magnetized lavas have ages between 1.79 and 1.96 m.y., which clearly brackets the original Olduvai event age. Sufficient data now existed for estimation of the duration of the Olduvai event, which was suggested to be about 0.1 m.y. Since three reversely magnetized lavas on St. George Island have ages indistinguishable from the four Olduvai event lavas, the authors introduced the possibility of a double or split event, where a short period of reversed polarity exists within the period represented by the event. Shortly afterward, Ninkovitch and others (1966, Fig. 11) detected evidence of a very short period of reversed polarity within the Olduvai event, which they dated (using sedimentation rates) as between 1.68 and 1.69 m.y. This is very close to the proposed age of 1.6 ± 0.05 m.y. of the Gilsa event (McDougall and Wensink, 1966).

6. In December, McDougall and others (1966) published results from the Newer Volcanics of Victoria, Australia (Fig. 2, xii). This paper was submitted for publication prior to the Jaramillo event discovery (Doell and Dalrymple, 1966). It was proposed that, because of the discovery of three reversely magnetized basalts with ages of about 0.8 m.y., the Brunhes/Matuyama boundary was now at t = 0.75 ± 0.07 m.y., and that a normal polarity event might exist at 0.81 m.y. Additional results showed that the Matuyama/Gauss boundary was at t = 2.35 ± 0.15 m.y. The concurrence of the first two of these suggestions with the Doell and Dalrymple result was mentioned in a note added to the paper in August, following review. A previously published (McDougall and Tarling, 1963) age of t = 0.86 m.y. for a normally magnetized basalt from the Kula Volcanic Series on the Island of Maui, representing possibly the first detec-

tion of what was now known as the Jaramillo event, was withdrawn from the polarity scale data, since the data sample was not from the same body as the paleomagnetically investigated lava.

7. At the end of 1966, McDougall and Chamalaun (1966) presented additional data from the island of Réunion, along with a revised polarity scale (Fig. 2, xiii). They included results from four normal polarity lavas with an age of 2.02 ± 0.07 m.y. The older limit of the Olduvai event was therefore proposed to be at t = 2.03 ± 0.05 m.y. They noted that this was slightly older than the age of the original Olduvai event detection of Evernden and Curtis (1965). As in the virtually simultaneous analysis of Cox and others (1966), they suggested that a period of geomagnetic field behavior including rapid changes of polarity might have occurred, and that the duration of Olduvai event was of the order of 0.05 to 0.1 m.y. Additional data acceptability criteria were proposed.

These identical conclusions about the detailed polarity history by two well-separated groups working on different sequences in different hemispheres constituted a most impressive verification of the reality of global polarity changes.

1967

1. Grommé and Hay (1967) published further analyses of the Olduvai Gorge K:Ar data (Evernden and Curtis, 1965) and new results from the two lavas of Bed I. They refer to the work of Evernden and Curtis (1965) showing "that the correct age to be assigned to the basalt of Bed I is 1.9 m.y." and conclude that the two closely spaced eruptions occurred during a period of normal polarity, at t = 1.9 m.y. The upper and lower age limits of the eruption are argued as being, depending on assumptions, between t = 1.75 and 1.9 m.y., or between 1.85 and 1.90 m.y. (Fig. 3, ii). Those age ranges were clearly still well separated from the age of the single detection of the Gilsa event, but very consistent with data from proposed detections of the Olduvai event in Alaska, Réunion Island, and Australia.

2. Cox and Dalrymple (1967a) presented statistical methods of evaluating the age of polarity boundaries, based largely on a new method for estimating the precision of K:Ar dating. The polarity scale due to the resulting analysis is given in Figure 3, iii.

3. In late 1967, Cox and Dalrymple (1967b)

confirmed the existence of the Gilsa event (t = 1.60 ± 0.05 m.y., according to McDougall and Wensink, 1966) by the discovery of a normal polarity lava of age 1.65 ± 0.09 m.y. on Nunivak Island, Alaska. They also identified two normal polarity events within the Gilbert Epoch (Fig. 1).

At this stage, with a total of 136 data points available for the entire polarity time scale, it became reasonably certain that the Matuyama epoch extended from about t = 0.7 to 2.4 m.y., with three recognized events (the Jaramillo, the Gilsa, and the Olduvai) occurring at about t = 0.9 m.y., 1.6 m.y., and 1.9 m.y., respectively (Fig. 3, iv).

1968

1. Hoare and others (1968) published fine details of the Nunivak Island study. This included a suggested polarity scale with slightly revised polarity boundaries (Fig. 3, v). The limits of the Olduvai event were placed at slightly older than t = 2.0 m.y. and just younger than 1.9 m.y., thereby accommodating results from Olduvai Gorge, the Pribiloff Islands, Réunion Island, and Victoria, Australia. The Gilsa event is still shown as well separated in time from the Olduvai.

2. Cox and others (1968) published a compilation of all polarity scale results. The preferred polarity scale (Fig. 3, vi) is slightly modified from that of Hoare and others (1968).

3. A new phase in the determination of the polarity time scale was initiated by Heirtzler and others (1968), using the polarity history inferred from magnetic profiles over oceanic ridge systems. It was assumed that recognition of the Brunhes/Matuyama and Matuyama/Gauss polarity boundaries in the mobile crust facilitates estimation, by extrapolation, of the ages of other magnetic polarity changes, manifested in details of the magnetic profiles. For the period under examination, Heirtzler and others (1968) suggested recognition of an Olduvai event within slightly younger limits than previously used, at t = 1.78 to 1.93 m.y. In addition, inconsistently appearing small anomalies in the profiles called anomaly "X" by Heirtzler and others (1968) suggested the possibility of hitherto unidentified polarity changes between the Olduvai event and Matuyama/Gauss boundary (Fig. 3, vii). They recognized no Gilsa event, but Vine (1968), using very similar methods, suspected that the Gilsa was in fact represented in marine magnetic profiles. Heirtzler and others (1968) predicted

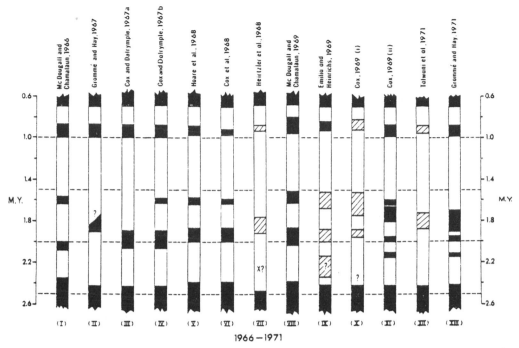

Figure 3.   Geomagnetic polarity time scale evolution between 1966 and 1971. Symbols as in Figure 2, except for shaded segments, which indicate polarity sequences inferred from analyses of marine magnetic profiles. See text for discussion of details.

Notes: (a) the upper and lower boundaries in scale (IX) are the standard deviation of the mean age of the central points of each event (Emilia and Heinrichs, 1969). (b) The younger boundary of the large event in scale (XIII) is taken at 1.7 m.y., since the authors describe an overlying volcanic unit of t = 1.71 m.y. age.

at least 171 reversals of polarity for the Tertiary. Results from a paleomagnetic survey of a thick sequence of Icelandic basalts (Dagley and others, 1967) covering the period t = 1 m.y. to 13 m.y. (Moorbath and others, 1968) showed that at least 58 polarity changes had occurred in 12 m.y. For the same period, Heirtzler and others (1968) predicted only 37 changes. This emphasizes the limited resolving power of the magnetic profile analytical method.

4. Cox (1968) introduced a statistical model employing the known polarity scale and polarity history inferred from marine profiles to evaluate and predict geomagnetic polarity behavior. He concluded that for the past 10 m.y., polarity events with durations less than 0.05 m.y. remained to be discovered. Shortly afterward, Watkins (1968b) presented evidence from several widely separated deep-sea sedimentary cores consistent with the occurrence of short events of 20,000 yrs duration or less at t = 0.82 and 1.07 m.y. Foster and Opdyke (1970) and Opdyke and Foster (1970) have

subsequently presented data to suggest that these events may indeed be real occurrences, but sound confirmation does not exist.

1969

1. McDougall and Chamalaun (1969) published a slightly refined polarity scale together with results from Mauritius and Rodriguez (Mascarene Islands). They show (Fig. 3, viii) an Olduvai event which extends from t = 1.9 to 2.05 m.y. This is a younger period than shown in their previous polarity scale (Fig. 3, i). The Olduvai event is still taken to be the largest event, because of the greater number of detections.

2. Emilia and Heinrichs (1969) analyzed many mid-ocean profiles in similar fashion to the previous work of Heirtzler and others (1968) and Vine (1968). They concluded (Fig. 3, ix) that the Gilsa and not the Olduvai was the major Matuyama event. With a Gilsa event at 1.66 ± 0.08 m.y. and an Olduvai event at 1.96 ± 0.076 m.y., a linear crustal spreading rate was accommodated by the

magnetic anomaly pattern over most active ridges. This was in contrast to the variable spreading rates which had to be invoked if the Olduvai event time was associated with the largest Matuyama event. This largest event was that most consistently identified in the mid-oceanic patterns and in those studies of deep-sea sedimentary cores which continued to identify only one event in the lower Matuyama. Emilia and Heinrichs also drew attention to more evidence of the existence of anomaly X of Heirtzler and others (1968) at about $t = 2.3 \pm 0.1$ m.y. (Fig. 3, vii) and an anomaly which they call "W" at 2.0 ± 0.06 m.y.

3. Earlier, Cox (1969) had published two polarity scales. One resulted from analysis of marine magnetic profiles (Fig. 3, x) with results very similar to those of Emilia and Heinrichs. The other scale (Fig. 3, xi) resulted from a combination of techniques: the epoch boundaries are due to results from K:Ar and paleomagnetic studies of igneous rocks, but the number, age and duration of events are determined in part from sedimentary core data, and in part from marine magnetic profiles. In this polarity scale, "split" Gilsa and Olduvai events are proposed. The nomenclature for the scale is included in Figure 1.

It appeared that by the end of 1969, resolution of all but the fine details of the Matuyama geomagnetic polarity history had taken place. Nevertheless, a major problem in the application of the polarity scale existed: in studies of deep-sea sedimentary cores, publications of substantial significance (Berggren and others, 1967; Glass and others, 1967; Hays and Opdyke, 1967; Ninkovitch, 1968; Ericson and Wollin, 1968; Phillips and others, 1968; Hays and others, 1969; Foster and Opdyke, 1970; Hays, 1970; Hays and Ninkovitch, 1970; Hays, 1971) continued to utilize that polarity scale proposed in early 1966, prior to discovery of the Gilsa event. The large event in the lower Matuyama is called the Olduvai in these publications, and the third event is not recognized. In contrast, another series of publications and presentations (Watkins and Goodell, 1967a, 1967b; Watkins, 1968b; Goodell and Watkins, 1968; Goodell and others, 1968; Steuerwald and others, 1968; Bandy and Casey, 1969, 1970, 1971; Bandy and others, 1969, 1971; Lamb, 1969: van Montfrans and Hospers, 1969; Clark, 1969, 1970; Opdyke and Glass, 1969; Kennett and Watkins, 1970; Herman, 1970; Fillon, 1970; Kennett and others, 1971; Birkeland and others, 1971; Watkins and Ken-

nett, 1971) recognized the three events, with the Gilsa event most commonly as the longest. Although no controversy could now reasonably exist regarding the minimum number of the three detected events within the Matuyama epoch (Fig. 3), there remained the problem of nomenclature, and (because of the unavoidable association of the event and age) the age of the large Matuyama event, and therefore the Pliocene-Pleistocene boundary.

1971

1. An incidental result of a geophysical survey of the Reykjanes ridge by Talwani and others (1971) included a suggested refinement of the polarity scale to one with only two events in the Matuyama (Fig. 3, xi). This is in sharp conflict with available data. The age of the major Matuyama event, not named by the authors, is given as 1.71 to 1.86 m.y. In their Figures 5, 13, and 14, however, Talwani and others (1971) show the large Matuyama event to include the 2-m.y. isochron of their magnetic profiles. The resolving power of the marine magnetic profiles in this part of the North Atlantic is low, because of the low spreading rates (1 cm/yr) involved. Because the large Matuyama event in the profiles of Talwani and others (1971) includes a 2-m.y. isochron derived from a constant spreading rate assumption and since the known large event is certainly younger than 2 m.y., non-linear spreading in the Reykjanes ridge area is a distinct possibility, and therefore the proposed Gilbert epoch chronology of Talwani and others (1971) must be of little value.

2. The most recent contribution to the polarity time scale definition is by Grommé and Hay (1971) in their third paper on Olduvai Gorge polarity:age results. One new date has been added to the 51 dates by Evernden and Curtis (1965), and new mapping has been used to clarify the stratigraphy of some of the previously dated samples. They suggest that the original dates can now be interpreted to be consistent with the age of the two normal polarity flows being about 1.79 m.y., with estimated standard deviations (from ages on minerals from tuffs overlying and underlying the basalt) of less than 0.1 m.y. The authors consider that the upper age limit is 2.0 m.y., and the minimum duration is 0.1 m.y. The best age of the overlying reversely magnetized body is taken as 1.71 m.y. Of major significance is a suggested resolution of the Olduvai versus Gilsa nomenclature problem for the major

30

event within the Matuyama epoch: Grommé and Hay propose that *if* the Olduvai and Gilsa are the same event, then the term Gilsa should be dropped, and the event at about t = 2.0 m.y. should be termed "Réunion" after the location of its initial unambiguous detection by McDougall and Chamalaun (1966). This proposal is discussed in further detail below in the section on magnetic polarity time scale nomenclature.

PROSPECTS FOR ADDITIONAL REFINEMENT OF THE POLARITY SCALE

In this section, consideration is given to the difficulties in further refinement of the polarity scale, particularly for the period t = 0 to 5 m.y. Discussion is best made in terms of the different experimental techniques involved.

Marine Magnetic Anomalies

One of the most fortunate circumstances in recent geophysical developments is that general midoceanic water depths (or the distances between magnetic sources and sea-level magnetic profiles) are of the order necessary for attenuation of the fine details in sea-level magnetic anomaly patterns to exactly that extent required for ready matching with the independently derived polarity scale, as known in 1966. This natural smoothing process was perhaps inadvertently enhanced in some early crustal spreading analyses (Vine, 1966) by the use of profiles derived from magnetic anomaly contour maps, rather than original profiles. If the available magnetic profiles at that time had been taken much closer or much farther from the magnetic sources, or if the polarity scale had not evolved to the extent where a Jaramillo and Olduvai event were known, it is conceivable that the great recent advances due to recognition of crustal spreading and subsequent development of plate tectonic models would have been long delayed, since the correspondence of the two polarity recording systems would inevitably have been obscured.

Vine (1966) has pointed out that crustal spreading rates and water depths are such that events occupying less than 5 to 6 km of crust (which at a spreading rate of 1 cm/yr represents periods shorter than about 0.05 m.y.) are unlikely to be reliably detected in marine magnetic profiles. This may be slightly improved upon by the use of stacking techniques on parallel profiles across the faster

spreading ridge systems (Blakely and Cox, 1971).

Larson and Speiss (1969) and Luyendyke (1969) believe that by analysis of short wavelength anomalies taken using a deep-tow magnetometer in the northeast Pacific, very fine details of the geomagnetic field behavior (particularly intensity variations) during the emplacement of the magnetic source materials, may be forthcoming. It is far more probable, however, that deviations from idealized dike-injection activity (Bott, 1967; Matthews and Bath, 1967; Harrison, 1968; Watkins and Richardson, 1971) and the inherent variations of magnetic properties in single igneous bodies (Watkins and Haggerty, 1967; Watkins, 1968a; Wilson and others, 1968; Watkins and others, 1970) which are so readily manifested in ground-level surveys over recent lavas, are likely to dominate over other factors in causing short wavelength anomalies. Peter (1970) has also commented on the interpretation of deep-tow marine magnetic data, preferring interpretations for the observed anomalies by means other than paleointensity variations. The spectacular successes of the interpretation of marine magnetic profiles in terms of varying polarity models (and polarity history sequences) may be tending to diminish, if not altogether remove, consideration of the various other possible causes of marine magnetic anomalies. Not all anomalies are of an amplitude which require polarity changes in the source material. The practice of predicting the polarity history from magnetic profiles, although perhaps successfully applied to the past 10 m.y., and even the past 50 m.y., when epoch-length periods are involved, can at best be encouraged as a prediction *only*. There now exists the possibility, for example, that a short period of reversed polarity may interrupt the predicted period of normal polarity from t = 8.79 to 9.94 m.y., according to paleomagnetic and K:Ar measurements on volcanic rocks in Australia (Evans, 1971) and Mexico (Watkins and others, 1971). The low-amplitude anomalies now associated with Cretaceous and Jurassic age oceanic crust (Helsley and Steiner, 1969; Einwich and others, 1970; Vine, 1970; Einwich and Vogt, 1971) may well be due, for example, to simple variations in the intensity of magnetization of the source rocks. Occurrence of an anomaly pattern over many separated parts of the oceanic crust is required before polarity history variations can be employed as a causative mechanism. Restraint on the use of the accommodating several degrees of freedom available (such as variable spreading rates and variable polarity history) is advisable.

It therefore appears that magnetic profiles are valuable, given sufficient numbers of profiles and consistency, in the confirmation of the broad features of the independently derived polarity history and for prediction of the broad features of the Cenozoic polarity history, but are to be treated with great caution when examined for confirmation of the reality of independently proposed events, especially when no sound age control is available. The resulting increase in amplitude (and decrease in wavelength) of anomalies when examined much closer to the submarine sources by deep-tow methods, is not likely to yield a corresponding increase in finer definition of the polarity history: rather, the various details which are undesirable in terms of the polarity history definition will become more evident in such profiles; but, again, if consistency of such data was forthcoming from several parts of the world, then polarity reversal mechanisms could be reasonably invoked.

K:Ar Dating and Paleomagnetism of Igneous Rocks

As discussed earlier, the present polarity scale (Fig. 1) has resulted almost entirely from potassium-argon dating of igneous rocks of known polarity. It is remarkable and perhaps fortunate that only two laboratories (U.S. Geol. Survey at Menlo Park, Calif., and the Dept. of Geophysics and Geochemistry at the Australian National Univ.) were involved in the pioneering stages of the polarity scale definition, although the data generated by Dalrymple, when associated with the laboratory of Evernden and Curtis, were used by the U.S. Geological Survey group. This is fortunate because the two groups involved were among the few isotope laboratories which used rigorous meaningful criteria, based largely on petrographic examination of materials prior to age analysis, to establish the suitability of young igneous rocks for dating (Curtis and Evernden, 1962; Dalrymple, 1963; McDougall, 1964, 1966; Cox and others, 1966). The present coherent state of the established polarity scale would not exist today without this early experimental control.

Cox and Dalrymple (1967a) recognized the limitations of the method, because of the approximate 3 to 5 percent precision limit in K:Ar dating. Given polarity epochs and events

of similar duration to those already recognized, they show that t = 6 m.y. is the probable limit of a definable polarity scale based only on K:Ar methods. Detailed studies of long, stratigraphically well-defined sequences of igneous rocks may assist in extending this range (Baksi and others, 1967; Evans, 1970), but such sequences would have to be very long and continuous indeed, to assist in any definition of the polarity scale prior to t = 10 m.y. K:Ar dating of sea-floor basalts, where ages can now be independently inferred from crustal spreading models, is not feasible because of problems of argon retention in quenched materials (Dalrymple and Moore, 1968). In any case, sea-floor basalts may not necessarily reflect the polarities inferred by the sea-level anomaly pattern (Watkins and Paster, 1971; Watkins and Richardson, 1971). It is conceivable that drilling into deeper unquenched parts of the ocean crust may provide materials more amenable to meaningful K:Ar analyses.

Despite the great limitations in the K:Ar technique, extension of the polarity scale to upper Miocene time and fine definition of the scale for the t = 0 to 5 or 6 m.y. period will depend ultimately on K:Ar studies. Magnetic profile analyses are, as discussed above, insufficiently precise and lacking in resolution for fine definition, and work on deep-sea sedimentary cores presents other difficulties, as discussed below.

Marine Sedimentary Cores

Mention has already been made of the problems created by the matching of paleomagnetic polarities in deep-sea sedimentary cores with an imperfectly defined polarity history, such as existed in 1966. This is largely the result of the difficulty involved in distinguishing between "noise" and that detrital remanent magnetism caused directly by geomagnetic field behavior. Watkins (1968b) has discussed some of the natural causes of spurious polarity reversals (or noise) in deep-sea sedimentary cores. These include faunal redeposition, various possible simple experimental errors during collection and laboratory measurements, and polarity recording lags resulting from delays between original deposition and final consolidation of deep-sea sediments (Dymond, 1969). Use of sedimentation rate extrapolation between ambiguously identified polarity epoch boundaries in cores (Watkins and Goodell, 1967a) may assist significantly in dating and therefore identifying independently proposed events for the Pliocene and Pleistocene, but apparent events in deep-sea sedimentary cores cannot be reliably attributed to "noise" in the absence of a perfectly defined polarity scale, because of the obvious circular reasoning involved. Similarly, the polarity predictions resulting from marine magnetic profile analyses cannot be used to distinguish between "noise" and true polarity history in deep-sea core data, because of the limited resolving power and extrapolated basis of the predictions. The follies of such an exercise would perhaps become apparent if a region of rapidly spreading oceanic crust (with a separation rate as great as 20 cm/yr) is discovered, with corresponding production of finely detailed polarity history predictions. It is reasonable to suspect that present data from deep-sea cores would need reinterpretation. It is clearly of value to have *all* paleomagnetic data from deep-sea cores published, rather than those in each core which match only the independently proposed polarity histories available at the time of publication (although interpretation of data is, of course, a privilege of any author). Otherwise, only reinforcement of established polarity scales will result, perhaps inhibiting further definition and correct resolution, since established polarity scales are not necessarily correct, particularly in early stages of their evolution.

It is of interest to note that although data from core RC 12–66 (Foster and Opdyke, 1970) may well have extended the established polarity scale back to $t = 9$ m.y. (and is in harmony with polarity scale derived from crustal spreading analyses), the established polarity scale for the past 4 m.y. or so (Fig. 1) owes relatively little to results from deep-sea sedimentary cores. Nevertheless, because of the limitation of precision of the K:Ar technique, it appears for all practical purposes, extension of the polarity scale to pre-upper Miocene times will be forced to depend on deep-sea sedimentary core results, coupled with the scale predicted from marine magnetic profiles. It must be realized that this predicted scale is based on a tenfold extrapolation, at least, so that the absolute ages involved in such a scale will have to be highly qualified. Very detailed K:Ar studies of critical polarity sequences no older than Miocene, will perhaps be of great value in controlling the quality of the scale. Dating of core materials such as volcanic ash may also assist in this context. The major acceptability criteria will have to be based on arguments involving identical results from various parts

of the world in both hemispheres, with micropaleontological age control.

In the author's opinion, the greatest danger of work on deep-sea sediments will be in attempting to resolve the fine structure of the geomagnetic polarity history. This is discussed in the following section.

PROBLEM OF THE DEFINITION OF SHORT POLARITY EVENTS

It may be instructive to realize that during the initial phases of work on the time scale (Cox and others, 1963a, 1963b; McDougall and Tarling, 1963; Grommé and Hay, 1963) periodicities of the order of millions of years were suspected for polarity changes. Shortly after Cox (1968) used statistical models to predict the existence of short polarity events (durations of fixed polarity of the order of 10^4 yrs), some evidence for such short polarity events was presented (Watkins, 1968b), but since the results were from a limited number of deep-sea cores, they could not be accepted as completely unambiguous until further independent results were forthcoming. Bonhommet and Babkine (1967) and Bonhommet and Zähringer (1969) have presented paleomagnetic, C^{14}, and K:Ar evidence for a short polarity event in French lavas at about $t = 0.01$ to 0.03 m.y. This has become known as the Laschamp event. Smith and Foster (1969) used data from deep-sea cores to suggest an event at $t = 0.108$ to 0.114 m.y. which they call the Blake event. Despite the relatively brief intervals of time which these proposed events represent, they are (if real) clearly of high potential value in future stratigraphic and even archeological studies, in addition to their relevance to an understanding of the polarity reversal frequency problem. But are these suggested events really due to global reversals of the Earth's magnetic field? Before additional short events are discovered, it is considered absolutely essential that much more consideration be given to the possibility of spurious data, and to the fundamental role of an inherent weakness involved in experimental approaches and publication. The fragile nature of the known polarity scale, particularly in its fine structure, is best demonstrated by such discussion.

Several curious interpretative inadequacies and experimental limitations emerge in examination of work on short polarity events. The most important interpretative aspect perhaps involves what can be termed the "reinforcement syndrome." There are several ways in which spurious data can be generated or can be thought to have been generated in paleomagnetic studies. For example, it is virtually certain that a recently discovered curious magnetic field behavior, where equatorial virtual geomagnetic poles (without evidence of association with a polarity reversal transition) occurred in the Cretaceous (Watkins and Cambray, 1971) and Tertiary (Cox, 1966; Doell, 1968; Lawley, 1970), was in fact discovered much earlier, within the vast volume of paleomagnetic data generated during the last 20 yrs, but was not recognized as such. Rejection of data may well have occurred, which would perhaps be a reasonable practice, if high-latitude geomagnetic poles were the focus of study; however, during early studies, spurious results were in all probability suspected and rejected. (This may have been easier to justify in these earlier studies, since the sampling of several parts of single lavas, as opposed to only one or two parts, is a relatively recent practice in paleomagnetic studies.) It is now probable, however, that many more such unusual pole positions will be discovered, since they are now understood to be a real magnetic field phenomenon. The first published description thus has great leverage: it enables workers pondering their own "curious" data to realize its real(?) meaning. But what if the behavior described in the initial publication is in fact erroneous? A substantial trap will have been laid. [Perhaps, the best known, or at least most significant, result of the "reinforcement syndrome" in the geological sciences is the very firmly established concept of four glacial periods during the last Ice Age. The initially defined system was confirmed by many different studies. It now appears that there may be seven or more separate glacial periods since t = 0.7 m.y. (Hays and others, 1969), and Antarctic glaciation, at least spasmodically, since the Eocene (Margolis and Kennett, 1971).] The nature of paleomagnetic data, which are not continuous in time, makes them particularly susceptible to this type of trap, since it is infinitely more difficult, if not impossible, to prove that a given magnetic field behavior has not taken place, than to demonstrate that it has occurred. Superimposed on this is an important human element: it is much more reasonable to be able to generate both the energy and belief (faith?) required for publication of data confirming a discovery than to publish much negative data of a pedestrian

nature. Thus the initial discovery is reinforced. This argument would appear to apply to any short polarity events which are suggested to have occurred: they are more likely to gain the status of a real phenomenon than to be proven nonexistent, regardless of the facts. This can happen even more convincingly, if theoretical arguments support the reality of the concept involved. Since deep-sea sedimentary cores which extend beyond the Brunhes/Matuyama (t = 0.69 m.y.) boundary are generally more difficult to obtain than cores extending to only t = 0.4 m.y. or so, Brunhes cores may be expected to be predominant in the materials available for study, so that the Laschamp or Blake events become strong candidates for correlation with, and therefore "confirmation" by, spurious data in the core. This may well have happened in a study of a short mid-Atlantic ridge core (Stesky and Strangway, 1970) where the Blake was interpreted to be represented. In this context, it is extremely interesting to note that of the seven cores used for the original identification of the Blake event (Smith and Foster, 1969), four have definite signs of faunal burrowing and disturbance close to or at the critical magnetic inclination changes, according to the associated published sedimentary logs. When the great difficulty involved in detection of burrowing in homogeneous muds and oozes is considered, and when it is realized that faunal redeposition may be obviously capable of distorting the magnetic signature, it is clearly not advisable to accept the Blake event as a real polarity change at this time. As far as the Laschamp event is concerned, sedimentary sequences of measured C^{14} age in California (Denham and Cox, 1970) and a Mediterranean deep-sea core of unusually high sedimentation rate (as much as 8 cm/1,000 yrs) studied by Opdyke and Ninkovitch (1971) yielded no evidence of the Laschamp event. The author has, similarly, found no consistent evidence of the Laschamp or Blake events in the southern hemisphere, despite much experimental effort involving measurement of the magnetism of several thousand specimens from six cores of high sedimentation rate. Of course, none of these negative results unequivocably proves that the Laschamp event did not occur. It may well be a real phenomenon. A series of papers on the "nonexistence" of an event within the decade following its proposed existence can nevertheless be visualized. The case for supercritical evaluation of data pertaining to polarity events is both obvious and strong. Additional aspects of the difficulties involved are discussed below in terms of the methods employed.

K:Ar and Paleomagnetism of Igneous Rocks

Despite the proposed existence of events within the Brunhes epoch, no reversely magnetized lava in a series of normal polarity lavas has yet been documented for the many Brunhes epoch lava sequences (and dated lavas) which are known to exist. This must be accepted as some (but not yet conclusive) evidence against the definite existence of events within the Brunhes.

Deep-Sea Sedimentary Cores

The shipboard (collection) and laboratory sources of possible spurious results have already been mentioned. Additional sources of misleading data include: (1) influxes of coarser (sand-sized and larger) sediments which are more likely to be randomly aligned, rather than the idealized alignment as a function of the ambient geomagnetic field direction, such as is readily accomplished by the finer sediments; (2) the restricting nature of the small diameter of piston cores, which does not allow lateral sampling of the critical horizon involved, such as would be the case in outcrops on land; and (3) the subtle nature of hiatuses which are likely to be present in many, perhaps most, cores (Watkins and Kennett, 1971).

It is also important to realize that correlation of any parameter within the cores can be a forced correlation, in that inferred sedimentation rates can be varied. This, of course, applies to any magnetic parameter (declination, inclination, magnetic intensity). The number of cores which have been examined is now sufficiently large to enable correlation of virtually any independent parameter (such as oxygen isotope, micropaleontological parameters, and the like) with magnetic data in *some* cores, somewhere, given the degree of freedom of varying the inferred sedimentation rate to accommodate different data sets. This emphasizes the huge problem of data selectivity in studies of correlations between magnetic and other parameters in cores. If, for example, "peaks" in paleotemperature data match peaks in magnetic parameter in one core, but not in several others, what can one conclude? Can it be concluded that the natural processes which tend to blur the detailed magnetic signature (if not the broader features such as major polarity changes) are not present in the one selected

core, but are present in others, where the "correlation" is not evident? In the author's opinion, such circular treatment of data is not permissible. To encourage this approach is to risk the fabrication of completely misleading models capable of spreading havoc (if taken seriously) in diverse independent disciplines. If under close and nonsubjective examination, correlations between magnetic and other parameters in sedimentary cores do emerge, careful consideration must be given to the various alternate models which could indirectly result in the given correlations. For example, the intensity of magnetization of sediments could conceivably vary as a function of a sediment type or critical particle size, the deposition of which is to some extent paleoclimatically controlled, resulting therefore in a potentially misleading correlation of intensity of magnetization (which is sometimes too simply assumed to be a function of geomagnetic paleointensity) and paleoclimate. Wollin and others (1971) may have detected such a correlation.

If polarity events are to be proposed on the basis of data from deep-sea sedimentary cores, it is considered essential that the event be detected in several cores, preferably of different sediment types which accumulated at different sedimentation rates, and at different water depths. It is also essential that, when feasible, independent dating (isotope or micropaleontological) methods are employed. Any evidence of sediment disturbance should result in data rejection. Data should also be obtained from several parts of both hemispheres, since it is entirely possible that an apparent reversal of the magnetic field could be observed in a restricted region (of continental dimension) as the result of non-dipole activity alone, during a period of subdued main dipole intensity, without any real reversal occurring. Such geomagnetic behavior would, by definition, not be a magnetic polarity event. Unless these minimum requirements are met, it is suggested that short inclination "reversals" from deep-sea sedimentary cores should not be given "event" status. If a term is needed for a short-period apparent movement of the magnetic field, "deviation" or "excursion" would be more appropriate.

Sediments Outcropping above Sea Level

Extension of the polarity history observations in deep-sea sediments to studies of upper

Cenozoic sediments outcropping on land has already begun (Nakagawa and others, 1969; van Montfrans and Hospers, 1969; Bucha, 1970; Pevzner, 1970; Bout, 1970; Kennett and others, 1971). Materials involved vary from marine mudstones, siltstones, and limestones (Kennett and others, 1971) to loam, loess, and fluvial deposits (van Montfrans and Hospers, 1969), and even cycles of "loess, forest soils, and water sediments" of old river beds (Bucha, 1970). Interdigitated volcanic rocks and sediments (with mammalian fauna) in France have been studied by Bout (1970).

Kennett and others (1971) took three samples per site in their study of New Zealand sediments. In some cases, two out of three samples have coherent results, while the third appeared to be quite spurious. Minimum scatter computational methods, involving removal of unstable components in several different demagnetizing fields (Watkins and Richardson, 1968), resulted in reduction of the significance of the spurious data point when data from the entire New Zealand sedimentary sequence were evaluated. It was clear, however, that if the sampling technique had involved only one sample per site, the resulting data set could certainly have included apparent "events." It appears that most other studies of sediments from above sea level may have employed only one specimen per site. Together with the virtually complete lack of knowledge and understanding of the magnetic mineralogy of the sediments, such sampling systems would suggest that any resulting evidence for short events in such materials is of minimal value. The data of Bucha (1970) in particular, which are interpreted to indicate polarity events with durations as short as a few hundred years during the Brunhes epoch, have the appearance of being spurious, at least in part.

In common with the requirements considered to be essential before polarity events which are apparently detected in deep-sea sedimentary cores can be accepted as real, results from continental sedimentary sequences should be from several sections, preferably of various sedimentary types with micropaleontological age control, from various parts of the world. This may appear to be a too rigorous set of acceptability criteria, but to invoke a globally synchronous reversal of the Earth's magnetic field involves undertaking a substantial responsibility to future investigators. If accepted, such proposals could produce long-lasting consequences in the geological sciences.

MAGNETIC STRATIGRAPHY NOMENCLATURE PROBLEM

As described above, there now exists a well-established nomenclature convention in magnetic polarity scale studies; periods of one polarity of the order of 1 m.y. are termed epochs, and those lasting the order of 50,000 to 100,000 yrs are termed events. The four epochs recognized for the Pleistocene and Pliocene (Brunhes, Matuyama, Gauss, and Gilbert) are named after deceased contributors to magnetic studies, and the events are named on a type-locality basis, after the site where the events are first recognized (Fig. 1). The term "short event" has only recently entered the literature and is taken to mean periods of fixed polarity of the order of 10,000 yrs. The type-locality concept has been applied to the short events so far proposed (the Laschamps and Blake, both within the Brunhes epoch). Most recently, McElhinny and Burek (1971) have used the terms "interval" and "zone" for some Mesozoic geomagnetic periods. The former is used as a suffix to the name of a contributing scientist (although Irving and Parry, 1963, used the same term as a suffix after a geographic location for the Permian geomagnetic polarity character) and the latter are added to terms which are geographic in origin.

In contrast, the detection of epochs in deep-sea sedimentary cores has resulted in proposal of a system calling for numbering of epochs (Foster and Opdyke, 1970) to parallel conventions in ocean magnetic profile analyses (Heirtzler and others, 1968; Talwani and others, 1971). Thus the known epochs (Fig. 1) are 1 through 4, and the epoch below the Gilbert is 5, and so on. Similarly, some events within epochs have been given alphabetical identification: Hays and Opdyke (1967) name three events in the Gilbert "a, b, and c." The problems with this type of nomenclature system are obvious: subsequent discovery of events between those recognized, will present an inelegant series of choices. In fact, there now appear to be four events recognized within the Gilbert epoch (Foster and Opdyke, 1970).

When considered relative to the time involved in defining conventional stratigraphic nomenclature on a global scale, the polarity time scale can be considered to be in its infancy. It is therefore perhaps surprising that conflicts or ambiguities already exist in the literature.

The conflict in nomenclature (within the $t = 0.6$ to 2.6 m.y. period under examination) involves, as described earlier in this paper, the name of the major event in the lower half of the Matuyama Epoch (Fig. 3): is it the "Olduvai" event or "Gilsa" event? The significance of this problem, in terms of the age of the Pliocene-Pleistocene boundary has also been discussed above. Grommé and Hay (1971) have recently suggested resolution of this controversy. The merits of this proposed resolution, and its implications to future nomenclature modifications or controversy is appropriately reviewed here.

The original K:Ar result, for this normal polarity unit, was quoted by Grommé and Hay (1963) as being approximately 1.8 m.y. After the realization that a polarity event was represented and publication of the detailed K:Ar results (Evernden and Curtis, 1965), a closer evaluation of the available data was made by Grommé and Hay (1967). Although it was concluded that the lavas involved are not older than 2.0 m.y., problems including possible contamination of ignimbrites and tuffs with older materials led to an ambiguity in the available interpretations: the age of the normal polarity basalt was thought to lie between t = 1.85 and 1.90 m.y., *or* between t = 1.75 and 1.9 m.y. Meanwhile, the Gilsa event, named for the 1966 detection in Iceland of a normal polarity lava of t = 1.60 ± 0.05 m.y. (McDougall and Wensink, 1966) was confirmed by the t = 1.65 ± 0.09 m.y. result for a normal polarity lava on Nunivak Island (Cox and Dalrymple, 1967b). As discussed previously, Cox (1969) later estimated the Gilsa event to be split and to extend from t = 1.61 to 1.79 m.y., and, largely on the basis of later apparent independent detection of the older event, the Olduvai was inferred to extend from t = 1.95 to 1.98 m.y., and from t = 2.11 to 2.13 m.y. Later, Watkins and Abdel-Monem (1971) detected a normal polarity lava at t = 1.64 ± 0.03 m.y. on the island of Madeira, above a reversed polarity lava of age 1.76 ± 0.07 m.y. This was interpreted to be a detection of the Gilsa event. The second re-evaluation of K:Ar ages for the Olduvai Gorge, together with some new data (Curtis and Hay, 1972), is quoted by Grommé and Hay (1971) to show that the best estimate for the age of a normal polarity tuff in the Olduvai section is now 1.79 ± 0.03 m.y. with the best estimate for the age of an overlying reversely magnetized tuff being 1.71 m.y. Dalrymple (1972) evaluates the same data to yield an age for the two normal polarity units as 1.82 m.y. In their abstract, Grommé and Hay

conclude that the Olduvai event is no older than 2.0 m.y., and ranges from 0.1 to 0.2 m.y. in length. It is therefore suggested that the Gilsa and Olduvai events may actually represent the same normal polarity event, and "if so the Gilsa should now be abandoned" (Grommé and Hay, 1971, p. 179). This is a very reasonable, and by previously established convention, undeniably correct conclusion. Grommé and Hay (1971) solve the resulting problem of the name of the normal polarity event at about t = 2 m.y., detected in several parts of the world by each of the three recording methods and previously called Olduvai, by naming it the Réunion event, after the locality of its first unambiguous detection by McDougall and Chamalaun (1966). The authors also summarize many data to show that their Olduvai event is the major event, in terms of duration, in the Matuyama epoch, and mention the fact that the Pliocene-Pleistocene boundary, previously thought to be between t = 1.8 and 2.0 m.y. would now appear to be 0.2 to 0.3 m.y. younger.

To support their proposal for this substantial change in nomenclature, Grommé and Hay (1971, p. 183) present an argument based on convention: "among workers dealing with paleomagnetism of deep-sea sedimentary cores, it has been nearly universal usage to give the name Olduvai to the earlier of the two prominent normal polarity events observed within the Matuyama epoch," and then they cite eight references in that context. Their argument would have some merit were it not for the fact that there existed a larger number of similar publications or presentations (Watkins and Goodell, 1967a, 1967b; Goodell and Watkins, 1968; Goodell and others, 1968; Watkins, 1968b; Steuerwald and others, 1968; Bandy and Casey, 1969, 1970; Bandy and others, 1969; Lamb, 1969; van Montfrans and Hospers, 1969; Clark, 1969, 1970; Opdyke and Glass, 1969; Kennett and Watkins, 1970; Herman, 1970; Fillon, 1970) not adopting the Olduvai convention, but using "Gilsa" for the large Matuyama event, after McDougall and Chamalaun (1966) and Cox (1969). Since Grommé and Hay's (1971) suggested revision, several more publications using "Gilsa" have appeared: Bandy and others (1971); Birkeland and others (1971); Kennett and others (1971); Ruddiman (1971); and Watkins and Kennett (1971). A more convincing argument is the reference to the Opdyke and Foster (1970) result from a study of the duration of the large

Matuyama event in a series of twelve North Pacific cores, where a range of t = 1.71 to 1.86 m.y. was inferred. As Figure 4 shows, however, using all the available published data through 1970 where both the Brunhes/Matuyama and Matuyama/Gauss boundaries are detected (other than those published by the groups firmly advocating three events in the Matuyama), the range of ages to which the large event can correspond, using sedimentation rates, shows that the method is not convincing.[1] In view of the virtual proof that three events occurred in the Matuyama, it is perhaps curious that only two of the 23 cores involved (Fig. 4) actually recorded three events. This could indicate that the third event is of very limited duration. Van Montfrans (1971) has recently analyzed an unspecified but large number of deep-sea sedimentary core paleo-

[1] Inspection of the Opdyke and Foster (1970) data, which were seen only after review of this manuscript, shows that although nine new estimates would be available to add to the twenty-three in Figure 4, this conclusion would not be affected.

magnetic results. He used cross-correlation methods to compare the published data with Cox's (1969) polarity time scale (Fig. 3, xi), and concluded that three detectable polarity events exist in the Matuyama epoch.

Grommé and Hay's proposed nomenclature revision, while of some merit, is unfortunately not definitive. As they stress, their suggestions are valid only if the Gilsa lava is the same age as the Olduvai Gorge material, and, as they infer, this is not definitely proven. What certainly emerges from the three analyses of the Olduvai Gorge result is the difficulty of K:Ar work on the Olduvai Gorge igneous rocks. Since it is far from impossible that both events will be found in one single lava or continental sedimentary sequence in the future, and far more reliable data be forthcoming, it would seem somewhat premature to adopt the proposed nomenclature change. Otherwise, the three separate proposed detections of the Gilsa event (on Iceland at t = 1.60 m.y., in Alaska at t = 1.65 m.y., and on Madeira at t = 1.64 m.y. which are *all* younger than the proposed large

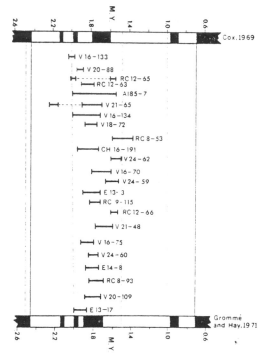

Figure 4.   Time spans of the large polarity event within the lower half of the Matuyama reversed polarity epoch, as inferred from paleomagnetic data in selected deep-sea sedimentary cores. The purpose of diagram is to illustrate the difficulty of exact determination of the

time spans of polarity events. The data are *all* the published results (at the time of writing) in which the given sedimentary core includes both the Brunhes/Matuyama boundary and the Matuyama/Gauss boundary (*see* Fig. 1). These two boundaries facilitate estimation of the age range of the largest normal polarity event time range by use of linear sedimentation rates assumed for the Matuyama epoch. The cores are selected as being only from those publications which recognized a polarity scale with only two major polarity events within the Matuyama in contrast to those which recognize three events (Fig. 1). Identification number on each data bar is the core number: the letters involved refer to the ship used (V = *Vema*, RC = *Robert Conrad*, CH = *Chain*, E = *Eltanin*, A = *Atlantis*). Polarity scales due to Cox (1969) and Gromme and Hay (1971) shown for comparison at left and right, respectively: black = normal polarity; clear = reversed polarity. Horizontal dashed lines at 0.5-m.y. intervals. The two vertical dashed lines (for cores RC12–65 and V21-65, respectively) join two observed time spans of normal polarity in each core, for the lower Matuyama epoch. Notes: (a) the normal polarity events under consideration are those published as interpreted data, and do not involve any data reinterpretation. (b) Short events associated with the large event are not included when doubt is expressed about the reliability of the data. For example, a second event is associated with V16-75 (Opdyke and Glass, 1969, Fig. 2) but is not regarded as reliable (Opdyke and Glass, 1969, Fig. 5) and is therefore not included in this compilation. (c) A similar diagram has now appeared (*in* Opdyke and Foster, 1970); this includes three of the above data, and involves nine additional points.

event duration of Grommé and Hay and Foster and Opdyke's preferred age of 1.71 to 1.86 m.y.) will *all* have to both be arbitrarily assumed to be "anomalous," as will the age of 1.66 m.y. inferred by Emilia and Heinrichs (1969) from oceanic magnetic profiles. This can be argued to be applying unreasonable data rejection. In addition, it is perfectly feasible that the Gilsa event may represent the upper part of a "split" event (Fig. 1), in which case both terms should be retained, as being discrete but close magnetic events. There can be little doubt that the term "Gilsa" cannot yet be dropped. Since a reversely magnetized tuff of 1.71 m.y. age at Olduvai Gorge overlies the critical normal polarity units, and since it is older than all the proposed Gilsa event K:Ar detections as discussed above, it would therefore appear most reasonable at this time to call the large event by two names: the Gilsa for the upper part, and the Olduvai for the lower part. Opdyke and Foster (1970, p. 102) favor retention of "Gilsa" for the upper part of a split "Olduvai" event. Whether or not conventional stratigraphic systems will allow this practice remains to be seen.

The Olduvai versus the Gilsa nomenclature controversy also suggests a hypothetical problem: if an event were named with an age of t = 6.5 m.y., but was subsequently shown by more accurate dating to actually be at t = 6.0 m.y., but in the meantime the 6.0 m.y. event had been discovered elsewhere and therefore given a different name, which event and name would have precedence? Would the first actual detection (even if erroneous) or first *accurate* detection carry to the event name? Perhaps a resolving precedent has already been established: a single reversed polarity lava of t = 2.95 ± 0.06 m.y. on Waianae (West Oahu, Hawaii) described by McDougall and Tarling (1963) might, in retrospect, have been the first detection of the Mammoth event. McDougall and Chamalaun (1966) did not include this determination in their summary because there was some doubt as to whether the age and polarity were determined on the same lava flow. Thus while justifiably dropped from consideration, the possibility of the initial Mammoth event detection being in Hawaii is still strong. Data which would have established the name of the event when the Mammoth results became known were therefore nullified by the existence of additional results, which rendered the age unreliable, despite the fact that the two older dates may well have been the poor

results or despite the fact that the results show local stratigraphic mapping imperfections. The first unambiguous *accurately* dated presentation of the event therefore should take precedence. This point is made to introduce the next section of this paper.

NEED FOR AN INTERNATIONAL SUBCOMMISSION OR PANEL ON THE GEOMAGNETIC POLARITY TIME SCALE

It is virtually certain that inconsistencies and nomenclature conflicts similar to those resulting during the first eight years of work on the polarity history will emerge in future studies and use of the polarity time scale.[2] As the major features of the polarity scale are now known for the past 6 m.y. and future work on this part of the scale will therefore amount to refinement (event or short event definition), more complex problems and arguments will, in all probability, occur. There can be no doubt that establishment of an international body with plenary powers to set up procedural guidelines and to consider claims for new terms would greatly assist in maintaining the desirable coherence of the polarity time scale. It is envisaged that such a group would mainly review (biannually?) the merits of any published nomenclature revisions or proposals. Without such control, the scale will, in all probability, lose part of its resolution and great potential in contributing substantially to understanding of Pliocene and Pleistocene phenomena. The older parts of the scale will be, of course, equally in need of a controlling body. This discussion of the polarity time scale evolution and its prospects for future fine definition is therefore appropriately concluded by considering the principles and state of established relevant commissions. The principles of a polarity nomenclature system will inevitably bear some resemblance to those incorporated into the International Code of Stratigraphic Nomenclature. Since it has not yet been completely agreed upon, Sylvester-Bradley (1967) has suggested that the code could be profitably modeled on the Code of Zoological Nomenclature. Sylvester-Bradley has written the following to the author, in response to a request for

---

[2] Creer (1971) has drawn attention to a great problem which now exists in the Mesozoic polarity scale nomenclature. McElhinny and Burek (1971) have proposed a system totally different than a version suggested earlier by Pecherski (1970).

comment on the desirability of a polarity time-scale commission:

The first essential is to distinguish those areas of nomenclature which can be precisely and objectively defined by edict and those which can allow for differences of opinion that result from disparate information or varied interpretation. In zoological nomenclature, objectivity is obtained by specifying every taxon in terms of a single type. In stratigraphic nomenclature, the "stratotype" concept is an attempt (only partially successful) to achieve similar objectivity. In zoological nomenclature, flexibility is achieved by declaring the *boundaries* of each taxon to be variable according to the subjective opinion of each taxonomist concerned.

In a polarity time-scale nomenclature, the only objectivity that I can see must come from a type section in a type locality. The name of an event would then be defined by the type section. The *age* of the event then becomes a matter of subjective opinion, varying according to the interpretation of laboratory and field observations. If two named events are claimed to be coincident in age, the two names become subjective synonyms. Which name of two rival synonyms should be applied can be decided by an arbitrary stipulation. This might be according to a Law of Priority or a Law of Prescription. In the former case, the first name to be published takes precedence. In the latter case, the name to be accepted is prescribed by some specific international commission.

I would myself favour a Law of Priority. This would give precedence to the name "Olduvai Event" over the name "Gilsa Event," if they are the same event. The two names would be regarded as *subjective* synonyms only, however. If at a subsequent time, new evidence were to lead a specialist to return to the belief that the events were not coincident, the name "Gilsa Event" would be resurrected.

R. M. Carter (University of Otago), who has recently been involved with Tertiary stratigraphic nomenclature in New Zealand (Carter, 1970) has made the following comments to the author concerning problems of nomenclature analogous to those that could occur in a polarity time scale: (1) Although the type of locality concept is widespread in stratigraphy, some geologists such as Scott (1960) think that it is partly unnecessary. (2) Although the principle of priority is often applied to stratigraphy, it is very much less rigorous than in the field of taxonomy, and is usually only invoked when other things are equal. (3) Oldroyd (1966) is of the opinion that the International Sub-Commission on Zoological Nomenclature spends most of its time passing numerous exceptions to its own unnecessarily rigorous rules, the worst

of which is that of strict priority.

Carter has also kindly provided the author with the following two examples of stratigraphic conflicts, which may be considered as analogous to either present or future polarity scale problems:

1. Two of the New Zealand Miocene stages are the Awamoan (proposed by Thompson in 1916) and the Altonian (proposed by Finlay and Marwick in 1947). Subsequent work in each of the type sections (Hornibrook, 1969; Scott, 1969) suggests that the latter may be equivalent at least in part to the former; thus the Altonian may be synonymous to the older Awamoan, and if a principle of priority is established, will not survive as a stage.

2. The Petanian and Nukumaruan are two stages that have been similarly discovered to partly cover the same time range within the New Zealand Pleistocene. The former was proposed by Thompson (1916) for beds in Hawkes Bay, and has clear priority over the latter, which was proposed by Finlay and Marwick (1940) for beds in the Wanganui area. The Nukumaruan is, however, in a continuous section with the stratotypes of the immediately older Waitotaran and immediately younger Castlecliffian stages. Nukumaruan has therefore been retained rather than the Petanian which was proposed earlier. This example serves to illustrate the considerable flexibility that stratigraphers retain, by refusing to be bound by an absolute law of priority.

In conclusion, in only nine years, the polarity time scale has evolved to an extent where it is clearly capable, with careful use, of being the foundation of an understanding of the diverse global phenomena represented in deep-sea sediments, and hitherto obscure fine time lines in volcanic and sedimentary sequences of the continents. As demonstrated in this summary of the polarity time-scale evolution and some of the problems inherent in some of the associated data acquisition, a certain vulnerability to circular reasoning and possibly unrealized natural complexities exist in evaluating the fine details of the scale. At this stage, it would therefore appear opportune to establish an international group to facilitate an unambiguous and continual development of the polarity history definition, as well as the associated nomenclature.

ACKNOWLEDGMENTS

My gratitude goes to J. P. Kennett, P. Sylvester-Bradley, and R. M. Carter for their

valuable comments on stratigraphic nomenclature convention. G. B. Dalrymple made
several valuable comments during review of the
manuscript, and kindly provided a reprint of
his 1972 publication, which also involves a
review of the polarity time-scale development.

REFERENCES CITED

Baksi, A. K., York, D., and Watkins, N. D., 1967,
Age of the Steens Mountain geomagnetic
polarity transition: Jour. Geophys. Research,
v. 72, p. 6299–6308.

Bandy, O. L., and Casey, R. E., 1969, Major late
Cenozoic planktonic datum planes, Antarctica
to the tropics: Antarctic Jour. U.S., v. 4, p.
170–171.

—— 1970, Quaternary paleoclimatic variations
[abs.]: Am. Assoc. Petroleum Geologists Bull.,
v. 54/55, p. 835.

—— 1972, Major Quaternary paleo-oceanographic
cycles, Antarctic to the tropics, in Adie, R. J.,
ed., Conf. Antarctic Geology and Geophysics,
2d, Oslo, Proc. (in press).

Bandy, O. L., Casey, R. E., and Wright, R. C.,
1969, Climatic deterioration near the Brunhes-
Matuyama boundary: INQUA Congress
(Paris) Session VIII (Symposium on deep-sea
sediments), Proc., p. 62–64.

—— 1971, Late Neogene planktonic zonation,
magnetic reversals, and radiometric dates,
Antarctic to the tropics: Am. Geophys. Union,
Antarctic Research Ser., v. 15, p. 1–26.

Berggren, W. A., Phillips, J. D., Bertels, A., and
Wall, D., 1967, Late Pliocene-Pleistocene
stratigraphy in deep-sea cores from the south-
central North Atlantic: Nature, v. 216, p.
253–255.

Birkeland, P. W., Crandell, D. R., and Richmond,
G. M., 1971, Status of correlation of Quaternary stratigraphic units in the western con-
terminous United States: Quaternary Research, v. 1, p. 208–227.

Blakely, R. J., and Cox, A., 1970, A search technique for modelling sea-floor magnetic lineations [abs.]: EOS (Am. Geophys. Union
Trans.), v. 51, p. 745.

Bonhommet, N., and Babkine, J., 1967, Sur la
présence d'aimanations inversées dans la
Chaîne des Puys: Acad. Sci. Comptes Rendus,
v. 264, p. 92–94.

Bonhommet, N., and Zähringer, J., 1969, Paleomagnetism and potassium argon age determinations of the Laschamp geomagnetic
polarity event: Earth and Planetary Sci.
Letters, v. 6, p. 43–46.

Bott, M. H. P., 1967, Solution of the linear inverse
problem in marine interpretation with application to oceanic magnetic anomalies: Jour.
Geophys. Research, v. 13, p. 313–323.

Bout, P., 1970, Absolute ages of some volcanic
formations in the Auvergne and Velay areas

and chronology of the European Pleistocene:
Paleogeography, Paleoclimatology, Paleoecology, v. 8, p. 95–106.

Brunhes, B., 1906, Recherches sur le direction
d'aimantation des roches volcaniques: Jour.
Physique, v. 5, p. 705–724.

Bucha, V., 1970, Geomagnetic reversals in Quaternary revealed from a paleomagnetic investigation of sedimentary rocks: Jour. Geomagnetism
and Geoelectricity, v. 22, p. 253–272.

Carter, R. M., 1970, A proposal for the sub-division
of Tertiary time in New Zealand: New Zealand
Jour. Geology and Geophysics, v. 13, p. 350–
363.

Chamalaun, F. H., and McDougall, I., 1966, Dating geomagnetic polarity epochs in Réunion:
Nature, v. 210, p. 1212–1214.

Clark, D. L., 1969, Paleoecology and sedimentation
in part of the Arctic Basin: Arctic Institute
North America Jour., v. 22, p. 233–245.

—— 1970, Magnetic reversals and sedimentation
rates in the Arctic Ocean: Geol. Soc. America
Bull., v. 81, p. 3129–3134.

Cox, A., 1966, Geomagnetic secular variation in
Alaska [abs.]: Am. Geophys. Union Trans., v.
47, p. 78.

—— 1968, Lengths of geomagnetic polarity intervals: Jour. Geophys. Research, v. 73, p. 3427–
3460.

—— 1969, Geomagnetic reversals: Science, v. 163,
p. 237–245.

Cox, A., and Dalrymple, G. B., 1967a, Statistical
analysis of geomagnetic reversal data and the
precision of potassium-argon dating: Jour.
Geophys. Research, v. 72, p. 2603–2614.

—— 1967b, Geomagnetic polarity epochs: Nunivak
Island, Alaska: Earth and Planetary Sci.
Letters, v. 3, p. 173–177.

Cox, A., Doell, R. R., and Dalrymple, G. B., 1963a,
Geomagnetic polarity epochs and Pleistocene
geochronometry: Nature, v. 198, p. 1049–1051.

—— 1963b, Geomagnetic polarity epoch: Sierra
Nevada II: Science, v. 142, p. 382–385.

—— 1964a, Geomagnetic polarity epochs: Science,
v. 143, p. 351–352.

—— 1964b, Reversals of the Earth's magnetic
field: Science, v. 144, p. 1537–1543.

—— 1965, Quaternary paleomagnetic stratigraphy,
in Wright, H. E., Jr., and Frey, D. G., eds.,
The Quaternary of the United States: Princeton, N. J., Princeton University Press, p. 817–
830.

—— 1968, Radiometric time scale for geomagnetic
reversals: Geol. Soc. London Quart. Jour.,
v. 124, p. 53–66.

Cox, A., Hopkins, D. M., and Dalrymple, G. B.,
1966, Geomagnetic polarity epochs: Pribilof
Islands, Alaska: Geol. Soc. America Bull., v.
77, p. 883–910.

Creer, K. M., 1971, Mesozoic palaeomagnetic reversal column: Nature, v. 233, p. 545–546.

Curtis, G. H., and Evernden, J. F., 1962, Age of

basalt underlying Bed I, Olduvai: Nature, v. 194, p. 610–612.

Curtis, G. H., and Hay, R. L., 1972, Further geologic studies and K-Ar dating of Olduvai Gorge and Ngorongoro Crater, in Bishop, W. W., and Miller, J. A., eds., Calibration of hominid evolution: Edinburgh, Scottish Academic Press (in press).

Dagley, P., Wilson, R. L., Ade-Hall, J. M., Walker, G.P.L., Haggerty, S. E., Sigurgeirsson, T., Watkins, N. D., Smith, P. J., Edwards, J., and Grasty, R. L., 1967, Geomagnetic polarity zones for Icelandic lavas: Nature, v. 216, p. 25–29.

Dalrymple, G. B., 1963, Potassium-argon dates of some Cenozoic volcanic rocks of the Sierra Nevada, California: Geol. Soc. America Bull., v. 74, p. 379–390.

—— 1972, Potassium-argon dating of geomagnetic reversals and North American glaciations, in Bishop, W. W., and Miller, J. A., eds., Calibration of hominid evolution: Edinburgh, Scottish Academic Press (in press).

Dalrymple, G. B., and Moore, J. G., 1968, Argon-40: Excess in submarine pillow basalts from Kilauea Volcano, Hawaii: Science, v. 161, p. 1132–1134.

Denham, C. R., and Cox, A., 1970, Paleomagnetic evidence that the Laschamp polarity event did not occur between 30,000 and 12,000 years ago [abs.]: EOS (Am. Geophys. Union Trans.), v. 51, p. 745.

Dickson, G. O., and Foster, J. H., 1966, Magnetic stratigraphy of a deep-sea core from the Pacific Ocean: Earth and Planetary Sci. Letters, v. 1, p. 458–462.

Doell, R. R., 1968, Paleomagnetic studies of lavas on the islands of Kauai and Oahu [abs.]: Am. Geophys. Union Trans., v. 49, p. 127.

Doell, R. R., and Dalrymple, G. B., 1966, Geomagnetic polarity epochs: A new polarity event and the age of the Brunhes-Matuyama boundary: Science, v. 152, p. 1060–1061.

Doell, R. R., Dalrymple, G. B., and Cox, A., 1966, Geomagnetic polarity epochs: Sierra Nevada Data, 3: Jour. Geophys. Research, v. 71, p. 531–541.

Doell, R. R., Dalrymple, G. B., Smith, R. L., and Bailey, R. A., 1968, Paleomagnetism, potassium-argon ages, and geology of rhyolites and associated rocks of the Valles Caldera, New Mexico: Geol. Soc. America Mem. 116, p. 211–248.

Dymond, J., 1969, Age determinations of deep-sea sediments: A comparison of three methods: Earth and Planetary Sci. Letters, v. 6, p. 9–14.

Einwich, A. M., and Vogt, P. R., 1971, Continued studies of the magnetic "smooth zone" in the western North Atlantic [abs.]: EOS (Am. Geophys. Union Trans.), v. 52, p. 195.

Einwich, A. M., Higgs, R. H., and Lowry, H. M., 1970, Magnetic anomalies in the "smooth zone" western North Atlantic [abs.]: EOS (Am. Geophys. Union Trans.), v. 51, p. 274.

Emilia, D. A., and Heinrichs, D. F., 1969, Ocean floor spreading: Olduvai and Gilsa events in the Matuyama epoch: Science, v. 166, p. 1267–1269.

Ericson, D. B., and Wollin, G., 1968, Pleistocene climates and chronology in deep-sea sediments: Science, v. 162, p. 1227–1234.

Evans, A. L., 1970, Geomagnetic polarity reversals in a late Tertiary lava sequence from the Akaroa Volcano, New Zealand: Geophysics, v. 21, p. 163–184.

Evernden, J. F., and Curtis, G. H., 1965, The potassium-argon dating of late Cenozoic rocks in East Africa and Italy: Current Anthropology, v. 6, p. 343–385.

Evernden, J. F., Curtis, G. H., and Lipson, J., 1957, Potassium-argon dating of igneous rocks: Am. Assoc. Petroleum Geologists Bull., v. 41, p. 2120–2147.

Evernden, J. F., Savage, D. E., Curtis, G. H., and James, G. T., 1964, Potassium-argon dates and the Cenozoic mammalian chronology of North America: Am. Jour. Sci., v. 262, p. 145–198.

Fillon, R. H., 1970, Paleoclimatic fluctuations in glacial-marine sediments of Antarctic deep-sea cores: Geol. Soc. America, Abs. with Programs (Ann. Mtg.), v. 2, no. 7, p. 550.

Finlay, H. J., and Marwick, J., 1940, The division of the Upper Cretaceous and Tertiary in New Zealand: Royal Soc. New Zealand Trans., v. 70, p. 77–135.

—— 1947, New divisions of the New Zealand Upper Cretaceous and Tertiary: New Zealand Jour. Sci., B 28, p. 228–236.

Foster, J. H., and Opdyke, N. D., 1970, Upper Miocene to Recent magnetic stratigraphy in deep-sea sediments: Jour. Geophys. Research, v. 75, p. 4465–4475.

Glass, B., Ericson, D. B., Heezen, B. C., Opdyke, N. D., and Glass, J. A., 1967, Geomagnetic reversals and Pleistocene chronology: Nature, v. 216, p. 437–442.

Goodell, H. G., and Watkins, N. D., 1968, The paleomagnetic stratigraphy of the Southern Ocean: 20° West to 160° East longitude: Deep-Sea Research, v. 15, p. 89–112.

Goodell, H. G., Watkins, N. D., Mather, T. T., and Koster, S., 1968, The Antarctic glacial history recorded in sediments of the southern oceans: Paleogeography, Paleoclimatology, Paleoecology, v. 5, p. 41–62.

Grommé, C. S., and Hay, R. L., 1963, Magnetization of basalt of Bed I, Olduvai Gorge: Nature, v. 200, p. 560–561.

—— 1967, Geomagnetic polarity epochs: New data from Olduvai Gorge, Tanganyika: Earth and Planetary Sci. Letters, v. 2, p. 111–115.

—— 1971, Geomagnetic polarity epochs: Age and duration of the Olduvai normal polarity event: Earth and Planetary Sci. Letters, v. 10, p.

42

179–185.

Harrison, C.G.A., 1966, The paleomagnetism of deep-sea sediments: Jour. Geophys. Research, v. 71, p. 3033–3043.

—— 1968, Formation of magnetic anomaly pattern by dike injection: Jour. Geophys. Research, v. 73, p. 2137–2142.

Hay, R. L., 1963, Stratigraphy of beds I through IV, Olduvai Gorge, Tanganyika: Science, v. 139, p. 829–831.

Hays, J. D., 1970, Stratigraphy and evolutionary trends of radiolaria in North Pacific deep-sea sediments: Geol. Soc. America Mem. 126, p. 185–218.

—— 1971, Faunal extinctions and reversals of the Earth's magnetic field: Geol. Soc. America Bull., v. 82, p. 2433–2448.

Hays, J. D., and Ninkovitch, D., 1970, North Pacific deep-sea ash chronology and age of present Aleutian underthrusting: Geol. Soc. America Mem. 126, p. 263–290.

Hays, J. D., and Opdyke, N. D., 1967, Antarctic radiolaria, magnetic reversals, and climatic change: Science, v. 158, p. 1001–1010.

Hays, J. D., Saito, T.. Opdyke, N. D., and Burckle, L. H., 1969, Pliocene-Pleistocene sediments of the equatorial Pacific: Their paleomagnetic, biostratigraphic, and climatic record: Geol. Soc. America Bull., v. 80, p. 1481–1514.

Heirtzler, J. R., Dickson, G. O., Herron, E. N., Pitman, W. C., and Le Pichon, X., 1968, Marine magnetic anomalies, geomagnetic field reversals and motions of the ocean floor and continents: Jour. Geophys. Research, v. 73, p. 2119–2136.

Helsley, C. E., and Steiner, M. B., 1969, Evidence for long intervals of normal polarity during the Cretaceous period: Earth and Planetary Sci. Letters, v. 5, p. 325–332.

Herman, Y., 1970, Arctic paleo-oceanography in late Cenozoic time: Science, v. 169, p. 474–477.

Hoare, J. M., Condon, W. H., Cox, A., and Dalrymple, G. B., 1968, Geology, paleomagnetism, and potassium-argon ages of basalts from Nunivak Islands, Alaska: Geol. Soc. America Mem. 116, p. 377–414.

Hornibrook, N. de B., 1969, in Woods, B. L., The geology of the Inatapere Subdivision, western Southland: New Zealand Geol. Survey Bull., v. 79, p. 98–102.

Irving, E., 1964, Paleomagnetism and its application to geological and geophysical problems: New York, John Wiley & Sons, Inc., 399 p.

Irving, E., and Parry, L. G., 1963, The magnetism of some Permian rocks from New South Wales: Royal Astron. Soc. Jour. Geophysics, v. 7, p. 395–411.

Jacobs, J. A., 1963, The Earth's core and geomagnetism: New York, Pergamon Press, 137 p.

Kennett, J. P., and Watkins, N. D., 1970, Geomagnetic polarity change, volcanic maxima, and faunal extinction in the South Pacific: Nature, v. 227, p. 930–934.

Kennett, J. P., Watkins, N. D., and Vella, P., 1971, Paleomagnetic chronology of Pliocene-early Pleistocene climates and the Plio-Pleistocene boundary in New Zealand: Science, v. 171, p. 276–279.

Lamb, J. L., 1969, Planktonic foraminiferal datums and late Neogene epoch boundaries in the Mediterranean, Caribbean, and Gulf of Mexico: Gulf Coast Assoc. Geol. Socs. Trans., v. 19, p. 559–578.

Larson, R. L., and Spiess, F. N., 1969, East Pacific Rise Crest: A near-bottom geophysical profile: Science, v. 163, p. 68–71.

Lawley, E. A., 1970, The intensity of the geomagnetic field in Iceland during Neogene polarity transitions and systematic deviations: Earth and Planetary Sci. Letters, v. 10, p. 145–149.

Luyendyke, B. P., 1969, Origin of short-wavelength magnetic lineations observed near the ocean bottom: Jour. Geophys. Research, v. 74, p. 4869–4881.

Margolis, S. V., and Kennett, J. P., 1971, Cenozoic paleoglacial history of Antarctica recorded in subantarctic deep-sea cores: Am. Jour. Sci., v. 271, p. 1–36.

Matthews, D. H., and Bath, J., 1967, Formation of magnetic anomaly pattern of mid-Atlantic ridge: Geophysics, v. 13, p. 349–357.

Matuyama, M., 1929, On the direction of magnetization of basalt in Japan, Tyosen and Manchuria: Japan Acad. Proc., v. 5, p. 203–205.

McDougall, I., 1964, Potassium-argon ages from lavas of the Hawaiian Islands: Geol. Soc. America Bull., v. 75, p. 107–128.

—— 1966, Precision methods of potassium-argon isotopic age determination on young rocks, in Runcorn, S. K., ed., Methods and techniques in geophysics II: New York, Interscience, p. 274–304.

McDougall, I., and Chamalaun, F. H., 1966, Geomagnetic polarity scale of time: Nature, v. 212, p. 1415–1418.

—— 1969, Isotopic dating and geomagnetic polarity studies on volcanic rocks from Mauritius, Indian Ocean: Geol. Soc. America Bull., v. 80, p. 1419–1442.

McDougall, I., and Tarling, D. H., 1963, Dating of polarity zones in the Hawaiian Islands: Nature, v. 200, p. 54–56.

—— 1964, Dating geomagnetic polarity zones: Nature, v. 202, p. 171–172.

McDougall, I., and Wensink, H., 1966, Paleomagnetism and geochronology of the Pliocene-Pleistocene lavas in Iceland: Earth and Planetary Sci. Letters, v. 1, p. 232–236.

McDougall, I., Allsop, H. L., and Chamalaun, F. H., 1966, Isotopic dating of the New Volcanic series of Victoria, Australia, and geomagnetic polarity epochs: Jour. Geopnys. Research, v. 71, p. 6107–6118.

McElhinny, M. W., 1971, Geomagnetic reversals during the Phanerozoic: Science, v. 172, p. 157–159.

McElhinny, M. W., and Burek, P. J., 1971, Mesozoic paleomagnetic stratigraphy: Nature, v. 232, p. 98–102.

McMahon, B. E., and Strangway, D. W., 1967, Kiaman magnetic interval in the western United States: Science, v. 155, p. 1012–1013.

Melloni, M., 1853, Sur l'aimantation des roches volcaniques: Acad. Sci. Comptes Rendus, v. 37, p. 229.

Mercanton, P. L., 1926, Inversion de l'inclinaison magnétique terrestre aux âges géologiques: Terrestrial Magnetism and Atmospheric Electricity, v. 31, p. 187–190.

Moorbath, S., Sigurdsson, H., and Goodwin, R., 1968, K-Ar ages of the oldest exposed rocks in Iceland: Earth and Planetary Sci. Letters, v. 4, p. 197–205.

Nagata, T., Akimoto, S., Uyeda, S., Shimizu, Y., Ozima, M., and Kobayashi, K., 1957, Palaeomagnetic study on a Quaternary volcanic region in Japan: Philos. Mag. Supp. Adv. Physics, v. 6, p. 255–263.

Nakagawa, H., Niitsuma, N., and Hayasaka, I., 1969, Late Cenozoic geomagnetic chronology of the Boso Peninsula: Geol. Soc. Japan Jour., v. 75, p. 267–280.

Ninkovitch, D., 1968, Pleistocene volcanic eruptions in New Zealand recorded in deep-sea sediments: Earth and Planetary Sci. Letters, v. 4, p. 89–102.

Ninkovitch, D., Opdyke, N. D., Heezen, B. C., and Foster, J., 1966, Paleomagnetic stratigraphy, rates of deposition, and tephrachronology in North Pacific deep-sea sediments: Earth and Planetary Sci. Letters, v. 1, p. 476–492.

Oldroyd, H., 1966, The future of taxonomic entomology: Systematic Zoology, v. 15, p. 253–260.

Opdyke, N. D., and Foster, J., 1970, The paleomagnetism of cores from the North Pacific: Geol. Soc. America Mem. 126, p. 83–120.

Opdyke, N. D., and Glass, B. P., 1969, The paleomagnetism of sediment cores from the Indian Ocean: Deep-Sea Research, v. 16, p. 249–261.

Opdyke, N. D., and Ninkovitch, D., 1971, The paleomagnetism of two deep-sea cores from the eastern Mediterranean [abs.]: EOS (Am. Geophys. Union Trans.), v. 52, p. 186.

Opdyke, N. D., and Runcorn, S. K., 1956, New evidence for reversal of the geomagnetic field near the Pliocene-Pleistocene boundary: Science, v. 123, p. 1126–1127.

Opdyke, N. D., Glass, B., Hays, J. D., and Foster, J., 1966, Paleomagnetic study of Antarctic deep-sea cores: Science, v. 154, p. 349–357.

Pecherski, D. M., 1970, Palaeomagnetism and palaeomagnetic correlation of Mesozoic formations of north-east USSR, in Paleomagnetic and biostratigraphic characteristics of some important Mesozoic and Cenozoic series from north and far-east USSR: Akad. Nauk SSSR, Sci. Works of North-East Complex Institute (SVKNEE), Magadan, v. 37, p. 58–114.

Peter, G., 1970, Discussion of a paper by B. P. Luyendyke, "Origin of short-wavelength magnetic lineations observed near the ocean bottom": Jour. Geophys. Research, v. 75, p. 6717–6720.

Pevzner, M. A., 1970, Paleomagnetic studies of Pliocene-Quaternary deposits of Pridniestrovie: Paleogeography, Paleoclimatology, Paleoecology, v. 8, p. 215–219.

Phillips, J. D., Berggren, W. A., Bertels, A., and Wall, D., 1968, Paleomagnetic stratigraphy and micropaleontology of three deep-sea cores from the central North Atlantic Ocean: Earth and Planetary Sci. Letters, v. 4, p. 118–130.

Pitman, W. C., and Heirtzler, J. R., 1966, Magnetic anomalies over the Pacific Antarctic ridge: Science, v. 154, p. 1164–1168.

Roche, A., 1951, Sur les inversions de l'aimantation remante des roches volcaniques dans les monts d-Auvergue: Acad. Sci. Comptes Rendus, v. 233, p. 1132–1134.

—— 1956, Sur la date de la dernière inversion du champ magnétique terrestre: Acad. Sci. Comptes Rendus, v. 243, p. 812–814.

Ruddiman, W. F., 1971, Pleistocene sedimentation in the equatorial Atlantic: stratigraphy and faunal paleoclimatology: Geol. Soc. America Bull., v. 82, p. 283–302.

Rutten, M. G., 1959, Paleomagnetic reconnaissance of mid-Italian volcanos: Geologie en Mijnbouw, v. 21, p. 373–374.

Rutten, M. G., and Wensink, H., 1960, Paleomagnetic dating, glaciations and the chronology of the Plio-Pleistocene in Iceland: Internat. Geol. Cong., 21st, Pt. IV, Proc., p. 62–70.

Scott, G. H., 1960, The type locality concept in time-stratigraphy: New Zealand Jour. Geology and Geophysics, v. 3, p. 580–584.

Scott, G. S., 1969, Awamoan Stage (lower Miocene, N.Z.): Implications of Allan's revision and relation to Altonian: New Zealand Jour. Geology and Geophysics, v. 12, p. 383–390.

Smith, J. D., and Foster, J. H., 1969, Geomagnetic reversals in the Brunhes normal polarity epoch: Science, v. 163, p. 565–567.

Stesky, R. M., and Strangway, D. W., 1970, Magnetic studies of deep-sea cores from the mid-Atlantic ridge [abs.]: EOS (Am. Geophys. Union Trans.), v. 51, p. 274.

Steuerwald, B. A., Clark, D. L., and Andrews, J. A., 1968, Magnetic stratigraphy and faunal patterns in Arctic Ocean sediments: Earth and Planetary Sci. Letters, v. 5, p. 79–85.

Sylvester-Bradley, P. C., 1967, Towards an international code of stratigraphic nomenclature, *in* Teichert, C., and Yochelson, E. L., eds., Essays in paleontology and stratigraphy: Univ. Kansas Spec. Pubs. Geol., v. 2, p. 49–56.

Talwani, M., Windisch, C. C., and Langseth, M. G., 1971, Reykjanes Ridge crest: A detailed geophysical study: Jour. Geophys. Research, v. 76, p. 473–517.

Thompson, J. A., 1916, On stage names applicable to the division of the Tertiary in New Zealand: New Zealand Inst. Trans., v. 48, p. 28–40.

van Montfrans, H. M., 1971, Palaeomagnetic dating in the North Sea Basin: Earth and Planetary Sci. Letters, v. 11, p. 226–235.

van Montfrans, H. M., and Hospers, J., 1969, A preliminary report on the stratigraphical position of the Matuyama-Brunhes geomagnetic field reversal in the Quaternary sediments of the Netherlands: Geologie en Mijnbouw, v. 48, p. 565–572.

Vine, F. J., 1966, Spreading of the ocean floor: New evidence: Science, v. 154, p. 1405–1415.

—— 1968, Magnetic anomalies associated with mid-ocean ridges, *in* Phinney, R. A., ed., The history of the Earth's crust: Princteon, N. J., Princeton University Press, p. 73–89.

—— 1970, Sea-floor spreading and continental drift: Jour. Geol. Education, v. 18, p. 87–90.

Vine, F. J., and Wilson, J. T., 1965, Magnetic anomalies over a young ocean ridge off Vancouver Island: Science, v. 150, p. 485–489.

Watkins, N. D., 1968a, Comments on the interpretation of linear magnetic anomalies: Pure and Appl. Geophysics, v. 69, p. 170–192.

—— 1968b, Short period geomagnetic polarity events in deep-sea sedimentary cores: Earth and Planetary Sci. Letters, v. 4, p. 341–349.

Watkins, N. D., and Abdel-Monem, A., 1971, Detection of the Gilsa geomagnetic polarity event on the island of Madeira: Geol. Soc. America Bull., v. 82, p. 191–198.

Watkins, N. D., and Cambray, F. W., 1971, Paleomagnetism of Cretaceous dikes from Jamaica: Royal Astron. Soc. Jour. Geophysics, v. 22, p. 163–179.

Watkins, N. D., and Goodell, H. G., 1966, The stratigraphic use of paleomagnetism in sedimentary cores from the Southern Ocean [abs.]: Am. Geophys. Union, v. 47, p. 478.

—— 1967a, Confirmation of the reality of the Gilsa geomagnetic polarity event: Earth and Planetary Sci. Letters, v. 2, p. 123–129.

—— 1967b, Geomagnetic polarity changes and faunal extinction in the Southern Ocean: Science, v. 156, p. 1083–1089.

Watkins, N. D., and Haggerty, S. E., 1967, Primary oxidation variation and petrogenesis in a single lava: Contr. Mineralogy and Petrology, v. 15, p. 251–271.

Watkins, N. D., and Kennett, J. P., 1971, Antarctic bottom water: Major change in velocity during the late Cenozoic between Australia and Antarctica: Science, v. 173, p. 813–818.

Watkins, N. D., and Paster, T. P., 1971, The magnetic properties of igneous rocks from the ocean floor: Royal Soc. London Philos. Trans., ser. A, v. 268, p. 507–550.

Watkins, N. D., and Richardson, A., 1968, Palaeomagnetism of the Lisbon Volcanics: Royal Astron. Soc. Jour. Geophysics, v. 15, p. 287–304.

—— 1971, Intrusives, extrusives, and linear magnetic anomalies: Royal Astron. Soc. Jour. Geophysics, v. 23, p. 1–13.

Watkins, N. D., Gunn, B. M., and Coy-Yll, R., 1970, Major and trace element variations during the initial cooling of an Icelandic lava: Am. Jour. Sci., v. 268, p. 24–49.

Watkins, N. D., Gunn, B. M., Baksi, A. K., York, D., and Ade-Hall, J., 1971, Paleomagnetism, geochemistry, and potassium-argon ages of the Rio Grande de Santiago Volcanics, central Mexico: Geol. Soc. America Bull., v. 82, p. 1955–1968.

Wilson, R. L., Haggerty, S. E., and Watkins, N. D., 1968, Variation of palaeomagnetic stability and other parameters in a vertical traverse of a single Icelandic lava: Royal Astron. Soc. Jour. Geophysics, v. 16, p. 179–192.

Wollin, G., Ericson, D. B., and Ryan, W.B.F., 1971, Variations in magnetic intensity and climatic changes: Nature, v. 232, p. 549–551.

MANUSCRIPT RECEIVED BY THE SOCIETY JUNE 23, 1971
REVISED MANUSCRIPT RECEIVED SEPTEMBER 20, 1971

3

Copyright © 1977 by the Geological Society of America

Reprinted from *Geol. Soc. America Bull.* **88**:1–15 (1977)

Extension of the geomagnetic polarity time scale to 6.5 m.y.: K-Ar dating, geological and paleomagnetic study of a 3,500-m lava succession in western Iceland

IAN McDOUGALL   *Research School of Earth Sciences, Australian National University, Canberra, A.C.T., 2600*
KRISTJAN SAEMUNDSSON }
HAUKUR JOHANNESSON   }   *National Energy Authority, Reykjavik, Iceland*
NORMAN D. WATKINS   *Graduate School of Oceanography, University of Rhode Island, Kingston, Rhode Island, 02881*
LEO KRISTJANSSON   *Science Institute, University of Iceland, Reykjavik, Iceland*

ABSTRACT

More than 400 successive lavas in Borgarfjördur, western Iceland, have been subjected to paleomagnetic and K-Ar age analysis. Volcanism in the region was virtually continuous between about 7.0 and 2 m.y. ago, during which time more than 3,500 m of volcanics and interbedded sediments accumulated. Regression analysis of the K-Ar age and aggregate thickness data demonstrates that the rate of growth of the lava pile was remarkably uniform at 730 m/m.y. throughout the whole period. The magnetostratigraphic and K-Ar data indicate a nearly complete record of the geomagnetic polarity history and extend the polarity time scale based upon data from subaerial volcanic rocks to about 6.5 m.y. ago. The boundaries of polarity epoch 5 are shown to be 5.34 and 5.83 m.y. Epoch 6 lies between about 6.54 and 5.83 m.y. ago, during which time three normal polarity events are recognized. The ages determined for the polarity-interval boundaries in this study confirm recent estimates derived from analyses of marine magnetic anomalies using sea-floor–spreading assumptions.

The results show that it is possible to obtain reliable K-Ar ages on lavas that have undergone zeolite facies metamorphism at temperatures up to about 150°C by careful selection of samples that are well crystallized and have their original high-temperature mineralogy preserved.

A revised estimate of 5.2 ± 0.1 m.y. is proposed for the age of the Miocene-Pliocene boundary.

INTRODUCTION

Beginning in 1963, K-Ar and paleomagnetic studies of basaltic rocks led to the delineation of the geomagnetic polarity history for the past 4.5 m.y. (summary by Watkins, 1972). The time scale which this polarity history represents has assumed diverse importance, the most spectacular of which is its application to analyses of linear marine magnetic anomalies, in terms of the sea-floor–spreading hypothesis (Vine, 1966). This in turn has led, by extrapolation, to a predicted polarity time scale for the Cenozoic (Heirtzler and others, 1968) and Mesozoic (Larson and Pitman, 1972). Attempts are currently being made to test the validity of this predicted scale by paleomagnetic and micropaleontological studies of deep-sea sediments (Opdyke and others, 1974; Theyer and Hammond, 1974) and paleomagnetic studies of selected outcrops, especially in the Mesozoic (Helsley and Steiner, 1974). Ambiguities in the predicted time scale (Baldwin and others, 1974) have drawn attention to the difficulties in extrapolating from the past 4.5 m.y. of polarity history, derived by K-Ar and paleomagnetic techniques, and applying this information to the Mesozoic. It has become clearly desirable to extend this scale to at least t = 7 m.y., to

minimize extrapolation errors, as well as to provide a detailed polarity scale for utilization back to 7 m.y. ago. The precision of K-Ar ages is insufficient to extend the polarity time scale using basalt extrusives, unless unusually long stratigraphically controlled sequences can be studied.

It is now well established that K-Ar ages can be obtained with an accuracy of a few percent on subaerial lavas that are holocrystalline and free of alteration. In contrast, lavas that are not well crystallized or contain glass or secondary minerals commonly yield K-Ar ages that are incorrect. In many cases, the measured ages are demonstrably too young, almost certainly the result of the poor argon retention properties of glass and alteration products. As shown by Baksi (1974), however, even slightly altered lavas, and particularly those containing clay minerals, may give K-Ar ages that are too old. Baksi presented convincing evidence that this can be caused by fractionation of atmospheric argon associated with the clay minerals in the sample during preheating under vacuum at moderate temperatures (200° to 300°C) prior to extraction of the argon in the high vacuum system. Problems of this kind relating to poor argon retention properties of the phases in a lava, and laboratory-induced fractionation effects, may be expected during the K-Ar dating of Icelandic lavas because of their somewhat altered nature.

In this paper, we shall present paleomagnetic and K-Ar data for a very thick sequence of lava flows from the Borgarfjördur region in western Iceland (Fig. 1, inset). This region is located to the west of the Reykjanes-Langjökull rift zone toward which the strata dip at an average of about 5°. The structural relations favor an origin of the rocks within this rifting zone and subsequent spreading away from it.

Volcanism in Iceland is limited to several zones of rifting and transcurrent faulting which are part of the Mid-Atlantic Ridge system. In contrast to the situation on the Reykjanes Ridge (Herron and Talwani, 1972), crustal spreading in Iceland may occur along two or more axes simultaneously (Walker, 1975), and it has been suggested that during the geological history of Iceland the spreading axes may have jumped repeatedly (Saemundsson, 1974). Paleomagnetic and K-Ar dating studies of stratigraphically controlled long sequences are obvious means of establishing Icelandic geology on a firm stratigraphic basis, so that ideas about tectonic history of Iceland may become less speculative. Thus, in addition to attempting to extend the polarity time scale back from t = 4.5 m.y., our data will assist in defining the regional geological history.

PREVIOUS WORK

Brunhes (1906) and Mercanton (1926) first recognized the potential value of magnetic polarity as a stratigraphic index, by virtue of the contrasting polarities between old and younger volcanic

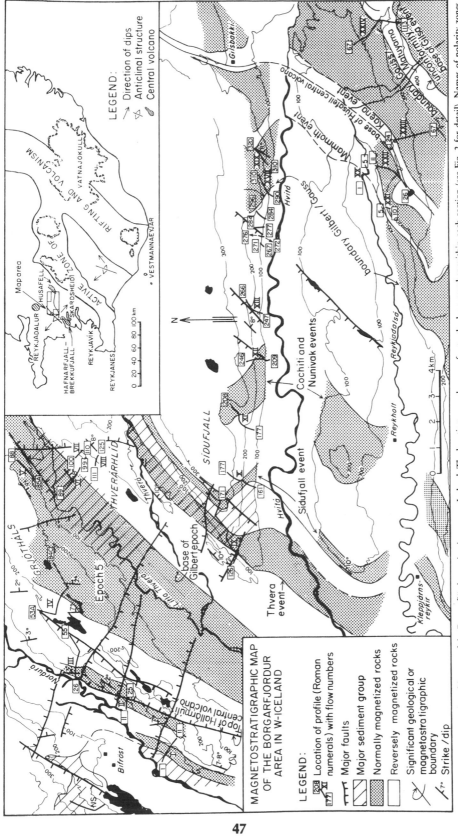

Figure 1. Magnetostratigraphic map of the Borgarfjordur region in western Iceland. The location of each profile is shown. Roman numerals (I to XXIV) refer to section number; numbers enclosed in boxes refer to the lava number within each section (see Fig. 2 for detail). Names of polarity zones added; for detail see text.

47

rocks in the Massif Central of France. Later, Matuyama (1929) proposed that the Pleistocene volcanoes in parts of both Japan and Korea could be distinguished from older rocks on a polarity basis. Roche (1951), Opdyke and Runcorn (1956), and Nagata and others (1957) made similar suggestions. It was Einarsson (1957), however, who first demonstrated the regional geological potential of this approach by the identification and mapping (using a simple compass) of 14 successive polarity zones in western Iceland. The map which he produced has not been improved upon, or matched in scale in other volcanic provinces. Piper (1971) has speculated upon the relationship between the polarity zones mapped by Einarsson (1957, 1962) and the established geomagnetic polarity history, but no reliable K-Ar data were available at that time to confirm his suggestions.

The clear need for a combined paleomagnetic and K-Ar dating approach to the unraveling of Icelandic geology (which cannot be carried out by conventional means, because of the restricted occurrence of sediments and the general lack of definitive fossils) was recognized nearly a decade ago (Dagley and others, 1967). Only recently, however, has this approach begun to yield reliable results; the main difficulty has been the problem of K-Ar dating of basaltic rocks that have undergone mild thermal metamorphism. This thermal metamorphism appears to be ubiquitous to virtually all volcanics in Iceland that have been buried by more than a few hundred metres, and it is expressed by the progressive development of zeolites and other secondary minerals with increasing depth in the lava piles (Walker, 1960). This regional alteration takes place below the water table by reaction of water with the basalts to produce secondary minerals in response to the rather high geothermal gradients observed in Iceland, which are commonly in the order of 50° to 160°C per kilometre (Pálmason, 1973; Pálmason and Saemundsson, 1974).

In the flood basalt sequences of eastern Iceland, Walker (1960, 1964, 1974) has identified and mapped a number of zeolite zones which show distinctive mineral assemblages. Lavas near the original top of the lava pile are free of zeolites, but at an estimated depth of about 150 m zeolites begin to appear. Walker designated the uppermost zeolite zone as the chabazite-thomsonite zone, which passes down into the analcite zone at a depth of about 600 m. The analcite zone passes successively at greater depths into the mesolite-scolecite and laumontite zones. Walker showed that these zeolite zones are nearly horizontal and that they crosscut lava sequences in which dips increase from a few degrees near the top of the piles to 10° or more at sea level. He presented a strong case for the zeolite-zone boundaries being more or less parallel to the top of the lava pile and thus related to depth of burial and increasing temperature. Similar relationships appear to apply in the flood basalt sequences elsewhere in Iceland. From the high geothermal gradients of up to 160°C/km encountered in the area bordering the Reykjanes-Langjökull volcanic zone (Pálmason, 1973), it can be inferred that at a depth of about 600 m where analcite begins to develop, the temperature is likely to be approximately 100°C. This value is intermediate between the estimates of Ade-Hall and others (1971) of 80°C for this boundary (obtained using known thermal gradients) and 140°C, derived from experimental results of analcite stability field studies (see also Coombs and others, 1959). We conclude that lavas in which the chabazite-thomsonite association is developed have been metamorphosed at temperatures not exceeding 140°C and perhaps as low as 100°C.

Two major problems arise in attempting to measure K-Ar ages on basalts so affected: the progressive hydrothermal alteration of the lavas results in degradation of the original pyrogenic minerals and glass, if present, to secondary minerals which may not retain radiogenic argon quantitatively, even at ambient temperature; and in addition it is apparent that the metamorphism has occurred sometime after the eruption of the lavas, so that it is conceivable that K-Ar ages may record approximately the time of alteration rather than the time of crystallization of the lavas. The approach which we have used in attempting to evaluate the possible magnitude of these problems was to measure ages on carefully selected samples from throughout the sequence, and then to compare the results with the known stratigraphy.

FIELD METHODS

Mapping

The geology of the area was previously mapped in detail by Johannesson (1972) and Saemundsson and Noll (1974). These results formed the basis for the present study, which began in the summer of 1973. Mapping involved the definition of lava types, measurement of the thicknesses of lava flows and sedimentary beds, the identification of zeolite zoning and magnetic polarity (using a portable fluxgate magnetometer), the measurement of the tectonic dip, and the mapping of intrusions. The entire stratigraphic column was assembled from 24 profiles which were linked together using suitable reference horizons.

The location of each of the 24 separate profiles is shown in Figure 1. The prefix NP was applied to the lower 320 lavas in profiles I to XIX, and NT was applied to the upper 120 bodies in profiles XX to XXIV. Thus the lavas are numbered consecutively from NP1 to NP320, followed upward by NT1 to NT112. The stratigraphy of each section is given, together with much additional information, including lava identification number, in Figure 2.

Most of the sections were measured along steep slopes or gullies where a continuous succession of rock bodies could be observed without much difficulty. This was not, however, always possible, and so doubts about the continuity do exist in the area of Grjóthals (profile IV) and in Thverárhlid (profile VI) which cross relatively level country, where hidden faults might exist. The sections as a rule do not overlap significantly, as the stratigraphic connections in most cases were sufficently well established by the geological mapping. An overlap by more than one unit was considered necessary in a few cases, however, where between-profile correlation proved difficult (for example, profiles IX to X and XXII to XXIII in Fig. 2). Several connections are uncertain, but we are fairly confident that only a few units are involved, either missed or double counted. Individual flows were labeled by successive numbers, using durable yellow paint and starting from the bottom of the column near the Nordura River (Fig. 1). In some cases, flow units within a single compound lava flow also were given numbers. The mapping of faults proved very time consuming because of their large number and unpredictable throws.

The thickness of the units was established both from altimeter readings on steep escarpments and from direct measurement at the outcrop using a tape measure. It is difficult to assess the error limits of the thickness measurements. Check measurements indicate that the thicknesses assigned tend to be too low rather than too high. An error of as much as 10 percent may thus be present. Our results are consistent with the thickness of about 3,750 m estimated by Einarsson (1957) for approximately the same sequence, but very different from a mere 1,500 m reported by Piper (1971).

Paleomagnetism

Following the mapping, gasoline-powered drills were used to take 3 or 4 separate cores, of 2.5-cm diameter and 12-cm average length, from each numbered lava. Cores were oriented in geographic coordinates while still attached to the outcrop. A total of 1,560 cores were obtained in this way. Sampling sites were so chosen to avoid proximity to dikes. A limited number of lavas proved to be impossible to sample, for various reasons.

K-Ar Dating

Each lava in the sequence was examined in the field. Samples were taken using a 3-kg hammer from the least altered, most massive parts of a total of 154 lavas which showed the least evidence of

Figure 2. Diagrams showing the stratigraphy of each profile sampled. Both altitude and aggregate thickness above base of section are indicated. Magnetic polarity shown by small circles; open-reversed polarity, closed-normal polarity. Correlation between profiles shown by dashed lines. Section numbers (I to XXIV) and lava numbers as in Figure 1.

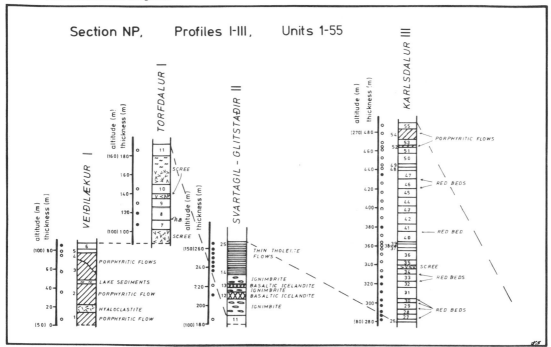

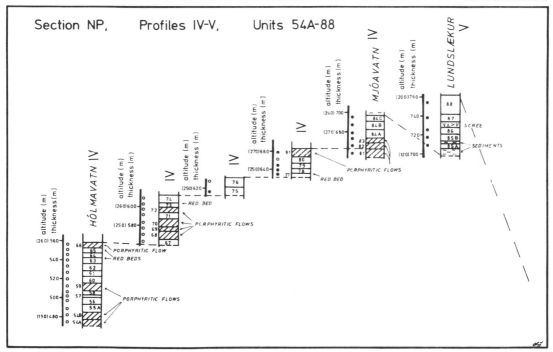

49

Figure 2. (*Continued*).

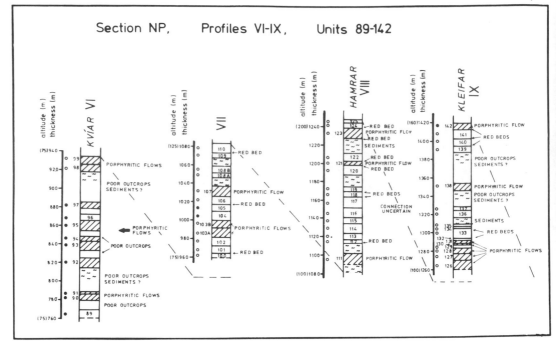

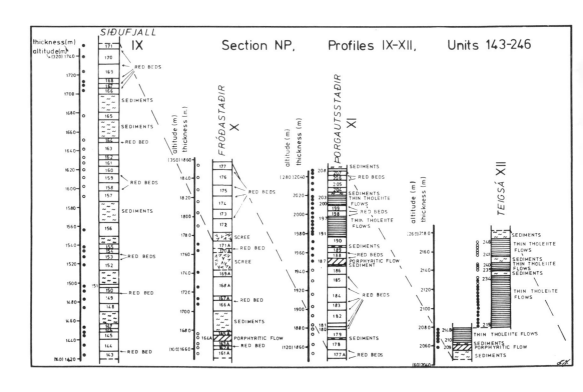

Figure 2. (*Continued*).

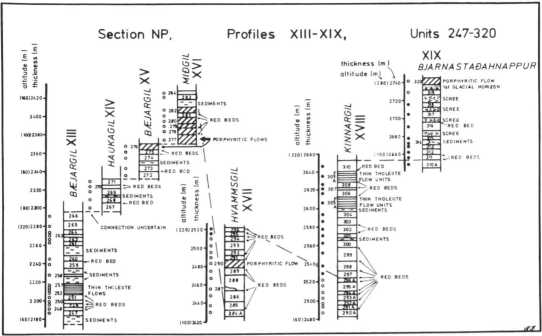

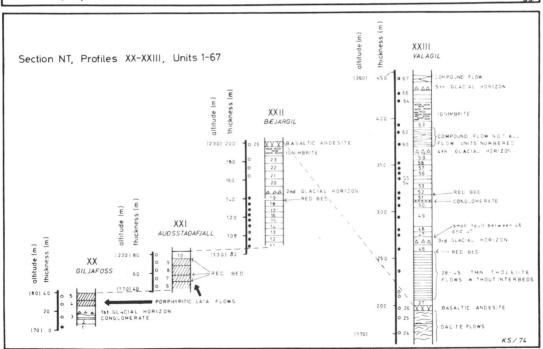

51

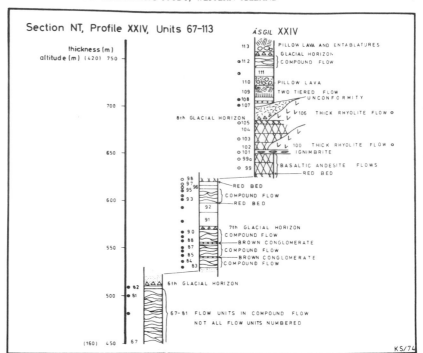

Figure 2. *(Continued)*.

secondary alteration. Thus more than one-half of the lavas were judged to be too altered to be worth collecting, primarily because of the extensive development of amygdales containing zeolite.

LABORATORY METHODS

Paleomagnetism

A 2.2-cm-long specimen was sliced from that part of each core which was farthest from the weathering surface. The direction and intensity of natural remanent magnetism (NRM) was measured using a 5-hertz spinner magnetometer, to a precision of ±2 percent. The unstable components in each specimen were then minimized by demagnetization to progressively higher peak alternating magnetic fields of 100, 200, and finally 300 oersteds, with remeasurement of the remanent magnetism direction and intensities being made after each treatment. Mean data for each lava were computed for the NRM, and following all three demagnetization treatments. The final mean direction of magnetization of each body was selected using the minimum scatter criterion: the demagnetizing field employed for each specimen is that which produces minimum within-lava variation of directions of remanent magnetism. All data were corrected for the small regional dip of the lavas, which were assumed to be originally horizontal, and were subjected to conventional Fisher (1953) statistical analysis. The data rejection criterion of Vincenz and Bruckshaw (1960) was employed. In this method, the resultant vector (**R**) obtained using unit vector per specimen must exceed a critical value so that the mean direction for a body can be interpreted as nonrandom at the 95 percent confidence level.

K-Ar Dating

A thin section made from each sample was examined under the microscope. It should be emphasized that the choice of samples petrographically suitable for dating is still a rather subjective procedure because of the lack of information about the location of the potassium in the lavas, and because of incomplete knowledge on the argon retention properties of glass, mesostasis, and incipiently altered minerals. As noted earlier, well crystallized, fresh lavas appear to retain radiogenic argon quantitatively, but such rocks are very rare in Iceland. It is not difficult to recognize fresh lavas, nor is it difficult to recognize strongly altered lavas which are unlikely to retain radiogenic argon quantitatively. Difficulties do arise, however, in deciding whether a slightly altered lava should be considered for dating or rejected. In the present study, the philosophy adopted was to choose samples that show the least alteration of the primary phases and minimum development of the secondary minerals. A total of 24 samples were considered to be suitable for K-Ar analysis.

Those samples chosen for dating were crushed to a size of 1.4 to 0.7 mm; an aliquot was taken, reduced to less than 0.1 mm and reserved for potassium analysis by a flame photometric technique similar to that given by Cooper (1963). Ten to 20 g of the coarser material was used for each argon extraction, carried out in a vacuum line according to procedures previously described (McDougall, 1966). In the present study, however, the bottle containing the sample was mounted outside the argon extraction line and baked overnight at less than 120°C to minimize the possibility of fractionation of atmospheric argon, a phenomenon observed by Baksi (1974) in some slightly altered lavas. During fusion of the sample by induction heating, a tracer of ^{38}Ar, prepared from a gas pipette system, was added to the evolved gases. After purification, the isotopic composition of the argon was measured on an extensively modified AEI MS10 mass spectrometer, which has a 4.2 kilogauss permanent magnet and a digital readout system. Data were reduced by computer, and errors were assigned according to the method described by McDougall and others (1969), in which the statistically determined uncertainties associated with the isotopic analysis, the calibration of the tracer, and the potassium determination are combined quadratically.

RESULTS

Local Geology

The stratigraphic column comprises a total thickness of 3.5 km of lava flows and interbasaltic detrital beds of volcanic origin. Central volcanoes, which form an integral part of the lava pile, were avoided, although near the base and top of the column some of the sections lie close to the central volcanoes of Hallarmúli (Johannesson, 1975) and Húsafell (Saemundsson and Noll, 1974). Except for a few intermediate and acid flows and rare ignimbrites, most of the units are basaltic lavas. Several different lava types occur among the basalts. These are olivine tholeiites which are usually strongly zeolitized and commonly consist of flow units; flows of olivine-poor or olivine-free tholeiite with no or negligible infillings of chalcedony or silica-rich zeolites; and flows with porphyritic plagioclase, which are normally very thick and massive. These different lava types tend to form mappable series of one dominant lava type in a manner similar to that found by Walker (1959) in eastern Iceland. The lateral extent of flows is unknown, although a single flow could occasionally be traced over a distance of several kilometres. Individual groups of flows persist along strike, however, for more than 10 km.

The thickness of flows varies considerably, depending on the lava type and probably also on the topography at the time of eruption. Intermediate and acid lava flows are always very thick, up to tens of metres in the latter case. Compound olivine tholeiite flows are also very thick and probably represent segments of lava shields. The olivine-free tholeiite flows usually have a thickness of 7 to 10 m, but occasionally groups occur which consist of only 2- to 4-m-thick flows. Most of the thin tholeiite flows originated from nearby central volcanoes.

Interbasaltic clastic beds make up a significant portion of the column (Fig. 2). There are three types of clastic beds. The most common are thin red beds, usually less than 1 m thick and regarded as soil of aeolian origin. The second type consists of sediments of fluvial and lacustrine origin. These are most common near the base of the section and at about 900 and 1,700 m above the base. The thick sedimentary horizons extend for considerable distances along strike, and perhaps represent periods of low lava extrusion rates, particularly as some of them are associated with unconformities. It

may be significant that the major sedimentary horizons, when traced along strike, are found to occur near the base or top of central volcanoes. They provide what might be considered as a floor or roof to a lenticular body of rocks produced by one central volcano and its associated fissure or dike swarm. The sedimentary horizons may therefore be merely the result of topographic changes and perhaps unequal subsidence following the extinction of one central volcano and the birth of another. The third type of clastic beds are the tillites and conglomerates of fluvioglacial origin which are frequently found to interfinger with hyaloclastite breccias. This type of sediment is typical of the uppermost 700 m of the section. The glacial horizons occur at fairly regular intervals separated by lava sequences about 100 m thick. A more detailed discussion of the glacial horizons is given by Saemundsson and Noll (1974).

The lava pile in the Borgarfjördur area has been affected by regional hydrothermal alteration. The degree of alteration increases with depth in the lava pile reaching the mesolite-scolecite zone of Walker (1960) at the valley bottoms in the western part of the area. Exposures in the Hvitá valley lie entirely in the chabazite-thomsonite zone. The analcite zone lies between the two and is represented on the higher ground in Grjótháls and Thverárhlíd (Fig. 3). The topmost zeolite-free zone is absent, probably having been removed by denudation. Mapping in the Borgarfjördur region shows (1) that the stratigraphically lowest 40 lavas (NP1-NP40) lie within the mesolite-scolecite zone, (2) that lavas from NP41 to NP160 lie in the analcite zone; and (3) that the remainder of the lavas are in the chabazite-thomsonite zone (Fig. 3). Thus the lowest lavas are likely to have been reheated to temperatures as high as 150°C, whereas those above the analcite zone probably have not been raised above 100°C since original crystallization and cooling.

Intrusives constitute only a small fraction of the total exposed rock volume. Thin sills and irregular veins occur near the section in Hallarmúli; elsewhere, subvertical dikes predominate. The predominant trend of dikes is northeast-southwest roughly parallel to the strike of the lavas. In the western part of the area, however, dikes trending northwest-southeast are also common. The dikes increase in number westward across the area. There is a sudden increase in the number of dikes west of Sidufjall where northwest-trending dikes add to the prevailing northeast-trending dikes. A close correspondence is observed between the intensity of zeolitization and the dike intensity (Fig. 3), as is the case in eastern Iceland (Walker, 1960).

The lavas of the area dip generally southeast between 2° and 8°. In Nordurárdalur, however, and in much of Grjótháls, the lavas dip toward the east. This local deviation of the dip is perhaps related to the activity of the Hallarmúli and Reykjadalur central volcanoes, as the sequence showing the anomalous dip is equivalent in time to the activity of these volcanoes. Locally, there is a slight unconformity between the Hallarmúli sequence, with its acid and intermediate rocks, and the overlying rocks which are related to the Reykjadalur central volcano.

Two sets of faults which are of different age and trends are rec-

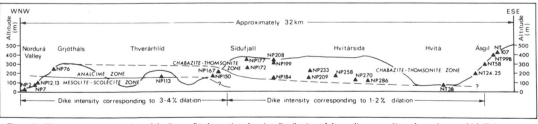

Figure 3. Diagrammatic cross section of the Borgarfjördur region showing distribution of the zeolite zones. Sites of samples on which K-Ar ages were measured are indicated together with sample number.

ognized. Most faults are normal with throws of a few metres up to several tens of metres. One set of faults is parallel to the strike of the lavas; it formed during and soon after the lavas were erupted. This is confirmed by the presence of dikes along the fault planes and more convincingly by a decrease in throw upward which is matched by lavas banking against the fault wall. The other set of faults has a northwest strike, perpendicular to the strike of the lavas. Most of these faults are young, forming faults scarps, sometimes in Holocene deposits. These young faults are thought to be related to the genesis of the Snaefellsnes Fracture Zone (Sigurdsson, 1970), which has an easterly trend and consists of west-northwest–trending *en echelon* segments.

Petrography

The lavas of the Borgarfjördur region are mainly tholeiitic basalts, although a few basaltic icelandites and dacitic lavas are found. Some of the basalts are strongly feldsparphyric. Petrographically, the lavas show considerable variations in grain size and texture, but except for dacite lavas (NT24, NT25), all have rather similar mineralogy. Most of the lavas are fine grained and intergranular, consisting of plagioclase, clinopyroxene, and iron oxide with or without olivine or its alteration products. Brown or green mineraloid and (or) clay commonly are present (as much as 10 percent), filling vesicles or occurring interstitially to the high temperature phases. In some rocks, pale brown, nearly isotropic mesostasis occurs in which microlites often are found. In the more vesicular lavas, infillings of zeolite are common. The basalts range from relatively fresh rocks with plagioclase and pyroxene and even olivine, virtually unaltered, through rocks in which plagioclase is cracked and shows incipient zeolitization into rocks wherein much of the plagioclase has been altered to zeolite.

Few of the lavas were both holocrystalline and free of incipient alteration and mineraloid and are thus not ideally suited for K-Ar dating. Samples accepted for dating showed no more than incipient alteration of plagioclase, pyroxene, and mesostasis, if present. Mineraloid was present in many of the retained samples, but samples showing obvious evidence of zeolitization were rejected.

The basaltic icelandites, NP12 and NP13, from near the bottom of the sequence in the mesolite-scolecite zone contain some fresh olivine in the groundmass and about 10 percent pale brown, remarkably fresh mesostasis. The dacite lavas, NT24 and NT25, in the Húsafell sequence are extremely fine grained to cryptocrystalline and glassy rocks that are relatively fresh.

For the samples collected, only a slight increase in the degree of alteration of the rocks from the lower part of the Borgarfjördur sequence is detected compared with the upper part of the sequence; far fewer samples, however, were regarded as worth collecting in the field from the lower part. Samples obtained from the more massive lavas commonly contained few amygdules, and some showed only incipient zeolitization in thin section, whereas adjacent vesicular lavas were strongly altered and zeolitized. It appears that the denser parts of the lavas were nearly impervious to ground water so that the original mineralogy was preserved, contrasting with the associated more vesicular, altered lavas. Nevertheless, even the relatively fresh lavas must have been at the same elevated temperatures as the surrounding altered lavas during the burial metamorphism.

Paleomagnetism

Using the statistical criterion of Vincenz and Bruckshaw (1960), 15 of the lavas sampled provided such a high scatter of directions between specimens that no reliable direction of remanent magnetism could be determined. The latitude of the virtual geomagnetic pole (VGP) for each body is plotted in Figure 4 against the height of the body above the base of the sampled section: the VGP latitude is the best index of polarity, which is considered as normal

polarity for VGP latitudes in the northern hemisphere and reversed polarity for VGP latitudes in the southern hemisphere. When a VGP latitude for a lava is between 45°N and 45°S, it may well represent either a transitional direction (if between lavas of opposite polarity) or an "excursion" (if not transitional between opposite polarities).

The paleomagnetic data are here discussed only in terms of the local stratigraphy and our attempt to obtain further definition of the polarity time scale. Detailed discussion of the results in terms of models of behavior of the geomagnetic field are published elsewhere (Watkins and others, 1976), together with a complete table of the mean directions of remanent magnetism and the relevant statistical information for each lava.

K-Ar Ages

The K-Ar data are given in Table 1, arranged in stratigraphic order. Errors are quoted at the level of one standard deviation. The duplicate potassium determinations generally agree to better than 1 percent. Most of the replicate argon measurements agree to within 2 percent, with a few disagreeing by more than 5 percent. A mean age is given for each sample, and the error assigned is either one standard deviation as calculated from the replicate ages or the standard deviation of the age that has the largest error, whichever is greater. This is done because in those cases where the replicates agree well, the calculated uncertainty for the mean age is unreasonably small when errors associated with potassium and tracer calibration measurements are taken into account.

DISCUSSION

K-Ar Ages

The measured K-Ar ages are remarkably consistent with the stratigraphy, suggesting that they record approximately the age of crystallization of the lavas rather than the age of subsequent zeolite facies (burial) metamorphism. This conclusion is supported strongly by the absence of any relation between the measured age of each sample and its position in the sequence with respect to the zeolite zones (Fig. 3), the boundaries of which are nearly horizontal and approximately parallel to the top of the lava pile. Even the samples from low down in the sequence (NP2, 7, 12, 13) within the mesolite-scolecite zone yield ages that are consistent with the stratigraphy. The wide range of potassium contents found in the samples (Table 1) taken together with the consistency of the results provides further evidence that the ages are meaningful. Thus despite the presence of incipient alteration in many of the samples dated, it would appear that loss of radiogenic argon during burial metamorphism has not been severe. Nevertheless, it would be unwise to assume that the measured ages are as accurate as would be expected from rocks that are completely unaltered and have not undergone burial metamorphism. Indeed, examination of Table 1 shows that there are a few anomalous ages, particularly toward the top of the NP section. The worst example is for sample NP209 from which at least 25 percent loss of radiogenic argon is indicated by comparison with ages on samples stratigraphically below and above it. Petrographically, sample NP209 is a very fine grained but well-crystallized basalt, with no identifiable glass or mesostasis, and it is virtually free of alteration and mineraloid. This demonstrates the difficulty of predicting with certainty from petrographic examination the likelihood that a particular sample will yield a reliable, accurate K-Ar age.

The present study is, however, most encouraging in that it has shown that well-crystallized, minimally altered lavas that have undergone zeolite facies metamorphism at temperatures up to 150°C in the mesolite-scolecite, analcite, and chabazite-thomsonite zones give K-Ar ages that are reasonably accurate. In the rest of this

TABLE 1. POTASSIUM-ARGON AGES ON WHOLE-ROCK SAMPLES, MAINLY BASALTS,
FROM THE BORGARFJÖRDUR AREA, WESTERN ICELAND

Field no.	Lab. no.	K (wt %)	Rad.^{40}Ar (10^{-12} mol/g)	$\frac{100 \text{ Rad.}^{40}\text{Ar}}{\text{Total }^{40}\text{Ar}}$	Calculated age (m.y.) ± 1 s.d.	Mean age (m.y.) ± 1 s.d.	Polarity[*]	Locality
NT107	73-1391	0.4192, 0.4161	1.185	23.8	1.59 ± 0.02	1.59 ± 0.02	R	Ásgil, Húsafell
NT99B	73-1388	0.6580, 0.6537	2.855	59.3	2.44 ± 0.03	2.42 ± 0.03	R	Ásgil, Húsafell
			2.818	61.7	2.41 ± 0.03			
NT58	73-1379	0.4730, 0.4725	1.955	24.1	2.32 ± 0.04	2.35 ± 0.05	N	Okvegúr, Húsafell
			2.003	13.8	2.38 ± 0.05			
NT25	73-1375	1.905, 1.907	9.988	55.0	2.94 ± 0.03	2.92 ± 0.04	R	Okvegúr, Húsafell
			9.815	91.8	2.89 ± 0.03			
NT24	73-1373	1.631, 1.635	8.439	40.2	2.90 ± 0.04	2.88 ± 0.04	R	Okvegúr, Húsafell
			8.283	37.3	2.85 ± 0.04			
NT3B	73-1361	0.1681, 0.1687	0.835	8.6	2.78 ± 0.09	2.77 ± 0.12	R	Reykjadalsá, Húsafell
			0.824	6.1	2.75 ± 0.12			
NP286	73-1332	0.2992, 0.2991	1.862	18.6	3.49 ± 0.07	3.46 ± 0.07	R	Hvammsloekur, Sidufjall
			1.831	18.3	3.44 ± 0.06			
NP270	73-1327	0.5607, 0.5577	3.623	25.9	3.64 ± 0.06	3.66 ± 0.06	R	Haukagil, Sidufjall
			3.665	24.8	3.68 ± 0.05			
NP258	73-1322	0.4860, 0.4819	2.897	19.1	3.36 ± 0.07	3.45 ± 0.13	R	Húsagil, Sidufjall
			3.051	20.3	3.54 ± 0.06			
NP233	73-1318	0.5871, 0.5843	3.673	14.2	3.52 ± 0.08	3.56 ± 0.08	?	Teigsá, Sidufjall
			3.758	14.4	3.60 ± 0.07			
NP209	73-1314	0.3027, 0.3051	1.500	7.3	2.77 ± 0.11	2.82 ± 0.11	N	Teigsá, Sidufjall
			1.561	6.7	2.88 ± 0.11			
NP208	73-1313	0.3220, 0.3218	2.364	13.6	4.12 ± 0.11	3.89 ± 0.20	?	Thorgautsstadagil, Sidufjall
			2.179	13.1	3.80 ± 0.08			
			2.146	16.0	3.74 ± 0.07			
NP199	73-1310	0.4285, 0.4295	2.780	18.2	3.64 ± 0.07	3.60 ± 0.07	N	Thorgautsstadagil, Sidufjall
			2.720	17.5	3.56 ± 0.06			
NP184	73-1304	0.5351, 0.5334	3.859	59.4	4.05 ± 0.06	4.08 ± 0.06	R	Thorgautsstadagil, Sidufjall
			3.919	57.7	4.12 ± 0.05			
NP177	73-1300	0.5665, 0.5659	4.162	57.1	4.12 ± 0.06	4.10 ± 0.06	R	South face of Sidufjall
			4.125	56.9	4.09 ± 0.05			
NP172	73-1297	0.5559, 0.5601	4.171	74.2	4.19 ± 0.06	4.20 ± 0.06	R	South face of Sidufjall
			4.192	74.6	4.22 ± 0.05			
NP162	73-1290	0.4120, 0.4171	3.453	22.5	4.67 ± 0.09	4.60 ± 0.10	R	Northwest face of Sidufjall
			3.347	19.9	4.53 ± 0.06			
NP150	73-1284	0.3830, 0.3839	3.439	55.8	5.03 ± 0.08	4.94 ± 0.13	N	Sidufjall
			3.314	64.1	4.85 ± 0.06			
NP113	73-1271	0.2441, 0.2444	2.371	19.6	5.45 ± 0.09	5.26 ± 0.16	R	Aesustigur, Thverárhlíd Valley
			2.246	19.3	5.16 ± 0.09			
			2.254	19.3	5.18 ± 0.09			
NP76	73-1252	0.3689, 0.3679	3.693	34.4	5.62 ± 0.09	5.78 ± 0.15	?	Grjótháls
			3.889	60.6	5.92 ± 0.07			
			3.805	64.8	5.79 ± 0.07			
NP13	73-1213	0.8672, 0.8671	9.551	88.4	6.18 ± 0.09	6.20 ± 0.09	R	Nordurárdalur, east of Bifrost
			9.611	89.4	6.22 ± 0.07			
NP12	73-1212	0.9452, 0.9504	10.486	86.6	6.20 ± 0.10	6.25 ± 0.10	N	Nordurárdalur, east of Bifrost
			10.645	75.8	6.30 ± 0.07			
NP7	73-1207	0.3934, 0.3956	4.749	43.0	6.75 ± 0.11	6.74 ± 0.11	N	Nordurárdalur, east of Bifrost
			4.734	45.1	6.73 ± 0.08			
NP2	73-1202	0.2565, 0.2546	3.238	30.9	7.11 ± 0.12	7.05 ± 0.12	R	Nordurárdalur, east of Bifrost
			3.188	30.3	7.00 ± 0.09			

Note: $\lambda_e = 0.585 \times 10^{-10}$ yr^{-1}; $\lambda_\beta = 4.72 \times 10^{-10}$ yr^{-1}; ^{40}K/K = 1.19×10^{-2} atom %.
[*] N = normal; R = reversed.

paper, we assume that these measured ages are valid estimates, to a first approximation, of the time since the lavas were erupted. Thus the sequence of lavas examined in the Borgarfjördur region is considered to have erupted over the period from about 7.0 to 1.6 m.y. ago, from the late Miocene, through the Pliocene and into the early Pleistocene.

Magnetostratigraphy

As shown in Figure 2, overlaps were employed between six of the separate profiles, because of some doubts in the between-profile correlations. The polarity data do not conflict with any of these correlations. The polarity log in Figure 4 excludes data from re-

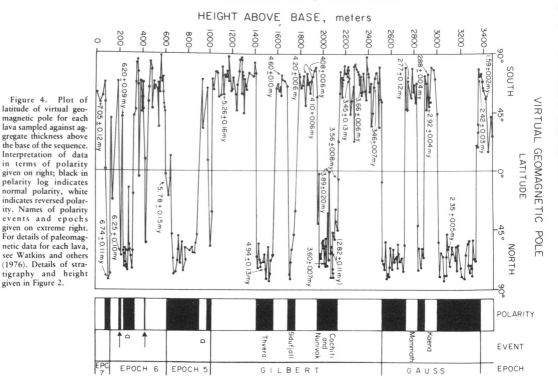

Figure 4. Plot of latitude of virtual geomagnetic pole for each lava sampled against aggregate thickness above the base of the sequence. Interpretation of data in terms of polarity given on right; black in polarity log indicates normal polarity, white indicates reversed polarity. Names of polarity events and epochs given on extreme right. For details of paleomagnetic data for each lava, see Watkins and others (1976). Details of stratigraphy and height given in Figure 2.

peated sampling of the same lava, and it is therefore presented as a magnetostratigraphic column for the Borgarfjördur region. With this, a magnetostratigraphic map can then be constructed for the area (Fig. 1): the identification and naming of the magnetostratigraphic units is discussed in the following section. It is important to realize that polarity stratigraphy, while most commonly based on measurements made at the outcrop, ideally must be based on polarities determined in the laboratory, because unstable normal polarity components, which can be removed only in the laboratory, often can mask an original reversed polarity.

It is suggested that the Borgarfjördur magnetostratigraphic column can be employed as a "type" section for western Iceland: those earlier paleomagnetic surveys of western Iceland involving laboratory analyses (for example, Wilson and others, 1972) did not attempt to compile a continuous column.

Polarity Time Scale

In Figure 5, the K-Ar and paleomagnetic polarity data are plotted against the stratigraphic thickness above the base of the section sampled in the Borgarfjördur region. The polarity time scale of Talwani and others (1971) is shown at the top of the diagram using the standard nomenclature extended to include epochs 5 to 7 after Opdyke (1972). In the following discussion, we adhere to the arbitrary selection of boundaries between polarity epochs that these authors have established, rather than attempt to select our own polarity patterns.

The combined data indicate that the sequence is virtually continuous from the early Matuyama to the base of epoch 6 and into epoch 7 and that there is a linear relation between measured age and stratigraphic position. Although the K-Ar ages generally are consistent with the stratigraphy, there are several cases where the measured ages conflict with the stratigraphy, the most obvious example of which is the anomalously young age for sample NP209. Clearly, individual ages are of insufficient accuracy for them to be used to estimate the ages of the boundaries of polarity intervals. As inspection of Figure 5 shows, however, these data are amenable to linear regression analysis, providing a means of using the results collectively to determine the age of the boundaries of the intervals of constant polarity without reference to existing time scales, in much the same way as the original polarity time scale was arrived at, by interpolation between data from diverse localities, but with the substantial advantage in this study of working with data from a single sequence.

A simple regression analysis approach was used in which the aggregate stratigraphic thickness (lavas and interbedded sediments) was assumed to be known without error and the variance of the age parameter was taken as constant. In effect, we are assuming that the thickness of section accumulated per unit time is constant; this is unlikely to be strictly correct, but other models involving much extrapolation are considered by us to be far more subjective. In the regression analysis, the clearly erroneous result from NP209 was omitted, as was the datum for NT107, because this sample was collected from a lava above an unconformity. Taking all the remaining data (22 points), the best fit line was:

$$y = 6.72 - 0.001372\,x \qquad (1)$$

where y is the age in million years and x is the height in metres above the base of the section. The correlation coefficient for this regression is 0.990, and the standard error of the slope is 3.1 percent.

This regression yields an age of 3.27 m.y. for the easily recognized Gilbert-Gauss boundary at 2,520 m (Fig. 4), compared with the well-determined age of 3.32 m.y. for this boundary based upon

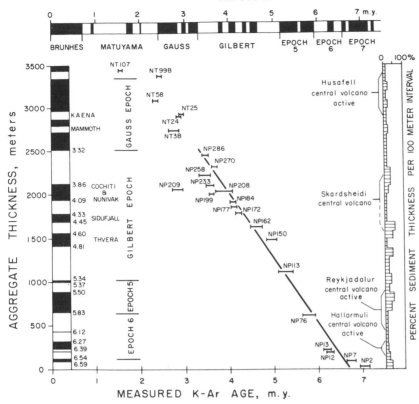

Figure 5. The K-Ar ages, with precision limits indicated, plotted against aggregate stratigraphic thickness above the base of the sequence. Polarity log for the sequence shown adjacent to the thickness axis; black = normal polarity, white = reversed polarity. Polarity time scale after Talwani and others (1971) given at top of diagram. Regression line through data is least-squares fit with Gilbert-Gauss epoch boundary fixed at 3.32 m.y. at 2,520 m (see text for discussion). Ages indicated for polarity interval boundaries on polarity log derived from regression. The proportion of sediment present for each 100 m of section shown on right, and period of activity of central volcanoes in the region also is indicated.

K-Ar dating of fresh lavas of known polarity from elsewhere (Cox, 1969; Dalrymple, 1972). The agreement is satisfactory and suggests that the assumptions made in the analysis are reasonable. Nevertheless it was decided to fix the Gilbert-Gauss boundary at 3.32 m.y. and to fit a regression to all data points from below this stratigraphic level in order to derive a polarity time scale that is directly comparable with time scales obtained by extrapolation using magnetic anomaly patterns and magnetostratigraphy in deep-sea cores, for which the 3.32 m.y. age for the Gilbert-Gauss boundary commonly is assumed. The regression which is shown in Figure 5, yields the equation:

$$y = 6.72 - 0.001348\,x \qquad (2)$$

based upon 17 data points in addition to that for the Gilbert-Gauss boundary. The ages for the boundaries between lavas of opposite polarity in the Borgarfjördur region as determined from this regression are given on the polarity log in Figure 6. The resulting polarity log is illustrated in Figure 6, where it is compared with the polarity time scales derived from marine magnetic anomalies and magnetostratigraphic data from deep-sea sedimentary cores. Standard errors for the intercept and slope parameters for the line shown in Figure 5 are less than 2.7 percent, so that uncertainties in the assigned ages of the polarity-interval boundaries must be of this order, that is, in the range of 0.1 to 0.2 m.y. Despite the relatively large uncertainties in the derived ages, the results from the Borgarfjördur region strongly support a reduction in the age of individual polarity boundaries as given by Vine (1968) and Heirtzler and others (1968) to values more nearly those proposed by Talwani and others (1971) and Klitgord and others (1975).

In the Borgarfjördur sequence, the Kaena and Mammoth reversed polarity events of the Gauss epoch are well represented. Three normal polarity events are recognized within the Gilbert Epoch rather than four, as have been identified in the magnetic anomaly patterns observed across mid-ocean ridges. The ages derived from the regression suggest that the Cochiti and Nunivak events are recorded as a single polarity interval without an intervening reversed interval, either because of a hiatus in the eruption of the lavas following the extinction of Skardsheidi central volcano (Fig. 5), or because of unrecognized difficulties in correlating between profiles XI and XII (Fig. 2) to the extent that some lavas have been omitted. The normal polarity events in the lower part of the Gilbert Epoch, called C_1 and C_2 by Opdyke (1972) and identified as 3.3 and 3.4 in magnetic anomaly analyses, are clearly distinguishable and have estimated ages similar to those proposed by Talwani and others (1971) and Klitgord and others (1975), as shown in Figure 6. By comparison with the polarity pattern observed elsewhere, we identify the Gilbert–epoch 5 boundary at 1,015 m above the base of the Borgarfjördur sequence, providing (with the use of equation 2) an age of 5.34 m.y., which is slightly older than recent estimates (Fig. 6). The upper normal polarity interval of epoch 5 is represented, however, by only two flows in the section (Fig. 4) and is thus much attenuated compared to previous estimates, suggesting that there was a reduced rate of eruption of lavas during this time. On the other hand, the age derived using equation 2 for the base of epoch 5 is within 0.1 m.y. of the independent estimates (Fig. 6). The base of epoch 5 is characterized by a "precursor" in which the VGP returns briefly to opposite but low latitudes, prior to completing the polarity reversal. We have defined the time of reversal to be when the VGP crosses the equator en route directly to high latitudes. This occurs at 650 m above the base of the section. Epoch 6 is shown as of wholly reversed polarity by Foster and Opdyke (1970), Opdyke (1972), and Opdyke and others (1974), although a normal polarity interval is indicated in the polarity log of one core (Conrad 72-66) studied by these workers. Theyer and Hammond (1974) recognized a normal polarity event in epoch 6 from their studies on deep-sea cores. Similarly, in the magnetic anomaly pattern, the equivalent of epoch 6 features a normal polarity event, according to Talwani and others (1971) and earlier results, as included in Figures 5 and 6. It is difficult to correlate unequivocally the polarity sequence found in the Borgarfjördur region with the deep-sea core and magnetic-anomaly polarity patterns, but we suggest that the base of epoch 6 is at 125 m above the base of the section with an estimated age of 6.54 m.y. This age agrees well with that assigned to the epoch 6–epoch 7 boundary by Talwani and others (1971) and Theyer and Hammond (1974), and it provides an upper segment to epoch 7 which features a short interval of normal polarity overlying a reversed event (Figs. 5 and 6). Our data indicate the presence of additional very short intervals of normal polarity within epoch 6 at about 6.12 m.y. (lava NP47, at an elevation of 440 m) and at 6.43 m.y. (lava NP12, at an elevation of 210 m). Low latitude VGP occurrences within epoch 6 (lavas NP10 and NP41 at 145 and 380 m, respectively) are not considered to be reversals, but probably only excursions of the geomagnetic field.

Thus the results from the Borgarfjördur sequence in western Iceland have enabled us to propose a polarity time scale for the interval between the Gilbert-Gauss boundary at 3.32 m.y. and the inferred base of epoch 6 at about 6.54 m.y. This extends the polarity time scale based upon direct K-Ar dating of subaerially erupted lavas to considerably earlier times than has been hitherto possible.

Nomenclature

As recently recommended by the Polarity Time Scale Subcommission of the International Union of Geological Sciences and International Union of Geodesy and Geophysics (*Geotimes*, June 1975, p. 26–27), it may be appropriate to name the polarity epochs and events recognized in studies of subaerial volcanics, which are then distinguished from the numbered epochs and events derived from deep-sea sedimentary core studies, and the "anomaly numbers" employed during analyses of marine magnetic profiles. We are very conscious of the need to avoid adding unnecessarily to the already burdensome number of names in the geological literature, however, and for the present at least, we propose to limit extending the nomenclature to the two events in the Gilbert Epoch. We therefore propose that the two normal events in the lower Gilbert are called the Sidufjall and Thvera events, and these are thought to correlate with events C_1 and C_2, respectively, of Opdyke (1972). Our suggestion would thus eliminate any numbering or lettering of events for the youngest four polarity epochs, but would employ the same numbering and lettering system used in deep-sea sediment studies for earlier epochs and events. Future work, particularly magnetostratigraphic mapping, may conceivably lead to revision of this proposal, in which case the polarity events identified within epochs 5 and 6 would require a nomenclature based on their type locations in the various sections shown in Figures 1 and 2.

Lava Extrusion Rates

The regression analyses discussed in the previous section indicate that the rate of accumulation of lavas and sediments in the Borgarfjördur region was nearly constant over the 5-m.y. interval studied. For the whole sequence, the average accumulation rate is 730 m/m.y.; and as the section contains about 430 lavas, the average rate of eruption is one lava per 11,150 yr. A more precise evaluation of the measured accumulation rate was carried out by using a regression procedure in which errors are allowed for in both parameters (McIntyre and others, 1966; York, 1966). The estimate

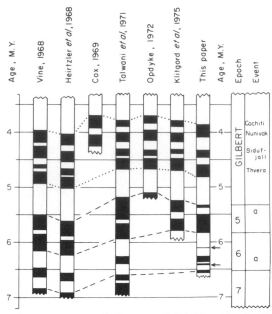

Figure 6. Comparison of polarity time scale derived from this study with scales proposed by other workers. Black = normal polarity, white = reversed polarity. See text for correlation with, and naming of, epochs and events. Arrows represent newly discovered events in epoch 6.

of precision of each K-Ar age as given in Table 1 was employed in the calculations and the error assigned to the aggregate stratigraphic thickness was permitted to change until the line fitted to within the errors at the 95 percent confidence level. This was found to be the case for a 5 percent coefficient of variation for the aggregate thickness parameter, at which value the modified regression equation yielded ages for the polarity boundaries which differed by less than one percent. The uncertainties in the K-Ar ages are likely to be underestimated in most cases because they are precision estimates and thus do not take into account geological errors, such as loss of radiogenic argon. This means that the 5 percent error assigned to the aggregate thickness may well be overestimated. These analyses further emphasize the remarkable uniformity in the rate of accumulation of lavas and sediments in the Borgarfjördur region of western Iceland.

The thickness of sediment in the section is about 15 percent of the total, although the sediments may make up as much as 50 percent of restricted intervals of the section (Fig. 5). In the lowest 225 m of the sequence, sediments make up about 20 percent of the thickness. The K-Ar ages on lavas from this part of the section suggest a lower than average extrusion rate (Fig. 5). An alternative possibility, however, is that the K-Ar ages on samples NP12 and NP13 are somewhat too low because of loss of radiogenic argon, as both samples contain about 10 percent of poorly crystallized mesostasis. Relatively high proportions of sediment occur in the sequence between about 700 m and 1,650 m above the base, but inspection of Figure 5 does not reveal any obvious decrease in the rate of accumulation of sediments and lavas. Thus it would appear that sediments are deposited at approximately the same rate as the lavas in terms of thickness per unit time; as might be expected during intervals of reduced rate of extrusion of lavas, the relative amount of sediment deposited therefore increases. We mentioned above, however, that the younger normal polarity interval in epoch 5 is represented by only two lavas in our section, possibly indicating a low extrusion rate or hiatus. Inspection of Figure 5 shows that the suggested gap coincides closely with the time of extinction of the Reykjadalur central volcano to the north of the area mapped (Fig. 1). During epochs 5 and 6, the two central volcanoes of Hallarmúli and Reykjadalur succeeded one another, possibly giving rise to an increased growth rate of the lava pile during their active periods. Subsequent to cessation of eruption from these volcanoes, lavas from the rift zone bank up against them only gradually, perhaps as a result of topographic effects, giving rise to the possible hiatus or slow rate of accumulation. Similar topographic effects may also explain the possible hiatus in the upper Gilbert epoch between the Cochiti and Nunivak events, following the extinction of the Skardsheidi central volcano. Figure 5 also shows that there may have been a more rapid growth rate from the middle of the Gauss epoch coinciding with the activity of Húsafell central volcano nearby. Again there is a hiatus separating the volcano from the on-lapping lavas, as shown by the difference in age between NT99B and NT107, and this may be at least partly explained by the original topographic high created by the volcano persisting for a long period, until lavas from the central axis piled against the flank, reducing the relief.

Although we have shown that the lava pile has grown at a rate of about 730 m/m.y., it is difficult to calculate the average volume of lava per unit time because of lack of knowledge as to the areal extent and volume of the lavas and the area of active volcanicity. Nevertheless, one comparison can be made. The volcanic production in Iceland during postglacial time is about 480 km³ per 10,000 yr (Jakobsson, 1972), equivalent to a layer about 20 m thick over the entire area of about 25,000 km² of the active volcanic zone each 10,000 yr. This compares with a value of 7.3 m per 10,000 yr for the growth of the lava pile in the Borgarfjördur region, so that the rate of accumulation in the Borgarfjördur region was only about 30 percent of that now occurring on the average in the active volcanic zone.

The average production rate of lavas in Iceland is about three to four times greater than that of comparable segments of the Mid-Atlantic Ridge (Jakobsson, 1973). This is expected because Iceland is a positive element on the sea floor, and the volume of volcanic material must therefore be greater than that involved in normal sea floor spreading at the submarine mid-ocean ridges. The sustained and uniform rate of volcanic activity in the Borgarfjördur region of western Iceland is the first well-documented example of its kind in a spreading-ridge environment, and despite the differences in production rates between the sea floor and the study area, the sustained volcanism which has been observed must strengthen conclusions reached by several earlier workers that the mode of formation of new crust in Iceland may serve as an excellent model for the generation of crust at active spreading centers in general.

Age of the Miocene-Pliocene Boundary

The Miocene-Pliocene boundary defined in Italy was thought to lie within planktonic foraminiferal zone N18 by Blow (1969). More recent studies indicate that the boundary is either at the base of zone N18 or within latest zone N17 (Berggren and Van Couvering, 1974). Data from deep-sea sediment cores indicate that zone N18 is within the early part of the Gilbert Epoch and therefore slightly younger than the top of epoch 5. Berggren and Van Couvering (1974) accept an age of 5.1 m.y. for the upper boundary of epoch 5 and suggest that the Miocene-Pliocene boundary has an age between about 4.9 and 5.1 m.y. Our results from the Borgarfjördur region indicate that the epoch 5–Gilbert boundary has an age of about 5.34 m.y., somewhat older than previous estimates, and thus the age of the Miocene-Pliocene boundary may be as great as 5.2 to 5.3 m.y. From K-Ar dating of volcanics in Fiji thought to lie astride the Miocene-Pliocene boundary, Gill and McDougall (1973) and McDougall and Page (1975) suggested that the boundary has an age equal to or slightly younger than 5.3 m.y. Thus these data seem to be convergent toward an age for the Miocene-Pliocene boundary of 5.2 ± 0.1 m.y. Nevertheless, there exist arguments for

the boundary having a younger age (for example, in New Zealand; Kennett and Watkins, 1974), and these differences remain to be reconciled.

CONCLUSIONS

All the lavas in the Borgarfjördur sequence of western Iceland have suffered burial metamorphism to zeolite facies grade, with temperatures in the deeper parts of the section probably rising to about 150°C. Samples which have their original mineralogy preserved and lack obvious evidence of zeolitization yield K-Ar ages that are stratigraphically compatible and that provide good estimates of the age of eruption of the lavas rather than the age of subsequent metamorphism. The results show that the sequence formed over a 5-m.y. interval from about 7.0 m.y. to about 2 m.y. ago, during late Miocene and Pliocene time.

Regression analysis of the ages and aggregate stratigraphic thickness reveals a remarkably uniform rate of growth of the lava pile of 730 m/m.y., providing on the average an eruption rate of one lava every 11,000 yr. Strong evidence for steady-state growth of the lava pile associated with an active spreading center in western Iceland over a period of about 5 m.y. is provided, and it is suggested that very similar processes may well be operating at sea-floor spreading centers.

The magnetostratigraphic studies, together with the K-Ar data have provided a revised history of geomagnetic polarity for the 5-m.y. eruptive interval. Ages for the boundaries of polarity intervals derived from the regression and which are virtually independent of previous estimates enable a polarity time scale to be proposed which is similar to, but which provides finer detail than, that derived by Talwani and others (1971) and Klitgord and others (1975) from analyses of marine magnetic anomalies.

ACKNOWLEDGMENTS

This research was supported by National Science Foundation Grant No. GA37178. Considerable support also was provided by the National Energy Authority of Iceland and by the Australian National University. The field work was carried out with the assistance of Brooks Ellwood of the University of Rhode Island and Olafur Flovenz. Able technical assistance in the K-Ar dating was provided by Z. Roksandic and Mrs. R. Maier, and in the paleomagnetic analyses by Ms. L. Steere and S. Keany.

REFERENCES CITED

Ade-Hall, J. M., Palmer, H. C., and Hubbard, T. P., 1971, The magnetic and opaque petrological response of basalts to regional hydrothermal alteration: Royal Astron. Soc. Geophys. Jour., v. 24, p. 137–174.
Baksi, A. K., 1974, Isotopic fractionation of a loosely held atmospheric argon component in the Picture Gorge basalts: Earth and Planetary Sci. Letters, v. 21, p. 431–438.
Baldwin, B., Coney, P. J., and Dickinson, W. R., 1974, Dilemma of Cretaceous time scale and rates of sea-floor spreading: Geology, v. 2, p. 267–270.
Berggren, W. A., and Van Couvering, J. A., 1974, The late Neogene: Biostratigraphy, geochronology, and paleoclimatology of the last 15 million years in marine and continental sequences: Palaeogeography, Palaeoclimatology, Palaeoecology, v. 16, p. 1–216.
Blow, W. H., 1969, Late middle Eocene to Recent planktonic foraminiferal biostratigraphy: Internat. Conf. Planktonic Microfossils, 1st, Proc., Brönnimann, P., and Renz, H. H., eds., Geneva, 1967: Leiden, E. J. Brill, p. 199–421.
Brunhes, B., 1906, Recherches sur la direction d'aimentation des roches volcaniques: Jour. Physique, v. 5, p. 705–724.
Coombs, D. S., Ellis, A. J., Fyfe, W. S., and Taylor, A. M., 1959, The zeolite facies, with comments on the interpretation of hydrothermal syntheses: Geochim. et Cosmochim. Acta, v. 17, p. 53–107.
Cooper, J. A., 1963, The flame photometric determination of potassium in

geological materials used in potassium-argon dating: Geochim. et Cosmochim. Acta, v. 27, p. 525–546.

Cox, A., 1969, Geomagnetic reversals: Science, v. 163, p. 237–245.

Dagley, P., Wilson, R. L., Ade-Hall, J. M., Walker, G.P.L., Haggerty, S. E., Sigurgeirsson, T., Watkins, N. D., Smith, P. J., Edwards, J., and Grasty, R. L., 1967, Geomagnetic polarity zones for Icelandic lavas: Nature, v. 216, p. 25–29.

Dalrymple, G. B., 1972, Potassium-argon dating of geomagnetic reversals and North American glaciations, in Bishop. W. W., and Miller, J. A., eds., Calibration of hominoid evolution: New York., Wenner-Gren Foundation, p. 107–134.

Einarsson, Tr., 1957, Magneto-geological mapping in Iceland with the use of a compass: Philos. Mag., v. 6, suppl. p. 232–239.

——1962, Upper Tertiary and Pleistocene rocks in Iceland: Soc. Scientiarum Islandica, v. 36, p. 1–196.

Fisher, R., 1953, Dispersion on a sphere: Royal Soc. London Proc., v. A217, p. 295–305.

Foster, J. A., and Opdyke, N. D., 1970, Upper Miocene to Recent magnetic stratigraphy in deep-sea sediments: Jour. Geophys. Research, v. 75, p. 4465–4473.

Gill, J. B., and McDougall, I., 1973, Biostratigraphic and geological significance of Miocene-Pliocene volcanism in Fiji: Nature, v. 241, p. 176–180.

Heirtzler, J. R., Dickson, G. O., Herron, E. M., Pitman, W. C., and Le Pichon, X., 1968, Marine magnetic anomalies, geomagnetic field reversals and motions of the ocean floor and continents: Jour. Geophys. Research, v. 73, p. 2119–2136.

Helsley, C. E., and Steiner, M. B., 1974, Paleomagnetism of the Lower Triassic Moenkopi Formation: Geol. Soc. America Bull., v. 85, p. 457–464.

Herron, E. M., and Talwani, M., 1972, Magnetic anomalies on the Reykjanes Ridge: Nature, v. 238, p. 390–392.

Jakobsson, S. P., 1972, Chemistry and distribution pattern of Recent basaltic rocks in Iceland: Lithos, v. 5, p. 365–386.

Johannesson, H., 1972, The Tertiary lava pile between Nordurardal and Hvitarsida, western Iceland [B.S. dissert.]:Reykjavik, Univ. Iceland, p. 1–57 [in Icelandic].

——1975, Structure and petrochemistry of the Reykjadalur central volcano and the surrounding areas, midwest Iceland [Ph.D. thesis]: Durham, England, Univ. Durham, p. 1–273.

Kennett, J. P., and Watkins, N. D., 1974, Late Miocene–early Pliocene paleomagnetic stratigraphy, paleoclimatology, and biostratigraphy in New Zealand: Geol. Soc. America Bull., v. 85, p. 1385–1398.

Klitgord, K. D., Huestis, S. P., Mudie, J. D., and Parker, R. L., 1975, An analysis of near-bottom magnetic anomalies: Sea-floor spreading and the magnetized layer: Royal Astron. Soc. Geophys. Jour. v. 43, p. 387–424.

Larson, R. L., and Pitman, W. C., III, 1972, World-wide correlation of Mesozoic magnetic anomalies, and its implications: Geol. Soc. America Bull., v. 83, p. 3663–3662.

Matuyama, M., 1929, On the direction of magnetization of basalt in Japan, Tyôsen and Manchuria: Japan Acad. Proc., v. 5, p. 203–205.

McDougall, I., 1966, Precision methods of potassium-argon isotopic age determination on young rocks, in Runcorn, S. K., ed., Methods and techniques in geophysics, Vol. 2: New York, Interscience, p. 279–304.

McDougall, I., and Page, R. W., 1975, Toward a physical time-scale for the Neogene: Data from the Australian region, in Saito, T., and Burckle, L. H., eds., Late Neogene epoch boundaries: New York, Micropaleontology Press, Am. Mus. Nat. History p. 75–84.

McDougall, I., Polach, H. A., and Stipp, J. J., 1969, Excess radiogenic argon in young subaerial basalts from the Auckland volcanic field, New Zealand: Geochim. et Cosmochim. Acta, v. 33, p. 1485–1520.

McIntyre, G. A., Brooks, C., Compston, W., and Turek, A., 1966, The statistical assessment of Rb-Sr isochrons: Jour. Geophys. Research, v. 71, p. 5459–5468.

Mercanton, P. L., 1926, Inversion de l'inclinaison magnétique terrestre aux âges géologiques: Terrestrial Magnetism and Atmospheric Electricity, v. 31, p. 187–190.

Nagata, T., Akimoto, S., Uyeda, S., Shimizu, Y., Ozima, M., and Kobayashi, K., 1957, Paleomagnetic study on a Quarternary volcanic region in Japan: Philos. Mag. Suppl. Adv. Physics, v. 6, p. 255–263

Opdyke, N. D., 1972, Paleomagnetism of deep-sea cores: Rev. Geophysic and Space Physics, v. 10, p. 213–249.

Opdyke, N. D., and Runcorn, S. K., 1956, New evidence for reversal of the geomagnetic field near the Pliocene-Pleistocene boundary: Science v. 123, p. 1126–1127.

Opdyke, N. D., Burckle, L. H., and Todd, A., 1974, The extension of the magnetic time scale in sediments of the central Pacific Ocean: Eart and Planetary Sci. Letters, v. 22, p. 300–306.

Pálmason, G., 1973, Kinematics and heat flow in a volcanic rift zone with application to Iceland: Royal Astron. Soc. Geophys. Jour., v. 3. p. 451–481.

Pálmason, G., and Saemundsson, K., 1974, Iceland in relation to the Mid Atlantic Ridge: Ann. Rev. Earth and Planetary Sciences, v. 2, p. 25 50.

Piper, J.D.A., 1971, Ground magnetic studies of crustal growth in Iceland Earth and Planetary Sci. Letters, v. 12, p. 199–207.

Roche, A., 1951, Sur les inversions de l'aimantation rémanente des roche volcaniques dans les monts d'Auvergne: Acad. Sci. Comptes Rendus v. 233, p. 1132–1134.

Saemundsson, K., 1974, Evolution of the axial rifting zone in Northern Iceland and the Tjörnes fracture zone: Geol. Soc. America Bull., v. 85 p. 495–504.

Saemundsson, K., and Noll, H., 1974, K/Ar ages of rocks from Húsafell western Iceland, and the development of the Húsafell central volcano Jökull, v. 24, p. 40–59.

Sigurdsson, H., 1970, Structural origin and plate tectonics of the Snaefellsnes volcanic zone, western Iceland: Earth and Planetary Sci Letters, v. 10, p. 129–135.

Talwani, M., Windisch, C. C., and Langseth, M. G., Jr., 1971, Reykjane ridge crest: A detailed geophysical study: Jour Geophys. Research v. 72, p. 473–517.

Theyer, F., and Hammond, S. R., 1974, Paleomagnetic polarity sequence and radiolarian zones, Brunhes to polarity epoch 20: Earth and Planetary Sci. Letters, v. 22, p. 307–319.

Vincenz, S.A., and Bruckshaw, J. McG., 1960, Note on the probability distribution of a small number of vectors: Cambridge Philos. Soc. Proc v. 56, p. 21–26.

Vine, F. J., 1966, Spreading of the ocean floor: New evidence: Science v. 154, p. 1405–1415.

——1968, Magnetic anomalies associated with mid-ocean ridges, in Phinney, R. A., ed., The history of the Earth's crust: Princeton, N.J Princeton University Press, p. 73–89.

Walker, G.P.L., 1959, Geology of the Reydarfjördur area, eastern Iceland Geol. Soc. London Quart. Jour., v. 114, p. 367–393.

——1960, Zeolite zones and dike distribution in relation to the structure o the basalts of eastern Iceland: Jour. Geology, v. 68, p. 515–528.

——1964, Geological investigations in eastern Iceland: Bull. Volcanol. v. 27, p. 351–362.

——1974, The structure of eastern Iceland, in Kristjansson, L., ed. Geodynamics of Iceland and the North Atlantic area: Dordrecht, Hol larn, D. Reidel Pub. Co., p. 177–188.

——1975, Excess spreading axes and spreading rate in Iceland: Nature v. 255, p. 468–471.

Watkins, N. D., 1972, Review of the development of the geomagnetic polarity time scale and discussion of prospects for its finer definition Geol. Soc. America Bull., v. 83, p. 551–574.

Watkins, N. D., McDougall, I., and Kristjansson, L., 1976, Upper Miocene and Pliocene geomagnetic secular variation in the Borgarfjördur area of western Iceland: Royal Astron. Soc. Geophys. Jour. (in press).

Wilson, R. L., Watkins, N. D., Einarsson, Tr., Sigurgeirsson, Th., Haggerty S. E., Smith, P. J., Dagley, P., and McCormack, A. G., 1972, Palaeo magnetism of ten lava sequences from south-western Iceland: Roya Astron. Soc. Geophys. Jour., v. 29, p. 459–471.

York, D., 1966, Least-squares fitting of a straight line: Canadian Jour Physics, v. 44, p. 1079–1086.

MANUSCRIPT RECEIVED BY THE SOCIETY JULY 21, 1975
REVISED MANUSCRIPT RECEIVED MARCH 1, 1976
MANUSCRIPT ACCEPTED APRIL 14, 1976

Part II

EXTENSION OF THE POLARITY SCALE TO DEEP-SEA SEDIMENTS: MULTICHANNEL RECORDERS

Editor's Comments
on Papers 4 Through 13

13   THEYER and HAMMOND
Abstract from *Paleomagnetic Polarity Sequence and Radio-larian Zones, Brunhes to Polarity Epoch 20*

Deep-sea sediments are multichannel recorders of the earth's history, particularly that which is related to paleoenvironmental conditions and the evolution of life in the oceans. By the middle 1960s, studies of deep-sea cores were beginning to accelerate rapidly, and although the sequence of events could often be arranged through knowledge of stratigraphic superposition, the chronology of events was still poorly known, and correlation between high and low latitude sequences was still in its infancy. With this background, the potential for applying the polarity time scale to stratigraphy was rapidly perceived by a few workers, and magnetostratigraphy was attempted on deep-sea sediments. A thorough and valuable review of the paleomagnetism of deep-sea sediments up until 1971 has been provided by Opdyke (Paper 4). This reviews important information relevant to several topics discussed in this volume; namely, the development of magnetostratigraphy in deep-sea sediments; the polarity time scale; extension of the time scale to sequences older than the latest Cenozoic; and the geologic applications of magnetostratigraphy in deep-sea sediments. Other topics not included in this volume are also discussed; namely, short paleomagnetic events; long-term behavior of the earth's magnetic field; and paleomagnetic intensity changes.

The first workers to recognize paleomagnetic reversals in deep-sea sediments were Harrison, a geophysicist, and Funnell, a paleontologist (Paper 5), previous fellow students at Cambridge University, England, who later collaborated in this joint study at Scripps Institution of Oceanography, California. This was to be one of the first collaborative efforts by a geophysicist and a paleontologist and fortunately was a successful one. Also this study was the first to demonstrate, from study of deep-sea cores, that the last reversal of the earth's field occurred within the Quaternary. The cores examined were rather short (<2 m), and thus the results were limited in scope. The work by Harrison and Funnell (Paper 5) had in turn been stimulated by previous studies by Dr. M. J. Keen (1963) who had carried out the first modern paleomagnetic study of deep-sea sediments as a graduate student at the Department of Geodesy and Geophysics, Cambridge University, working under the supervision of Dr. M. N. Hill. Keen studied piston cores mostly collected from the Iberian Abyssal Plain, although a few were from shallower depths on topographic prominences. The sedimentary sequences collected from the abyssal plain region were mostly turbidites of very high sedimentarion rate and even the longest piston core of 6 m did not penetrate to the Brunhes/Matuyama paleomagnetic boundary. Even the shallower cores, perhaps because of their shorter lengths ($<$3 m), did not penetrate this boundary. Thus Keen (1963) was not able to detect a paleomagnetic reversal himself, but he did demonstrate that some of the sediments were stabley magnetized. This discovery itself was encouraging enough for Dr. J. C. Belshé, a professor at Cambridge University, to suggest to C. Harrison to work on the problem of detecting paleomagnetic reversals in deep-sea sediments. Dr. C. Harrison had joined the department in 1960 as a graduate student, overlapping with M. Keen who was still a graduate student there. However, it was necessary for Harrison to do most of his work at Scripps Institution of Oceanography, California, where a large collection was available of gravity and short piston cores mostly from the Pacific Ocean. While at Scripps, Harrison worked under the auspices of Dr. V. Vacquier. Most of the piston cores in the Scripps collection were too disturbed for paleomagnetic work, and of the short gravity cores examined, two had fortunately penetrated the Brunhes/Matuyama boundary. This boundary in these cores was very close to the level where the radiolarian *Pterocanium prismatium* became extinct. Brian Funnell noticed a correlation between the paleomagnetic reversal and extinction of this species and suggested the writing of a joint paper, which is reproduced as Paper 5. Harrison (1966) provided more detailed paleomagnetic results for these two gravity cores in a later paper. He recognized that deep-sea sediments possess a relatively stable natural remanent magnetism (NRM) and that cores provide a good opportunity for the study of rather complete sequences laid down over several million years. This allows an ideal opportunity for the examination of both magnetic reversals and the shorter term variations of the magnetic field. In this initial study, the paleomagnetic nomenclature of Cox et al. (1964, 1965) was utilized, and although the cores were short, the boundary between the Brunhes and Matuyama Epochs and the Olduvai Event within the Matuyama Epoch were recognized. From a stratigraphic perspective, the most important observation made by Harrison (1966), was that an isochronous correlation was demonstrated in sections separated by more than 3000 km. By the middle 1960s, deep-sea cores were beginning to play some role in the reinforcement of the polarity history established from sequences of volcanic rocks on land. The work, however, contributed little toward the development of the polarity time scale itself.

The most dramatic demonstrations of the importance of magnetostratigraphy in deep-sea sediments, and those that made the greatest impact on the development of this discipline, were yet to come in the middle 1960s. There existed at this opportune time a small team of scientists at the Lamont-Doherty Geological Observatory who were all in communication with each other and were about to capitalize on the initial pioneering efforts of Harrison and Funnell. The combined efforts of this group were to advance magnetostratigraphy rapidly to the point where it would become a central and eventually a relatively routine procedure, in the stratigraphy of marine sediments. Dr. Neil Opdyke, a student of Dr. Keith Runcorn, of the University of Newcastle on Tyne, England, had recently joined the staff at the Lamont Geological Observatory to continue his paleomagnetic research. In the middle 1960s his interest was still largely concerned with Paleozoic and Mesozoic paleomagnetism. However, in relation to this work, Opdyke employed a demagnetization technique that proved important in providing clearer and hence more readily interpretable paleomagnetic records. A large collection of piston cores was already stored at the Lamont Observatory, having been accumulated during many years of largely geophysical exploration of the oceans under the direction of Dr. Maurice Ewing. Dr. Ewing had a keen sense of the potential value of large collections of geophysical and sedimentary data for future oceanographic studies. Other large core collections were also being built up at this time, the most notable of which resulted from expeditions of the U.S.N.S. *Eltanin* in the Southern Ocean. Both collections were soon to provide an important basis for the extension of magnetostratigraphy to deep-sea sediments by the Lamont group and another group at Florida State University, Tallahassee. At the same time the development of the polarity history continued, based on terrestrial volcanic sequences, by Cox and his associates at the U.S. Geological Survey in California and by McDougall and his associates at the Australian National University, Canberra. That the earth's magnetic field has changed through time was by now widely accepted by geophysicists. It was also opportune that Dr. James Hays, of Lamont Geological Observatory, had very recently completed his Ph.D. dissertation (with Bruce Heezen as senior advisor) on

65

the latest Cenozoic radiolarians of Antarctic piston cores collected during *Vema* and *Eltanin* expeditions (Hays, 1965), and he had developed a biostratigraphy that was to be a critical component in the demonstration of the validity of magnetostratigraphy as a tool in correlation and chronology of deep-sea sediments. Also at Lamont at this time were two graduate students working on their dissertations, Dr. Billy Glass, working on deep-sea sediments and their paleomagnetic record, and Dr. John Foster, who had developed a slow-spinner magnetometer ideal for measuring the paleomagnetism of deep-sea sediments. These two students together conducted some initial experiments on the paleomagnetism of deep-sea sediments.

The Lamont group decided to initially study the paleomagnetism of a number of deep-sea cores from the Antarctic region for several reasons: First, the cores were long ($\sim$5–12 m) and hence contained good potential for a relatively long paleomagnetic record; second, a radiolarian biostratigraphy had already been established by James Hays on the cores, which would allow independent verification of a magnetostratigraphy and of intercore correlations. The previous radiolarian work had also shown that these cores were devoid of unconformities. Of primary importance in the selection of these cores, however, was their geographic location at high latitudes where the inclination of the magnetic field is high. Hence a magnetic vector pointing up relative to the horizontal indicates an opposite polarity to the magnetic vector pointing downward in the same core. Thus it is necessary to know only which part of the core being measured is to the top and which is down. In equatorial areas the inclination of the magnetic field in contrast approaches zero and hence was more difficult to interpret in unoriented cores during the earlier days of less sensitive instrumentation. The combination of the talent and expertise of the group paid off handsomely with the very elegant presentation of the paleomagnetic correlation of 7 long piston cores up to 4 m.y. from different segments of the Southern Ocean (Paper 6). The scale of Doell and Dalrymple (1966) was used, which for the first time enabled the rather accurate dating of a late Cenozoic biostratigraphic scheme, the radiolarian zonation of Hays (1965). These results were quickly accepted by the scientific community and were to have major repercussions in the understanding of late Cenozoic paleoceanography. The new chronology enabled these investigators to make some initial evaluations on the timing of Antarctic glaciation from the age of icerafted debris; the history of diatom ooze deposition and hence of upwelling of Antarctic waters and paleoproductivity and of possible links between extinction of radiolarian species and polarity reversals.

It was also during 1966 that a similar Lamont group led by Dr. Dragoslav Ninkovich, a Yugoslavian scientist with particular interest in volcanic ash distributions in deep-sea cores, produced a magnetostratigraphy of a set of deep-sea cores from the high latitude North Pacific, most of which were of Brunhes and Matuyama age (Paper 7). Ninkovich had immediately recognized the potential importance of magnetostratigraphy in enhancing tephrachronological research. This contribution also contained some other important new aspects since it was the first to demonstrate clearly the presence of the Jaramillo Event in deep-sea sediments; it was among the first to estimate the duration of the Jaramillo and Olduvai Events and to estimate rates of sediment deposition. It was also the first to intercalibrate correlations using both ash layers and paleomagnetic reversals, in turn providing a firm chronology for major explosive volcanic activity in the region.

These initial successes created a momentum that in turn led to other rapid advances. The most important of these that shortly followed was the publication in 1967 by Hays and Opdyke (Paper 8) of a chronological extension of the magnetostratigraphy of Antarctic sediments to about 4.5 m.y. ago, using three long *Eltanin* piston cores. The *Eltanin* oceanographic expeditions in the Southern Ocean were by this time obtaining some of the finest long piston cores in existence, thus allowing the magnetostratigraphy to be extended beyond the 3.5-m.y. maximum age in the first such study by the Lamont group (Paper 6). The longer stratigraphy allowed the definition and dating of new polarity events and radiolarian zones. This was the first contribution to document the paleomagnetic stratigraphy, in deep-sea sequences, of the Gilbert Reversed Epoch (t = 3.32 to 5.1 m.y. B.P.) providing strong evidence for the existence of three short polarity events within the Gilbert Epoch. A radiolarian biostratigraphy, based on the upward sequential disappearance of several radiolarian species, was dated using the polarity scale; a major radiolarian faunal change was discovered at about 2.5 m.y. ago, and further evidence was developed for a possible linkage between times of polarity change and radiolarian extinctions.

While the Lamont group was extending magnetostratigraphy in deep-sea sediments, independent paleomagnetic work was being conducted on cores from the Southern Ocean at Florida State University by Norman Watkins and his collaborators, including Dr. Grant Goodell. These workers also recognized the great potential of magnetostratigraphic studies of marine sediments, and in rather remarkably short time, determined the magnetostratigraphy of about 160 cores from the South Pacific sector of the Southern Ocean (Goodell and Watkins, 1968). Probably the most important aspect of their study was their convincing demonstration that large numbers of cores from wide areas of the oceans could be dated rapidly, providing a major new approach in the mapping of sedimentary deposits. Isopach maps based on isochrons provided by the paleomagnetic reversals allowed new insights to be developed on regional sediment patterns and processes, including erosion. The study of Goodell and Watkins (1968), however, suffered from a lack of an independent biostratigraphy that led to the incorrect magnetostratigraphic assignment of a number of cores. As a result, this work focused on the importance of collaboration between paleomagnetists and paleontologists to determine jointly a polarity stratigraphy, as each approach suffers from its own limitations. Later similar studies (Goodell et al., 1968) resulted in additional knowledge of the Antarctic glacial development and of Southern Ocean sedimentary history.

In these early studies the broader aspects of the polarity scale were readily verified, but differences of opinion emerged. These differences were often hotly debated between the different groups over the recognition and confirmation of short-term events, particularly those occurring within the Matuyama Epoch. These ambiguities became even more critical because of the association of the Pliocene-Pleistocene boundary with the major normal event in the Middle Matuyama. Details on the history of these discussions have been summarized in Watkins (Paper 2), Opdyke (Paper 4), and McDougall (1977). Disagreements largely concerned whether three events occur within the Matuyama Epoch (Olduvai, Gilsa, and Jaramillo Events: McDougall and Chamalaun, 1966; Watkins and Goodell, 1967) or whether there are only two events (Olduvai and the Jaramillo Events: Ninkovitch et al., Paper 7; Opdyke, Paper 4). This problem has not yet been completely resolved and, in fact, in one of the latest compilations of the polarity time scale, McDougall (1977) has suggested the presence of five events in the Matuyama, although the longest and most conspicuous event is identified as the Olduvai Event. The problem of identification of polarity events in deep-sea sediments contributed to additional thought about the character of the preservation of short-term events in deep-sea sediments and the natural and unnatural causes of spurious polarity reversals (Watkins, 1968). These include faunal redeposition, error in technique, and the polarity recording lags resulting from delays between original deposition and final consolidation of deep-sea sediments (Dymond, 1969).

Correlation over long distances using fossils has always been difficult even when planktonic groups are used. This is because different water masses within the oceans are marked by distinct planktonic assemblages, and species that do range over wide areas often exhibit different stratigraphic ranges in different oceanic regions. It was precisely because of these long-standing problems that magnetostratigraphy was grasped so enthusiastically by paleontologists and stratigraphers as a new correlation tool over vast distances, especially between the tropics and polar regions. Among the first such attempts in magnetostratigraphic correlation between high and low latitude planktonic microfossil zonations were those of Berggren et al. (1967), Phillips et al. (1968), and Berggren (Paper 9). Multiple microfossil criteria were utilized to recognize the Pliocene-Pleistocene boundary in various parts of the world's oceans. In tropical areas, for instance, Berggren et al. (1967), Berggren (Paper 9), and Phillips et al. (1968) were able to date the Pliocene-Pleistocene boundary in deep-sea sediments as about 1.8 m.y. because of the association of this boundary with the Olduvai Event (Gilsa of some workers). The Pliocene-Pleistocene boundary in the deep-sea sediments was based on the first appearance of the planktonic foraminifer *Globorotalia truncatulinoides*, which also first appears within the type section of the Pliocene-Pleistocene boundary in Calabria, southern Italy. Likewise, Glass et al. (1967) conducted studies on cores previously paleontologically studied by Ericson et al. (1963) and also showed an association of the Pliocene-Pleistocene boundary with the Olduvai Event. In contrast, in the Antarctic area, where calcareous microfossils are of much less importance in deep-sea sediments, the Pliocene-Pleistocene boundary can be distinguished on the extinction of the radiolarian *Eucyritidium calvertense* (Hays and Opdyke, 1967). Thus as a result of magnetostratigraphy, more detailed intra- and interoceanic correlations over great distances were possible for biostratigraphic schemes of latest Cenozoic age.

Correlation of the polarity time scale with multiple microfossil biostratigraphies continued, and its value was further elegantly demonstrated in another classical contribution by the Lamont group lead by Hays (Paper 10) in which they dated and correlated a large number of microfossil events in eastern equatorial Pacific deep-sea cores of Quaternary and Pliocene age ($t = 4.5$ m.y. to P.D.). This was the first such comprehensive study involving planktonic foraminifera, radiolaria, diatoms, discoasters, silicoflagellates, and calcium carbonate cycles. Such a study was expedited by the presence at Lamont at this time of a group of experts in micropaleontology including James Hays (radiolaria), Lloyd Burckle (diatoms), and Tsunemasa Saito (foraminifera). The paleomagnetic dating of these various microfossil sequences produced a major surprise that was to demonstrate further the strength of magnetostratigraphy in providing a chronological framework for late Cenozoic sediments. It was discovered that changes in the fossils used for the differentiation of the Miocene-Pliocene boundary were only 4.5 to 5 m.y. old and that they occurred near the base of the Gilbert Epoch. Up until this time, rather slim and often contradictory radiometric evidence had suggested a 9 m.y. ago for this boundary, and thus the Pliocene Epoch was suddenly discovered to be only about half the duration previously believed. For instance, incorrect radiometric ages derived from ash layers in the marine sedimentary sequence of California (Bandy, 1967) had supported the longer duration of the Pliocene and rates of geological processes calculated from this spurious data were too low by a factor of about 2. Thus our concepts of the rates of Late Cenozoic geological process were substantially transformed by the application of magnetostratigraphy to microfossil biostratigraphy. The paleomagnetic stratigraphy of Hays et al. (Paper 10) was similar to the established polarity time scale, in addition to confirming the validity of the Kaena Reversed Event (McDougall and Chamalaun, 1966) in the Gauss Epoch and the identification of three normal events in the Gilbert Epoch. Quaternary climatic cycles were dated rather accurately, and further tantalizing relationships emerged between several microfossil extinctions and paleomagnetic reversals.

The interrelationships that were being developed between planktonic microfossil biostratigraphy and the polarity time scale were, by the late 1960s, being increasingly utilized for establishing the chronology of Cenozoic stratigraphy. Nevertheless, the research was still limited to the Pliocene and Quaternary (the last 5 m.y.). In 1968 the Deep Sea Drilling Project began its operations, providing for the first time, long, and often continuous marine sedimentary sequences for the entire Cenozoic and Late Cretaceous. It was largely because of this program that a satisfactory chronology became essential for application to a large diversity of historical geological studies resulting from the collection of this exciting core material. The prospects of directly establishing a magnetostratigraphy on most of these sedimentary cores was small because of substantial drilling disturbance except, in some cases, of more consolidated, older Tertiary material (Allis et al., 1975). Initial attempts (Ryan and Flood, 1973) demonstrated the difficulties of establishing magnetostratigraphy of such material (Kennett and Watkins, Paper 15). Thus calibration of the polarity history with biostratigraphy of deep-drilled sedimentary sequences has had to be carried out indirectly by radiometric dating of planktonic microfossil datums of marine sections exposed on land. This task has been carried out by Berggren in a sequence of papers (1969; 1972), one of which is included in this volume (Paper 11). The ages of Cenozoic microfossil zones were important for establishing ages of marine magnetic anomaly patterns of the ocean floor. During the early stages of drilling, much emphasis was placed on the determination of ages from fossils of sediments immediately overlaying oceanic basement rock of inferred magnetic anomaly age. The ages from magnetic anomaly patterns have been assumed using constant rates of sea-floor spreading. This assumption has been shown to be at times partly incorrect, as suggested by recent work such as that of Schlich (1975). He clearly shows a systematic offset between Cretaceous ages derived from magnetic anomaly patterns and those from the biostratigraphy of basal sediments. Nevertheless, the sea-floor magnetic anomaly patterns have formed a basis for the extension of the polarity history into much older Tertiary sediments. Two different approaches have been used, both of which require the study of piston cores. Because of their relatively short lengths, piston cores represent only limited amounts of time. This restriction has been partly alleviated by the use of long piston cores marked by low sedimentation rate. Such sequences often extend well into the Miocene. Using this approach, Foster and Opdyke (1970) were able to extend the magnetic stratigraphy in deep-sea sediments to about 12 m.y. B.P., using two of the longest (24 and 25 m) piston cores recovered from the equatorial Pacific. The magnetostratigraphy is in substantial agreement with the pattern of polarity epochs predicted by sea-floor spreading. These authors were able to subdivide the polarity history into seven new magnetic epochs numbered from Epoch 5 to Epoch 11. The paleomagnetic stratigraphy was not intercalibrated with any biostratigraphic scheme, perhaps an acceptable procedure in this case, because of a lack of evidence of stratigraphic hiatuses and of similarities of magnetostratigraphic patterns between the two cores.

At Lamont-Doherty Geological Observatory and at the University of Hawaii, a second approach for extending the polarity time scale was begun more recently by two groups working independently on equatorial Pacific cores. Opdyke et al. (Paper 12) recognized nineteen magnetic epochs from the earliest Miocene (t = 20 m.y. B.P.) to the Recent. Theyer and Hammond (Paper 13) established a polarity record from the earliest Miocene to the Present Day. These workers recognized twenty magnetic epochs and possibly as many as 30 polarity events. Both studies correlated cores using a large number of microfossil datums and established ages for these datums by intercalibration with the radiometric time scale of Berggren (Paper 11). The limitations of this approach have been fully appreciated by these workers, but they began the research because it is one of the few available approaches to extension of the polarity time scale through the Cenozoic.

REFERENCES

Allis, R. G., P. J. Barrett, and D. A. Christoffel. 1975. A paleomagnetic stratigraphy for Oligocene and Early Miocene marine glacial sediments at site 270, Ross Sea, Antarctica. In *Initial Reports of the Deep Sea Drilling Project*, Vol. XXVIII, edited by D. E. Hayes et al. Washington, D.C.: U.S. Government Printing Office, pp. 879–884.

Bandy, O. L. 1967. Foraminiferal definition of the boundaries of the Pleistocene in southern California, U.S.A. In *Progress in Oceanography*, Vol. 4, edited by M. Sears. Oxford: Pergamon Press, pp. 27–49.

Berggren, W. A. 1969. Cenozoic chronostratigraphy, planktonic foraminiferal zonation and the radiometric time scale. *Nature* 224:1072–1075.

Berggren, W. A., J. D. Phillips, A. Bertels, and D. Wall. 1967. Late Pliocene–Pleistocene stratigraphy in deep sea cores from the south-central North Atlantic. *Nature* **216**:253–255.

Cox, A., R. R. Doell, and G. B. Dalrymple. 1964. Reversals of the earth's magnetic field. *Science* **144**:1537–1543.

Cox, A., R. R. Doell, and G. B. Dalrymple. 1965. Quaternary paleomagnetic stratigraphy. In *The Quaternary of the United States*, edited by H. E. Wright, Jr., and D. G. Frey. Princeton: Princeton University Press, pp. 817–830.

Doell, R. R., and G. B. Dalrymple. 1966. Geomagnetic polarity epochs: A new polarity event and the age of the Brunhes-Matuyama boundary. *Science* **152**:1060–1061.

Dymond, J. 1969. Age determinations of deep-sea sediments: A comparison of three methods. *Earth and Planetary Sci. Letters* 6:9–14.

Ericson, D. B., G. Wollin, and M. Ewing. 1963. Pliocene–Pleistocene boundary in deep-sea sediments. *Science* **139**:727–737.

Foster, J. H., and N. D. Opdyke. 1970. Upper Miocene to Recent magnetic stratigraphy in deep-sea sediments. *Jour. Geophys. Research* **75**:4465–4473.

Glass, B., D. B. Ericson, B. C. Heezen, N. D. Opdyke, and J. A. Glass. 1967. Geomagnetic reversals and Pleistocene chronology. *Nature* **216**:437–442.

Goodell, H. G., and N. D. Watkins. 1968. The paleomagnetic stratigraphy of the Southern Ocean: 20° West to 160° East longitude. *Deep-Sea Research* **15**:89–112.

Goodell, H. G., N. D. Watkins, T. T. Mather, and S. Koster. 1968. The Antarctic glacial history recorded in sediments of the Southern Ocean. *Palaeogeography, Palaeoclimatology, Palaeoecology* **5**:41–62.

Harrison, C. G. A. 1966. The paleomagnetism of deep sea sediments. *Jour. Geophys. Research* **71**:3033–3043.

Hays, J. D. 1965. Radiolaria and Late Tertiary and Quaternary history of Antarctic seas. In *Biology of the Antarctic Seas II, Antarctic Research Series 5*. Washington, D.C.: American Geophysical Union, pp. 125–184.

Hays, J. D., and W. A. Berggren. 1971. Quaternary boundaries and correlations. In *Micropalaeontology of the Oceans*, edited by B. M. Funnell and W. Riedel. Cambridge: Cambridge University Press, pp. 669–691.

Keen, M. J. 1963. The magnetization of sediment cores from the eastern basin of the North Atlantic Ocean. *Deep-Sea Research* **10**:607–622.

McDougall, I. 1977. The present status of the geomagnetic polarity time scale. In *The Earth: Its Origin, Structure and Evolution (A volume in honor of J. C. Jaeger and A. L. Hales)*, edited by M. W. McElhinny. London: Academic Press, 34p.

McDougall, I., and F. H. Chamalaun. 1966. Geomagnetic polarity scale of time. *Nature* **212**:1415–1418.

Phillips, J. D., W. A. Berggren, A. Bertels, and D. Wall. 1968. Paleomagnetic stratigraphy and micropaleontology of three deep sea cores from the central North Atlantic Ocean. *Earth and Planetary Sci. Letters* **4**:118–130.

Ryan, W. B. F., and J. D. Flood. 1973. Preliminary paleomagnetic measurements on sediments from the Ionian (site 125) and Tyrrhenian (site 132) Basins of the Mediterranean Sea. In *Initial Reports of the Deep Sea Drilling Project*, Vol. XIII, edited by W. B. F. Ryan et al. Washington, D.C.: U.S. Government Printing Office, pp. 599–603.

Saito, T., L. H. Burckle, and J. Hays. 1975. Late Miocene to Pleistocene biostratigraphy of equatorial Pacific sediments. In *Late Neogene Epoch Boundaries*, edited by T. Saito and L. H. Burckle. New York: Mircopaleontology Press, pp. 226–244.

Schlich, R. 1975. Structure et age de L'Océan Indien Occidental. *Soc. Géol. France Mém.* 6, 104p.

Watkins, N. D. 1968. Short period geomagnetic polarity events in deep-sea sedimentary cores. *Earth and Planetary Sci. Letters* **4**:341–349.

Watkins, N. D., and H. G. Goodell. 1967. Confirmation of the reality of the Gilsa geomagnetic polarity event. *Earth and Planetary Sci. Letters* **2**:123–129.

ADDITIONAL READINGS

Berggren, W. A. 1977. The Pliocene/Pleistocene boundary in deep-sea sediments: Status in 1975. *Gior. Geologia* **41**:375–384.

Berggren, W. A. 1977. Late Neogene planktonic foraminiferal biostratigraphy of the Rio Grande Rise (South Atlantic). *Marine Micropaleo.* **2**:265–313.

Burckle, L. H. 1977. Pliocene and Pleistocene diatom datum levels from the equatorial Pacific. *Quaternary Research* **7**:330–340.

Burckle, L. H., and N. D. Opdyke. 1977. Late Neogene diatom correlations in the Circum-Pacific. In *Proceedings of the First International Congress on Pacific Neogene Stratigraphy.* Tokyo: Kaiyo Shuppan Co., pp. 255–284.

Clark, H. C., and J. P. Kennett. 1973. Paleomagnetic excursion recorded in latest Pleistocene deep-sea sediments, Gulf of Mexico. *Earth and Planetary Sci. Letters* **19**:267–274.

Couvering, J. A. van. 1978. Status of Late Cenozoic boundaries. *Geology* **6**:169.

Couvering, J. A. van, and W. A. Berggren. 1977. Biostratigraphical basis of the Neogene time scale. In *Concepts and Methods of Biostratigraphy,* edited by E. G. Kauffman and J. E. Hazel. Stroudsburg, Pa.: Dowden, Hutchinson & Ross, pp. 283–306.

Dickson, G. O., and J. H. Foster. 1966. The magnetic stratigraphy of a deep sea core from the North Pacific Ocean. *Earth and Planetary Sci. Letters* **1**:458–462.

Gartner, S. 1977. Calcareous nannofossil biostratigraphy and revised zonation of the Pleistocene. *Marine Micropaleo* **2**:1–25.

Harrison, C. G. A. 1974. The paleomagnetic record from deep-sea sediment cores. *Earth-Sci. Rev.* **10**:1–36.

Harrison, C. G. A., and B. L. K. Somayajulu. 1966. Behaviour of the earth's magnetic field during a reversal. *Nature* **212**:1193–1195.

Hays, J. D. 1970. Stratigraphy and evolutionary trends of radiolaria in North Pacific deep-sea sediments. *Geol. Soc. America Mem. 126,* pp. 185–218.

Heye, D. 1971. Correlation of sediment cores on the basis of their magnetization. *Sedimentology* **16**:111–117.

Johnson, H. P., and J. M. Ade-Hall. 1975. Magnetic results from basalts and sediments from the Nazca Plate. *Nature* **257**:471–473.

Johnson, H. P., H. Kinoshita, and R. T. Merrill. 1975. Rock magnetism and paleomagnetism of some North Pacific deep-sea sediments. *Geol. Soc. America Bull.* **86**:412–420.

Keen, M. J. 1963. The magnetization of sediment cores from the eastern basin of the North Atlantic Ocean. *Deep-Sea Research* **10**:607–622.

Kellogg, D. E., and J. D. Hays. 1975. Microevolutionary patterns in Late Cenozoic radiolaria. *Paleobiology* **1**:150–160.

Kent, D. V. 1973. Post-depositional remanent magnetisation in deep-sea sediment. *Nature* **246**:32–34.

Kent, D. V., and W. Lowrie. 1974. Origin of magnetic instability in sediment cores from the central North Pacific. *Jour. Geophys. Research* **79**:2897–3000.

Kent, D. V., and N. D. Opdyke. 1977. Palaeomagnetic field intensity variation recorded in a Brunhes Epoch deep-sea sediment core. *Nature* **266**:156–159.

Kobayashi, K., K. Kitazawa, T. Kanaya, and T. Sakai. 1971. Magnetic and micropaleontological study of deep-sea sediments from the west-central equatorial Pacific. *Deep-Sea Research* **18**:1045–1062.

Louden, K. E. 1977. Paleomagnetism of DSDP sediments, phase shifting of magnetic anomalies, and rotations of the west Philippine Basin. *Jour. Geophys. Research* **82**:2989–3002.

Molyneux, L., and R. Thompson. 1973. Rapid measurement of the magnetic susceptibility of long cores of sediment. *Royal Astron. Soc. Geophys. Jour.* **32**:479–481.

Montfrans, H. M. van, and J. Hospers. 1969. A preliminary report on the stratigraphical position of the Matuyama-Brunhes geomagnetic field reversal in the Quaternary sediments of the Netherlands. *Geologie en Mijnbouw* **48**:565–572.

Opdyke, N. D., and B. P. Glass. 1969. The paleomagnetism of sediment cores from the Indian Ocean. *Deep-Sea Research* **16**:249–261.

Opdyke, N. D., D. V. Kent, and W. Lowrie. 1973. Details of magnetic polarity transitions recorded in a high deposition rate deep-sea core. *Earth and Planetary Sci. Letters* **20**:315–324.

Rees, A. I. 1971. The magnetic anisotropy of samples from the Deep Sea Drilling Project Leg I, Orange, Texas, to Hoboken, N. J. *Marine Geology* **11**:M16–M23.

Rees, A. I., U. von Rad, and F. P. Shepard. 1968. Magnetic fabric of sediments from the La Jolla submarine canyon and fan, California. *Marine Geology* **6**:145–178.

Saito, T. 1976. Geologic significance of coiling direction in the planktonic foraminifera Pulleniatina. *Geology* **4**:305–309.

Saito, T. 1977. Late Cenozoic planktonic foraminiferal datum levels: The present state of knowledge toward accomplishing pan-Pacific stratigraphic correlation. In *Proceedings of the First International Congress on Pacific Neogene Stratigraphy*. Tokyo: Kaiyo Shuppan Co., pp. 61–80.

Saito, T., L. H. Burckle, and J. D. Hays. 1975. Later Miocene to Pleistocene biostratigraphy of equatorial Pacific sediments. In *Late Neogene Epoch Boundaries*, edited by T. Saito and L. H. Burckle. New York: Micropaleontology Press, pp. 226–244.

Smith, J. D., and J. H. Foster. 1969. Geomagnetic reversal in Brunhes normal polarity epoch. *Science* **163**:565–567.

Theyer, F. 1973. Reply to N. D. Watkins, J. P. Kennett, and P. Vella, 1973. *Nature Phys. Sci.* **244**:47–48.

Theyer, F., and S. R. Hammond. 1974. Cenozoic magnetic time scale in deep-sea cores: Completion of the Neogene. *Geology* **2**:487–492.

Watkins, N. D., J. P. Kennett, and P. Vella. 1973. Palaeomagnetism and the *Globorotalia truncatulinoides* datum in the Tasman Sea and Southern Ocean. *Nature Phys. Sci.* **244**:45–47.

4

Reprinted from Rev. Geophysics and Space Physics **10**:213–249 (1972)

Paleomagnetism of Deep-Sea Cores

Neil D. Opdyke

Lamont-Doherty Geological Observatory of Columbia University
Palisades, New York   10964

This review is intended to cover the principal developments that have occurred within the last six years in the paleomagnetic study of marine sediments. Recent work utilizing the reflecting-light microscope indicates that detrital high-temperature Fe-Ti oxides are probably responsible for most of the magnetic remanence in marine sediments. These minerals possess a spectrum of coercivities that makes it necessary to use alternating-field–demagnetization techniques to isolate stable components. It is possible to use the standard magnetic stratigraphy for the last 4 m.y. of earth history derived from terrestrial lavas. Using the ages of the magnetic boundaries from this time scale it is possible by extrapolation and interpolation to better determine the ages of the major events. The ages of these events in increasing age are Jaramillo, 0.87 to 0.92 m.y.; Olduvai, 1.71 to 1.86 m.y.; Kaena, 2.82 to 2.90 m.y.; Mammoth, 3.0 to 3.085 m.y.; Cochiti, 3.72 to 3.82 m.y.; Nunivak, 3.97 to 4.14 m.y.; 'c' event of the Gilbert series, 4.33 to 4.65 m.y. Through the use of long cores from the central Pacific and through correlation using fossil datums, it has been possible to extend the magnetic stratigraphy back to the upper middle Miocene to magnetic epoch 5. It is concluded that very short magnetic events are probably short-term excursions of the field and not true magnetic events. It is shown that the field of the earth averages to an axial-dipole field within a period of 27,000 years and that the field over the last two million years has acted as a geocentric axial dipole. The evidence shows that when reversals of the dipole occur, the values of the reversed inclination are not significantly different from the normal values. The use of magnetic stratigraphy in marine geology has opened up a new era in study of sedimentary processes and evolution of marine organisms.

Paleomagnetic studies of deep-sea cores began with the work of *Johnson et al.* [1948], who found that the directions of magnetization of red-clay cores from the Antarctic region gave results less scattered than those obtained from glacial varves. The next significant study of marine sediments was that of *Keen* [1963], who showed that the inclination observed in cores in the region of the Iberian abyssal plain was about the same as that of the ambient field at the sampling site, which happened to be close to that of the axial-dipole field.

The major impetus to the paleomagnetic study of marine cores was provided by the demonstration by *Harrison and Funnell* [1964] that reversals of magnetization were preserved in cores from the central Pacific, and by *Lin'kova* [1965] that reversals were detectable from the Arctic Ocean sediments. *Opdyke et al.* [1966] showed that long-distance core-to-core correlation of marine sediments was possible by using the magnetic stratigraphy worked out on lava sequences on land.

METHODS AND TECHNIQUES

The sediment used for magnetic studies from oceanic areas has been obtained through the use of gravity cores [*Harrison*, 1966], piston cores [*Opdyke et al.*, 1966], or by the Joides drilling ship [*Opdyke and Phillips*, 1969]. All these techniques possess some undesirable features. Gravity cores are usually too short to allow sampling of several reversals of the earth's field and sometimes suffer internal disturbance. The piston-coring technique [*Ewing et al.*, 1967] has provided by far the largest amount of data presently available; however, if the scope is not properly adjusted, the top of the core is sometimes missed. Stretching of the core also occasionally occurs. The deep-sea drilling cores often suffer from severe disturbance caused by drilling and are usually discontinuous. *Harrison* [1966] has stated that greater accuracy is obtained if the specimens are taken and measured while moist, and this method has been used by several workers [*Goodell and Watkins*, 1968; *Phillips et al.*, 1968]. However, *Opdyke et al.* [1966] report satisfactory results from dried sediments, although it seems certain that measurement while moist is probably the better procedure in cores with a large clay content, which shrink a great deal on drying. Nevertheless, usable data have been obtained from these cores even after drying [*Opdyke and Foster*, 1970]. Best results are obtained from cores with reasonably high contents of biogenic material (Foraminifera, Radiolaria, and Diatomaceae) with a maximum contraction of 10%.

At the Lamont-Doherty Geological Observatory, a 10-cm sampling interval has proved to be satisfactory for initial survey work, with closer spacing to delineate boundary positions. After removal from the core, specimens can be measured on any magnetometer with sufficient sensitivity. The magnetometer used at Lamont-Doherty Geological Observatory is a modified low-speed type, at present widely used for marine paleomagnetism [*Foster*, 1966].

MAGNETIC PROPERTIES OF MARINE SEDIMENTS

Marine sediments have magnetic characteristics similar to continental igneous and sedimentary rocks with intensities characteristically less than 1×10^{-4} gauss and present the same types of problems to the investigators. The minerals responsible for the remanent moment of marine sediments are the same as those present in terrestrial rocks.

Haggerty [1970] has developed a technique for impregnating marine sediments by using a dental cement to produce polished sections. The opaque minerals can then be identified and studied with a reflected-light microscope. Haggerty has found detrital titano-magnetites with varying degrees of oxidation (as revealed by the relative abundance of exsolved ilmenite lamellae), as well as members of the hematite-ilmenite solid solution series, along with other oxides and sulfides. These minerals are present in various grain sizes down to sizes below the resolving power of the microscope. Haggerty believes that these minerals are detrital, probably being carried from the continents by winds. The magnetization of the sediments probably resides in particles less than 10 microns in size. That these minerals are responsible for the magnetization of marine sediments is borne out by Curie point determinations on central and north Pacific cores, which show that both titano-magnetites and hematite are present [*Dickson and Foster*, 1966; *Opdyke and Foster*, 1970]. These findings agree with the study of *Keen* [1963] on North Atlantic cores.

STABILITY OF MAGNETIZATION

Opdyke et al. [1966] showed that both normal and reversed magnetized specimens from cores of the southern hemisphere behave in a stable or partially stable fashion (Figure 1). Most marine sediments behave in a partially stable fashion, and in most cases unstable secondary components are either present in the direction of the earth's field at the sampling site or are acquired during storage. It has therefore proved necessary to partially demagnetize all specimens in peak fields of from 50 to 150 oersteds. Figure 2 shows how the internal consistency of the data is substantially improved after alternating-field partial demagnetization. However, this method is not uniformly successful in removing unwanted secondary components. Figure 3 shows the change in inclination down core V20-98, a red-clay core from the central Pacific. It can be seen that at the point where the magnetic inclination becomes internally inconsistent at 470 cm, the intensity curves diverge markedly, indicating that a large secondary component is being removed in low fields. In certain specimens the intensity drops almost an order of magnitude, indicating the presence of a very large unstable component. Evidently the cleaning techniques being employed are not adequate to recover the original direction of magnetization. This type of behavior is common in cores from red-clay areas of the central Pacific [*Opdyke and Foster*, 1970]. In some cases the marked increase in instability in the cores coin-

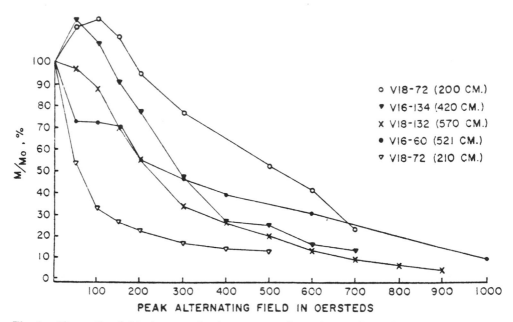

Fig. 1. Alternating-field demagnetization curves of specimens from five cores. M/M_0 is the ratio of remaining to initial intensity after treatment in the corresponding field.

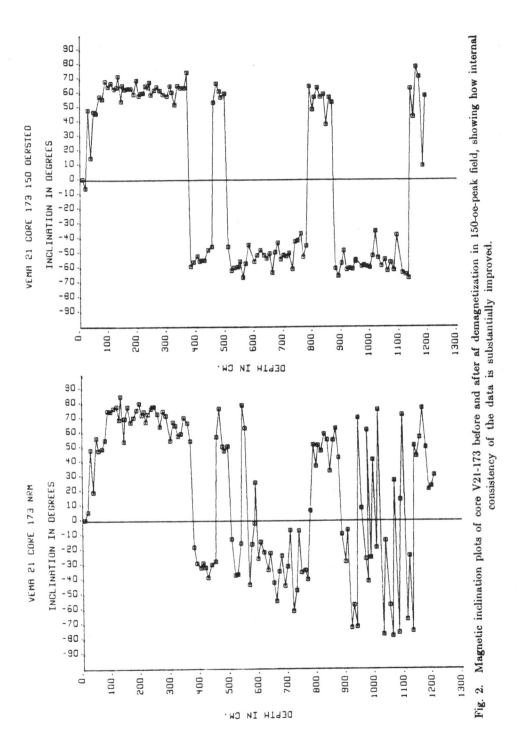

Fig. 2. Magnetic inclination plots of core V21-173 before and after af demagnetization in 150-oe-peak field, showing how internal consistency of the data is substantially improved.

cides with a lithologic color change, which is apparently due to an increase in manganese content with depth [*Haggerty*, 1970].

GEOMAGNETIC TIME SCALE

The time scale used in this paper is modified after *Cox* [1969] and follows the terminology of *Cox et al.* [1964] for the last 4.5 m.y. of the reversal history of the earth's field. The major modification to this time scale is the duration and position of the Olduvai gorge as determined by *Grommé and Hay* [1971].

All studies of cores that are presumed to cover the last 4 m.y. of earth history are interpreted with respect to the model presented above. Figure 4 shows inclination plots for three cores from the North Pacific. The rate of sedimentation decreases from top to bottom, so that longer and longer portions of the geomagnetic time scale are seen. The upper portions of all these cores are normally magnetized, and it is reasonable to believe that the upper normally magnetized section of each core represents the Brunhes normal epoch, defined on the basis of potassium-argon dating by *Cox et al.* [1963]. It should be emphasized that the magnetic stratigraphy obtained from marine sediments is based on

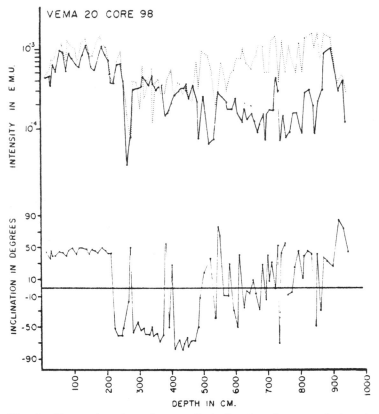

Fig. 3. Composite plots of magnetic inclination (bottom plot) and intensity of magnetization (top plot) for core V20-98. The dotted line indicates the sample NRM intensity in emu; the solid line shows the intensity after af demagnetization in 50 oe.

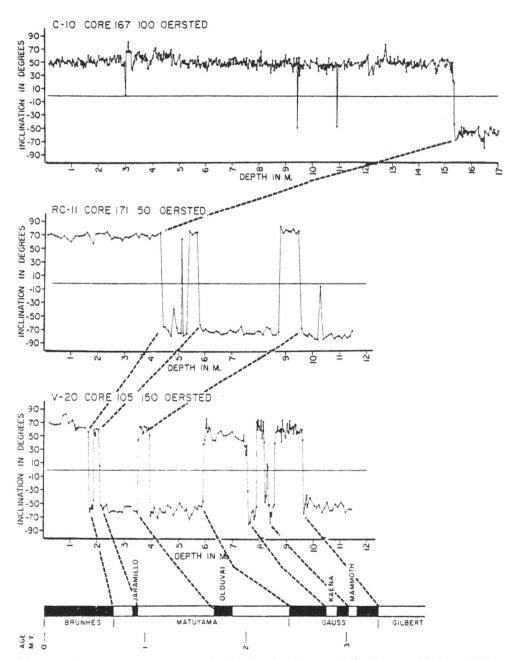

Fig. 4. Magnetic inclination versus depth plots for three north Pacific cores (C1O-167, RC11-171, and V20-105) with time scale and correlations.

the classical method of stratigraphy, the law of superposition. *Opdyke et al.* [1966], to make a distinction between magnetic stratigraphy based on superposition and that based on ^{40}K-^{40}Ar dating, referred to sediments that they believed represented the Brunhes normal-polarity epoch as the Brunhes normal-polarity series. Similarly, the reversely magnetized sediments below the Brunhes series, which are presumed to correlate with the Matuyama reversed-polarity epoch, are called the Matuyama reversed-polarity series, and so on, back to the Gilbert reversed-polarity epoch. If unconformities are present in the sediments, however, serious mistakes can be made. To obviate this possibility, some independent method of telling relative time must be applied. The most widely used method to date is the evolution or extinction of planktonic microfossils such as Radiolaria, diatoms, or Foraminifera [*Opdyke et al.*, 1966; *Hays and Opdyke*, 1967; *Hays et al.*, 1969]. Once the fossil zonation has been tied to the paleomagnetic time scale, the paleontology of the cores is of inestimable value in aiding in the interpretation of the reversal stratigraphy.

It is highly desirable to have an independent test of the assumption that the last reversal identified in oceanic cores can be correlated with the Brunhes-Matuyama boundary, as determined from work on lava flows and dated by the ^{40}K-^{40}Ar method. One way of making such a direct test is to compare the rate of sedimentation averaged over the last 700,000 years with rates of sedimentation derived from an independent method such as radiometric dating using ^{230}Th, ^{231}Pa, ^{14}C. This comparison was made by *Harrison and Somayajulu* [1966], who used ^{230}Th for one core from the eastern Pacific, with good agreement. A more extensive comparison [*Ku et al.*, 1968] on fourteen cores from the Indian and Pacific oceans indicated a high degree of correlation between the various methods of determining rates of deposition. As an example, core V16-75 (22°S, 58°E) from the Indian Ocean yielded a paleomagnetically determined rate of 6.2 mm per 10^3 years for 700,000 years, a rate derived from excess ^{231}Pa of 7.2 mm per 10^3 years over 70,000 years, an excess ^{230}Th rate of 5.8 mm per 10^3 years for 300,000 years, and a radiocarbon rate of 8.4 mm per 10^3 years for 20,000 years. The high $CaCO_3$ content near the top of the core may account for the somewhat higher rate obtained by ^{14}C; nevertheless, the agreement is good considering the different lengths of time over which the rates are determined. In view of this study it seems reasonable to conclude that the Brunhes series in oceanic cores is time-synchronous with the Brunhes polarity epoch as determined from terrestrial lava flows.

A further test of the paleomagnetic time scale in marine cores was carried out by *Dymond* [1969], who dated volcanic ashes in cores on which the paleomagnetic stratigraphy was known by using the ^{40}K-^{40}Ar method. With this technique it was possible to date horizons within the Matuyama series. In two of the cores studied, V21-145 and V21-173, the ^{40}K-^{40}Ar ages are in very good agreement with the ages of the ashes extrapolated from the magnetic data. The 95% confidence limits on the age of the ash in the third core, V20-108, make it just significantly different from the magnetic reversal age. All three of these cores are from the north-central Pacific [*Opdyke and Foster*, 1970]. The ages on eight ash layers from the fourth core, which was raised from the Indian Ocean

south of the Java trench, are significantly older than the ages postulated on
the basis of the magnetics. Dymond suggested that the magnetic age is not the
true age and that the discrepancy has been caused by a delay in the magnetization being locked into the core. He argues that the magnetization does not become
stable in this core until the magnetic particles have been buried to a depth of
1.5 meters.

The core in question, V19-153, has been used by *Glass* [1967] in the study
of microtectites and their distribution in the Indian Ocean. The microtectites in
six of the cores in the southeast Asia strewn field occur at or just above the
Brunhes/Matuyama boundary [*Glass*, 1967]. *Dymond* [1969] has suggested
that the glassy objects in V19-153 might be volcanic glass; however, recent
geochemical analysis clearly shows that they are microtectites [*Cassidy et al.*,
1969]. They have recently been dated by using fission-track techniques, and
the results are in agreement with the paleomagnetic age. The explanation of
why the ^{40}K-^{40}Ar dating method is in conflict with other dating techniques
within this particular core will have to await further research.

MAGNETIC EVENTS

Magnetic events were first recognized as single lava flows of opposite
polarity within sequences of dominantly opposite sign. Lava flows that have
varying rates of extrusion are difficult to use for studies of the duration of
events, particularly when the errors on the ^{40}K-^{40}Ar dates are larger in some
cases than the duration of the event itself. Sedimentary sequences in areas that
have a more or less continuous rate of sedimentation across events can be used
to date their duration and time of occurrence. This has proved a successful technique in the study of the Jaramillo event [*Opdyke*, 1969] and the Olduvai event
[*Opdyke and Foster*, 1970].

Magnetic events that span a length of time of the order of 50,000 years
and more are often found in piston cores; thus within the Matuyama series, the
Jaramillo and Olduvai events can be discriminated. The Jaramillo [*Doell and
Dalrymple*, 1966] is the event directly preceding the Brunhes/Matuyama
boundary. *Opdyke* [1969] was able to date it by extrapolation from the Brunhes/
Matuyama boundary and to place the beginning of the event at 916×10^3
years and its termination at 870×10^3 years, with a standard deviation of $33
\times 10^3$ years. The average duration of the event was found to be 54×10^3 years,
with a standard deviation of 15×10^3 years. This estimate is in good agreement
with a date on a lava flow with a transition direction from the type area in New
Mexico of 88×10^3 years and with the estimate of *Cox and Dalrymple* [1967],
who gave a mean age of 91×10^3 years for the event.

The Olduvai event. The most important event from a stratigraphic viewpoint occurs in the lower part of the Matuyama series, which in previous studies
on deep-sea sediments has been correlated with the Olduvai event [*Opdyke et al.*,
1966; *Dickson and Foster*, 1966; *Hays and Opdyke*, 1967; *Berggren et al.*, 1967;
Hays et al., 1969]. It is the event in which the Pliocene/Pleistocene boundary
occurs [*Berggren et al.*, 1967]. Confusion has arisen in that some workers believe
this large event is not the Olduvai but is, instead, the younger Gilsa [*Cox*, 1969;

Vine, 1968]. In an effort to more closely delineate its age as seen in deep-sea cores, a time-versus-depth plot was constructed from the cores that contain the large event and also pass through to the upper-Gauss normal series. In this study 13 cores are available that have these qualifications. An age of 0.69 m.y. has been used for the age of the Brunhes/Matuyama boundary, and an age of 2.43 m.y. has been used for the Gauss Matuyama boundary. Assuming that the rate of sedimentation is constant between two points, a mean age for the event can be obtained (Figure 5). The age of the upper boundary of the event ranges between 1.59 and 1.90 m.y., with an average age of 1.71 ± 0.26 m.y. The lower boundary ranges in age between 1.75 and 1.98 m.y., with an average of 1.86 ± 0.19 m.y. The average length of the event would be 150,000 ± 47,000 years.

The correlation of this large event with that seen in Olduvai gorge has been confirmed by a restudy of Olduvai gorge by *Grommé and Hay* [1971], who show that the Olduvai normal-geomagnetic-polarity event is represented in its type area by five normally magnetized volcanic-rock units whose age range may be as great as 1.6 to 1.9 m.y. It can be seen that this interval spans the time obtained from the cores. There can be no doubt that the large event seen in marine sediments in which the Pliocene/Pleistocene boundary occurs is correlative with the Olduvai event as seen in Olduvai gorge.

The Gilsa normal-polarity event was named by *McDougall and Wensink* [1966] for a normally magnetized lava flow in Iceland that was dated at 1.60 ± 0.05 m.y. *Grommé and Hay* [1971] have suggested that this lava was probably erupted during the Olduvai event and that therefore the name Gilsa could be dropped, since Olduvai has clear priority. There is evidence from three deep-sea cores (V20-109, E13-3, and V12-18) that a short normal-polarity event occurred just after the Olduvai event. It has been suggested by *Ninkovich et al.* [1966] that this could be identified with the Gilsa event.

Evidence also exists for normally magnetized lavas in the 0.3 m.y. preceding the Olduvai event. *Cox* [1969] lists seven normally magnetized lavas with ages ranging from 1.95 to 2.09 m.y., which he included in the 'Olduvai' event. It seems probable that a short event or events are present in this part of the time scale. Evidence for short events at this time has been pointed out by *Heirtzler et al.* [1968], *Cox* [1969], and *Emilia and Heinrichs* [1969] on the basis of small anomalies that appear on sea-floor–spreading profiles. Short events also appear in the magnetic record from several cores (E13-3, V16-34, V21-65, and V11-75). *Stacey* [1969] and *Grommé and Hay* [1971] have suggested they be named the Reunion events after work done on Reunion Island by *McDougall and Chamalaun* [1966], and this usage is followed here. Although it seems that these events are real, their importance for stratigraphic work in marine sediments is minimal principally because they have so short a duration that they are not seen in many cores that pass through the Matuyama series in which the rates of sedimentation are of the order of 5 mm per 1000 years. It seems possible, however, that these short events will be more important in studies of expanded sections on land or in deep-sea drilling cores where rates of sedimentation are higher. This has already been shown to be true in Japan [*Nakagawa et al.*, 1969] and New Zealand [*Kennett et al.*, 1971].

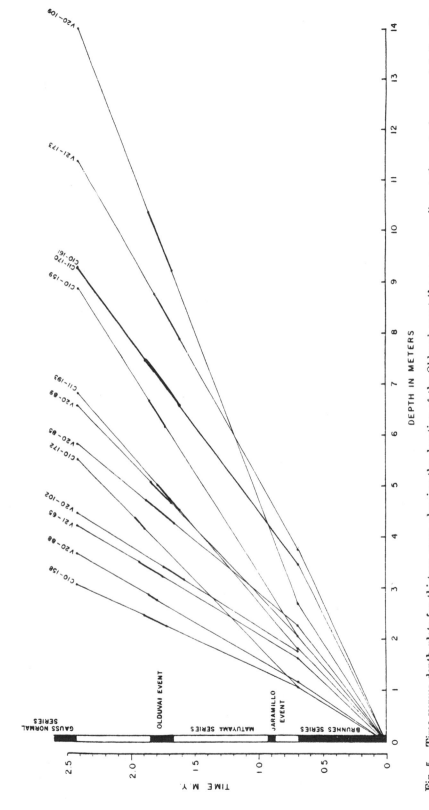

Fig. 5. Time versus depth plots for thirteen cores, showing the duration of the Olduvai event if an average sedimentation rate is assumed during the Matuyama epoch based on the Brunhes/Matuyama boundary at 0.69 m.y. and the Matuyama/Gauss boundary at 2.43 m.y. The average duration of the Olduvai event figured in this manner is 150,000 years, from 1.71 to 1.86 m.y. B.P.

85

Mammoth and Kaena events. During the Gauss normal epoch are two prominent events. One was originally dated at 3.06 m.y. and called the Mammoth event by *Cox et al.* [1964]; the second, which is younger, has been named the Kaena event by *McDougall and Chamalaun* [1966]. These events are often seen in the Gauss series in cores that penetrate through to sediments of Gilbert-series age. Figure 6 shows the Gauss series in four cores plotted against time; dates used are 2.43 m.y. for the Gauss/Matuyama boundary and 3.32 m.y. for the Gauss/Gilbert boundary. The average lengths of the Kaena and Mammoth events are 0.085 m.y. and 0.087 m.y., respectively. The boundaries of the Kaena event are 2.82 and 2.90 m.y.; boundaries of the Mammoth event are 3.0 and 3.085 m.y. These dates are in close agreement with the original K/Ar dates for the Kaena event, 2.8 ± 0.06 m.y., and the Mammoth event, 3.06 ± 0.02 m.y. The figures obtained are also closely comparable with those of *Cox* [1969] and *Talwani et al.* [1971], who used interpolation techniques derived from sea-floor–spreading profiles (Table 1). It is impressive that these different studies yield closely similar results.

From the magnetic stratigraphy deduced from the research vessel *Eltanin* 19 profile over the East Pacific rise, *Pitman and Heirtzler* [1966] postulated the existence of three major events and one minor event during the Gilbert reversed-magnetic epoch. Three of these new events were found in Antarctic cores by *Hays and Opdyke* [1967] and were designated as events 'a,' 'b,' and 'c' of the Gilbert reversed epoch. Events 'a' and 'b' were subsequently named the Cochiti and Nunivak events by *Cox and Dalrymple* [1967] on the basis of lava flows, dated by the potassium-argon method. Event 'c' has not yet been named, but its resolution into a small upper and a longer lower portion has been demonstrated by *Foster and Opdyke* [1970]. The apparent dates for these events as obtained by extrapolation indicate that they are younger, however, than predicted from the *Eltanin* 19 profile. V20-88 has a remarkably constant rate of sedimentation through to the base of Gauss series sediments (Figure 7). Assuming that the core has a continuous rate of deposition throughout, one obtains dates for the events in the Gilbert given in Table 2. These values are similar to those obtained previously by *Hays and Opdyke* [1967] for core E13-17. The best estimate of the age of the Gilbert/epoch 5 boundary is that of *Foster and Opdyke* [1970], who suggest a date for this transition of 5.1 m.y. *Talwani et al.* [1971] have recently proposed an alteration in the sea-floor–spreading time scale covering this period of time based on data from the Reykjanes ridge. This alteration brings the sea-floor–spreading time scale into close coincidence with dates determined from sedimentary studies. Figure 8 gives the author's synthesis of the magnetic-polarity time scale from the present to epoch 5.

EXTENSION OF MAGNETIC TIME SCALE

Recently it has become possible to extend the reversal sequence seen in marine sediments to the upper middle Miocene [*Foster and Opdyke*, 1970]. This is owing to the recovery of two long piston cores from the central Pacific, 26 and 28 meters in length, which penetrate through normal magnetic epoch '5' (Figures 9 and 10). RC12-65, although shorter in length than RC12-66, pene-

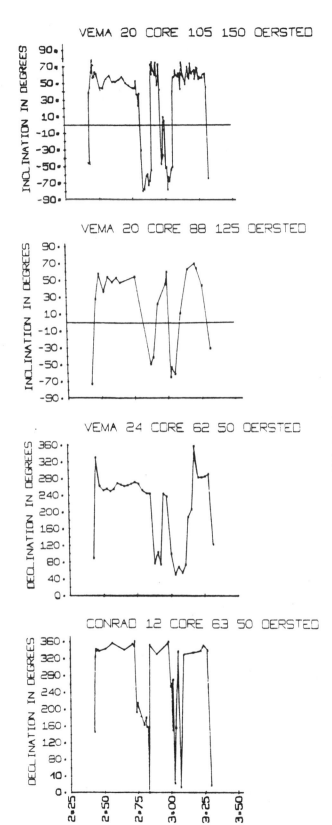

Fig. 6. Magnetic inclination versus depth for cores V20-105 and V20-88 (top) and declination versus time plots for V24-62 and RC12-63 (bottom), showing the Gauss series using a date of 2.43 m.y. for the Gauss/Matuyama boundary and a date of 3.32 m.y. for the Gauss/Gilbert boundary.

TABLE 1. Gauss Epoch Events

Event		This Paper	Cox, 1969	Talwani, 1971
Kaena	Upper boundary	2.82 m.y. B.P.	2.80 m.y. B.P.	2.84 m.y. B.P.
	Lower boundary	2.90 m.y. B.P.	2.90 m.y. B.P.	2.94 m.y. B.P.
	Total time	85,000 years	100,000 years	100,000 years
Mammoth	Upper boundary	3.000 m.y. B.P.	2.94 m.y. B.P.	3.04 m.y. B.P.
	Lower boundary	3.085 m.y. B.P.	3.06 m.y. B.P.	3.10 m.y. B.P.
	Total time	87,000 years	140,000 years	60,000 years

The original ^{40}K-^{40}Ar dates that defined these events were 2.8 m.y. B.P. for the Kaena event and 3.06 m.y. B.P. for the Mammoth.

trates much further in time. The reason for this is that RC12-65 is largely carbonate free, whereas RC12-66 contains appreciable amounts of carbonate. Fortunately RC12-65 contains large numbers of siliceous microfossils that allow a correlation to previous studies in the region [*Hays et al.*, 1969] and to other areas. The core contains 23 changes of polarity based on changes of declination below the Gilbert reversed-polarity epoch. *Foster and Opdyke* [1970] have arbitrarily divided these magnetic polarity changes into magnetic epochs and events following the proposed nomenclature of *Hays and Opdyke* [1967]. In this stratigraphic scheme the magnetic epochs are numbered in order of increas-

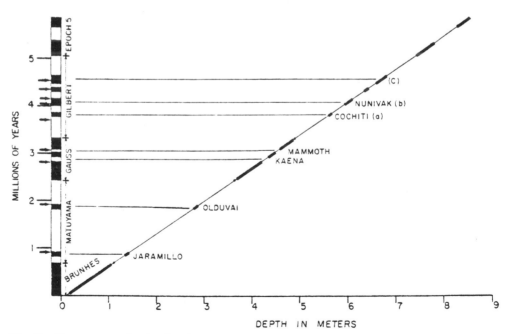

Fig. 7. Time versus depth plot for core V20-88. The slope of the line was determined by using 0.69 m.y. for the Brunhes/Matuyama boundary, 2.43 m.y. for the Gauss/Matuyama boundary, and 3.32 m.y. for the Gauss/Gilbert boundary. The normal (black) and reversed (white) portions of the core were projected onto the time axis. The original dates from which the magnetic events were established are shown by the solid arrows on the left.

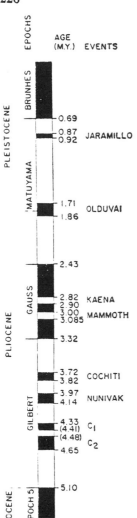

Fig. 8. Revised magnetic stratigraphy of marine sediments based on K-Ar ages and extrapolation of magnetic data for cores.

ing age, with the Brunhes normal magnetic epoch being magnetic epoch 1; Matuyama, epoch 2; Gauss, epoch 3; and Gilbert, epoch 4. In this system all even-numbered epochs will be predominantly reversely magnetized, and odd-numbered epochs will be normally magnetized. The events will be of opposite polarity and are designated as events 'a,' 'b,' 'c,' etc., in order of increasing age. RC12-65 has been divided into eleven magnetic epochs, most of which contain magnetic events. The major uncertainty is the possibility that major unconformities exist, causing a distortion in the magnetic record. The fauna and flora of these cores have been carefully analyzed, and no large hiatuses have been detected; hence it seems unlikely that important segments of the record are missing. However, distortions may occur in the relative lengths of magnetic epochs and events. Indeed, such a distortion in RC12-65 occurs during the Matuyama reversed-polarity epoch, where the magnetic stratigraphy is relatively

TABLE 2. Gilbert Epoch Events

Event		Heirtzler et al., 1968	Cox, 1969	Hays and Opdyke, 1967	Talwani, 1971	Opdyke, 1971
Cochiti	Upper boundary	4.04 m.y. B.P.	3.70 m.y. B.P.	3.82 m.y. B.P.	3.78 m.y. B.P.	3.72 m.y. B.P.
	Lower boundary	4.22 m.y. B.P.	3.92 m.y. B.P.	3.91 m.y. B.P.	3.88 m.y. B.P.	3.82 m.y. B.P.
	Total time	180,000 years	220,000 years	90,000 years	100,000 years	100,000 years
Nunivak	Upper boundary	4.35 m.y. B.P.	4.05 m.y. B.P.	4.08 m.y. B.P.	4.01 m.y. B.P.	3.97 m.y. B.P.
	Lower boundary	4.53 m.y. B.P.	4.25 m.y. B.P.	4.20 m.y. B.P.	4.17 m.y. B.P.	4.14 m.y. B.P.
	Total time	180,000 years	200,000 years	120,000 years	160,000 years	170,000 years
C_1	Upper boundary	4.66 m.y. B.P.	4.38 m.y. B.P.	4.34 m.y. B.P.	4.31 m.y. B.P.	4.33 m.y. B.P.
	Lower boundary	4.77 m.y. B.P.	4.50 m.y. B.P.		4.41 m.y. B.P.	
	Total time	110,000 years	120,000 years		100,000 years	
C_2	Upper boundary	4.80 m.y. B.P.			4.48 m.y. B.P.	
	Lower boundary	5.01 m.y. B.P.		4.53 m.y. B.P.	4.66 m.y. B.P.	4.65 m.y. B.P.
	Total time	200,000 years			180,000 years	
	Total time, C	310,000 years		190,000 years	350,000 years	320,000 years

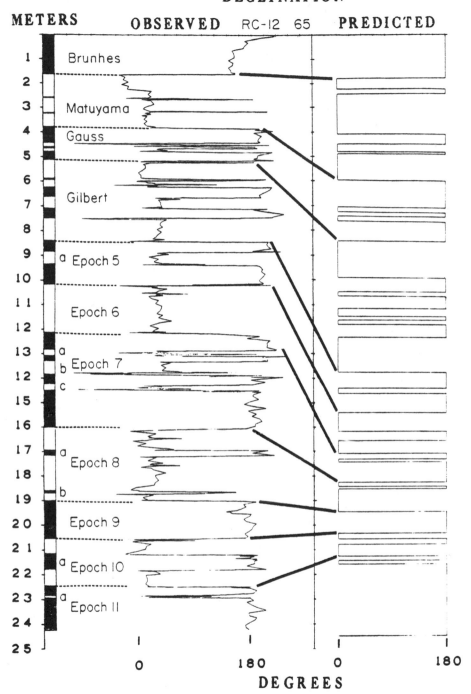

Fig. 9. The change of magnetic declination in RC12-65 after af partial demagnetization of 50 oe with respect to the split face of the core as a function of depth, compared with the reversal sequence predicted by the sea-floor-spreading time scale of *Vine* [1968]. The black (normal) and white (reversed) bar diagram on the left indicates the proposed extension of the geomagnetic time scale based on this study.

CONRAD 12 CORE 66

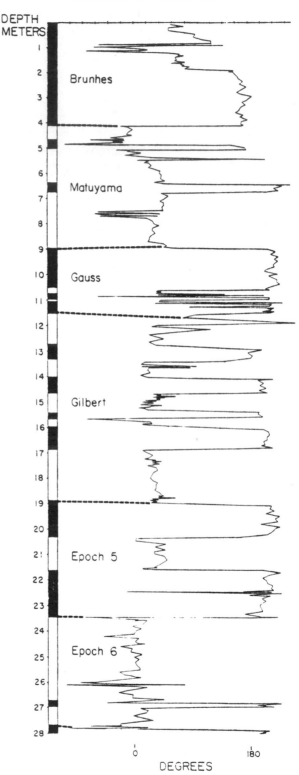

Fig. 10. The change of magnetic declination in core RC12-66 after af partial demagnetization of 50 oe with respect to the split face of the core as a function of depth. Predicted time scale on the left. The black indicates normal and the white reversed magnetics.

well known. Until several cores have been obtained that confirm the magnetic stratigraphy beyond epoch 5, the relative sequence must remain tentative; however, the relative position of the reversed and normal sections is secure, since they occur in stratigraphic superposition. The magnetic stratigraphy is known through magnetic epoch 5, since we have at least 5 cores that penetrate to this point in time.

Extension of magnetic time scale beyond epoch 11. Recently workers at Lamont-Doherty Geological Observatory have been attempting to extend the magnetic sequence beyond magnetic epoch 11 by correlating the older portion of RC12-65 to other cores that have older sections but have unconformities at the top. This has been accomplished by correlating important floral and faunal datums between the cores and then matching the reversal patterns. By using this technique the reversal sequence has been extended to magnetic epoch 15 in sediments of upper middle Miocene age (Figure 11). This age would be about 15 m.y. on the *Berggren* [1969] time scale [*Opdyke et al.*, 1970]. By using this technique it will probably be possible in the near future to extend the magnetic

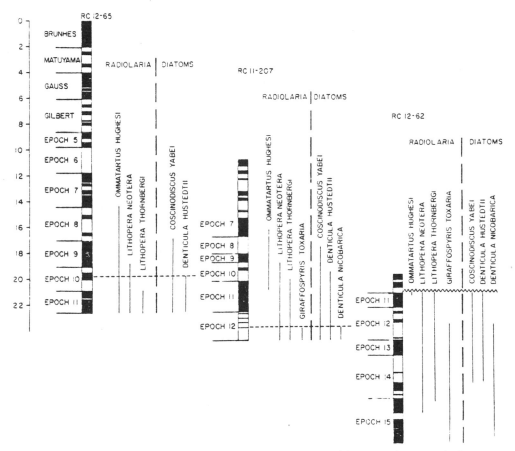

Fig. 11. Correlation of important floral and faunal datums with magnetic reversals for cores RC12-65, RC11-207, and RC12-62.

sequence in marine cores back in time to at least the Oligocene, although the reliability is correspondingly reduced. It has proved possible to correlate epoch 11 to the *Discoaster hamatus* zone, which is in turn correlative to the *Globorotalia menardii* foraminiferal zone. Epoch 12 can be correlated on the basis of Foraminifera to the *Globorotalia meyeri* zone, which is the highest zone in the middle Miocene. The boundary between the middle Miocene and the late Miocene must lie between magnetic epochs 11 and 12. Epoch 15 can be correlated, in part, to the foraminiferal *Globorotalia periferoacuta* zone. *Berggren* [1969] estimates this zone to be approximately 13.5 m.y. B.P.

Figure 11 shows the magnetic stratigraphy of marine sediments as far back in time as it is presently known. The magnetic stratigraphy has been graded into four categories of reliability. The highest level of reliability is that in which the ages of the magnetic boundaries have been fixed by ^{40}K-^{40}Ar dating. The second level of reliability is the sector of time where the sequence of reversals has been fixed by superposition and dated by extrapolating rates of sedimentation, and where at least three or more cores are seen to possess the same magnetic sequence. The third level of reliability is where the magnetic sequence is fixed by stratigraphic superposition in one core. The fourth, and last, level of reliability is where the magnetic sequence has been extended by overlap of cores possessing unconformities at the top, with the correlation between cores being affected by using floral and faunal datums. Obtaining absolute dates on the magnetic sequence of third- and fourth-level reliability that are older than 5 m.y. B.P. is difficult; however, the dates that are possible to obtain have been determined by the correlation of datum levels from the cores containing magnetic sequences to sections that have been dated radiometrically.

SHORT EVENTS AND ASSOCIATED PROBLEMS

During the Brunhes normal epoch two short magnetic events have been postulated: one called the Laschamp event has been observed in recent lavas from France by *Bonhommet and Babkine* [1967], and the second, called the Blake event, has been observed in cores by *Smith and Foster* [1969]. The existence of short events has been suggested on theoretical grounds by *Cox* [1969]. *Smith and Foster* [1969] postulated that the Blake event was about 5000 years in length and occurred within the X zone of *Ericson* [1961]. One of the questions that has not been properly answered is how short a change of polarity must be in order to qualify as an event as opposed to a short-term fluctuation of the field. Marked rapid excursions of the magnetic field apparently occur where either the inclination or declination shows serious departures from dipolarity. Excursions of this sort have been noted from lava flows on land. *Evans* [1970], in a study of upper Cenozoic lavas from New Zealand, illustrates two cases in which the declination reverses, but the inclination remains stable, and another case in which the inclination changes, but the declination does not alter. These deviations of the field do not appear as true reversals, since the direction of magnetization does not change through 180° with respect to adjacent lavas. It is suggested that such excursions of the field are aborted reversals. The Laschamp event may be such an excursion, since the direction observed is not truly reversed. It has proved impossible to identify this event in cores with high rates of sedimentation that cover the proper time interval. True magnetic events should be identified by complete reversal of the field and not by just a change in one of the components. As will be shown later, a complete field reversal appears to take about 10,000 years to complete, and hence we should expect that the shortest complete event would require about 20,000 years to take place. One might therefore conclude that the Blake event, which is estimated to be 5,000 years in length, is not a true short event but is more probably a magnetic excursion. The Blake event is very difficult to identify, even in cores that cover the correct interval of time, and it appears to be present in only 10% of these cores. It may therefore be concluded that the events that have been postulated to be short events within the Brunhes epoch are not true short events but are field fluctuations with periods of the order of the time taken for field reversal. They should therefore be classed as magnetic excursions. In sediments where the rate of sedimentation is high, they may be of some value for correlation. Magnetic excursions of this type seem more likely to happen when the field is disturbed during reversals of the field, and such short excursions have been reported at the Brunhes/Matuyama boundary by N. Niitsuma (unpublished data, 1971) and by *Bucha* [1970] from sections exposed on land. *Opdyke and Foster* [1970] also show such short excursions that may be associated with the reversals of the Jaramillo event. Watkins [1969], in a Miocene reversal sequence from Oregon, also records an excursion that is associated with a reversal. Possible excursions of the field at times other than major reversals have been documented by *Watkins* [1968] within Matuyama series sediments. However, in the absence of some independent very fine correlative techniques it is very difficult to unambiguously confirm their existence, since rates of sedimentation are low in cores that penetrate to Matuyama series age sediments. Future work on exposed sedimentary sequences with high rates of deposition may provide a better understanding of this kind of behavior.

INVESTIGATION OF LONG-TERM BEHAVIOR OF THE EARTH'S FIELD BY USING THE MAGNETISM OF DEEP-SEA SEDIMENT CORES

The hypothesis that the earth's magnetic-dipole field closely approximates a geocentric axial dipole and has done so throughout geologic time is the basis for calculation of all paleomagnetic pole positions. At the present time the axial-dipole field is inclined at an angle of 11.5° to the axis of rotation. Worldwide averages of paleomagnetic data from rocks of Pleistocene age consistently yield a mean pole position that is not significantly different from the present axis of rotation. *Wilson* [1970], however, has recently argued that although the main dipole field is axial, it is not centered but is displaced into the northern hemisphere by 160 km. It has also been suggested that the dipole field is able to maintain persistent offsets of up to 5° over lengths of time up to one million years [*Doell and Cox*, 1965].

Unlike a lava flow that records the earth's magnetic field at a virtual instant of time, each sample from a sediment core is an average of the field over a period of several thousands of years, depending on the rate of sedimentation. Thus secular variations and other nondipole field effects with periods of 10^3 years or less are averaged out. Therefore one could expect that possible variations of the dipole field, with periods of 10^4–10^6 years, could be found in the inclination values of cores.

Harrison [1966] made a study of the dipole field from sediment-core inclinations without coming to a strong conclusion. His analysis, in which he used the methods of *Cain et al.* [1965], showed that the fit to his data was about equal for an axial dipole, a nonaxial dipole, and an inclined dipole. The internal consistency of these data was not good; however, Harrison employed no partial demagnetization techniques, which have been shown to remove unstable magnetic components in cores. Also, Harrison was using data from gravity cores, necessarily less than 200 cm in length, and was analyzing only 5 to 16 samples per core. The study therefore lacks definition although 52 cores from all oceans of the world were examined.

Opdyke and Henry [1969] utilized a larger set of data, all of which represent specimens that were af demagnetized in fields of 50 to 150 oe, and selected the cores that showed the highest internal consistency. The criteria used for inclusion in the study were (1) that each core show at least one reversal of the field and (2) that the standard deviation of the inclination values about the arithmetic mean be less than 15°. The 52 cores that met these criteria come from all the world's oceans (Figure 12). Magnetic samples were usually taken at 10-cm intervals and represented time spans of from 1000 to 9000 years. The cores were taken from geographic latitudes between 55°N and 62°S, and the average inclinations varied between zero and 73°. An example of the latitude dependence of inclination within these cores for Brunhes-age sediments is shown in Figure 13.

Even in cores taken near the equator, the data are often good enough to recognize reversals from shifts of inclination of 10°. These shifts have been confirmed as reversals by the synchronous shift of 180° in the declination (Figure 14). Figure 15 shows the plot of inclination versus latitude for the 52 cores in this study. It can be seen that the fit of the inclination data to these curves is good and that no major divergences were found in the time intervals represented by the cores. To test more vigorously for an angular offset of the dipole from the rotational axis, a simple axis rotation was employed. If the magnetic pole was rotated away from the rotational pole by a certain angle, there would be a corresponding shift of the expected magnetic inclination, or if the inclination is considered fixed, this would correspond to a shift of the original latitude θ to a new apparent dipole latitude θ'. A root-mean-squares value of the difference between observed and apparent dipole latitudes $[\Sigma(\theta - \theta')^2]^{1/2}$ was calculated for all the cores and minimized. This was done by calculating apparent dipole latitudes for pole positions incremented by 5° intervals outward from the rotational pole and then by 1° intervals in the minimal area. The best-fit pole for the inclination values from the cores occurred at 89°N, 211°E. Figure 16 is a plot of the variation with latitude of $[\Sigma(\theta - \theta')^2]^{1/2}$ along this line of longitude. The minimum is shown as an offset of only 1° from the rotational pole. Such a small offset cannot be considered as significant. Standard deviations of this pole position and a similar comparison with the rotational pole, the present dipole

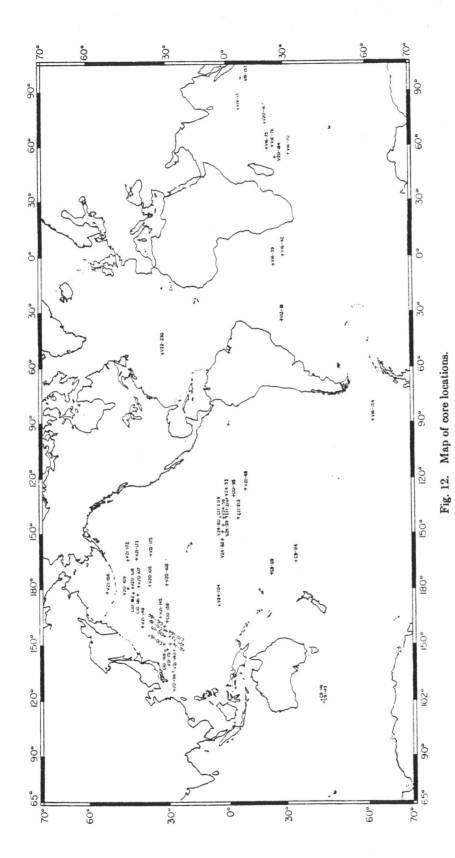

Fig. 12.   Map of core locations.

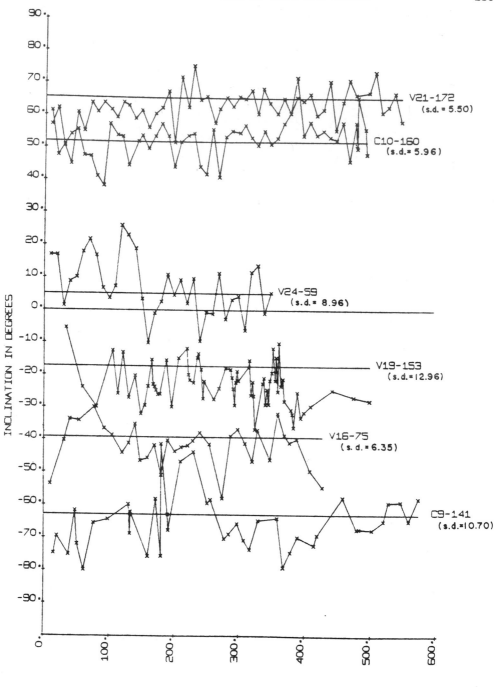

DEPTH IN CM.

Fig. 13. Composite computer plot of magnetic inclination versus depth in core for six cores from different latitudes showing latitude dependence of inclination with these cores for Brunhes age sediments. Theoretical axial-dipole inclinations are shown by the heavy lines through each core. The standard deviation of each core from its arithmetic mean is indicated in parentheses.

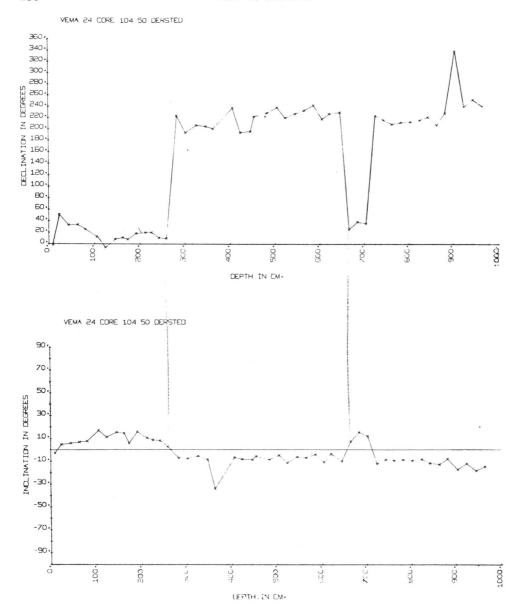

Fig. 14. Computer plots of V24-104 showing the close correlation between reversals derived from the declination and the change in inclination values for a core taken from a very low latitude.

field, and a pole calculated from Hawaiian data that *Doell and Cox* [1965] believe represents the dipole are shown in Table 3. It can clearly be seen that the data support the hypothesis of a geocentric axial-dipole field. *Kobayashi et al* [1971] have recently obtained data for a set of cores from the western equatorial Pacific that support this conclusion.

Since it appears that inclination values average to those that can be expected from an axial geocentric dipole, it would be useful to know how much time is necessary to sample to obtain this average. In order to try to answer this question and other questions related to the problem of dipole movement, two well-dated cores with high rates of sedimentation from the eastern Mediterranean have been studied (*Opdyke et al.*, 1971). One of these cores, V10-50, is only 3.5 meters in length but terminates in a prominent ash layer that has been correlated to the Santorini eruption, dated archeologically and by ¹⁴C at 3500 years B.P. V10-50 can be correlated to a second core, V10-58, through this ash, which occurs at one-meter depth. Core V10-58 has two further dated horizons, the first a sapropel at 200 cm dated by ¹⁴C at 8000 years B.P. and the second a volcanic ash at 550 cm that has been correlated to an ash dated at 25,000 years B.P. The dominant lithology in the two cores is a calcareous lutite. The rate of sedimentation within

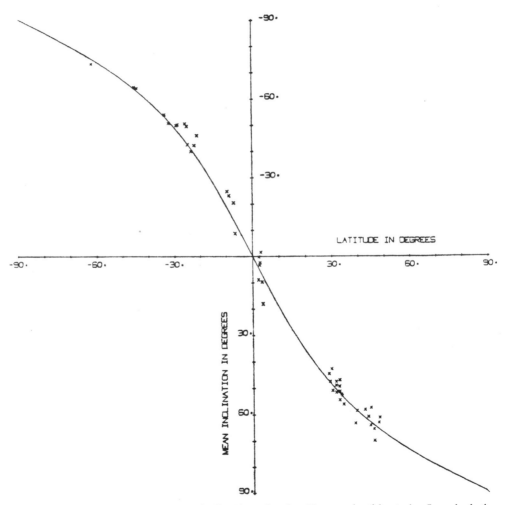

Fig. 15. Core latitude versus mean inclination plot for 52 cores in this study. In calculating mean, the sign of reversed specimens was changed, and total means were calculated having the sign of the normal field direction. Expected geocentric axial-dipole inclination is indicated by heavy line.

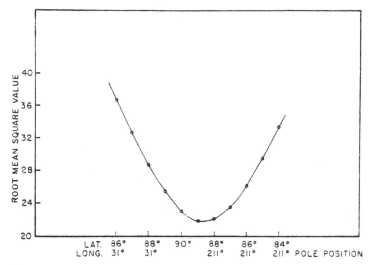

Fig. 16. Plot of variation with latitude of root-mean-squares sum $[\Sigma(\theta-\theta')^2]^{1/2}$ along longitude 211°E. The minimum value occurs at pole position 89°N, 211°E.

the lutite is 18.2 cm/1000 years in V10-58 and 75.5 cm/1000 years in V10-50. These cores therefore provide an opportunity to study in some detail the behavior of the field over the last 30,000 years; to this end, the lutite sections of the core were sampled continuously and measured after af demagnetization in a peak field of 100 oe. Figure 17 shows the variation of magnetic inclination with depth in the cores. The average inclination in V10-50 is 45.8° with a standard deviation of 12.6°, which is well within one standard deviation of the dipole field of 55° for the locality. Core V10-58, which has a lower rate of sedimentation, has an average value for the inclination of 54.7° (s.d. = 14.3°), which is identical to the inclination of the geocentric axial dipole. These data therefore indicate that the dipole field averages to a geocentric axial dipole within approximately 25,000 years. The data have been smoothed using a five-point running mean and show a period of 6000 years, which may be a result of the movement of the dipole field about the axis of rotation. It should be noted that the span of time covered

TABLE 3. Comparison of the Root-Mean-Squares Values for
Various Important Quaternary Poles

	Pole Position		
	Latitude, deg	Longitude, deg	Standard Deviation
Rotation axis	90	0	3.36
Dipole that best fitted inclination data	89	211	3.10
Present earth dipole	79	290	7.35
Pole from Hawaiian data [Doell and Cox, 1965]	84.2	312.9	5.52

by V10-58 should cover the interval in which the Laschamp event is known to occur. No clear inversion of the inclination is present; however, at about 340 cm the inclination reaches a low value of 10°, corresponding to an abrupt drop in intensity. It may be that this excursion of the field, which has an age from 1500 to 1600 years, represents the Laschamp event.

NORMAL- AND REVERSED-FIELD ANALYSIS

It has been suggested [*Wilson*, 1970] that reversals of the field are not truly 180° opposed but have a persistent offset. Both normally and reversely magnetized data from the 50 cores analyzed by *Opdyke and Henry* [1969] do not support this idea. To determine whether a significant difference in inclination exists, the individual differences $|I_n| - |I_r|$ were computed for each core, and the average difference was found to be $+0.78°$, with a standard deviation of 5.43°. The 95% confidence limit for the mean is $\pm10°$. It can be concluded that

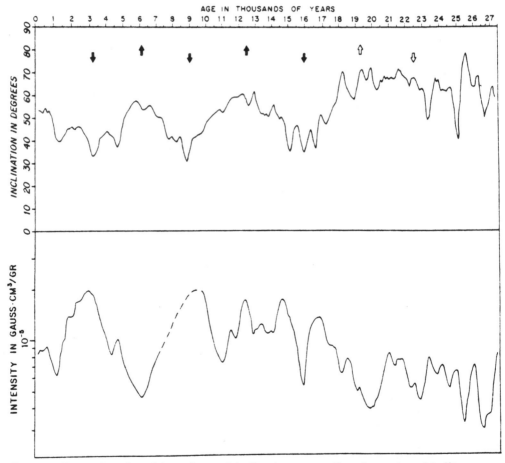

Fig. 17. Composite plot of intensity and inclination versus time for eastern Mediterranean over the last 27,000 years obtained by combining results from cores V10-50 and V10-58.

there is no significant difference between the normal and reversed inclinations in the cores.

Using data from cores, it is more difficult to tell whether changes of declination are exactly 180° because of greater problems with orientation; however, in cores where the data are of high quality, the change in declination is not significantly different from 180°.

INTENSITY CHANGES

Changes of intensity of magnetization within cores are common and occur at reversals of the field and at lithologic breaks. Changes of intensity with field reversal have been noted by *Opdyke et al.* [1966], *Harrison* [1966], *Ninkovich et al.* [1966], *Opdyke and Foster* [1970], and *Kobayashi et al.* [1971]. The change of intensity at field reversals has been ascribed to a decrease of intensity of the earth's field by *Ninkovich et al.* [1966] and *Harrison and Somayajulu* [1966]. *Kobayashi et al.* [1971] have suggested three possible mechanisms: (1) the decrease in intensity at reversals is due to the true drop of intensity during a polarity change of the earth's field; (2) the observed intensity decrease is due to the overlapping influence of the normal and reversed field on uncompacted sediment in which the fine magnetic particles are still somewhat mobile; or (3) the intensity change is completely attributable to variation in the content of magnetic minerals. The third possibility can be eliminated by the use of other magnetic parameters such as susceptibility or saturation remanence, which show that in most cases no striking changes in magnetic mineral content occur across reversals. *Kobayashi et al.* [1971] consider that the second of the two possibilities is the most reasonable on the basis of the fact that the sediment-water interface within 30 cm of the surface is often very fluid, with high porosity, allowing the rotation or partial rotation of grains already deposited. They point out that if this process is operating, the record of polarity events shorter than, say, 30 cm would be lost by remagnetization of the sediment before the sediment is compacted. However, it is well documented that the intensity of the dipole field decreases at reversals [*Van Zijl et al.*, 1962; *Dagley and Lawley*, 1971]. It seems reasonable to believe, therefore, that the sediment is accurately recording a variation in field intensity. Figure 18 shows the variation of intensity over a series of reversals through to the middle Miocene showing that at each reversal a matching decrease in intensity occurs. If it is assumed that a reversal of the field is being recorded, then the time taken for a reversal of the field can be estimated. *Harrison* [1966], using overlapping samples and normalizing the variations of intensity by the ratio NRM/K, estimated that the time taken for a reversal to occur is 10,000 years. *Ninkovich et al.* [1966], utilizing the decrease in intensity and assuming no variation of the content of magnetic minerals, found that the field took only 2000 years to alter direction, but that the time taken for the intensity to decrease and rebuild was 20,000 years. *Foster and Opdyke* [1970], using a similar technique on the reversals of the Jaramillo event, got values of 38,000 years for the complete inversion of the field. *Cox and Dalrymple* [1967], using data derived from potassium argon dating on magnetically studied lavas, obtained an estimate of 4600 years for the

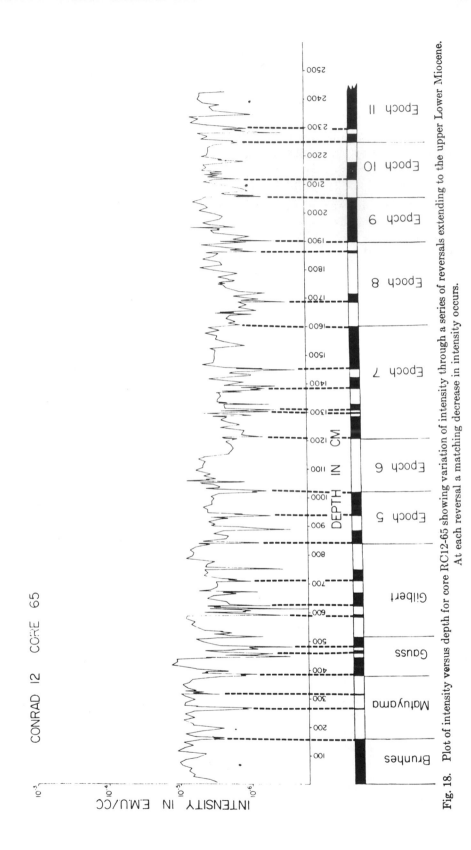

Fig. 18. Plot of intensity versus depth for core RC12-65 showing variation of intensity through a series of reversals extending to the upper Lower Miocene. At each reversal a matching decrease in intensity occurs.

104

magnetic vector to reverse, with a 95% confidence limit of from 1600 to 2100 years. It can be seen that the times taken for reversals are in general agreement, lending support to the idea that the reversal of the field is being recorded in the cores.

Not all intensity changes that occur in marine sediments can be ascribed to changes in the magnetic field, since it is clear that changes in lithology can also bring about changes in magnetic intensity. Figure 19 shows the variation of lithology, magnetic susceptibility, and magnetic intensity in core V10-58 from the eastern Mediterranean. This core has three distinct lithologies: volcanic ash, which has the highest intensities of magnetization; calcareous lutite, which is well-magnetized but has intensities less than that of the volcanic ash; and a sapropel layer, which is weakly magnetized with intensities of magnetization one order of magnitude less than that of the calcareous lutite. The specific susceptibility varies in the same manner with the highest values in the ash and the lowest in the sapropel. It is clear that the use of variation of intensity of magnetization to study the intensity variations of the earth's field is feasible only if variations due to changes in lithology can effectively be eliminated. This can be accomplished through using the Koenigsberger ratio $Q_n' = M/K$, as done by *Harrison* [1966], or by using saturation magnetization, or perhaps by anhys-

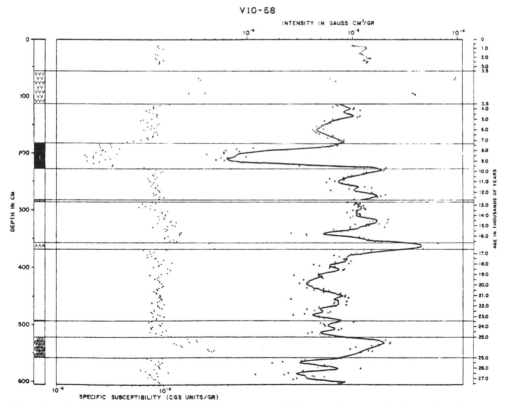

Fig. 19. Variation of lithology, magnetic susceptibility, and magnetic intensity in core V10-58 from the eastern Mediterranean.

teretic magnetization. In the case of the cores from the eastern Mediterranean, the specific susceptibility was measured [Opdyke et al., 1971]. The sections containing volcanic ash and the sapropel were eliminated, since the mode of magnetization in these sections is probably different from the calcareous lutite. A curve of changing intensity of magnetization against time is then made by combining data from V10-50 and V10-58. This curve matches well what we know of the variation of intensity over the last 8000 years [Bucha, 1970], which gives confidence that the change in Figure 17 represents the change in intensity of the earth's field over the last 28,000 years.

It also becomes clear that magnetic susceptibility offers a means of correlation in cores within individual basins where rates of sedimentation are high. The use of relative susceptibility as a correlative tool has also been demonstrated by Radhakrishnamurty et al. [1968].

GEOLOGIC APPLICATIONS OF MAGNETIC STRATIGRAPHY

Paleomagnetic stratigraphy has proved to be a development of major importance in marine geology. It is the perfect stratigraphic tool, since it permits the stratigrapher to draw time-synchronous horizons with known dates for sediments of the last 3.5 m.y. of earth history. This can be done between stratigraphic sections with no common fossils and of different facies [Goodell and Watkins, 1968; Hays et al., 1969; Opdyke and Foster, 1970].

Figure 20 shows the paleomagnetic correlation of cores from the Arctic, Pacific, Indian, and Atlantic oceans, all of which contain different fossil assemblages and have varying lithologies. All these cores contain a complete record of the Pleistocene and are correlated via the magnetics to the North Atlantic core of Berggren et al. [1967], in which the Pliocene/Pleistocene boundary has been identified on the basis of the evolution of Gl. truncatulinoides from Gl. tosaensis.

It is also possible to use magnetic stratigraphy to date evolution, any transitions [Berggren et al., 1967; Hays et al., 1969] and extinction [Opdyke et al., 1966; Watkins and Goodell, 1967; Hays, 1971].

Figure 21 demonstrates how it is possible to use magnetic stratigraphy to correlate radiolarian faunal zones from the Antarctic, Arctic, and tropical Pacific oceans [Hays, 1970]. These in turn may be correlated to the standard pelagic foraminiferal sequence [Hays et al., 1969]. The magnetic stratigraphy thus allows a secure correlation to be made between the different fossil groups even though they do not occur in the same regions and are preserved in different sediment types. It has also been used to date climatically significant sedimentary parameters such as ice-rafted debris [Opdyke et al., 1966; Conolly and Ewing, 1970] and volcanic ash [Ninkovich et al., 1966; Ninkovich, 1968; Kennett and Watkins, 1970]. It has also been possible to determine rates of sedimentation over large areas of the ocean floor. Figure 22 shows an isopach of rates of sedimentation within the basin of the North Pacific [Opdyke and Foster, 1970].

It seems clear that the magnetic study of marine cores will be an increasingly important tool in marine geology. It is certain that continued study will gradually reveal a detailed record of the behavior of the earth's magnetic field.

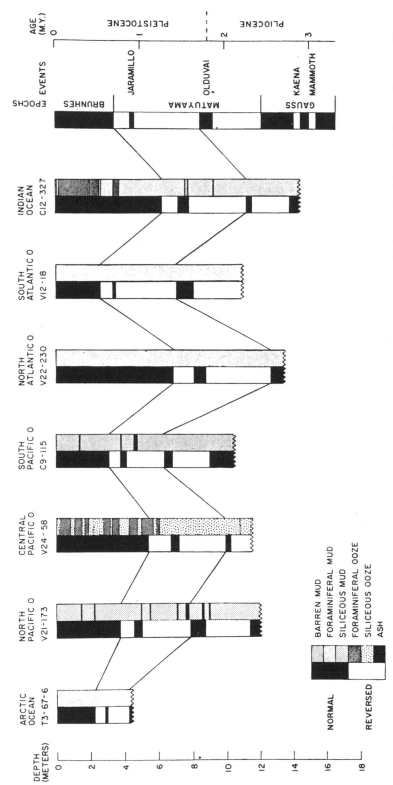

Fig. 20. Paleomagnetic correlation of cores from the Arctic, Pacific, Indian, and Atlantic oceans, all of which contain different fossil assemblages and have varying lithologies.

107

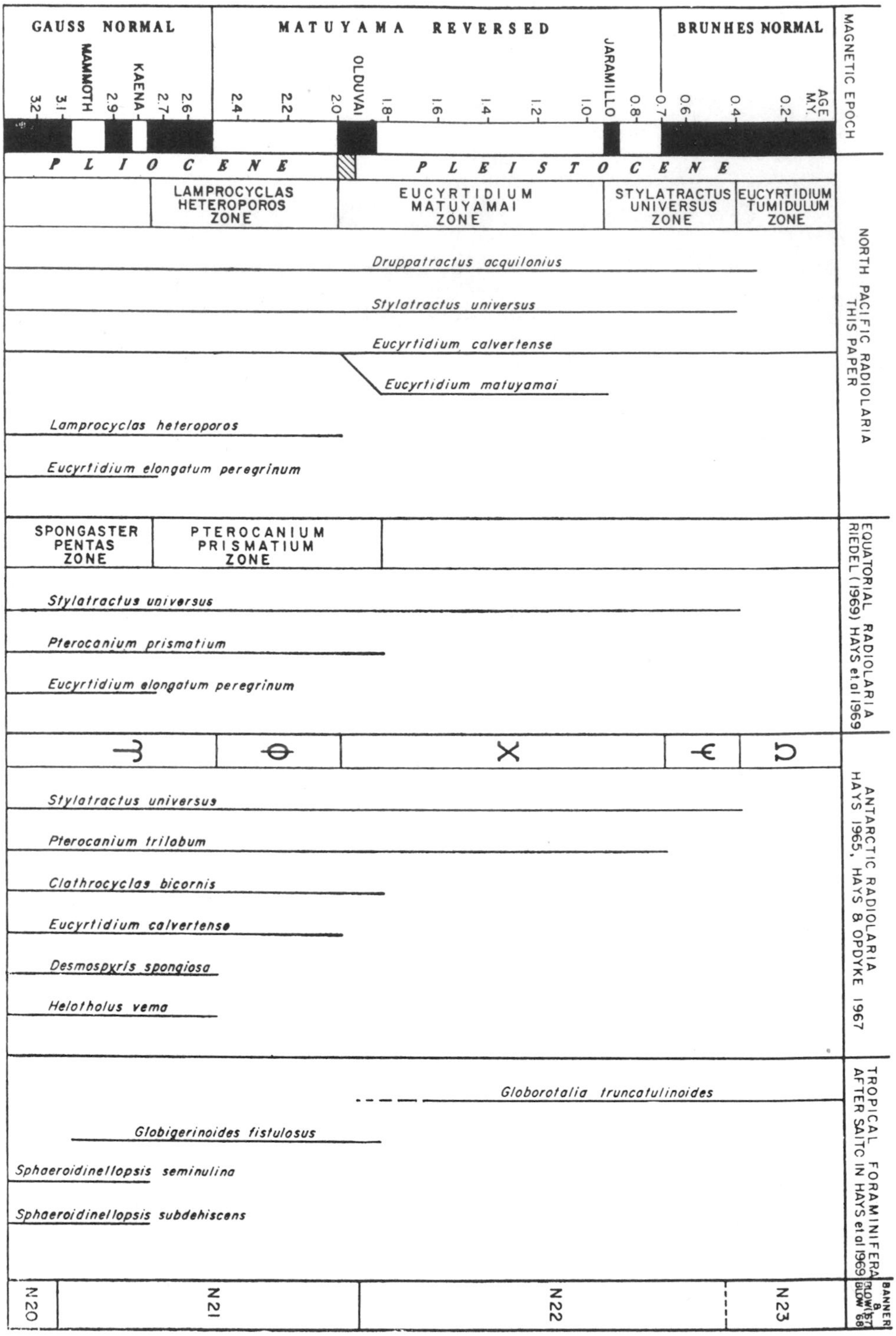

Fig. 21. Correlation of radiolarian faunal zones from the Antarctic, Arctic, and tropical Pacific oceans with magnetic stratigraphy.

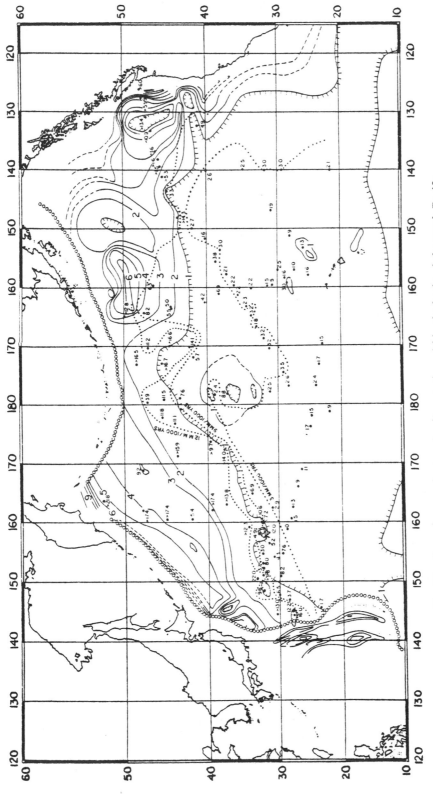

Fig. 22. Isopach map of rates of sedimentation within the basin of the north Pacific.

109

Acknowledgments. The writer would like to thank Drs. W. Lowrie, L. Burckle, and A. McIntyre, who critically read the manuscript and made many helpful comments.

This study has been supported by the National Science Foundation under grant GA-11050 and by Naval Research grant N00014-67-A-0108-0004.

Lamont-Doherty Geological Observatory Contribution 1734.

REFERENCES

Berggren, W. A., Cenozoic chronostratigraphy, planktonic foraminiferal zonation and the radiometric time scale, *Nature, 224*, 1070–1075, 1969.

Berggren, W. A., J. D. Phillips, A. Bertels, and D. Wall, Late Pliocene-Pleistocene stratigraphy in deep-sea cores from the south-central north Atlantic, *Nature, 216*, 253–255, 1967.

Bonhommet, N., and J. Babkine, Magnetisme terrestre—Sur la presence d'aimantations inversees dans la Chaine des Puys, *C.R.H. Acad. Sci., Ser. B, 264*, 92–94, 1967.

Bucha, V., Geomagnetic reversals in Quaternary revealed from a paleomagnetic investigation of sedimentary rocks. *J. Geomagn. and Geoelec., 22*, 253–272, 1970.

Cain, J. C., W. E. Daniels, S. J. Hendricks, and D. C. Jensen. An evaluation of the main geomagnetic field, 1940–1962. *J. Geophys. Res., 70*, 3647–3674, 1965.

Cassidy, W. A., B. Glass, and B. C. Heezen, Physical and chemical properties of Australasian microtektites, *J. Geophys. Res., 74*, 1008–1025, 1969.

Conolly, J. R., and M. Ewing, Ice-rafted detritus in northwest Pacific deep-sea sediments, in *Geological Investigations of the North Pacific, Mem. 126*, edited by J. D. Hays, pp. 219–231, Geological Society of America, Boulder, Colo., 1970.

Cox, A., Geomagnetic reversals, *Science, 163*, 237–245, 1969.

Cox, A., and G. B. Dalrymple, Statistical analysis of geomagnetic reversal data and the precision of potassium-argon dating, *J. Geophys. Res., 72*, 2603–2614, 1967.

Cox, A., R. R. Doell, and G. B. Dalrymple. Geomagnetic polarity epochs: Sierra Nevada II, *Science, 142*, 382–385, 1963.

Cox, A., R. R. Doell. and G. B. Dalrymple. Reversals of the earth's magnetic field, *Science, 144*, 1534–1543, 1964.

Dagley, P., and E. Lawley, Anomalous paleomagnetic directions and transition zones (abstract), *Eos Trans. AGU, 52*, 186, 1971.

Dickson, G. O., and J. H. Foster, The magnetic stratigraphy of a deep sea core from the north Pacific Ocean, *Earth Planet. Sci. Lett., 1*, 458–462, 1966.

Doell, R. R., and A. Cox, Paleomagnetism of Hawaiian lava flows. *J. Geophys. Res., 70*, 3377–3405, 1965.

Doell, R. R., and G. B. Dalrymple, Geomagnetic polarity epochs: A new polarity event and the age of the Brunhes-Matuyama boundary, *Science, 152*, 1060–1061. 1966.

Dymond, J., Age determinations of deep-sea sediments: A comparison of three methods, *Earth Planet. Sci. Lett., 6*, 9–14, 1969.

Emilia, D. A., and D. F. Heinrichs, Ocean floor spreading: Olduvai and Gilsa events in the Matuyama epoch, *Science, 166*, 1267–1269. 1969.

Ericson, B. D., M. Ewing, G. Wollin, and B. C. Heezen, Atlantic deep-sea sediment cores, *Geol. Soc. Amer. Bull., 72*, 193–286, 1961.

Evans, A. L., Geomagnetic polarity reversals in a late Tertiary lava sequence from the Akaroa Volcano, New Zealand, *Geophys. J., 21*, 163–183, 1970.

Ewing, M., D. E. Hayes, and E. M. Thorndike, Corehead camera for measurement of currents and core orientation, *Deep-Sea Res., 14*, 253–258, 1967.

Foster, J. H., A paleomagnetic spinner magnetometer using a fluxgate gradiometer, *Earth Planet. Sci. Lett., 1*, 463–466, 1966.

Foster, J. H., and N. D. Opdyke, Upper Miocene to Recent magnetic stratigraphy in deep-sea sediments, *J. Geophys. Res., 75*, 4465–4473, 1970.

Glass, B., Microtektites in deep-sea sediments, *Nature, 214*, 372–374, 1967.

Goodell, H. G., and N. D. Watkins. The paleomagnetic stratigraphy of the Southern Ocean: 20° west to 160° east longitude. *Deep-Sea Res., 15*, 89–112, 1968.

Grommé, C. S., and R. L. Hay, Geomagnetic polarity epochs: Age and duration of the Olduvai normal polarity event, *Nature, 10,* 179–185, 1971.

Haggerty, S. E., Magnetic minerals in pelagic sediments, *Carnegie Inst. Wash. Yr. Book 68,* 332–336, 1970.

Harrison, C. G. A., The paleomagnetism of deep-sea sediments, *J. Geophys. Res., 71,* 3033–3043, 1966.

Harrison, C. G. A., and B. M. Funnell, Relationship of paleomagnetic reversals and micro-palaeontology in two late Cenozoic cores from the Pacific Ocean, *Nature, 204,* 566, 1964.

Harrison, C. G. A., and B. L. K. Somayajulu, Behaviour of the earth's magnetic field during a reversal, *Nature, 212,* 1193–1195, 1966.

Hays, J. D., Stratigraphy and evolutionary trends of radiolaria in north Pacific deep-sea sediments, in *Geological Investigations of the North Pacific, Mem. 126,* edited by J. D. Hays, pp. 185–218, Geological Society of America, Boulder, Colo., 1970.

Hays, J. D., Extinctions and reversals of the earth's magnetic field, *Geol. Soc. Amer. Bull., 82,* 2433–2448, 1971.

Hays, J. D., and N. D. Opdyke, Antarctic Radiolaria, magnetic reversals and climatic change, *Science, 158,* 1001–1011, 1967.

Hays, J. D., T. Saito, N. D. Opdyke, and L. H. Burckle, Pliocene-Pleistocene sediments of the equatorial Pacific: Their paleomagnetic, biostratigraphic, and climatic record, *Geol. Soc. Amer. Bull., 80,* 1481–1514, 1969.

Heirtzler, J. R., G. O. Dickson, E. M. Herron, W. C. Pitman III, and X. Le Pichon, Marine magnetic anomalies, geomagnetic field reversals, and motion of the ocean floor and continents, *J. Geophys. Res., 73,* 2119–2136, 1968.

Johnson, E. A., T. Murphy, and O. W. Torreson, Pre-history of the earth's magnetic field, *Terr. Magn. Atmos. Elec., 53,* 349–372, 1948.

Keen, M. J., The magnetization of sediment cores from the eastern basin of the north Atlantic Ocean, *Deep-Sea Res., 10,* 607–622, 1963.

Kennett, J. P., and N. D. Watkins, Geomagnetic polarity change, volcanic maxima and faunal extinction in the south Pacific, *Nature, 227,* 930–934, 1970.

Kennett, J. P., N. D. Watkins, and P. Vella, Paleomagnetic chronology of Pliocene-early Pleistocene climates and the Plio-Pleistocene boundary in New Zealand, *Science, 171,* 276–279, 1971.

Kobayashi, K., K. Kitazawa, T. Kanaya, and T. Sakai, Magnetic and micropaleonotological study of deep-sea sediments from the west-central equatorial Pacific, *Deep-Sea Res.,* in press, 1971.

Ku, T., W. Broecker, and N. D. Opdyke, Comparison of sedimentation rates measured by paleomagnetic and the ionium methods of age determination, *Earth Planet. Sci. Lett., 4,* 1–16, 1968.

Lin'kova, T. I., Some results of paleomagnetic study of Arctic Ocean floor sediments, translated by E. R. Hope from *Natoyascheye i Proshloye Magnitnogo Polia Zemli,* pp. 279–281, Nauka, Moscow, 1965, Directorate of Scientific Information Services, *Publ. T 463 R,* Canada, 1966.

McDougall, I., and F. H. Chamalaun, Geomagnetic polarity scale of time, *Nature, 212,* 1415–1418, 1966.

McDougall, I., and H. Wensink, Paleomagnetism and geochronology of the Pliocene-Pleistocene lavas in Iceland, *Earth Planet. Sci. Lett., 1,* 232–236, 1966.

Nakagawa, H., N. Niitsuma, and I. Hayasaka, Late Cenozoic geomagnetic chronology of the Boso Peninsula, *J. Geol. Soc. Jap., 75,* 267, 1969.

Ninkovich, D., Pleistocene volcanic eruptions in New Zealand recorded in deep-sea sediments, *Earth Planet. Sci. Lett., 4,* 89–102, 1968.

Ninkovich, D., N. D. Opdyke, B. C. Heezen, and J. H. Foster, Paleomagnetic stratigraphy, rates of deposition and tephrachronology in north Pacific deep-sea sediments, *Earth Planet. Sci. Lett., 1,* 476–492, 1966.

Opdyke, N. D., The Jaramillo event as detected in oceanic cores, in *The Application of*

Modern Physics to Earth and Planetary Interiors, edited by S. K. Runcorn, pp. 549–552, John Wiley, New York, 1969.

Qpdyke, N. D., and J. H. Foster, Paleomagnetism of cores from the north Pacific, in *Geological Investigations of the North Pacific, Mem. 126*, edited by J. D. Hays, pp. 83–119, Geological Society of America, Boulder, Colo., 1970.

Opdyke, N. D., and K. W. Henry, A test of the dipole hypothesis, *Earth Planet. Sci. Lett., 6*, 139–151, 1969.

Opdyke, N. D., and J. D. Phillips, Paleomagnetic stratigraphy of sites 1–7 (leg 1), preliminary report, in *Initial Reports of the Deep Sea Drilling Project*, vol. 1, edited by M. Ewing et al., pp. 501–519, U.S. Government Printing Office, Washington, D.C., 1969.

Opdyke, N. D., B. Glass, J. D. Hays, and J. Foster, Paleomagnetic study of Antarctic deep-sea cores, *Science, 154*, 349–357, 1966.

Opdyke, N. D., L. H. Burckle, J. D. Hays, and T. Saito, Extension of the magnetic stratigraphy to the middle Miocene in deep-sea sediments (abstract), *Abstr. Program Geol. Soc. Amer., 2*, 642, 1970.

Opdyke, N. D., D. Ninkovich, and W. Lowrie, The paleomagnetism of two deep-sea cores from the eastern Mediterranean (abstract), *Eos Trans. AGU, 52*, 186, 1971.

Phillips, J. D., W. A. Berggren, A. Bertels, and D. Wall, Paleomagnetic stratigraphy and micropaleontology of three deep-sea cores from the central north Atlantic Ocean, *Earth Planet. Sci. Lett., 4*, 118–130, 1968.

Pitman, W. C. III, and J. R. Heirtzler, Magnetic anomalies over the Pacific-Antarctic ridge, *Science, 154*, 1164–1171, 1966.

Radhakrishnamurty, C., S. Likhite, B. S. Amin, and B. L. K. Somayajulu, Magnetic susceptibility stratigraphy in ocean sediment cores, *Earth Planet. Sci. Lett., 4*, 464–468, 1968.

Smith, J. D., and J. H. Foster, Geomagnetic reversal in Brunhes normal polarity epoch, *Science, 163*, 565–567, 1969.

Stacey, F. D., *Physics of the Earth*, 324 pp., John Wiley, New York, 1969.

Talwani, M., C. C. Windisch, and M. G. Langseth, Jr., Reykjanes ridge crest: A detailed geophysical study, *J. Geophys. Res., 76*, 473–517, 1971.

van Zijl, J. S. V., K. W. T. Graham, and A. L. Hales, Evidence for a genuine reversal of the earth's field in Triassic-Jurassic times, 1, *Geophys. J., 7*, 23–39, 1962.

Vine, F. J., Magnetic anomalies associated with mid-ocean ridges, in *The History of the Earth's Crust*, edited by R. A. Phinney, pp. 73–89, Princeton University Press, Princeton, N.J., 1968.

Watkins, N. D., Short-period geomagnetic polarity events in deep-sea sedimentary cores, *Earth Planet. Sci. Lett., 4*, 341–349, 1968.

Watkins, N. D., Non-Dipole behaviour during an upper Miocene geomagnetic polarity transition in Oregon, *Geophys. J., 17*, 121–149, 1969.

Watkins, N. D., and H. G. Goodell, Geomagnetic polarity change and faunal extinction in the Southern Ocean, *Science, 156*, 1083–1087, 1967.

Wilson, R. L., Permanent aspects of the earth's non-dipole magnetic field over upper Tertiary times, *Geophys. J., 19*, 417–437, 1970.

Reprinted from *Nature* **204**:566 (1964)

RELATIONSHIP OF PALAEOMAGNETIC REVERSALS AND MICROPALAEONTOLOGY IN TWO LATE CAENOZOIC CORES FROM THE PACIFIC OCEAN

C. G. A. Harrison and B. M. Funnell

RECENT investigation has revealed palæomagnetic reversals preserved in the clayey radiolarian ooze of two deep-sea cores from the equatorial Pacific Ocean. The cores are:

MSN 12G; position 3° 01′ N., 174° 02′ W.; depth, 5,230 m.

MSN 142G; position 5° 20′ N., 146° 13′ W., depth, 5,089 m.

The cores are illustrated, their lithologies briefly described, and Radiolaria listed by Riedel and Funnell[1].

Palæomagnetism. The uppermost parts of both cores are thought to be normally magnetized with respect to the present Earth's field. As the cores come from equatorial positions and were not orientated on collection it is not possible to be certain that this is so. There is, however, no evidence of erosion or non-deposition at or near the tops of either sequence. The first reversal occurs in *MSN 12G* between 74 and 95 cm, and in *MSN 142G* between 78 and 87 cm. Other reversals occur below these levels in both cores. Assuming that the uppermost magnetic vector is normal, the magnetization of the cores is as follows:

MSN 12G; 6–74 cm (N, 5 samples), 95–116 cm (*R*, 7 samples), 119, 123 cm (N), 133 cm (*R*), 135 cm (N), 138–146 cm (*R*, 7 samples).

MSN 142G; 15–78 cm (N, 8 samples), 87–107 cm (*R*. 3 samples), 116, 125 cm (N), 134, 142 cm (*R*).

The palæomagnetic results below the first reversed zone (*R*1) do not seem to be consistent with the hypothesis that reversals of the Earth's magnetic field have occurred about every million years, unless extremely slow ratios of deposition are assumed for the lower portions of these cores. However, recent results from continental rocks[2-5] are also inconsistent with this hypothesis, and it seems likely that below the *R*1 zone reversals occurred more rapidly than was formerly suggested.

Micropalæontology. In *MSN 12G*, Radiolaria between 0 and 43 cm are similar to Recent assemblages and indicate a Quaternary age. Specimens of *Pterocanium prismatium* occur rarely at 90–92 cm, and commonly from 127 cm to the bottom of the core. Some re-worked Eocene, Miocene and possibly Oligocene forms are present throughout. In *MSN 142G*, a Quaternary age is indicated by the Radiolaria between 0 and 76 cm, but from 94 cm to the bottom of the core *Pterocanium prismatium* is rather common.

In this core some re-worked Miocene forms are present throughout.

In both cores examined the first reversal appears to correspond more or less exactly with the upper limit of occurrence of *Pterocanium prismatium*. Lithologically there does not seem to be any break in accumulation at this level in *MSN 12G*, which consists throughout of a vaguely mottled, brown clayey siliceous ooze, but in *MSN 142G* there is a burrow-mottled gradation, between buff or brown ooze above and grey-brown ooze below, at approximately 85 cm.

In recent investigations of the palæomagnetic polarity of basalts from the continents and oceanic islands, normal magnetization has been found to characterize all rocks up to 0·98 m.y. B.P., prior to which a period of reversed magnetism (*R*1) is found extending back to 1·9 m.y.[2]. The reversal at 0·98/0·99 m.y. would appear to fall within the Quaternary, as defined in accordance with the recommendations of the International Geological Congress 1948 (refs. 4 and 6). If, as seems most likely, the first reversal recorded in the cores corresponds to that found on land, an intra-Quaternary age of 0·98/0·99 m.y. is implied for the level of the reversal in the cores.

The upper limit of occurrence of *Pterocanium prismatium* has hitherto been regarded as approximating to the Tertiary–Quaternary boundary[1,7]. In the Pacific it corresponds rather closely with the disappearance of *Discoaster* spp. from the calcareous nannoplankton, a phenomenon which has also been attributed to the Tertiary–Quaternary boundary in the Atlantic[8].

The association of a first magnetic reversal with the upper limit of occurrence of *Pterocanium prismatium* in the two Pacific cores suggests that this level has an age of very slightly less than 1 m.y. and that it is intra-Quaternary in terms of current stratigraphical definitions.

The work of one of us (C. G. A. H.) was supported by American Chemical Society petroleum research grant No. 700 A.

[1] Riedel, W. R., and Funnell, B. M., *Quart. J. Geol. Soc. Lond.*, **120**, 305 (1964).
[2] Cox, A., Doell, R. R., and Dalrymple, G. B., *Science*, **142**, 382 (1963).
[3] Cox, A., Doell, R. R., and Dalrymple, G. B., *Science*, **143**, 351 (1964).
[4] Evernden, J. F., Savage, D. E., Curtis, G. H., and James, G. T., *Amer. J. Sci.*, **262**, 145 (1964).
[5] McDougall, I., and Tarling, D. H., *Nature*, **200**, 54 (1963).
[6] Funnell, B. M., *Quart. J. Geol. Soc.*, *Lond.* (*Suppl. Vol.*), **120S** (in the press).
[7] Riedel, W. R., Bramlette, M. N., and Parker, F. L., *Science*, **140**, 1238 (1963).
[8] Ericson, D. B., Ewing, M., and Wollin, G., *Science*, **139**, 727 (1963).

6

Reprinted from *Science* 154:349–357 (1966)

Paleomagnetic Study of
Antarctic Deep-Sea Cores

N. D. Opdyke, B. Glass, J. D. Hays, J. Foster

That the earth's magnetic field has changed polarity is now accepted as a fact by most geophysicists. Self reversal of certain magnetic minerals have also been shown to be possible, but instances in which self-reversal mechanisms have been proven continue to be rare. Cox *et al.* (*1*) and McDougall and Tarling (*2*), by carrying out simultaneous paleomagnetic and radiometric studies of Pliocene and Pleistocene lava flows, have been able to trace the changes in polarity of the earth's field through the last 4 million years. These they have named in order of increasing age (in millions of years): the Brunhes normal epoch, 0 to 0.7; the Matuyama reversed epoch, 0.7 to 2.4 ± 0.1; the Gauss normal epoch, 2.4 ± 0.1 to 3.35. The Gilbert reversed epoch ended at 3.35 million years and its beginning is not known. Within the Matuyama reversed epoch, two short periods of normal polarity occurred at about 0.9 and 1.9 million years, and these have been termed the Jaramillo and the Olduvai events, respectively. Within the Gauss normal epoch there was a short period of reversed polarity at 3.0 million years, called the Mammoth event. This magnetic stratigraphy is shown schematically in the left-hand column of Fig. 1.

Although most paleomagnetic studies have been of continental rocks, a few have been made of oceanic sediments by McNish and Johnson (*3*), Johnson *et al.* (*4*), Keen (*5*), Harrison and Funnell (*6*), and Fuller *et al.* (*7*). Of these only Harrison and Funnell and Fuller *et al.* found reversals of polarity.

The purpose of this study is to determine whether the direction of remanent magnetism in deep-sea cores can be used for purposes of stratigraphic correlation. Several long cores from high

Dr. Opdyke is senior research associate, Mr. Glass and Mr. Foster are research assistants, and Dr. Hays is a research associate at Lamont Geological Observatory of Columbia University, Palisades, New York.

southern latitudes were chosen because at these latitudes the inclination of the magnetic field is sufficiently steep that cores oriented in azimuth are unnecessary; all one needs to know is which end of the core is the top. Also the paleontological stratigraphy based on Radiolaria is known, and this gives no indication of large hiatuses.

All cores used in this study are from the Antarctic region (Fig. 1; Table 1). Three of them were taken in the Bellingshausen Basin (V16-132, -133, and -134), one from near Drake Passage (V18-72), and three from the south Indian Ocean (V16-57, -60, and -66). Three of the cores (V16-133 and -134 and V18-72) were taken in an area of very moderate relief, whereas V16-57, -60, -66, and -132 were taken in areas of rugged relief. V16-60 and -132 were taken from the flanks of hills; V16-57 and -66, near the crests of hills.

Four of these cores (V16-132, -133, -134, and -60) consist of diatomaceous silty lutites in their upper portions and are silty lutites below (Fig. 1). V16-57, on the other hand, is more diatomaceous near the bottom, in fact the lowest 650 centimeters is a diatom ooze. V18-72 is only slightly diatomaceous in its upper part and contains silt and sand laminae, especially near the bottom. One core (V16-66) is a foraminiferal ooze. All the cores contain some ice-rafted material, and all except V16-60 and -66 contain manganese micronodules.

Laboratory Procedure

The cores were taken with a piston corer. Since they were cut into 10-foot (3.05-meter) sections aboard ship and the sections are not oriented relative to each other, the variation of the declination is not significant.

The cores were sampled every 10 centimeters, where possible, for a total of about 650 samples. The samples were cut into rough cubes measuring

approximately 2 centimeters. If one assumes sedimentation rates between 2 and 10 millimeters per 1000 years, a sample of this size covers a period of from 10,000 to 2000 years. Each was measured on a 5-cycle-per-second spinner magnetometer, an instrument that uses a commercially available two-channel, phase-sensitive, synchronous detector. For each of six spins a measurement is made on two orthogonal components of the projection of the magnetic moment upon a plane normal to the spin axes. The time for a complete six-spin measurement of a 10 cubic-centimeter specimen runs from 2 minutes, for a specimen with a moment of 1×10^{-5} electromagnetic units per cubic centimeter at a 1-second output-meter time constant, to 20 minutes for a moment of 1×10^{-7} electromagnetic units per cubic centimeter at a 24-second time constant. The 5-cycle-per-second rotation speed is very advantageous when one is working with mechanically weak specimens, which do not disintegrate at such speed.

In order to test for magnetic stability, alternating-field partial demagnetization was carried out in alternating-current fields of increasing strength (*8*). When a normal and a reversed specimen from core V16-133 were subjected to this treatment, the inclination in both specimens became steeper on application of low fields and remained stable even in high fields (Fig. 2).

The shapes of the alternating-field demagnetization curves (Fig. 3) vary widely. In all specimens there was an unstable component that was removed by fields of up to 150 oersteds. The rate of decrease of intensity with fields above this value decreases sharply, indicating that a stable component is present in most specimens. In some specimens having natural remanent magnetism directions that were almost horizontal, the sign of the inclination changes on alternating-field demagnetization. It seems probable that these unstable horizontal components were often acquired when the core was drying in a horizontal position; the direction of magnetization of specimens is known to change on drying, but it seems unlikely that this factor causes specimens that are steeply magnetized to change sign.

Each specimen from cores V16-57 and -134 was partially demagnetized in an alternating field of 150 oersteds; as a result of magnetic cleaning, the dispersion of the inclination in both cores decreased (Fig. 4).

Selected specimens from the other cores also were magnetically cleaned, particularly specimens from near reversals and specimens having shallow inclinations. In several instances these specimens changed sign on alternating-field demagnetization; in some cores this change served to move the boundaries between normal and reversed sections of the core (usually about 10 or 20 centimeters).

Magnetic Stratigraphy

A zone of normally magnetized sediment is present at the top of all cores studied (Fig. 1); we believe this represents the Brunhes normal epoch, which has been defined on the basis of K-A dating by Cox et al. (9). The K-A based stratigraphy is shown in the left-hand column of Fig. 1. We emphasize that the magnetic stratigraphy obtained from deep-ocean cores is based on the classical methods of stratigraphy—the law of superposition.

We shall hereafter refer to the sediments that we believe represent the Brunhes normal–polarity epoch as the Brunhes normal–polarity series. In all the cores a zone of reversed polarity occurs below the Brunhes normal series; we assume that it represents the Matuyama reversed epoch, and we shall refer to it as the Matuyama reversed

series. The change in polarity between the two zones is very abrupt in all cores (Fig. 4), with the possible exception of V16-66 in which a zone is very weakly magnetized between 300 and 400 centimeters; unfortunately the change in magnetic polarity takes place within this interval, so that the exact

Table 1. Sources of cores. Be, Bellingshausen; Ba, Basin; P, passage; AIR, Atlantic-Indian Ridge; MR, Madagascar Ridge.

Core	Coordinates		Water depth (m)	Region
V16-132	60°44.5′S,	107°29′W	4898	BeBa
V16-133	61°56.5′S,	95°03′W	5062	BeBa
V16-134	61°54.0′S,	91°15′W	5138	BeBa
V18-72	60°29.0′S,	75°57′W	4695	DrakeP
V16-57	45°14.0′S,	29°29′E	5289	Agulhas Ba
V16-60	49°59.5′S,	36°45.5′E	4574	AIR, S. flank
V16-66	42°39.0′S,	45°40′E	3072	S. MR

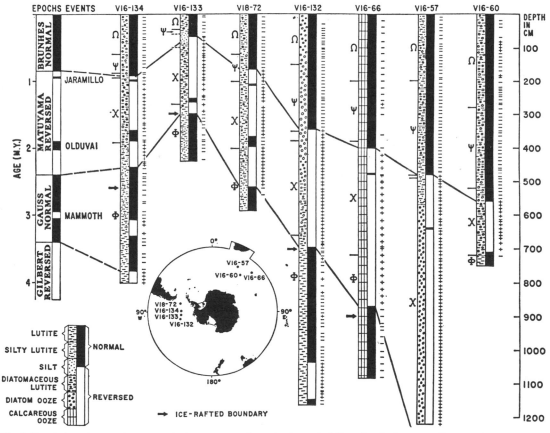

Fig. 1. Correlation of magnetic stratigraphy in seven cores from the Antarctic. Minus signs indicate normally magnetized specimens; plus signs, reversely magnetized. Greek letters denote faunal zones (17). Magnetic stratigraphy in left-hand column is from (1, 2). Inset: sources of cores.

position of the change in this core is in some doubt.

Within the Matuyama series, short intervals of normal polarity occur both at the top and toward the bottom of the series. The short normal interval near the top in V16-134, V18-72, V16-66, and V16-57 is presumed to be the Jaramillo event—recently named by Doell and Dalrymple (*10*). The interval of normal polarity toward the bottom of the series in cores V16-134, V16-133, and V18-72 we correlate with the Olduvai event.

None of these zones of normal polarity in the Matuyama series have been found in V16-132. In V16-60 the bottom of the core is normally magnetized, and it is uncertain whether this normal polarity represents the Olduvai event or the beginning of the Gauss normal epoch.

All the cores with the exception of V16-57, and possibly V16-60, have a normal interval below the Matuyama series, which we believe is the Gauss normal series. Within the Gauss normal series in V16-134 there is a short interval of reversed polarity, between 615 and 665 centimeters, which we correlate with the Mammoth event. In V16-132 the zone of reversed polarity between 1035 and 1145 centimeters we also interpret as correlative with the Mammoth event. V16-134 is the only core to pass through the Gauss normal series to an underlying reversed interval that we correlate with the Gilbert reversed epoch.

Two cores, V16-60 and -132, have intervals in which the polarity data are inconsistent. V16-60 has a long, dominantly normal zone from the top to 500 centimeters, but single specimens or groups of specimens have reversed directions at 10, 100, and 180 centimeters. In core V16-132 the directions are consistent down to 700 centimeters; lower, the specimens are weakly magnetized and individual specimens are polarized reversely to those immediately above and below them. Originally the change from the Matuyama to the Gauss series was placed at 800 centimeters.

On alternating-field demagnetization the polarity of this zone changed from dominantly positive to dominantly negative, moving the boundary up to 700 centimeters. In general, after alternating-current cleaning, the results were more consistent, but all anomalous determinations were not removed. The meaning of these apparently false re-

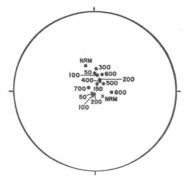

Fig. 2. Stereographic projection of directions of magnetization of two specimens from V16-133. Crosses represent negative inclinations; solid circles, positive. Numbers represent demagnetizing fields in oersteds, *NRM*, natural remanent magnetism.

versals is not clear, but they may be caused by one or more of the following factors: (i) magnetic instability (which in the case of V16-132 may be caused by partially decayed manganese micronodules), (ii) physical disturbance by the coring operation or extrusion of the core, (iii) errors in orienting sections of the core or in labeling samples, (iv) slumping or other bottom disturbances, or (v) the presence of previously unrecorded events.

The magnetic inclination in V16-57 and -134 was examined in some detail, and all specimens from these cores were partially demagnetized in a field of 150 oersteds. The magnetic inclination in V16-57 is very consistent (Fig. 4); the top 480 centimeters is normally magnetized, with no exceptions; at 480

centimeters a sharp reversal of sign occurs, and from there to the bottom of the core the specimens are reversely magnetized, with the exception of one normally magnetized specimen from 640 centimeters. The mean inclination of the top 480 centimeters, which is normally magnetized, is 65 degrees. In the lower part of the core, which is reversely magnetized, the mean inclination is 63 degrees; the ambient field at this location is about 66 degrees. Creer (*11*), averaging the earth's field around lines of latitude, found the mean field at 45 degrees south latitude to be 63 degrees. The inclination of the axial dipole field at this latitude is 63 degrees. Therefore the mean observed inclination in the core has a value that agrees with its present magnetic and geographic latitude.

In V16-134, as in V16-57, the dispersion of the inclination decreased on magnetic cleaning. The mean inclination of all the normally magnetized specimens is 71 degrees, and the same is true of the reversed specimens. The core was taken from 63 degrees south latitude where the ambient field is about 61 degrees. Creer (*11*) averaged the inclination of the ambient field around latitude 60 degrees south and obtained a value of 71.8 degrees, to which the numerical average obtained from both the normal and reversed sections of V16-134 is very close. The mode of the measured inclination is nearly 75 degrees, very close to the inclination of the axial dipole field—73.9 degrees at 60 degrees south. This fact suggests that the dipole field may average to an axial dipole field.

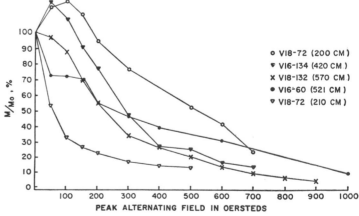

○ VI8-72 (200 CM)
▼ VI6-134 (420 CM)
✕ VI8-132 (570 CM)
● VI6-60 (521 CM)
▽ VI8-72 (210 CM)

Fig. 3. Alternating-field demagnetization curves of specimens from five cores. *M/Mo* is the ratio of remaining to initial intensity after treatment in the corresponding field.

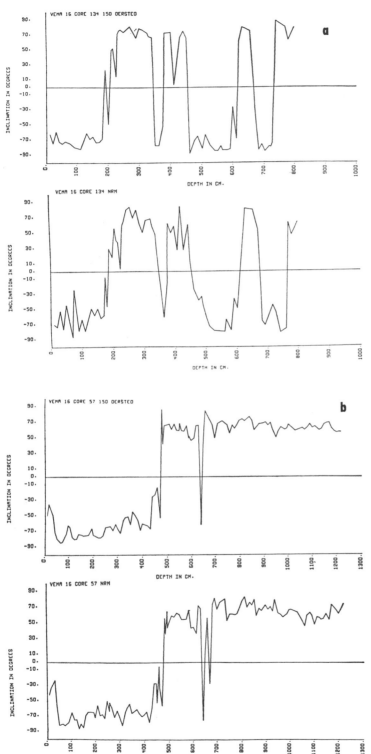

Fig. 4 (left). Plot of inclination versus depth in cores V16-134 (a) and V16-57 (b). *NRM*, natural remnant magnetism.

Earlier workers found that the inclination of the magnetic vector in sediments was consistently less than the inclination of the ambient field (4), but their samples were not treated in alternating fields. Our untreated samples also showed a relatively low inclination, but after treatment in alternating fields the inclination was in good agreement with the average field for the latitude of the core site. Keen (5), in his study of North Atlantic cores, observed a decrease of inclination with depth, which he attributed to compaction; we found no systematic decrease of inclination with depth in Antarctic cores.

The intensity of magnetization of our cores is usually in the range of 10^{-5} or 10^{-6} electromagnetic units per cubic centimeter. A plot of intensity versus depth, after partial alternating-current demagnetization at 150 oersteds, in cores V16-134 and V16-57 appears in Fig. 5; each specimen was weighed and the intensity is given in electromagnetic units per gram. There is no apparent correlation between small changes in lithology and intensities of magnetization, but there is some indication that the diatomaceous ooze in the bottom of V16-57 has higher intensities than the diatomaceous lutite near the top. Also there is a difference in intensity between cores of very different lithology. The core with the lowest intensities of magnetization is V16-66, a foraminiferal ooze; the other cores are dominantly diatomaceous lutites and have similar intensities.

Positive correlation exists between the reversals and low magnetic intensities. At every reversal in cores V16-57 and -134 there is a drop in intensity (Fig. 5). The lowest intensity measured in V16-57 occurs near the polarity change at 475 centimeters. In core V16-134 also, the lowest measured intensities are near the points of reversal, as one may expect if it is true (as has been postulated by use of the dynamo theory) that when reversal of the field occurs the intensity of the dipole field decreases to zero and builds again with the opposite polarity (*12*).

Keen (5) ascribed the magnetization of cores studied by him from the eastern basin of the North Atlantic to the presence of detrital magnetite. Harrison and Peterson (*13*) after study of

117

a magnetic mineral found in two cores from the Indian Ocean concluded that it was between magnetite and maghemite in structure, and that the magnetic properties of the sediment suggested formation *in situ*. When magnetic separates made from our cores were examined with an x-ray difractometer, magnetite lines were present in all. Curie-point determinations were made on specimens from several of these cores by Dickson (*14*), and were found to be in the region of about 600°C, close to the curie point of magnetite; however, other magnetic minerals may be present. Some of the magnetite occurs as inclusions in fragments of igneous rocks and therefore appears to be dominantly detrital in character, although the presence of authigenic magnetite cannot be excluded.

Radiolarian Stratigraphy

Four faunal zones have been established for Antarctic deep-sea cores on the basis of Radiolaria (*15*). These zones are designated, from oldest to youngest, Φ, X, Ψ, and Ω and are based primarily on the upward sequential decrease of radiolarian species (Fig. 6, a and b). This sequence has been observed in more than 80 Antarctic deep-sea cores. It has been considered that the upper three zones (X, Ψ, Ω) represent Quaternary sedimentation and that the Φ zone represents the late Tertiary (*15*).

Teh-lung Ku determined the average rate of sedimentation for the Ω zone in an Antarctic core (Eltanin 11-11) by measuring excess Th^{230} (*16*). Thus it was possible to determine the age of the Ψ-Ω boundary as about 400,000 years. Extrapolation through the Ψ and X zones, on the assumption that these zones were similar in rate of sedimentation to the Ω zone, gave estimated ages for the X-Ψ and Φ-X boundaries of 0.9 ± 0.18 and 1.6 ± 0.32 million years, respectively.

In the seven cores studied the magnetic reversals and faunal boundaries are consistently related to each other, an indication that they are both time-dependent phenomena (Fig. 1). By use of the dates of changes in polarity given by Cox *et al.* (*1*), rates of sedimentation can be calculated (Table 2), and from them the age of the Ψ-Ω boundary is calculated as between 0.4 and 0.5 million years. The X-Ψ boundary would have an age of about 0.7

million years; the Φ-X boundary, about 2.0 million years. These ages agree well with the estimates for these boundaries from excess Th^{230}. The fact that two independent methods of dating give similar ages for the faunal zones lends credence to both methods.

The Φ-X boundary was tentatively correlated by Hays (*15*) with the Pliocene-Pleistocene boundary of Ericson *et al.* (*17*), which is based on several paleontological criteria. Ericson *et al.* (*18*) estimated the age of the Φ-X boundary at about 1.5 million years. Riedel *et al.* (*19*) approximately correlated a faunal boundary in the Pacific,

based on the last appearance of the Radiolaria *Pterocanium prismatium* and various nanoplankton, with the boundary of Ericson *et al.* (*17*). The work of Harrison and Funnel (*6*) suggests that the last appearance of *P. prismatium* occurred at or about the time of the polarity change from the Matuyama reversed epoch to the Brunhes normal, about 0.7 million years ago. If this was so, the faunal boundary of Riedel *et al.* (*19*) would correspond in time with the X-Ψ boundary in Antarctic cores—not with Φ-X boundary as suggested by Hays (*15*).

To finally resolve the problem of

Table 2. Sedimentation rates of seven cores.

Interval (10^6 yr)	Rates (mm/10^3 yr)						
	V16-134	V16-133	V18-72	V16-132	V16-66	V16-57	V16-60
0.7 to 0.0	2.6	1.1	2.4	5.0	5.8	6.8	8.0
Olduvai to 0.7	1.5	1.6	1.7				
2.4 to 0.7	1.5	1.3	1.9	1.9	2.6	4.2	
Mammoth to 0.7	2.5						
3.5 to 2.4	2.9			3.0			
3.5 to 0.7	2.2			2.5			
Averages							
	2.3	1.2	2.1	3.2	4.1	6.6	8.0

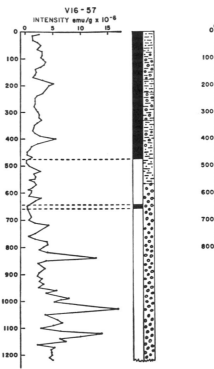

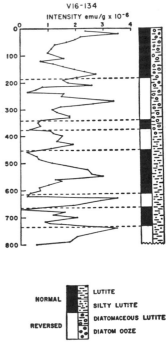

Fig. 5. Plot of intensity versus depth in cores V16-57 and -134 after alternating-current demagnetization at 150 oersteds. *emu*, Electromagnetic units.

the relation of various deep-sea stratigraphic boundaries, the paleomagnetic stratigraphy will have to be worked out for several cores containing the boundaries of Ericson *et al.* (*17*) and of Riedel *et al.* (*19*), which may then be compared with the Antarctic magnetic and faunal sequences.

The lithologic change from tan lutite below to diatomaceous sediments above, which occurs in Antarctic cores at about the same level as the Φ-X boundary, has an age of about 2 million years. This change was interpreted by Hays (*15*) to indicate the time when the vertical circulation of the Antarctic Ocean was stimulated by the initiation of large-scale freezing of sea ice around Antarctica, associated with glaciation of the continent.

Connolly and Ewing (*20*) found the first occurrence of ice-rafted debris in five Antarctic cores just below the Φ-X boundary, which they interpreted as indicating the earliest Pleistocene glaciation in the southern hemisphere. Ice-rafted detritus, as described by Connolly and Ewing (*20*), occurs in all our cores and a lower limit can be recognized in four (Fig. 1). Two (V16-66 and -132) of the four cores in which a lower limit of ice-rafted detritus can be recognized were studied by Connolly and Ewing (*20*); the lower limit of ice-rafted detritus occurred at 900 and 700 centimeters, respectively. In V16-134 the lower limit of ice-rafted debris is indistinct, but no such debris was observed below 500 centimeters. V16-133 has an abrupt lower limit of ice-rafted material between 300 and 310 centimeters. V16-57 and -60 contain ice-rafted detritus throughout. The sandy nature of core V18-72 prevented positive recognition of ice-rafted particles.

The lower limit of ice-rafted debris in cores that contain it occurs at or just below the lower boundary of the reversed interval that we interpret to represent the Matuyama reversed epoch, indicating an age of about 2.5 million years for the initiation of ice rafting. If Connolly and Ewing's (*20*) interpretation is correct, southern-hemisphere glaciation was initiated about 2.5 million years ago.

From the known dates of reversals, rates of sedimentation have been calculated on the assumption that the core tops are recent, there are no hiatuses, and the dating of the reversals is correct. These assumptions appear to be met in all cores, with the possible exception of V16-60. This point is illustrated by cores V16-134 and V18-72,

in which a time-versus-depth plot shows no marked deviations from the average sedimentation rate (Fig. 7). Regardless of the accuracy of the absolute ages of the reversals, it is clear that the average rate of sedimentation in the Indian Ocean cores is relatively higher than in the Bellingshausen Sea cores, whose rates are lower than about 3 millimeters per 1000 years (Table 2).

In cores for which a sedimentation rate during the Matuyama reversed epoch can be calculated, one finds that for all cores except V16-133 this rate is lower than that for the Brunhes normal epoch. In two cores the rate for the Gauss normal can be calculated, and in both it is higher than the rate for the Matuyama reversed epoch (Table 2).

Of the several horizons in Antarctic cores at which radiolarian species disappear, three have been chosen to mark stratigraphic zonal boundaries (*15*). Just above one of these (X-Ψ), several

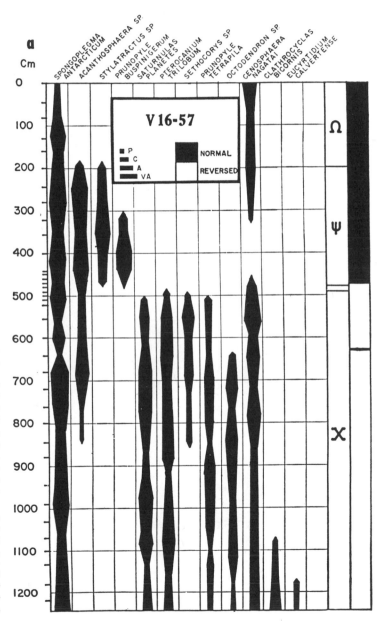

species appear for the first time. The possibility that abrupt appearances and disappearances of species reflect climatic changes can not be ruled out (*15*). Reversals occur within 30 centimeters of two of the boundaries (Φ-X and X-Ψ), but there is no evidence of a reversal near the Ψ-Ω boundary.

In the seven cores studied, the reversal that marks the boundary between the last reversed epoch (Matuyama) and the present normal epoch (Brunhes) falls within 30 centimeters of the first appearance of *Stylatractus* sp. and other species restricted to the zone (*Larcopyle* sp., Fig. 6a; and *Prunopyle buspinigerum*, Fig. 6b). In four cores the first appearance of these species comes within 10 centimeters of the reversal (V16-133, -134, -132, and -57). In the remaining three cores the reversal is below the appearance of these

species in two cores (V16-60 and -66) and above it in one (V18-72, Fig. 1).

Several of the Ω-zone species are morphologically similar to but distinguishable from species that are present in the X zone. Although they are morphologically distinct from their possible ancestors, *P. buspinigerum* and *Stylatractus* sp. may not be different species in a strict biologic sense.

The upper limit of the X zone (in Antarctic cores) is marked by the upper limit of both *Saturnulus planetes* and *Pterocanium trilobum*, but both species now live further north. This limit falls below the last reversal, but within 30 centimeters of it, in all cores but V16-60 and -66; in these two the reversal occurs below the top of the X zone (Fig. 1). In some cores (V16-134, -133, -132, and -57) there is a gap between the last occurrence of

X-zone species and the first appearance of species diagnostic of the Ψ zone. The particularly long gap (60 cm) between the X and Ψ zones in V16-133 exists because the paucity of Radiolaria in this interval makes it difficult to draw the boundaries.

The Φ-X boundary is based on the last common occurrence of two species: *Clathrocyclas bicornis* and *Eucyrtidium calvertense* (*15*); the former always ranges higher in the cores than the latter (Fig. 6, a and b). Although the number of species disappearing at this boundary is less than the number that disappear at the X-Ψ boundary, the aspect of the faunal change is more striking because of the great abundance of *E. calvertense* below the Φ-X boundary.

Six of the seven cores studied penetrate the Φ zone; in three of these (V16-134 and -133 and V18-72) the Φ-X boundary occurs within 10 centimeters of the base of the Olduvai event (Fig. 1). In two of the remaining three cores (V16-132, -66) the Olduvai event is not observed and the Φ-X boundary falls some 50 to 100 centimeters above the reversal between the Gauss normal series and the Matuyama reversed series. The magnetic stratigraphy in the lower part of V16-60 is so unclear that one cannot relate the position of the Φ-X boundary to it.

The coincidence or near coincidence of faunal changes with reversals in these cores suggests a causal relation. It has been suggested (*12*) that at the time of a reversal the intensity of the earth's dipole field must have decreased to zero; as a result, the earth's surface would have been subjected to a higher incidence of cosmic radiation than normal, which may have caused a higher mutation rate that strongly affected the evolutionary process (Uffen, *21*). Simpson (*22*) has recently presented paleomagnetic and paleontological evidence from the Cambrian to the Recent that, he believes, supports the thesis that reversals have in fact profoundly affected evolution. Both Uffen's and Simpson's views on accelerated mutation ignore the consensus among geneticists that there is no recognizable relation between evolutionary rate and mutation rate (*23*).

Our data are still insufficient to lend much support to Uffen's hypothesis, but more-detailed work now in progress at Lamont (on these and other cores) ultimately may serve to test it. We wish only to draw attention to the fact that our study showed two reversals and two faunal changes closely associated

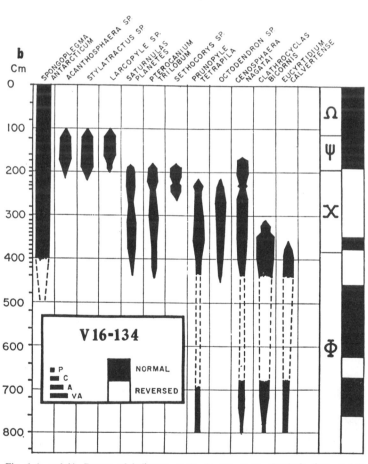

Fig. 6 (a and b). Ranges of indicator species in two cores. Graticule lines at left indicate sample locations in cores. *P*, present, *C*, common, *A*, abundant, *VA*, very abundant. Magnetic stratigraphies shown in right-hand column.

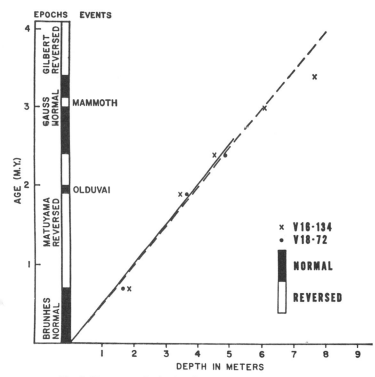

Fig. 7. Time-versus-depth plots for cores V16-134 and V18-72.

field of 150 oersteds. The average inclination in these two cores was then in good agreement with the average inclination of the ambient field for the latitude of the core site. It was also found that the intensities of the samples decreased at the points of reversal; this finding is to be expected if, as has been postulated by the dynamo theory, the intensity of the dipole field decreases to zero and builds again with opposite polarity. We believe that the magnetization of the cores results from the presence of detrital magnetite, although other magnetic minerals also may be present.

Four faunal zones (Φ, X, Ψ, and Ω) have been recognized in these Antarctic cores on the basis of upward sequential disappearance of Radiolaria. The faunal boundaries and reversals consistently have the same relations to one another, indicating that they are both time-dependent phenomena.

Using previously determined times of reversal, one may date the following events in the cores:

1) Radiolarian faunal boundaries: Φ-X, 2 million years; X-Ψ, 0.7 million years; Ψ-Ω, 0.4 to 0.5 million years. These dates are in good agreement with ages previously extrapolated from radiometric dates.

2) Initiation of Antarctic diatom-ooze deposition, approximately 2.0 million years ago.

3) First occurrence of ice-rafted detritus, approximately 2.5 million years ago.

One can also calculate rates of sedimentation, which vary in the cores studied from 1.1 to about 8.0 millimeters per 1000 years. Sedimentation rates for the Indian Ocean cores are higher than for the Bellingshausen Sea cores. The near-coincidence of faunal changes and reversals in the cores suggests but does not prove a causal relation.

We conclude from this study that paleomagnetic stratigraphy is a unique method for correlating and dating deep-sea cores, and that future work with such cores may provide a complete or nearly complete record of the history of the earth's magnetic field beyond 4 million years.

in several Antarctic cores; we hope other workers may have something to contribute to this problem. Harrison and Funnel's (7) finding that a reversal and extinction of the radiolarian species *Pterocanium prismatium* occur at about the same level in equatorial Pacific cores supports a possible causal relation.

Since reversals are almost instantaneous (geologically speaking), and because the polarity of the field is a global phenomenon, paleomagnetic stratigraphy is a powerful tool for correlating and dating deep-sea cores. Studies of North Pacific cores at Lamont show that reversals occurring in them can be correlated with those in Antarctic cores.

Deep-sea cores also provide a powerful tool for studying paleomagnetic stratigraphy. This study shows that some deep-sea cores contain a complete or nearly complete record of the history of the earth's magnetic field back to about 3.5 million years. There is striking agreement between the continuous record of deep-sea cores and the earlier discontinuous land record (1). Future work

on cores with high sedimentation rates will provide detailed geomagnetic histories of short intervals of time, while work on long cores with low sedimentation rates will extend the continuous record beyond 4 million years.

Summary

The magnetic inclinations and intensities of about 650 samples from seven deep-sea cores taken in the Antarctic were measured on a spinner magnetometer. This series of measurements provided a magnetic stratigraphy, based on zones of normally or reversally polarized specimens for each core, which was then correlated with the magnetic stratigraphy of Cox et al. (1). One core (V16-134) gave a continuous record of the paleomagnetic field back to about 3.5 million years.

When selected samples were subjected to alternating-field demagnetization, most were found to have an unstable component that was removed by fields of 150 oersteds; all samples from two cores were partially demagnetized in a

References and Notes

1. A. Cox, R. R. Doell, G. Dalrymple, *Nature* **198**, 1049 (1963); *Science* **142**, 382 (1963); R. R. Doell, G. B. Dalrymple, A. Cox, *J. Geophys. Res.* **71**, 531 (1966).
2. I. McDougall and D. Tarling, *Nature* **200**, 54 (1963).
3. A. G. McNish and E. A. Johnson, *Terrest. Magn. Atoms Elec.* **43**, 401 (1938).

4. E. A. Johnson, T. Murphy, O. W. Torreson, *ibid.* **53**, 349 (1948).
5. M. J. Keen, *Nature* **187**, 220 (1960); *Deep-Sea Res.* **10**, 607 (1963).
6. C. G. A. Harrison and B. M. Funnell, *Nature* **204**, 566 (1964).
   M. D. Fuller, C. G. A. Harrison, Y. R. Nayudu, *Amer. Assoc. Petrol. Geologists Bull.* **50**, 566 (1966).
8. F. Rimbert, *Rev. Inst. Franç. Petrole Ann. Combust. Liquides* **14**, 1 (1959); J. A. As and J. D. A. Zijderveld, *Geophys. J.* **1**, 308 (1958).
9. A. Cox, R. R. Doell, G. B. Dalrymple, *Science* **144**, 1537 (1964).
10. R. R. Doell and G. B. Dalrymple, *ibid.*, **152**, 1060 (1966).
11. K. M. Creer, *J. Geophys. Res.* **67**, 3461 (1962).

12. E. Bullard, *Proc. Cambridge Phil. Soc.* **51**, 744 (1955); T. Rikitake, *ibid.* **54**, 89 (1958); D. Allan, *Nature* **182**, 469 (1958).
13. C. G. A. Harrison and M. N. A. Peterson, *Amer. Mineralogist* **50**, 704 (1965).
14. G. Dickson, personal communication.
15. J. D. Hays, *Amer. Geophys. Union Antarctic Res. Ser.* **5**, 125 (1965).
16. ——, in *Progress in Oceanography* (Pergamon, London, in press).
17. D. B. Ericson, M. Ewing, G. Wollin, *Science* **139**, 727 (1963).
18. ——, *ibid.* **146**, 723 (1964).
19. W. R. Riedel, F. L. Parker, M. N. Bramlette, *ibid.* **140**, 1238 (1963).
20. J. R. Connolly and M. Ewing, *ibid.* **150**, 1822 (1965).
21. R. J. Uffen, *Nature* **198**, 143 (1963); in *Proc.*

   Intern. Geol. Congr. 22nd New Delhi (1964).
22. J. F. Simpson, *Geol. Soc. Amer. Bull.* **77**, 197 (1966).
23. N. D. Newell, personal communication.
24. Work supported by ONR contract Nonr 266(48) and NSF grants GP 4004 and GA-178. We thank the staff of Lamont Geological Observatory who contributed to the success of this investigation. In particular we thank James Heirtzler, Maurice Ewing, Wallace Broecker, and Allen Be for suggestions; N. D. Newell and Roger Batten for reading the manuscript and for comments; and Drs. Larson and Strangway of M.I.T. for allowing G. Dickson of the Observatory to make curie-point determinations on their equipment. Lamont Geological Observatory contribution 956.

7

Reprinted from *Earth and Planetary Sci. Letters* **1**:476 (1966)

PALEOMAGNETIC STRATIGRAPHY, RATES OF DEPOSITION AND TEPHRACHRONOLOGY IN NORTH PACIFIC DEEP-SEA SEDIMENTS*

Dragoslav NINKOVICH, Neil OPDYKE, Bruce C. HEEZEN and John H. FOSTER

Lamont Geological Observatory of Columbia University, Palisades, N.Y., USA

The paleomagnetic stratigraphy of 12 North Pacific deep-sea sediment cores has been investigated and has been used to date volcanic eruptions and to determine rates of deposition of pelagic sediments. Only four of the cores penetrated sediments deposited before the last reversal of the earth's magnetic field (0.7 m.y.). Of these, one penetrated to the Gauss series, two to sediments deposited during the Olduvai event and one penetrated to the middle of the Matuyama series. Eight other cores, 10-16 meters long, taken within 1000 km of the Japan-Kuril-Kamtchatka arc failed to reach the Matuyama series. The rate of deposition in North Pacific pelagic sediments vary from > 2 cm/1000 y in the area east of the Asiatic continent to < 0.8 cm/1000 y in the mid Pacific. Assuming continuous deposition, the length of the Jaramillo event can be established as 50 000 y and the Olduvai event as 14 000 y.

The apparent length of time during which the dipole field of the earth was reduced during reversals of the earth's magnetic field is approximately 20 000 y. In one of the cores the top of the Olduvai event is split. This may represent the Gilsa event.

The brown volcanic ash present in three of the cores apparently originated in an eruption 1.2 m.y. ago in the Aleutian Arc near the Andreanof Islands.

[*Editor's Note:* The article itself has not been reproduced here.]

---

* Lamont Geological Observatory Contribution.

Antarctic Radiolaria, Magnetic Reversals, and Climatic Change

James D. Hays and Neil D. Opdyke

Earlier (*1*) we reported the magnetic and radiolarian stratigraphy of seven cores from the Antarctic Ocean; the study established a paleomagnetic stratigraphy for antarctic deep-sea cores back to about 3.5 million years ago and verified the previously established (*2*) radiolarian stratigraphy. We now report the results of a study of three long antarctic cores containing records of continuous or nearly continuous sedimentation back to more than 4 million years ago; one of them contains a record probably longer than 5 million years. This longer stratigraphy makes possible the definition and approximate dating of new polarity events and radiolarian zones; it also contains evidence of a major faunal change about 2.5 million years ago that appears to correlate with evidence of cooling in other parts of the world.

Two (E13-3 and E13-17) of the three cores were taken (*3*) from the Bellingshausen basin; one (E14-8), from the western flank of the Mid-Pacific Ridge (Fig. 1 and Table 1). Each core contains an upper layer of siliceous sediment: diatom ooze in E13-17 (0 to 1020 centimeters) and E14-8 (0 to 1710 centimeters), and radiolarian clay in E13-3 (0 to 1000 centimeters). This siliceous sediment overlies a tan clay that is barren in E13-3 (1000 to 1603 centimeters) and E14-8 (1710 to 1830 centimeters) but interbedded in E13-17

with layers of radiolarian and diatomaceous clay (Figs. 2–4).

Magnetic Stratigraphy

Samples (8 cubic centimeters) were taken from all three cores at 10-centimeter intervals. The lower end of each sample was marked, and all were partially demagnetized in alternating fields (E14-8, 50 oersteds; E13-3 and E13-17, 150 oersteds). The direction and intensity of the remnant magnetism of each sample were measured on a magnetometer (5 cycles per second) described by Foster (*4*).

Inclination versus depth in the cores is shown in Fig. 5; the internal consistency is reasonably good in E13-17 and E13-3, but in E14-8 the quality of the data is poor, especially above 800 centimeters.

Correlation of the magnetic stratigraphy between these three cores and their correlation with the magnetic stratigraphy of Cox, Doell, and Dalrymple (*5*) are evident down to the Gilbert reversed-polarity epoch (Fig. 1 and Table 2).

The magnetic stratigraphy of E13-17 has been reported by Watkins and Goodell (*6*), who determined the inclination only upward or downward with an astatic magnetometer; the amount of inclination was not determined, and only one specimen in seven was demagnetized in alternating fields. Watkins and Goodell chose the Brunhes-Matuyama boundary at 500 centimeters, which is preceded by a long normal interval that they identified as

the Jaramillo event. The inclination plot for this core (Fig. 5a) shows that, at the point chosen by Watkins and Goodell as the Brunhes-Matuyama boundary, the inclination becomes less steep but never crosses the axis to become positive. It is probably because all specimens were not demagnetized that Watkins and Goodell misidentified the Brunhes-Matuyama boundary. We place the boundary at 615 centimeters, which position accords with the paleontologic zonation since the X-Ψ boundary occurs at 690 centimeters (Fig. 4). The relation between the paleontologic and magnetic stratigraphy is well documented (*1*); therefore the Jaramillo event as we identify it is the event that Watkins and Goodell called the Gilsa. Clearly there is in this core no evidence of the Gilsa event as postulated by MacDougall and Wensink (*7*).

Hitherto details of the magnetic stratigraphy within the Gilbert and below have remained unknown. A predominantly reversed section of core which presumably represents the Gilbert reversed-polarity series, lies below the Gauss normal series in all cores. Within this reversed interval are three prominent normal events in E13-3 and E13-17. The lower two events in E13-17 appear to be split, but the resolution of the data is insufficient for determination of this point without question. In the sediment below the Gauss normal series in E14-8 two distinct normal events are defined, and the upper part of a third is present. The core bottoms at 1830 centimeters, but the magnetic intensity drops to too low

Dr. Hays is a research associate and Dr. Opdyke is a senior research associate at Lamont Geological Observatory, Columbia University, Palisades, New York.

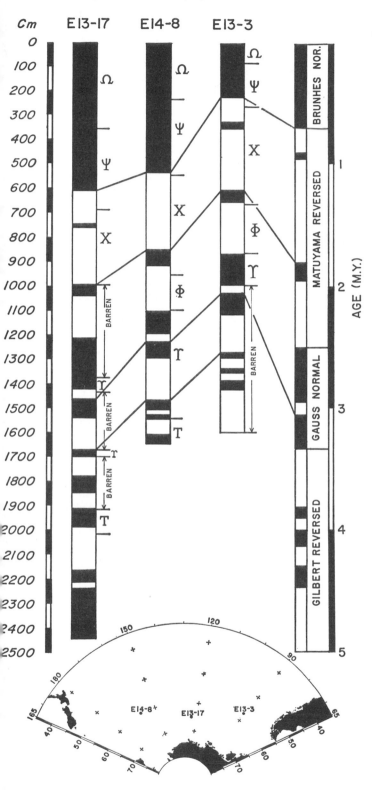

a level for reliability below 1650 centimeters.

Below the three events in E13-17 is an interval from 2000 to 2070 centimeters that is reversely magnetized. A void in the core extends from 2090 to 2180 centimeters, and the magnetic determinations from the specimens immediately above and below this void must be regarded as suspect. From 2200 to 2500 centimeters the core is predominantly normally magnetized, with one short reversed section at about 2225 centimeters.

These cores provide strong evidence of the existence of three short polarity events in the Gilbert reversed-polarity epoch; their existence has been confirmed in cores from other regions. Below these events in E13-17 are a reversed interval and then a long normal interval of epoch proportions. We propose establishment of the beginning of the Gilbert reversed-polarity series at the base of the reversed section of core that precedes the three newly established events.

The JOIDES deep-sea drilling program will provide for study cores that span significantly longer periods than do cores obtained by piston-coring techniques. Thus the present system of naming epochs of magnetic polarity after great magneticians, and events after geographic localities, will lead to great proliferation of names, since one may obtain a single deep-sea core spanning possibly 60 million years of the earth's magnetic history. It seems reasonable to restrict the present systems of naming (5) to epochs and events that have already been defined on the basis of correlation by potassium-argon decay. Barring a major breakthrough in the precision of potassium-argon dating techniques, it seems unlikely that the present system can be extended much beyond the Gilbert reversed-polarity epoch.

We therefore propose a numbering system that could be tied eventually to classical stratigraphic nomenclature and to the absolute time scale; it has the advantage of simplicity and allows a worker to recognize the positions of events and epochs in the sequence without need to memorize a long list of

Fig. 1. Correlation of magnetic stratigraphy in three antarctic cores. Solid black represents normally magnetized core: white, reversely magnetized. Greek letters denote faunal zones (2). Magnetic stratigraphy, in right-hand column down to the top of the Gilbert reversed, is from Cox et al. (5). Inset: sources of cores.

names. We propose to label the polarity epochs with Arabic numerals; events are labeled with letters from youngest to oldest within an epoch. By this system, for example, the Brunhes polarity epoch would be 1; the Matuyama, 2; and the Jaramillo and Olduvai events, 2*a* and 2*b*, respectively. If, after the system is established, new events are found, they may be inserted between the established events. If a new event were found between the Jaramillo and Olduvai events it could be designated event 2*a-b*. Thus a long set of magnetic stratigraphy can be usefully designated

and eventually, through absolute or paleontologic dating, incorporated into normal stratigraphic nomenclature. We do not intend this system to supersede the one now used for the last 3.5 million years of the earth's history, but rather intend it to be used to describe the sequential arrangement of epochs and events that will be found in the near future in oceanic cores. In this article we propose to designate the three new events in the Gilbert *a, b,* and *c,* from youngest to oldest, and the long normal interval lying below the Gilbert as magnetic epoch 5.

Rates of Sedimentation

Rates of sedimentation, with paleomagnetic stratigraphy used as a reference, have been published for seven antarctic cores (*1*). Generally the rates in these cores vary with time. Our E13-3 seems to have a remarkably constant rate of sedimentation. By use of 0.7 million years as the age of the Brunhes-Matuyama boundary, 2.5 million years for the base of the Matuyama, and 3.35 million years for the Gauss-Gilbert boundary (*8*), a time versus depth plot for this core produces

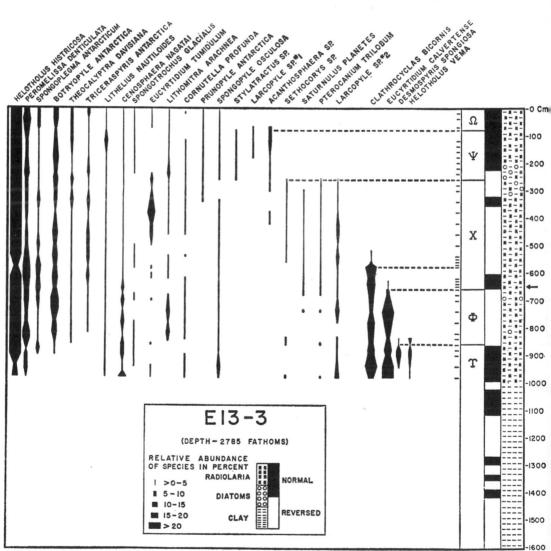

Fig. 2. Ranges of radiolarian species in core E13-3. Right-hand columns show lithology, magnetic stratigraphy, and faunal zones. Graticule at right indicates radiolarian samples. Arrow at far right shows lower limit of ice-rafted material.

a nearly straight line (Fig. 6). The rate of accumulation varies between 2.94 millimeters per 1000 years for the Gauss normal series and 3.56 millimeters per 1000 years for the Matuyama series.

The relative amounts of biogenic silica in these cores can be reliably estimated by inspection of the cores and study of the washed coarse fractions in samples taken at closely spaced intervals. Core E13-3 has a low content of biogenic silica throughout, consisting primarily of the shells of Radio-

laria, although diatoms occur in subordinate amounts.

The longest core studied, E13-17, does not have a constant rate of sedimentation (Fig. 6). The rate (millimeters per 1000 years) in the Brunhes series is 8.8; in the Matuyama series,

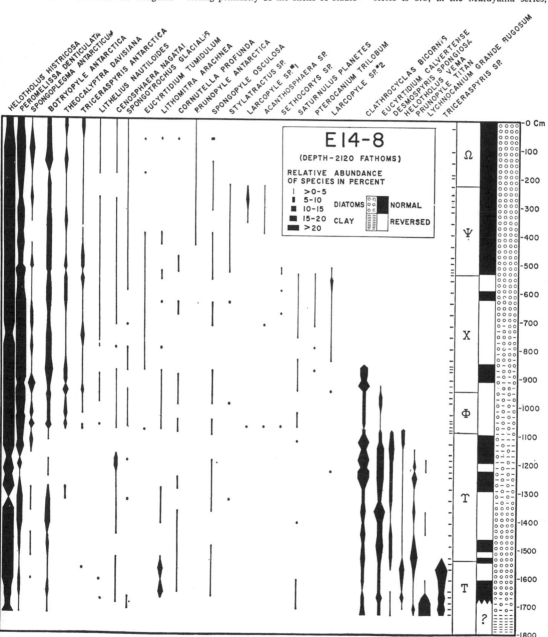

Fig. 3 (above). Ranges of radiolarian species in core E14-8. Right-hand column shows lithology, magnetic stratigraphy, and faunal zones. Graticule at right indicates radiolarian samples.　　Fig. 4 (right). Ranges of radiolarian species in core E13-17. Right-hand columns show lithology, magnetic stratigraphy, and faunal zones. Graticule at right indicates radiolarian samples. Arrow at far right shows lower limit of ice-rafted material. Hachures show where corrosion makes estimates of abundance difficult.

127

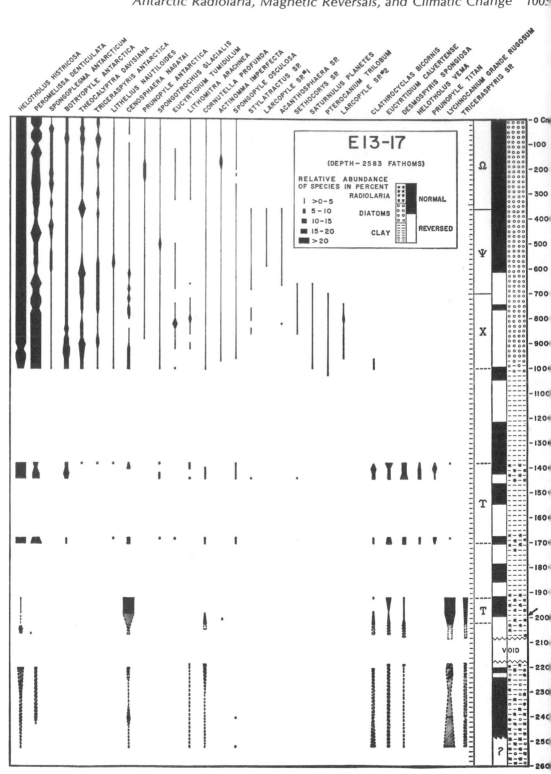

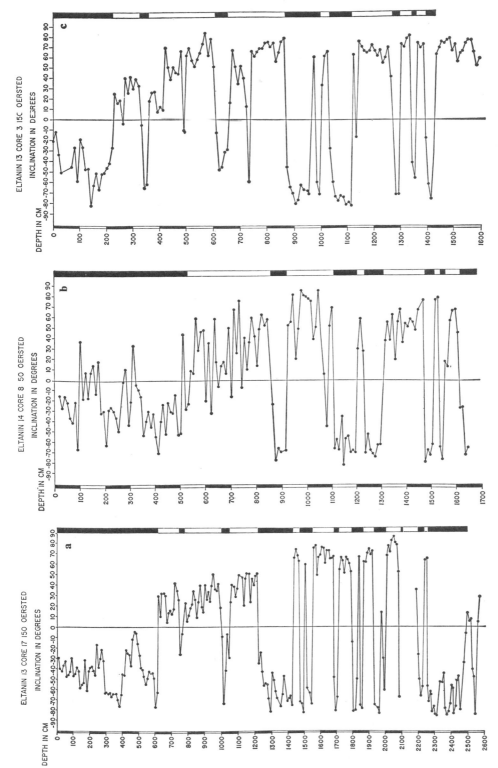

Fig. 5. Plots of inclination versus depth in cores E13-17 (a), E14-8 (b), and E13-3 (c).

129

3.34; and in the Gauss normal series 3.88. The top 950 centimeters of this core is rich in diatoms and Radiolaria, but these fossils are rare below 10 meters, occurring only in thin layers (Fig. 4). The higher rate of accumulation during the Brunhes than during the Matuyama or Gauss can be accounted for by the relatively large number of diatoms and Radiolaria present in the top 1000 centimeters.

Core E14-8, like E13-17, has a higher sedimentation rate (millimeters per 1000 years) in the Brunhes (7.7) than in the Matuyama (3.1) or Gauss (2.4); this variation also can be explained by the content of biogenic silica, which is high in the top 700 centimeters; below this depth, diatoms and Radiolaria decrease in abundance to 1085 centimeters (just above the Gauss normal series). The biogenic silica is least abundant between 1085 and 1230 centimeters, an interval falling within the Gauss normal series. Between 1230 and 1650 centimeters the diatom content increases to a diatom ooze.

Our data indicate that the rate of accumulation of sediment containing little biogenic silica is fairly constant in all three cores, ranging between 2.4 and 3.9 millimeters per 1000 years, and that the high rates occurring in the top portions of E13-17 and E14-8 can be accounted for by the high content of biogenic silica.

In discussing the radiolarian stratigraphy of antarctic cores, Hays (2) suggested that the change from highly diatomaceous sediments in the tops of these cores to sediments containing few siliceous fossils below was caused by recently increased productivity in antarctic waters due to intensification of vertical oceanic circulation associated with ice formation around the continent. Our study supports this hypothesis, in showing that the clay sediments represent slow sedimentation rather than rapid clay deposition masking the biogenic silica.

Using the rates of sedimentation determined for the Gauss normal series, one can extrapolate to the three new events in the Gilbert reversed-polarity series to obtain rough estimates of their ages (Table 3). In E13-3, where extrapolation is probably most significant, the three events would span the time between 3.86 and 4.39 million years ago. In E13-17 the three events end at 3.79 million years ago, while they would begin at 4.67 million years. The variable content of biogenic silica in the lower part of E14-8 makes ages extrapolated from this core subject to large errors.

In a recent paper Dalrymple et al. (9) have dated a reversal which occurs at 3.7 ± 0.1 million years ago (W10R/N). This probably represents the upper boundary of event "a," the age of which, by our extrapolation, is very close to the age they obtained. Ozima et al. (10) record magnetic reversals in New Mexico at 3.7 and 4.5 million years ago which are also probably associated with these events; however, because of the errors involved, exact correspondence is difficult to es-

Table 1. Sources of cores.

Core	Latitude (S)	Longitude (W)	Water depth (m)	Length (cm)	Region
E13-3	57°00'	89°29'	5090	1603	Bellingshausen basin
E13-17	65°41'	124°06'	4720	2642	Bellingshausen basin
E14-8	59°40'	160°17'	3875	1830	Flank of Mid-Pacific Ridge

Table 2. Intervals of magnetic series and events in three cores.

Series	Event	Core		
		E13-17	E14-8	E13-3
Brunhes		0–615	0–535	0–225
Matuyama		615–1215	535–1100	225–865
	Jaramillo	745–765	?	325–355
	Olduvai	995–1045	845–915	605–655
Gauss		1215–1545	1100–1300	865–1115
	Mammoth	1425–1465	1195–1225	995–1025

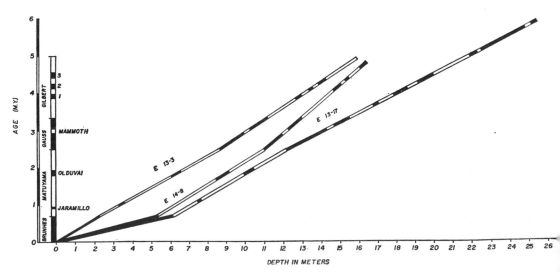

Fig. 6. Time versus depth plots of magnetic stratigraphy for three cores.

tablish. For example, the reversal dated by Cox et al. is from reversed polarity below to normal polarity above, at 3.7 million years ago, and the one by Ozima et al. is from normal polarity below to reversed polarity above, with the same age. Nevertheless, evidence for the existence of the events is present in lava flows on land.

One can also estimate the date of the beginning of the Gilbert epoch. Core E13-3 ends at 1600 centimeters and is reversed to its end; the extrapolated age of its base is 4.8 million years, so the beginning of the

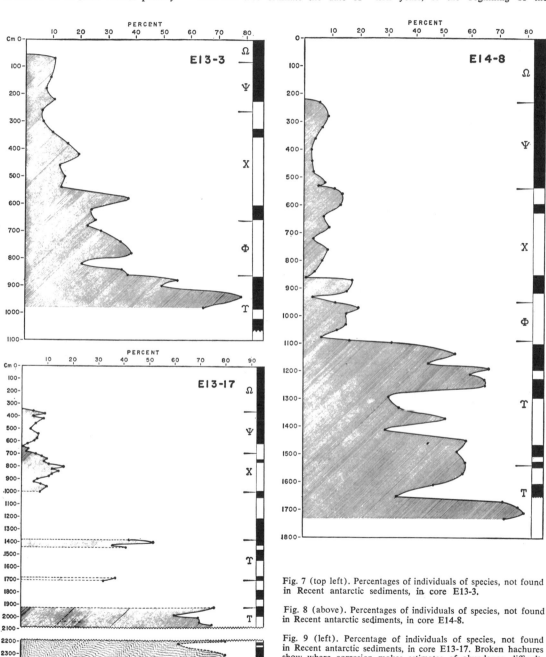

Fig. 7 (top left). Percentages of individuals of species, not found in Recent antarctic sediments, in core E13-3.

Fig. 8 (above). Percentages of individuals of species, not found in Recent antarctic sediments, in core E14-8.

Fig. 9 (left). Percentage of individuals of species, not found in Recent antarctic sediments, in core E13-17. Broken hachures show where corrosion makes estimates of abundance difficult.

Gilbert is unlikely to be younger. It is difficult to pick the beginning of the Gilbert in E13-17 because of a gap in the core between 2100 and 2200 centimeters; however, it is definitely normally magnetized below 2200 centimeters. If we extrapolate from the base of the Gauss normal series, using the rate of accumulation within the Gauss, we obtain an age of 5.25 million years for the top of epoch 5, which is probably a maximum age. Therefore an estimated age of 5 million years for the beginning of the Gilbert is reasonable but may be in error by as much as 200,000 years.

Radiolarian Stratigraphy

Four stratigraphic zones have been established in antarctic sediments on the basis of the upward sequential disappearance of species of Radiolaria (2); they have been named Φ (oldest), X, Ψ, and Ω (youngest). The base of the Ω zone has been dated radiometrically at about 400,000 years (11). The base of the Ψ zone (0.7 million years) and the base of the X zone (about 2 million years) have been dated paleomagnetically (1), while the base of the Φ zone has not yet been defined. Previously the faunal work ended near the top of the Φ zone, for in antarctic sediments south of the Antarctic Polar Front the sediments in many areas become barren below the Φ-X boundary (2).

The three cores selected for detailed study by us contain Radiolaria older than any previously reported from the floor of the Antarctic Ocean. Cores E13-3 and E14-8 have continuous radiolarian-bearing sequences back to about 3 and 4 million years ago, respectively, while E13-17 contains Radiolaria intermittently back to probably more than 5 million years ago.

In previous work (1, 2) certain species having restricted stratigraphic ranges were selected for study. We include 32 species constituting, in the time period studied, more than 90 percent of radiolarian fauna; the taxonomy of most has been discussed (2), and a report on the others (*Larcopyle* sp. No. 1, *Larcopyle* sp. No. 2, and *Triceraspyris* sp.) is in preparation. Altogether 296 slides were examined for Radiolaria; counts of at least 300 (usually 500) individuals were made on the 180 slides bearing sufficient Radiolaria. The results of these counts were

reproducible within 6 percent. From these data the stratigraphic ranges of the species were determined (Figs. 2–4), as well as the percentages of individuals in the sample belonging to species not found in the most recent antarctic sediments (Figs. 7–9).

The top four faunal zones in these cores bear the same relation to magnetic reversals as in cores previously reported (Figs. 2–4, Table 4) (1). The bottom of the Ω zone falls in the central portion of the Brunhes series. The X-Ψ boundary falls near but below the base of the Brunhes series, and the Φ-X boundary occurs just below the Olduvai event. The upper limit of the range of *Clathrocyclas bicornis* is near the top of the Olduvai event.

Both E13-3 and E14-8 contain a considerable thickness of radiolarian-bearing sediments below the Φ-X boundary. The next-older major faunal change occurs at the upper limit of two species (*Desmospyris spongiosa* and *Helotholus vema*) having similar upper limits (E13-3, 840 centimeters; E14-8, 1085 centimeters) just above the Gauss-Matuyama boundary (Figs. 2 and 3). This faunal change permits more precise definition of the Φ zone as the interval in antarctic deep-sea sediments containing individuals of *Clathrocyclas bicornis* and *Eucyrtidium calvertense* above the last common occurrence of *D. spongiosa* and *H. vema*. The sediments underlying the Φ zone and commonly containing *D. spongiosa* and *H. vema* will be designated the upsilon zone (Y). Near the Y-Φ boundary there is a sharp upward decrease in the occurrence of species that no longer live in antarctic waters, and a concomitant increase in abundance of species that comprise the Recent Antarctic assemblage, making the Y-Φ boundary the most striking faunal boundary yet studied. In fact the Y zone in both E13-3 and E14-8 is characterized by high percentages of extinct species (Figs. 7 and 8). Although the Y-Φ boundary is not present in E13-17, the interval containing Radiolaria within the Gauss normal series (1380 to 1440 centimeters) contains Y-zone species (*D. spongiosa* and *H. vema*) as well as relatively high percentages of species not found in the most recent antarctic sediments (Fig. 9).

Core E13-3 becomes barren shortly below the Y-Φ boundary, so only in E14-8 can one determine the upper boundary of the next-lower faunal change. Below 1540 centimeters in

E14-8, an undescribed spyroid (*Triceraspyris* sp.) is abundant; it is probably closely related to *Desmospyris spongiosa*, since the latter is abundant only above the upper limit of *Triceraspyris* sp. Since the upper limit of the range of *Triceraspyris* sp. is based solely on its range in E14-8, only tentatively do we designate the upper limit of this species the base of the Y zone and the top of the T zone. The upper limit of the T zone falls within event *a* of the Gilbert polarity series.

Relations of Faunal Boundaries to Temperature Changes

The only reliable evidence of past warmer conditions in antarctic sediments is the deep occurrence in cores of species that are normally restricted to Recent sediments north of the Antarctic Polar Front. Evidence (2) has been presented of a change from warmer to cooler conditions across the

Table 3. Apparent dates (by extrapolation from two cores) of beginnings and ends of three normal-polarity events in Gilbert reversed-polarity epoch.

Event	Date (× 10⁶ years ago)	
	Beginning	End
Core E13-3		
a	3.86	3.95
b	4.08	4.15
c	4.25	4.39
Core E13-17		
a	3.79	3.87
b	4.08	4.26
c	4.44	4.67

Table 4. Depths in two cores to faunal boundaries and nearest reversals.

Faunal boundary	Depth (cm)	
	Boundary	Nearest reversal
Core E14-8		
Ψ-Ω	230	No reversal
X-Ψ	530–540	530–540
Upper limit of *C. bicornis*	860–870	840–860
Φ-X	950–970	910–920
Y-Φ	1085–1095	1095–1105
T-Y	1550–1570	1540–1550
Core E13-3		
Ψ-Ω	50–70	No reversal
X-Ψ	240–250	220–230
Upper limit of *C. bicornis*	560–570	600–610
Φ-X	650–660	650–660
Y-Φ	860–870	860–870

X-Ψ boundary, but the other faunal boundaries down to the Φ-X boundary have shown no evidence of change in temperature.

Both E13-17 and E13-3 show definite evidence of warmer conditions at depth. In E13-17, below 2000 centimeters (4.5 million years) are various warm-water radiolarian species, including *Axoprunum stauraxonium* and *Heliodiscus asteriscus*. In all samples below 2180 centimeters (5.0 million years) are fragments of the diatom species *Ethmodiscus rex*; although it has not been reported south of 30°S (12), Burckle has seen fragments of this distinctive species in samples of Recent sediment from a core raised from the southern part of the Argentine basin (48°08'S,40°31'W). The occurrence of *E. rex* in E13-17 is therefore nearly 20 degrees farther south than the southernmost known occurrence of this species in Recent sediments.

In E13-3, the northernmost of the three cores, several species of warm-water Radiolaria occur commonly between 860 centimeters (Y-Φ; 2.4 million years) and 980 centimeters, where the core becomes barren.

There is no strong evidence in E14-8 of warmer conditions at depth, but below 1100 centimeters (Y-Φ boundary; 2.4 million years) several warm-water Radiolaria occur intermittently down to the point where the core becomes barren, at 1710 centimeters.

Evidence of change from warmer to cooler conditions exists in all three cores; the change took place approximately 2 million years earlier in E13-17 than in the two northern cores.

In spite of the differences in latitude between these cores and of the corresponding temperature differences, the faunal boundaries occur at about the same time in all three cores. Changes in temperature may have caused the disappearance of some of the species marking faunal boundaries, but this evidence opposes the possibility. Even more convincing is the fact that in a core taken north of the Polar Front (RC8-52; 41°06'S,101°25'E), in which the vast majority of species are warm-water species, the ranges of *Eucyrtidium calvertense* and *Clathrocyclas bicornis* are not significantly extended; in fact their upper limits bear a relation to the lower and upper boundaries of the Olduvai event resembling that found in cores taken from beneath the colder waters to the south of the Polar Front (Fig. 10).

Discussion

The limited geologic ranges of some planktonic organisms and their widespread distribution have made them excellent guide fossils, yet no completely satisfying explanation of the factor limiting their ranges has been proposed. Our data show with hitherto-unequaled precision that the upper limits of the ranges of certain radiolarian species are nearly isochronous over a broad area. In fact the magnetic-polarity epochs, down to the base of the Gauss normal-polarity series, can be characterized by their content of Radiolaria.

Of the multitude of factors that may extinguish pelagic organisms, only a few leave decipherable records in sediments. One factor that is frequently credited with causing extinctions is change in temperature; our data show that, at least for two species *Eucyrtidium calvertense* and *Clathrocyclas bicornis*), it is unlikely that change in temperature was responsible. This probability does not preclude the possibility that temperature change was involved in the disappearance from antarctic sediments of other radiolarian species such as those that disappear at the X-Ψ boundary, several of which are still living north of the Polar Front.

It has been suggested (1) that, because some antarctic Radiolaria disappeared near reversals, the reversals in some way influenced their disappearance. The faunal boundaries are drawn at the last common occurrences of the indicator species. The Ψ-Ω boundary has no corresponding reversal, although the other four faunal zones are usually associated with reversals (Table 4). Since nearly all species used in the definition of faunal zones survived a number of reversals before disappearing near one, we can only assume that, if a reversal did affect a species, either the effects of reversals differed or the reversal was a contributor to other environmental stresses that tended to weaken the species. In several instances a faunal boundary and a reversal coincide within the sampling interval (Table 4). The lack of coincidence in other instances may be due to postdepositional displacement of either the magnetic or the faunal boundary. It is possible, however, that the reversals and faunal limits are separated by thousands of years. Nevertheless there remain tantalizingly close correlations between four of the five faunal boundaries and reversals (Table 4). We hope that other students of pelagic organisms will examine their data in the light of these results.

Pliocene-Pleistocene Boundary

Stratigraphic boundaries traditionally have been drawn on the basis of faunal changes, regardless of whether these changes were climatically induced. It has been recommended (13) that "In order to avoid ambiguities, the Lower Pleistocene should include as its basal member in the type-area the Calabrian formation (marine) together with its terrestrial (continental) equivalent the Villafranchian. . . . [According] to evidence given, this usage would place the boundary at the horizon of the first indication of climatic deterioration in the Italian Neogene succession." The cooling trend in the Calabrian, evidenced by the first appearance of several cold-water species, has been taken to signal the initiation of glaciation in the Alps (14).

While direct faunal comparison between our three cores and the type locality of the Pliocene-Pleistocene boundary is difficult, indirect correlation is now possible. In the Antarctic the base of the Pleistocene was tentatively drawn by Hays (2) at the Φ-X boundary. Recently Banner and Blow (15) have established a planktonic foraminiferal zonation of the Neogene.

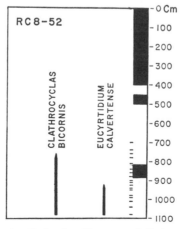

Fig. 10. Stratigraphic ranges of *Clathrocyclas bicornis* and *Eucyrtidium calvertense* compared with magnetic stratigraphy. Graticule at right indicates radiolarian samples.

The boundary between their zones N21 and N22, based on the evolutionary transition from *Globorotalia (G) tosaensis* to *Globorotalia (G) truncatulinoides*, occurs near the base of the stratotype Calabrian at Santa Maria di Catanzaro (*16*), the agreed earliest Quaternary (*13*). Berggren (*17*) has shown that this same evolutionary transition, marking the boundary between zones N21 and N22, occurs within the Olduvai event in a core from the North Atlantic. Since the Pliocene-Pleistocene boundary falls within the Olduvai event, the Φ-Χ boundary, falling near the base of the Olduvai event, represents a close approximation of the boundary in antarctic sediments.

Although the faunal change across the Φ-Χ boundary is significant, the greatest faunal change in these cores occurs at about 2.5 million years ago; evidence in E13-3 and E14-8 indicates a change from warmer to cooler conditions near this boundary.

Mathews and Curtis (*18*) have dated a so-called Pliocene-Pleistocene boundary in New Zealand at "more than 2 million years and probably nearer to 2.5 million years ago." This boundary is placed at the first indication of cooling conditions in a New Zealand pollen sequence (*19*); such cooling in New Zealand probably correlates with cooling indicated at the Υ-Φ boundary in antarctic cores E13-3 and E14-8.

Rutten and Wensink (*20*) have found that the first glacial till in Iceland falls in the middle of the Gauss normal-polarity series, so that it is about 3 million years old; this paleomagnetic age has recently been confirmed by potassium-argon dating (*7*). A similar age of 2.7 million years for a lava flow overlying the oldest till yet found in the Sierra Nevada of California has been reported (*21*).

Conolly and Ewing (*22*) determined that the first appearance of ice-rafted debris in several antarctic cores fell somewhat below the Φ-Χ boundary; this level was dated (*1*) at about 2.5 million years ago. Thus it appears that the major faunal change in antarctic cores during the last 5 million years occured at about the same time as evidence of deteriorating climatic conditions in other parts of the world. The time between 3 and 2.5 million years ago seems to have been critical in Neogene history and may have been the prelude to widespread glaciation in more temperate regions.

Antarctic Glaciation

Evidence is now available that glacial conditions existed on Antarctica well before the 3-million-year date reported (*20*) for Iceland. There is evidence (*23*) of glacial action probably older than 10 million years in the Jones Mountain area.

Preliminary examination (*24*) of our three antarctic cores shows that the age of the lower limit of ice-rafted debris varies from core to core. Core E13-3, the northernmost of the three, has ice-rafted grains in all samples above 650 centimeters (about 2 million years ago). In E13-17, the southernmost ice-rafted debris occurs to a depth of 1970 centimeters (more than 4 million years ago). Core E14-8 has ice-rafted material throughout; the age of its base is uncertain but probably exceeds 4 million years.

In age of its ice-rafted grains, E13-3 resembles nearby cores (*22*). The earlier appearance of glacial debris in the two western cores (E13-17 and E14-8) may indicate earlier glaciation of East than of West Antarctica. In any event our data indicate widespread glaciation on Antarctica at least 4 million years ago.

Summary

Our study extends the continuous record of the earth's magnetic field to more than 5 million years ago; it defines three new magnetic normal-polarity events and one new normal-polarity epoch. One core (E13-3) has a sufficiently constant rate of sedimentation to enable one to make reliable estimates of the ages of the three new events. The age of the upper boundary of the new polarity epoch is estimated at 5 million years. Because of the many new magnetic events and epochs that will be found through study of deep-sea sediments, we propose a numbering system that may simplify designation and increase the usefulness of marine magnetic stratigraphy.

The rate of deposition of clay in the cores is relatively constant, fluctuations being largely due to variation in amount of biogenic silica.

The cores contain a radiolarian stratigraphy similar to that reported (*1, 2*), and they also record disappearances of other species that make possible the definition of two new faunal zones.

The upper boundary of one of these zones (Υ) is the most striking faunal change yet encountered, occurring at about the same time as evidence of climatic deterioration in other parts of the world.

All the faunal boundaries are time-dependent; four coincide with reversals, or almost so. Change in temperature may have been responsible for some of the boundaries, but not apparently, for at least one. The close correspondence between reversals and faunal boundaries, whether or not changes in temperature were involved, adds a new dimension to the perplexing question of planktonic extinctions.

References and Notes

1. N. D. Opdyke, B. Glass, J. D. Hays, J. Foster, *Science* 154, 349 (1966).
2. J. D. Hays, *Amer. Geophys. Union Antarctic Res. Ser. No. 5* (1965), p. 125.
3. The cores were taken by U.S.N.S. *Eltanin* on cruises 13 and 14.
4. J. Foster, *Earth Planetary Sci. Letters* 1, 463 (1966).
5. A. Cox, R. R. Doell, G. B. Dalrymple, *Science* 144, 1537 (1964).
6. N. D. Watkins and H. G. Goodell, *Earth Planetary Sci. Letters* 2, 123 (1967).
7. T. MacDougall and H. Wensink, *ibid.* 1, 232 (1966).
8. A. Cox, R. R. Doell, G. B. Dalrymple *Quaternary of the United States*, H. E. Wright and D. G. Frey, Eds. (Princeton Univ. Press Princeton, N.J., 1965).
9. G. B. Dalrymple, A. Cox, R. R. Doell, C. S. Gromme, *Earth Planetary Sci. Letters* 2, 163 (1967).
10. M. Ozima, M. Kono, I. Kaneoka, H. Kinoshita, K. Kobayashi, T. Nagata, E. E. Larson D. Strangway, *J. Geophys. Res.* 72, 2615 (1967).
11. J. D. Hays, in *Progress in Oceanography* (Pergamon, London, in press).
12. J. D. H. Wiseman and N. I. Hendey, *Deep Sea Res.* 1, 47 (1953).
13. The recommendation was made by the Commission on the Pliocene-Pleistocene Boundary International Geological Congr., 18th, 1948 (1950).
14. C. I. Migliorinini, in *Proc. Intern. Geol. Congr., 18th, Gt. Brit. 1948* (1950), sect. H pt. 9, p. 66.
15. F. T. Banner and W. H. Blow, *Nature* 208 1164 (1965).
16. The stratotype is described by M. Gignoux [*Ann. Univ. Lyon* 36, 35 (1913)] as typical of his Calabrian stage, as proposed i Compt. Rend. 150, 841 (1910).
17. W. Berggren, *Proceedings 4th Session of the Committee on Mediterranean Neogene Stratigraphy*, Bologna, 1967, in press.
18. W. H. Mathews and G. H. Curtis, *Nature* 212, 979 (1966).
19. H. S. Gair, *New Zealand J. Geol. Geophys.* 4, 93 (1961).
20. M. G. Rutten and H. Wensink, in *Intern Geol. Congr. 21st* (1960), pt. 4, p. 62.
21. R. R. Curry, *Science* 154, 770 (1966).
22. J. R. Conolly and M. Ewing, *ibid.* 150, 1822 (1965).
23. C. Craddock, T. W. Bastien, R. H. Rutford *Antarctic Geology* (Interscience, New York 1964), p. 171.
24. The examination was made by D. Horn, Lamont Geological Observatory.
25. The work described was supported by grant from the National Science Foundation (GA 4004, GA 824, GA 558, and GA 861) and the Office of Naval Research (N00014-67-A-0108 0004). We thank M. Ewing and W. Broecker who read the manuscript and made helpful suggestions, and G. Goodell and N. Watkins who assisted when the cores were sampled at Florida State University. This article is Lamont Geological Observatory contribution No. 1122.

9

MICROPALEONTOLOGY AND THE PLIOCENE/PLEISTOCENE BOUNDARY IN A DEEP-SEA CORE FROM THE SOUTH-CENTRAL NORTH ATLANTIC

W. A. Berggren

ABSTRACT · Micropaleontologic and paleomagnetic investigations on a 7 m deep-sea core from the south-central North Atlantic (26°41.5' N Lat.; 39°23'W Long.) have allowed the recognition of the Pliocene/Pleistocene boundary based on criteria relatable to the holostratotype of the Calabrian Stage and an estimation of its « absolute » date. The Pliocene/Pleistocene boundary (Zone N21/N22) occurs within the Olduvai Normal Event and is dated at about 1.85 my. The major extinction of discoasters is found to occur about 100 cm above the base of the Pleistocene; a brief reappearance of discoasters was found in a sample within the Jaramillo Normal Event at about 0.9 my. A general cooling trend is observed in the lower half of the Pleistocene based on the relative ratios of *Pulleniatina obliquiloculata* and *Sphaeroidinella dehiscens* to *Globorotalia inflata* and *G. hirsuta*. A significant (stronger) cooling is suggested above a level within the Jaramillo Event (about 0.9 my) on the basis of a sharp reduction (and subsequent local disappearance) of the former two and a sharp increase in the latter two species. The suggestion is made that this stronger cooling may have been related to the onset of Pleistocene continental glaciation and that the « pre-glacial » and « glacial » Pleistocene may thus have an approximate ratio of 1 : 1.

[*Editor's Note:* The article itself has not been reproduced here.]

10

Reprinted from Geol. Soc. America Bull. 80:1481–1513 (1969)

Pliocene-Pleistocene Sediments of the Equatorial Pacific: Their Paleomagnetic, Biostratigraphic, and Climatic Record

JAMES D. HAYS
TSUNEMASA SAITO
NEIL D. OPDYKE
LLOYD H. BURCKLE

Lamont Geological Observatory, Columbia University, Palisades, New York 10964

Abstract: Magnetic stratigraphy of 15 oriented cores from the equatorial Pacific was determined as far back as the Gilbert reversed-polarity epoch.

Ranges of selected species of four major microfossil groups (diatoms, silicoflagellates, foraminifers and Radiolaria) are compared with the record of geomagnetic reversals during the last 4.5 m. y. in eastern equatorial Pacific deep-sea cores. Characteristics of the fossil assemblages are used as criteria for recognition of most of the paleomagnetic reversals that occurred during this interval. Two zones of major paleontological change occur characterized by extinctions of several species and coiling direction changes in some foraminifers. The first change comes in the middle of the Gauss normal magnetic series (about 3 m.y. B.P.) and the second near the Olduvai magnetic event (about 2.0 m.y. B.P.). Seven equatorial foraminiferal species, two radiolarian species, and two diatom species become extinct near reversals.

The establishment of the true chronostratigraphic relationships of these selected microfossil species allows us to date zonations of previous authors and provides absolute dates that can be used in worldwide correlation of marine sediments.

The percentage of calcium carbonate was determined throughout the lengths of four cores. Eight distinct carbonate cycles are present in the Brunhes series, having periodicities of about 75,000 years in the upper Brunhes to over 100,000 years in the lower Brunhes. It is possible to correlate these carbonate cycles among our cores and also to correlate them with the previous work of Arrhenius who equated the carbonate peaks with glacial stages and the troughs with interglacial stages. This interpretation is supported by paleomagnetic and C^{14} dating of the last carbonate high which is synchronous with the Wisconsin glaciation (80,000 to 11,500 years B.P.). It, therefore, is probable that there were eight major glacial fluctuations during the last 700,000 years.

During the last 400,000 years there is good correlation between the carbonate cycles of the Pacific and evidence of climatic fluctuations in the Atlantic established by Ericson and Wollin (1968) and Emiliani (1966) based on fossil abundances and oxygen isotope ratios, respectively.

The rates of sedimentation during the Brunhes series range between 3.5 mm/1000 years for siliceous ooze to 17.5 mm/1000 years for highly calcareous sediment.

CONTENTS

INTRODUCTION

This paper presents a combined paleontologic and paleomagnetic stratigraphy for
lower Pliocene through Pleistocene sediments
of the equatorial Pacific and relates it to the
carbonate cycles in sediments. All cores in the
Lamont collection from the area of study
(over 200, Fig. 1) have been examined to
determine the age and lithology of the sediment.

The sediment types from this area have been
thoroughly described by previous authors
(Revelle, 1944; Arrhenius, 1952; Revelle and
others, 1955; Arrhenius, 1963). The pattern
of surface sediment distribution (Fig. 2)
includes all previous work plus Lamont data
and does not depart strikingly from results of

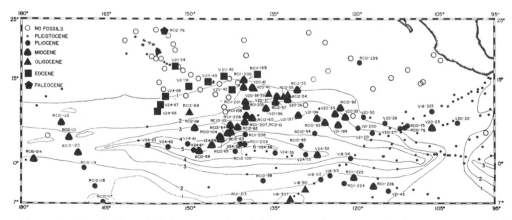

Figure 1.   Distribution of Pleistocene and pre-Pleistocene sediment cores in the eastern equatorial
Pacific. Sediment isopachs drawn from Ewing and others (1968) have a 100-m interval.

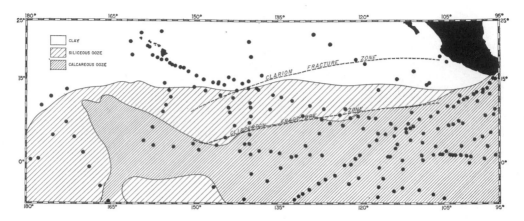

Figure 2. Distribution of sediment cores and sediments in the eastern equatorial Pacific.

earlier investigators. Calcareous sediments are being deposited beneath the equator to depths as great as 4800 m and at shallower depths north and south of the equator.

Siliceous oozes with little or no carbonate occur in a narrow belt to the north, west, and southwest of the calcareous sediments. These siliceous oozes accumulate at depths generally greater than 5,000 m along the equator and at somewhat shallower depths to the north. To the north of the siliceous ooze belt are clay sediments with few microfossils. The boundary between calcareous ooze and siliceous ooze north of the equator closely follows the Clipperton fracture zone (Menard, 1967). Since the Clipperton fracture zone is down-dropped on the north, the sediment change is probably due to greater solution on the north.

Ewing and others (1968) show in an extensive seismic reflection survey that a belt of thick sediments underlies the equatorial region of the Pacific. In the eastern half of the Pacific, the crest of this belt lies slightly north of the equator (4 degrees at 180°) and trends toward the equator to the east (Fig. 1). East of 170° W., these sediments are highly stratified while to the west they are acoustically transparent (Ewing and others, 1968). The upper layers in the highly stratified section thin to the north of the crest while the deeper layers appear to be continuous from south to north, coming to the surface north of the Clipperton fracture zone. A similar thinning of layers is evident in some areas on the southern flank of the thick sediments west of the East Pacific Rise (Ewing and others, 1968). This pattern of outcropping sediments results in the

common occurrence of Tertiary sediments within easy reach of the piston corer a few degrees north of the equator (Fig. 1).

The broad occurrence of outcrops or near outcrops of pre-Pleistocene sediments on the deep-ocean floor makes this region unique. Outcropping early and middle Tertiary sediments are commonly encountered associated with submarine topographic highs, but broad areas of outcropping Eocene to Miocene sediments are rarely found in the deep ocean. Lamont expeditions recovered 32 Pliocene, 28 Miocene, 4 Oligocene, 9 Eocene, and 1 Paleocene cores from this area (Fig. 1). All 64 cores from this region reported by Riedel and Funnell (1964) are either Tertiary in age or contain abundant, reworked Tertiary microfossils indicating the presence of Tertiary outcrops or nearby outcrops.

The pattern of core ages clearly shows two trends: first, an increasing age of sediments cored from the equator northward and a less distinct pattern of increasing age from the equator southward; second, increasing age of oldest sediments cored from the crest of the East Pacific Rise westward. The second trend has been discussed by Riedel (1967) and Burckle and others (1967) and shown to conform to the general theory of ocean-floor spreading (Hess, 1962; Dietz, 1961). The ages of these cores are compatible with estimated rates of spreading (Pitman and Heirtzler, 1966).

From the more than 200 cores in this area, we have selected 15 for paleomagnetic and paleontological study. Of these, four were selected for detailed paleontologic investigation

TABLE 1. LOCATIONS, DEPTHS, LENGTHS, AND AGES OF FOUR CORES

Core No.	Position		Depth (m)	Length (cm)	Age
V24–58	02°16′N	141°40′W	4490	1140	Pliocene
V24–59	02°34′N	145°32′W	4662	1250	Pliocene
V24–60	02°48′N	149°00′W	4859	1256	Pliocene
V24–62	03°04′N	153°35′W	4834	1334	Pliocene

(Table 1) because: (1) They are internally oriented for paleomagnetic measurements; (2) they are all more than 11 m long; (3) they contain complete or nearly complete sections extending into the Pliocene; (4) they all contain diatoms and Radiolaria while two also contain nannoplankton and foraminifera.

ACKNOWLEDGMENTS

We wish to thank Doctors W. Donn, D. B. Ericson, A. Gordon, and A. McIntyre for critically reviewing the manuscript and making numerous helpful suggestions. We are especially grateful to Dr. W. Broecker for carefully reading the manuscript and for several invaluable discussions of Pleistocene chronology. Thanks are also extended to Dr. D. Thurber for carbon-14 determinations, G. Cavallero for carbonate analyses, S. Turner for radiolarian illustrations, D. Gillanders and P. Epstein for typing and proofreading the manuscript, and M. Seely and D. Johnson for drafting.

We are particularly grateful to Dr. Maurice Ewing, Director of the Lamont Geological Observatory, who organized the cruises which collected the cores in this study. We are indebted to the crews of R/V Conrad and R/V Vema and Mr. Roy Capo for collecting and storing the cores.

We wish also to thank W. R. Riedel for providing us with two samples from Core LSDH 78P of the Scripps Institution of Oceanography collection.

The work was supported through grants GA 1193, GA 558, GA 861, GA 824 of the National Science Foundation and grant N-00014-67-A-0108-0004 of the Office of Naval Research.

LITHOLOGY OF SELECTED CORES

Core V24–58 consists largely of alternating layers of nearly white (5 YR 8/1)[1] calcareous ooze and very pale orange (10 YR 8/2) siliceous-calcareous ooze. The core is more calcareous above 550 cm than below (Fig. 12),

and is mottled throughout (probably because of burrowing organisms), but more intensely so in the low-carbonate sections. V24–59 is similar to V24–58, but has through much of its length a lower concentration of calcium carbonate (Fig. 13). The low-carbonate sections (pale yellowish brown (10 YR 5/2)) are darker than those in V24–58. The core is more intensely burrowed than V24–58. V24–60 is a pale yellowish-brown (10 YR 6/2) mottled radiolarian ooze alternating in the upper 4 m with lighter layers (pinkish gray, 5 YR 8/1). V24–62 is a moderate brown (5 YR 3/4) radiolarian ooze alternating in the upper 230 cm with pale yellowish-brown (10 YR 6/2) siliceous-calcareous ooze.

V20–163 was taken in the Indian Ocean from the eastern scarp of the Ninetyeast Ridge, at 17°12′ S. and 88° 41′ E. in 2706 m of water. This core is included to supplement foraminiferal ranges that are altered by solution in V24–59. It is 648 cm long and consists entirely of very compact pinkish-gray (5 YR 8/1) foraminiferal sand. The sediment is very slightly mottled, and no lithological breaks were observed. Holocene foraminiferal species are present only in the top few centimeters. No siliceous microfossils nor evidence of corrosion of calcareous foraminiferal tests were observed.

PALEOMAGNETIC STRATIGRAPHY

Recent paleomagnetic studies of deep-sea piston cores have relied on interpretation of magnetic inclinations to deduce the history of magnetic reversals in the cores. However, due to the low latitude sites and correspondingly low magnetic inclinations of our equatorial

---

[1] For two additional tables on locations, depth, and ages of cores, and zones of major floral and faunal changes, together with a photograph of core V24–59 (Pl. 1) order NAPS Document 00371 from ASIS National Auxiliary Publications Service, c/o CCM Information Sciences, Inc., 22 West 34th Street, New York, New York 10001; remitting $1.00 for microfiche or $3.00 for photocopies.

Pacific cores, interpretation must be based on the magnetic declination alone. Determination of relative magnetic declination is made possible by cutting a groove down the length of the core when the piston core is extruded on board ship. Into this groove a string is laid which serves to retain the orientation during shipment. Splitting the core in the laboratory along this string preserves its internal relative orientation making possible the detection of reversals as 180° shifts in the magnetic declination.

Starting with the 24th cruise of VEMA (1967) and the 11th cruise of CONRAD (1967), all Lamont cores have been oriented as described. Specimens were taken from split cores at intervals of from 10 to 20 cm (Opdyke and others, 1966) and measured on a 5-cps spinner magnetometer (Foster, 1966) after partial demagnetization by alternating fields.

The tight faunal control in this study makes misinterpretation of the magnetic stratigraphy very unlikely. The magnetic declination of the four cores which were given detailed paleontological study is shown in Figure 3. The reversal marking the base of the Brunhes normal series is easily identified in all cores, as are two normal events in the Matuyama reversed series, the upper event correlating with the Jaramillo event and the lower one with the Olduvai event. Core V24–58 bottoms in the lower Matuyama while the other cores pass into or through the Gauss normal series. V24–60 is difficult to interpret below 1150 cm, but it probably terminates in the Gauss normal series in the region of the Mammoth event. V24–62 bottoms in the Gilbert reversed series perhaps as low as the Gilbert "a" event while V24–59 ends below the Gilbert "c" event.

Some changes in the direction of magnetization are interpreted as being caused by the coring operation. For instance, the magnetic vector spirals in the top portion of three of the cores (clockwise in V24–60 and counter-clockwise in V24–62 and V24–59), which is probably a result of twisting of the coring apparatus upon entering the sediment.

In two of the cores there are also unexplained and sudden changes of direction. In V24–60 at a depth of 435 cm a change of 90° occurs, and a similar change is found in V24–62 at 530 cm, directly below the Olduvai event. These sudden shifts of direction are not understood, but probably are engendered by either the coring or handling operations.

The VEMA 24 cruise followed the crest of

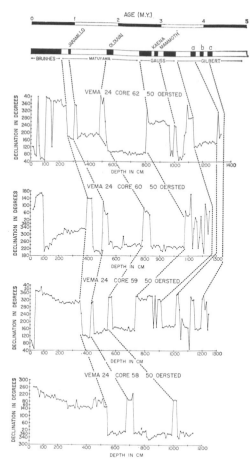

Figure 3. Magnetic correlation of sediment cores V24–58, 59, 60, and 62.

the thick equatorial sediments described by Ewing and others (1968). All cores with the exception of V24–52 are Plio-Pleistocene in age, as will be shown in a later section. Therefore, the last reversal is correlated with the Brunhes-Matuyama boundary. All cores east of 143° W. end in sediments of the Matuyama series. The three cores west of 143° end in older sediments (Fig. 4).

The CONRAD 11 and VEMA 24 tracks cross at right angles at 140° W. longitude. North of the Clipperton fracture zone, the cores are Miocene or older. The cores south of the Clipperton fracture zone show a clear trend of increasing rates of sedimentation toward the equator during the Pleistocene (Fig. 4).

140

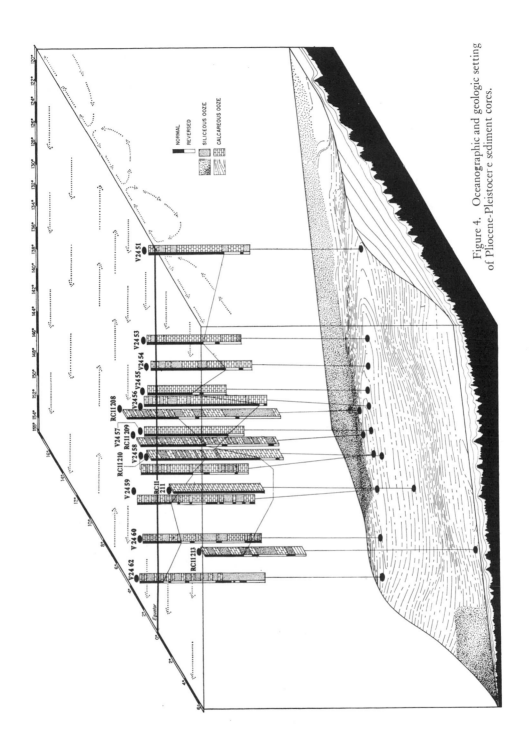

Figure 4. Oceanographic and geologic setting of Pliocene-Pleistocere sediment cores.

141

MICROPALEONTOLOGY

Planktonic Foraminifera

All microfossils were separated from the same sediment samples used for the paleomagnetic measurements. Foraminifers were washed on a 200 mesh (74-micron opening) sieve. More than 400 specimens were randomly separated from each sample; their species and coiling direction were determined. Samples were usually examined at intervals of 40 cm; after inspection, some intervals were studied every 10 cm if a marked faunal or coiling change was noticed.

In core V24–59, corrosion of foraminiferal tests were noted below 860 cm, with solution particularly evident in the interval between 870 cm and 1070 cm. According to Ruddiman and Heezen (1967) and Berger (1968), dis-

solution acts selectively on tests of different species. Those having thin test walls, such as *Candeina* and *Hastigerina*, are completely dissolved; those with a porous test, such as *Globigerinoides*, are partially corroded, while *Sphaeroidinella*, *Sphaeroidinellopsis*, and *Pulleniatina* are least affected. In the Sphaeroidinellids this may be due to a cortical covering of the test. Therefore, it is necessary to compare the ranges of planktonic foraminifera in V24–59 with data obtained from other cores containing better preserved faunas, such as core V20–163 (Figs. 6 and 7).

Planktonic foraminiferal assemblages from cores V24–58, V24–59, and V20–163 are, in general, similar to those described from corresponding horizons of the tropical Pacific and Indian Oceans (Parker, 1967; Figs. 5, 6 and 7). In our cores *Globorotalia tumida flexuosa*

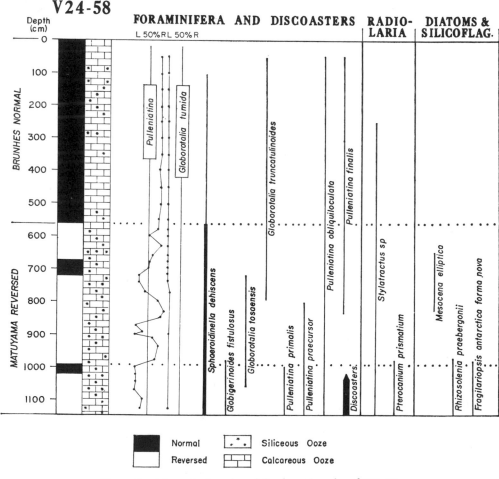

Figure 5.   Magnetic, faunal, and floral stratigraphy of V24–58.

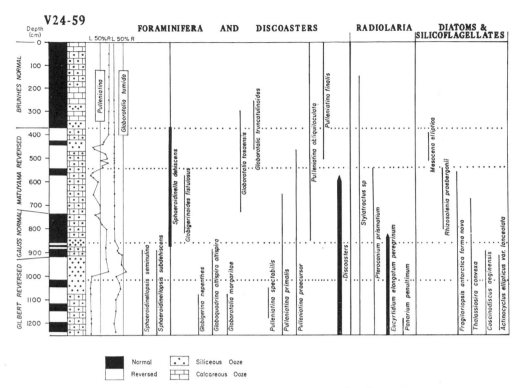

Figure 6.   Magnetic, faunal, and floral stratigraphy of V24–59.

(Koch) is the most dominant species below the middle Matuyama, while *Globoquadrina dutertrei* (d'Orbigny) is the most abundant above the middle Matuyama, amounting in places to as much as 80 percent of the total population. Among the many planktonic foraminifers in these cores, only those species which are stratigraphically important are discussed.

The core-to-core correlation of the ranges of solution-resistant species of *Sphaeroidinella*, *Sphaeroidinellopsis*, and *Pulleniatina* is very good, while ranges of less resistant species tend to diverge in response to the varying carbonate content of the cores.

A profound faunal change occurs in cores V24–59 and V20–163 near the middle of the Gauss series where the subspecies of *Globoquadrina altispira* (Cushman and Jarvis) and *Sphaeroidinellopsis*, all known since early Miocene, disappear, and the modern species *Sphaeroidinella dehiscens* (Parker and Jones) becomes one of the most common. In carbonate-rich core V20–163 the extinction of *Sphaeroidinellopsis* takes place at the top of the Mammoth event, but *G. altispira* continues slightly higher up to near the base of the Kaena

event (Fig. 7). The simultaneous disappearance of *G. altispira* and *Sphaeroidinellopsis* in V24–59 is ascribed to solution.

The successive evolutionary development of *G. tosaensis* Takayanagi and Saito from *Globorotalia crassaformis* (Galloway and Wissler) is observed in V20–163 in the lower part of the Gauss series, and typical *G. tosaensis* makes its first appearance just below the Mammoth event. The ranges of *G. tosaensis* and *G. altispira*, therefore, overlap 1 m in the region of the Mammoth event, as previously observed by Parker (1967).

From the middle of the Gauss epoch to the Matuyama-Brunhes boundary, *S. dehiscens* is a common species in all cores constituting nearly 4 percent of the total population, but becomes rare (less than 1 percent) in the Brunhes series. A similar reduction in abundance of *S. dehiscens* at the base of the Brunhes series has been reported in a number of other cores by Glass and others (1967).

Within the geologic section covered by our cores, there are four major intervals dominated by sinistrally coiled specimens of the genus *Pulleniatina* (Pl. 1), and they are useful in cor-

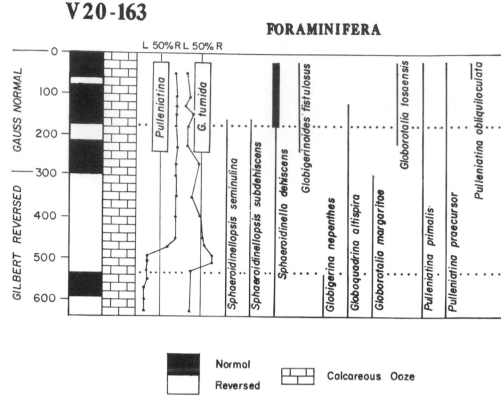

Figure 7.   Magnetic and faunal stratigraphy of V20–163.

relation. The oldest coiling change of the *Pulleniatina* group is very sharp, occurring just above the Gilbert "a" event. Nearly coincident with this oldest *Pulleniatina*-coiling change is a preference of *G. tumida flexuosa* for right coiling. The third horizon characterized by left-coiling *Pulleniatina* in core V24–58 between Olduvai and Jaramillo events is missing in V24–59, probably because of a disconformity. This interpretation is supported by the fact that the section between the Olduvai and Jaramillo events in core V24–59 is relatively short compared with the length of the Brunhes.

Pulleniatina spectabilis Parker is a short-ranging species found only in core V24–59. Its first appearance is just above the Gilbert "c" event, but well-developed specimens abound only in a short section between the Gilbert "a" and "b" events.

Globorotalia margaritae Bolli and Bermudez, *Globigerina nepenthes* Todd, *Globoquadrina altispira* (Cushman and Jarvis), and *Globigerinoides fistulosus* (Schubert) are, in our study,

considered to be solution-susceptible species since they are absent or rare in carbonate-poor sections. Therefore, ranges of these species in carbonate-poor cores are, in general, much restricted. In carbonate-rich cores, the extinction of *G. nepenthes* takes place at the top of the Gilbert "a" event just before the first coiling change in the *Pulleniatina* group (Fig. 2). *G. margaritae* becomes extinct at the Gilbert-Gauss boundary (Fig. 7). The last occurrence of *G. fistulosus*-rich zone is near the top of the Olduvai event in cores V24–58 and V24–59.

In our cores both *Globorotalia tosaensis* and *G. truncatulinoides* (d'Orbigny) are rare representing only 1 to 2 percent of the total fauna throughout their observed ranges. The sparseness of these two species is ascribed to the tropical location of our cores. According to Bradshaw (1959), *G. truncatulinoides* is primarily a central-water species in the Pacific and is most abundant between 20° N. and 40° N.

Discoasters

Discoasters were observed in three of the four cores studied. In two of them (V24–58 and 59), discoasters are reduced sharply in number just above the base of the Olduvai event. In the third core (V24–60), they disappear well below the Olduvai event, probably because of solution. The relationship of discoasters to the Pliocene-Pleistocene boundary has been discussed by a number of workers (Ericson and others, 1963; Riedel and others, 1963; Akers, 1965; Wray and Ellis, 1965; Glass and others, 1967; Berggren and others, 1967, 1967; and Bolli and others, 1968). The main problem encountered in trying to relate these tiny microfossils to the boundary is the ease with which they are reworked (McIntyre and others, 1967). Therefore, a number of workers are in disagreement as to the position of discoaster extinction in relation to the Olduvai event. While Glass and others (1967) place the sharp discoaster reduction in the lower part of the Olduvai, and Berggren and others (1967) report abundant discoasters some 80 cm above the top of the Olduvai, we note a sharp decrease in discoaster abundance within the Olduvai event.

Since V24–59 contains a long late Tertiary section, a study was made of the relative abundance of several species of discoasters. Below about 820 cm, a number of species are present including *Discoaster pentaradiatus*, *D. brouweri*, *D. surculus*, *D. exilis*, and *D. challengeri*. *D. brouweri* is the most common species just below 820 cm, but its abundance is variable and falls off steadily deeper in the core. Between 820 cm and the Olduvai event, *D. brouweri* is the most abundant species (always exceeding 90 percent), and in some samples it was the only species observed.

Radiolaria

A radiolarian stratigraphy for equatorial Pacific sediments was first defined by Riedel (1957) and later refined by him (Riedel, 1959; Riedel and Funnell, 1964; and Riedel, 1969). No attempt is made in this study to examine the stratigraphic ranges of all equatorial Radiolaria. Our purpose is rather to relate the ranges of a few stratigraphically significant radiolarian species to the paleomagnetic record and to compare these ranges with those of foraminifer and diatom species.

The Radiolaria included in this study are, in order of increasing upward range: *Panarium*

penultimum Riedel, *Eucyrtidium elongatum peregrinum* Riedel, *Pterocanium prismatium* Riedel, and *Stylatractus* sp. of Hays (1965; Figs. 5, 6, 8, and 9).

To delimit species ranges, samples were taken at 40-cm intervals with the sampling interval reduced to 10 cm to determine the upper limits of species. Counts of 1000 individuals of the total assemblage were made on several samples from each core to determine relative abundance.

Stylatractus sp. and *Pterocanium prismatium* are never abundant, each constituting 1 percent or less of the total assemblage. *Eucyrtidium elongatum pereqrinum* rarely constitutes more than 6 percent of the total assemblage. *Stylatractus* sp. and *Pterocanium prismatium* have sharp upper limits usually disappearing within 10 cm. Because these species are rare,

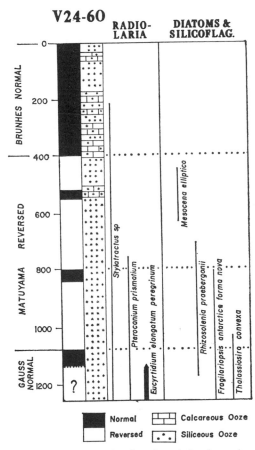

Figure 8. Magnetic, faunal, and floral stratigraphy of V24–60.

the upward extension of their occurrence through mixing is limited.

The disappearance of the abundant *Eucyrtidium elongatum peregrinum* is not as abrupt. To determine its upper limit in cores V24–59, 60, and 62, counts were made of 1000 individuals in samples taken at 10-cm intervals through a range of 70 cm. In core V24–59, this species declines from 18 percent at 840 cm to zero at 770 cm (Fig. 10) possibly due to a declining population, but more likely due to upward mixing.

Panarium penultimum shows a gradual decrease in the size of its polar columns and is transitional with the modern species *Penartus tetrathalamus* (Fig. 11). Picking the upper limit of *Panarium penultimum* is thus somewhat arbitrary since the transition occurs through some 100 cm in V24–59. For the purpose of this study, we have restricted the definition of *Panarium penultimum* to those individuals with well-developed polar columns (Fig. 11a) be-

cause these individuals most closely resemble the holotype (Riedel, 1957, Fig. 1) and because at levels where the polar columns are less well developed (Fig. 11b), many individuals have no columns at all and are indistinguishable from *Penartus tetrathalamus* (Fig. 11c). The last occurrence of forms with well-developed columns is fairly abrupt.

The upper limit of *Stylatractus* sp. falls in the central portion of the Brunhes normal series. The estimated age of the upper limit of *Stylatractus* sp. is between 300,000 and 380,000 years B.P. (Table 2). The two cores that have higher and more uniform carbonate content (V24–58 and 59, *see* Figs. 12 and 13) show younger ages for the upper limit of *Stylatractus* sp. than do the two cores having large fluctuations in carbonate content (V24–60 and 62, *see* Fig. 13). Since rates of sedimentation are apt to be more constant in the cores with more uniform carbonate content, the figure of 320,000 years from core V24–58 is probably the best estimate of the age of the upper limit of this species. Work on North Pacific cores (above 40° N.) has shown that this species disappears in this region about 400,000 years B.P. It disappears from Antarctic sediments at about the same time (Ku, 1966; Opdyke and others, 1966).

The age of the upper limit of *Pterocanium prismatium* can be roughly estimated by calculating the rate of sedimentation between the Olduvai and Jaramillo events, using 0.87 m.y.

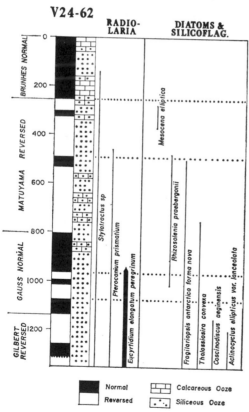

Figure 9. Magnetic, faunal, and floral stratigraphy of V24–62.

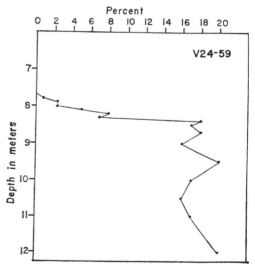

Figure 10. Abundance of *Eucyrtidium elongatum peregrinum* in V24–59.

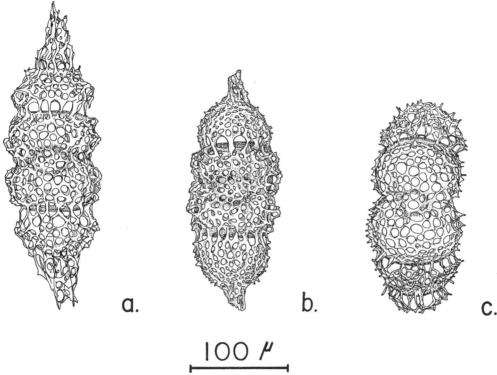

$\underset{\text{100 }\mu}{\longmapsto\!\!\!\longmapsto}$

100 μ

Figure 11.   Evolutionary transition from *Panarium penultimum* (a) to *Penartus tetrathalamus* (b and c).

for the top of the Jaramillo event (Opdyke, 1969) and 1.85 m.y. for the top of the Olduvai event. The age of the upper limit of *P. prismatium* is similar in cores V24–58 and V24–59 (Table 2). These represent probably the most reliable estimates since the higher rates of accumulation in these cores would reduce the effect of upward mixing. The estimated age of about 1.8 m.y. for the upper limit of *Pterocanium prismatium* is considered rather reliable.

Eucyrtidium elongatum peregrinum was originally reported to have an upper limit similar to that of *Pterocanium prismatium* (Riedel, 1957), but was later shown to become extinct

somewhat below *Pterocanium prismatium* (Riedel and others, 1963). *Eucyrtidium elongatum peregrinum* always disappears within the Gauss series in cores that we have studied. V24–62 has the most complete magnetic stratigraphy for the Gauss epoch showing clearly both the Mammoth and Kaena reversed-polarity events. In this core *Eucyrtidium elongatum peregrinum* has an upper limit 30 cm above the Kaena event giving an age for its disappearance of about 2.7 m.y.

The upper limit of *Panarium penultimum*, as defined here, falls just above the Gilbert "c" event giving it an age of about 4.5 m.y. B.P.

TABLE 2.   AGE OF UPPER LIMIT OF SELECTED RADIOLARIAN SPECIES

	V24–58	V24–59	V24–60	V24–62	Av.
Stylatractus sp.	324,000	300,000	368,000	377,000	341,000
Pterocanium prismatium	1,802,000	1,808,000	1,688,000	1,665,000	1,741,000
Eucyrtidium elongatum peregrinum		2,700,000		2,700,000	
Panarium penultimun		4,500,000			

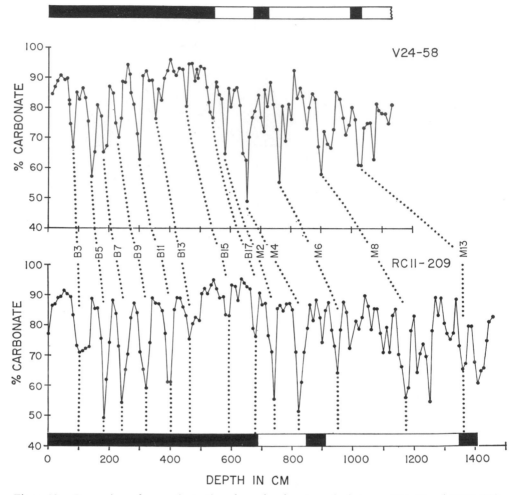

Figure 12.   Comparison of magnetic stratigraphy and carbonate cycles between V24–58 and RC11–209.

Diatoms and Silicoflagellates

All cores studied contain abundant diatoms. A few fairly abundant species having restricted stratigraphic ranges have been selected for study. Their ranges are compared with those of the foraminifera and Radiolaria, and with the paleomagnetic stratigraphy (Figs. 5, 6, 8, and 9).

Kolbe (1954) made a taxonomic stratigraphic study of the diatoms in cores recovered by the Swedish Deep-Sea Expedition. Muhina (1963) studied and described diatoms from a single core in the equatorial Pacific. Her study is of special interest because she described several new forms which were restricted in time. These forms, along with a few others, have been used in this paper.

The diatoms included in this study in order of increasing upward range are: *Actinocyclus ellipticus* var. *lanceolata* Kolbe, *Coscinodiscus aeginensis* Schmidt, *Thalassiosira convexa* Muhina, *Fragilariopsis antarctica forma nova* of Muhina, and *Rhizosolenia praebergonii* Muhina. *Actinocyclus ellipticus* var. *lanceolata* and the typical form (*Actinocyclus ellipticus*) have been observed in Miocene cores in this area. *A. ellipticus* var. *lanceolata* is common to abundant in Miocene and lower Pliocene sediments and disappears near the base of the Gauss series.

Coscinodiscus aeginensis disappears at about the same time as *A. ellipticus* var. *lanceolata*. In studying several other Miocene and lower Pliocene cores from the equatorial Pacific, it was discovered that this is a recurring species.

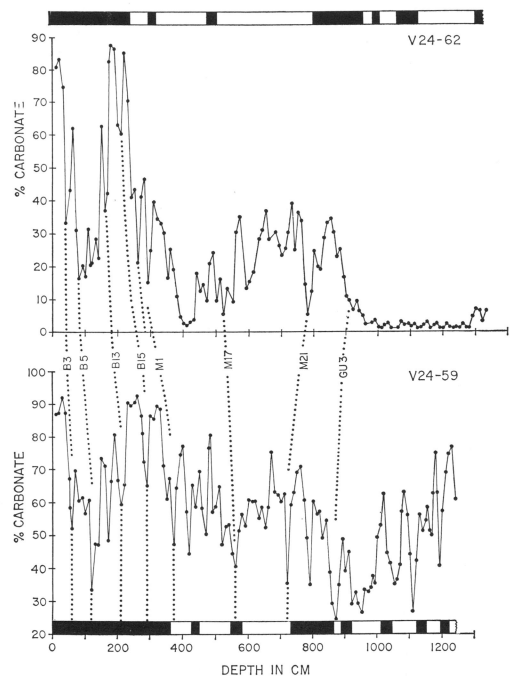

Figure 13. Comparison of magnetic stratigraphy and carbonate cycles between V24–59 and V24–62.

It is present in fair abundance in Miocene sediments and then disappears only to reappear below the base of the Gilbert "c" event. Its last occurrence is usually slightly above the last occurrence of *A. ellipticus* var. *lanceolata*. Kolbe (1954) noted two occurrences of *C. aeginensis* in Swedish Deep-Sea Expedition core 76, one at 23 cm and one in the deeper layers of the core.

Thalassiosira convexa extends up into the lower part of the Matuyama series below the Olduvai event. It has the same stratigraphic range in the North Pacific (J. Donahue, personal commun., 1968) and in a mid-latitude core from the Indian Ocean (V16–66), thus making it a good late Tertiary stratigraphic indicator.

F. antarctica forma nova disappears above the Olduvai event. It is interesting that similar species of *Fragilariopsis* in North Pacific (J. Donahue, personal commun., 1968) and Antarctic sediments also disappear above the Olduvai event.

Rhizosolenia praebergonii first appears in the middle part of the Gauss series, usually above the Mammoth event, and extends to above the Olduvai event. Generally, this species and *F. antarctica forma nova* disappear about the same time. *R. praebergonii* becomes progressively larger and more massive through its range.

A silicoflagellate, *Mesocena elliptica*, is stratigraphically useful, its range bracketing the Jaramillo event for which it is a useful guide fossil (Fig. 5, 6, 8 and 9).

Zones of Floral and Faunal Change

The combined nature of this study permits us to compare the ranges of species from four separate groups of micro-organisms and determine those intervals where changes occur in all groups. These intervals then can be related to the magnetic stratigraphy. Two zones of major faunal and floral change and several minor ones can be defined within the last 4.5 m.y.

The oldest zone of major change (foraminifera Datum V) occurs in the middle of the Gauss normal series. At this time (around 3 m.y. B.P.), a number of species of foraminifera, diatoms, and Radiolaria become extinct. The foraminiferal changes that occur within this interval have been used by earlier workers to define the Miocene-Pliocene boundary (Bandy, 1963; Bolli, 1966b; Bandy and Wade, 1967). Ericson and others (1963) reported the disappearance of *S. multiloba* (=*S. seminulina*) and *G. altispira* at the Pliocene-Pleistocene boundary. However Glass and others (1967) have shown that these two species disappear within the Pliocene about 3 m.y. B.P., a result with which our data agree (Pl. 1).

The next younger zone of major faunal change occurs in the vicinity of the Olduvai event (Datums IV and III). Here again, species of foraminifera, diatoms, and Radiolaria become extinct at about the same time.

In addition to the two zones of major faunal and floral changes, there are a number of other less profound changes. Three of these are worthy of mention because of their importance in correlation. First, at the top of the Gilbert "a" event, the first coiling change from left to right of both *Pulleniatina* and *G. tumida flexuosa* occurs. These changes are accompanied by the extinction of *G. nepenthes* and *P. spectabilis*. The second is at the base of the Brunhes series where the abundance of *S. dehiscens* is greatly reduced; the third is the extinction of *Stylatractus* sp. within the Brunhes series (Figs. 5, 6, 8, and 9).

RELATIONSHIP BETWEEN MAGNETIC REVERSALS AND FAUNAL AND FLORAL EXTINCTIONS

During the past several years there has been much discussion concerning the possible relationship between reversals of the earth's magnetic field and extinctions of planktonic organisms.

It has been suggested (Bullard, 1955) that at the time of a reversal the intensity of the earth's dipole field decreased to zero. Recent studies have presented data supporting this idea (Ninkovich and others, 1966; Harrison and Somayajulu, 1966). If this is true, then the earth's surface may have been subjected to a higher incidence of radiation during a reversal resulting in higher mutation rates that affected evolutionary processes (Uffen, 1963). Recently, several authors (Opdyke and others, 1966; Hays and Opdyke, 1967; Glass and others, 1967; and Watkins and Goodell, 1967) have presented data which show a close correlation between some extinctions of planktonic organisms and reversals of the earth's magnetic field. Subsequently, several articles have appeared presenting theoretical considerations that argue against a significantly higher dose of cosmic radiation at the earth's surface during a reversal (Waddington, 1967; Black, 1967; Harrison, 1968).

We present our data as a means of establishing the degree of correlation between reversals of the earth's magnetic field and extinction of planktonic organisms (Table 3). Of the three species of Radiolaria that became extinct during the interval of time covered by our cores, two (*Pterocanium prismatium* and *Eucyrtidium elongatum peregrinum*) disappear a few tens of centimeters above a reversal. The third (*Stylatractus* sp.) disappears in the middle of the Brunhes series well away from the last reversal of the earth's field.

All but one of the foraminifera that become extinct during the time we are considering have their last occurrence within a few tens of centimeters from a reversal.

Two out of five diatom species disappear a few tens of centimeters above a reversal (*Rhizosolenia praebergonii, Fragilariopsis antarctica*). The last occurrence of two of the other three species of diatoms (*Coscinodiscus aeginensis, Actinocyclus ellipticus* var. *lanceolata*) are not consistent from core to core, probably because of reworking. The final extinction of discoasters

TABLE 3. RELATION OF FLORAL AND FAUNAL EXTINCTIONS TO GEOMAGNETIC REVERSALS

Species	Core	Depth (cm)	Nearest Reversal (cm)	Magnetic Epoch and Event	Distance (cm) to Nearest Reversal (+) Above (−) Below
RADIOLARIA					
Stylatractus sp.	V24–58	250	545		Middle of
	V24–59	140	365	Top of Matuyama	Brunhes
	V24–60	210	395		
	V24–62	140	255		
Pterocanium prismatium	V24–58	970–80	990–1000		+20
	V24–59	530–40	540–50		+10
	V24–60	750–60	780–90	Top of Olduvai	+30
	V24–62	460–70	480–90		+20
	RC11–208	1120–30	1140–50		+20
Eucyrtidium elongatum peregrinum	V24–59	840–50	850–60		+10
				Top of Kaena	
	V24–62	930–40	965–75		+35
FORAMINIFERA					
Sphaeroidinellopsis seminulina	V20–163 (Indian O.)	170–80	170–80	Top of Mammoth	0
Sphaeroidinellopsis subdehiscens	V24–59 (Pacific O.)	880–90	870–80	Top of Mammoth	−10
Globigerinoides fistulosus	V24–58 (Pacific O.)	970–80	900–1000	Top of Olduvai	+20
	V24–59 (Pacific O.)	540–50	540–50	Top of Olduvai	0
	RC11–252 (Atlantic O.)	190–200	200–10	Top of Olduvai	+10
Globigerina nepenthes	RC11–252 (Atlantic O.)	440–50	450–60	Top of Event A	+10
	V20–163 (Indian O.)	530–40	530–40	Top of Event A	0
	V24–59 (Pacific O.) Solution effect	1070–80	1040–50	Base of Event A	−30
Globoquadrina altispira	V20–163	110–120	170–80	Top of Mammoth	+60
	V24–59 (Pacific O.)	880–90	870–80	Top of Mammoth	−10
Globorotalia margaritae	V24–59 (Pacific O.) Solution effect	1070–80	1040–50	Base of Event A	−30
	V20–163 (Indian O.)	290–300	290–300	Top of Gilbert	0
Pulleniatina spectabilis	V24–59 (Pacific O.)	1040–50	1040–50	Base of Event A	0

151

TABLE 3   (Continued)

Species	Core	Depth (cm)	Nearest Reversal (cm)	and Event Mgnetic Epoch	Distance (cm) to nearest Reversal (+) Above (−) Below
DIATOM					
Mesocena elliptica	V24–62	270–80	300–310	Top of Gilbert	+30
	V24–60	430–440	500–510	Top of Jaramillo	+70
	V24–58	640–50	570–580	Top of Jaramillo	+70
	V24–59	380–390	420–430	Base of Brunhes	+40
Rhizosolenia praebergonii	V24–62	470–80	480–490	Top of Olduvai	+10
	V24–60	690–700	780–790	Top of Olduvai	+90
	V24–58	960–970	990–1000	Top of Olduvai	+30
	V24–59	530–40	540–550	Top of Olduvai	+10
Fragilariopsis antarctica forma nova					
	V24–62	490–500	480–490	Top of Olduvai	−10
	V24–60	790–800	790–800	Top of Olduvai	0
	V24–58	970–80	990–1000	Top of Olduvai	+20
	V24–59	550–60	540–550	Top of Olduvai	−10
Thalassiosira convexa	V24–62	740–50	800–810	Top of Gauss	+60
	V24–60	1020–30	1070–80	Top of Gauss	+50
	V24–59	660–70	740–50	Top of Gauss	−80
Coscinodiscus aeginensis	V24–62	1140–50	1130–40	Base of Gauss	−10
	V24–59	880–90	880–90	Base of Mammoth	0
Actinocyclus ellipticus var. *lanceolata*					
	V24–62	1190–1200	1130–40	Base of Gauss	−60
	V24–59	900–910	900–910	Base of Gauss	0

in our cores occurs in the neighborhood of the Olduvai event.

It is possible that burrowing organisms may depress the position of a reversal in a core by a few to a few tens of centimeters. In our judgment, there is no way in which burrowing could raise the position of a reversal in a core. Burrowing, on the other hand, can extend upward the last occurrence of a species. Burrowing action, therefore, may have increased the distance of the last occurrence of species above reversals in our cores. Extinction of foraminifera show the closest correlation with reversals, followed by Radiolaria and diatoms. The foraminifera—larger and heavier—are less easily mixed than the smaller and lighter Radiolaria and even smaller diatoms.

While foraminifera, as already discussed, show some solution no such effects were noted for either the Radiolaria or diatoms. The correlation between last occurrences and reversals of solution-resistant species (*S. seminulina, S. subdehiscens*) is very good in all cores. Some solution-susceptible species (*G. nepenthes, G. margaritae*), on the other hand, show good correlation with reversals only in cores that show no evidence of solution.

Those species of foraminifera that show close correlation of their last occurrence with a re-versal show morphological peculiarities that may be indicative of a high degree of specialization. A few examples are the strongly serrated margin of the chambers of *G. fistulosus*, the trochospiral form of *G. altispira* and *G. nepenthes*, and the unique nature of the narrowly angled periphery of *P. spectabilis*.

The close correlation between extinctions and reversals in the cores studied here and in Antarctic cores (Opdyke and others, 1966; Hays and Opdyke, 1967; Watkins and Goodell, 1967) suggests a causal connection. Since only a few organisms become extinct near a reversal, and some have survived many reversals before becoming extinct, it is clear that a reversal alone is probably insufficient to cause an extinction. However, highly specialized species may be vulnerable to slight changes in the environment and for these, small direct or indirect effects of a reversal may be sufficient to cause their extinction.

It is clear that detailed morphological studies of those species that become extinct near reversals are now in order to determine whether their specific variability is changed toward the end of their range.

Our zones of major floral and faunal change occur at times of frequent reversals. The central part of the Gauss series contains at least four reversals within about 300,000 years. There is now evidence that the Olduvai event contains a short reversed interval, thereby having four reversals in about 150,000 years (Ninkovich and others, 1966).

From this study and previous work (Opdyke and others, 1966; Hays and Opdyke, 1967), most reversals since the Gilbert "a" event have one or more species disappearing near them. Reversals that have extinctions associated with them in the Antarctic (top of Gauss, top and bottom of Olduvai) are, with the exception of the top of the Olduvai, not the reversals that have extinctions associated with them in our equatorial cores. Therefore, nothing appears peculiar about individual reversals that makes some more likely than others to have species extinctions associated with them, but, rather, it is probably the degree of specialization of individual members of the population that make some species prone to extinction at different times in different areas. Because of specialization these species might have become extinct in any case, but the reversal may have hastened their extinction.

CORRELATION OF OUR PALEOMAGNETICALLY DATED BIOSTRATIGRAPHY WITH PREVIOUS WORK

In the past few years, much has been learned about the taxonomy and stratigraphic ranges of Pliocene-Pleistocene marine planktonic microfossils. Despite this knowledge, biostratigraphic zonations and inter-regional correlations still remain in dispute.

For the identification of Miocene-Pliocene and Pliocene-Pleistocene boundaries, we follow the planktonic-foraminiferal criteria proposed by Banner and Blow (1965, 1967) and Blow (1968).

The type Pliocene in Italy consists primarily of marginal marine facies which yield only few species of globorotaliids (Selli, 1967). In Sicily, however, sediments of the Messinian stage of uppermost Miocene are overlain by fossiliferous "Trubi" beds of the Zanclian stage of lowest Pliocene (Selli, 1967; Tongiorgi and Tongiorgi, 1964). According to Blow (1968) the development of *S. dehiscens* occurs about 40 ft above the base of the exposed Trubi marl. Therefore, the Miocene-Pliocene boundary can be recognized outside of the stratotype section as a horizon shortly below the first appearance of *S. dehiscens*.

The 18th International Geological Congress in Great Britain, Section H (1950), recommended that the lower Pleistocene should include the marine Calabrian formation in the type area as its basal member. Banner and Blow (1965) showed that the base of the type-Calabrian stage approximates the first evolutionary appearance of planktonic foraminifer *G. truncatulinoides* from *G. tosaensis*. Berggren and others (1967) showed that this evolutionary transition occurs within the Olduvai event. Hays and Berggren (1970) reviewed the status of the Pliocene-Pleistocene boundary in deep-sea sediments and concluded that the boundaries defined by Ericson and others (1963) in the Atlantic, Riedel and others (1963) in the Pacific, and Hays (1965) in the Antarctic were correlative with the boundary in the type section and dated it at about 1.8 m.y. B.P.

A comprehensive review of late Cenozoic foraminiferal biostratigraphy has been presented by Parker (1967). Among the authors reviewed by Parker, a marked disparity exists concerning the placement of the foraminiferal zones with reference to the European-type sections, although, in general, the authors have used the same kind of foraminiferal criteria in establishing their biostratigraphic subdivisions. From our study it is possible to correlate the criteria of various authors with the paleomagnetic record which provides a worldwide absolute time scale. Plate 1 summarizes our interpretation of the late Cenozoic biostratigraphy of various workers. The following criteria establish useful time planes for worldwide correlation.

Useful Criteria in Order of Increasing Age

Datum (I). A marked upward decrease in abundance of *Sphaeroidinella dehiscens* at the Brunhes-Matuyama boundary.

Datum (II). The first evolutionary appearance of *Pulleniatina finalis* from *P. obliquiloculata* midway between the Jaramillo and Olduvai events.

Datum (III). The upper limit of *Globigerinoides fistulosus* near the top of the Olduvai event.

Datum (IV). The first evolutionary appearance of *Globorotalia truncatulinoides* near the lower boundary of the Olduvai event. (This datum is poorly defined in the tropics where temperate-water species are rare.)

Datum (V). The extinction of all species of *Sphaeroidinellopsis* coincident with increased upward abundance of *Sphaeroindinella dehiscens* at the top of the Mammoth event. (This is the same as Bandy's (1963) *Sphaeroidinella dehiscens* datum and is the most easily recognized foraminiferal datum.)

Datum (VI). The extinction of *Globorotalia margaritae* at the Gauss-Gilbert boundary.

Datum (VII). A sharp left-to-right coiling change in *Pulleniatina*, primarily of *P. primalis*, just above the Gilbert "a" event.

Datum (VIII). The extinction of *G. nepenthes* at the top of the Gilbert "a" event.

Datum (IX). The first evolutionary appearance of *Pulleniatina spectabilis* from *P. primalis* near the top of the Gilbert "c" event. (This datum may be applicable only in the Indo-Pacific region to which this species is apparently restricted.)

Bandy (1963), working on late Cenozoic sediments of the Philippines, recognized a profound faunal change across what he believed to be the Miocene-Pliocene boundary, where *Globoquadrina altispira globosa*, *Sphaeroidinella* (=*Sphaeroidinellopsis*) *seminulina* vars., *S. seminulina subdehiscens*, and the "right-coiling *Globorotalia menardii* complex" became extinct at the same level. This boundary also coincides with the first appearance in abundance of *Sphaeroidinella dehiscens dehiscens*, which he called the *Sphaeroidinella dehiscens* datum. Within his upper Miocene he observed that the extinction of *G. nepenthes* was approximately coincident with a marked coiling change in *Pulleniatina obliquiloculata* from sinistral to dextral.

The faunal boundaries recognized by Bandy are present in our cores and occur in an almost identical succession (Pl. 1). The extinction of *G. nepenthes* (our Datum VIII) closely followed by an abrupt left-to-right coiling change in the species of *Pulleniatina* (our Datum VII) occurs at the top of the Gilbert "a" event at about 3.7 m.y. This confirms the observation of Ericson and others (1963) that *G. nepenthes* ranges up into the Pliocene. The extinction of all the species of *Sphaeroidinellopsis* and the increased upward abundance of *S. dehiscens* occur in our cores at the top of the Mammoth event dated at approximately 3 m.y. The excellent agreement of these faunal changes between our cores and the Philippines section indicate that Bandy's Miocene-Pliocene boundary is 3 m.y. old and is equivalent to the top of the Mammoth event of the Gauss normal series. Parker (1967) also concluded that Bandy's boundary lies within the Pliocene be-

tween Zones N 19 and N 21 of Banner and Blow (1965).

Bandy (1967) has suggested an age of about 9 m.y. for the point of initial appearance of *S. dehiscens* on the basis of radiometric dating of a vitric tuff collected from the middle part of the Malaga mudstone in the Malaga Cove section of California. According to Ingle (1967), this section consists largely of foraminiferal barren, radiolarian-rich mudstone. Therefore, the 9 m.y. date for the first appearance of *S. dehiscens* must have been based on the correlation of this date with other sections which yielded *S. dehiscens*. From Bandy's paper (1967) it is not clear whether the initial appearance of *S. dehiscens* is equivalent to the *S. dehiscens* datum as defined in the Philippines. Our data indicate that Bandy's *S. dehiscens* datum is about 3 m.y. old, but the earliest occurrence of *S. dehiscens* is older (at least 4.5 m.y.).

Parker (1967) has suggested that a part of the section which corresponds to Banner and Blow's Zone N 20 may be missing in the Philippines at Bandy's Miocene-Pliocene boundary. In the Philippines a short interval characterized by the sinistrally coiled forms of *Pulleniatina* is present immediately above the boundary. Since left-coiling forms of *Pulleniatina* do not recur in our cores until the upper Gauss normal series, it is possible that an interval is missing from the Philippine section. *G. altispira* has been found in our cores to range slightly higher than the last occurrence of *S. seminulina*. Therefore, the simultaneous disappearance of *G. altispira* and *S. seminulina* in the Philippines may be additional evidence of a stratigraphic break.

Bandy and Wade (1967) applied Bandy's foraminiferal criteria established in the Philippines to a South Atlantic core (V15–164). This core was earlier investigated by Ericson and others (1963) who placed the Plio-Pleistocene boundary at around 170 cm on the basis of planktonic foraminifera. Bandy and Wade (1967) noted patterns of foraminiferal ranges closely comparable with those observed in the Philippines and concluded that Ericson's Plio-Pleistocene boundary represents a stratigraphic hiatus in this core with Pleistocene sediments directly overlying lowermost Pliocene *Globigerina* ooze. We would agree that this core has a hiatus, but the sediment underlying the hiatus is middle Pliocene in age (Pl. 1). Our attempt to date this core paleomagnetically failed because of weakly magnetized sediments.

The foraminiferal zonation of Bolli and

Bermudez (1965) for middle Miocene to Pliocene warm-water sediments has been applied to a Neogene Indonesian section (Bolli, 1966). Bolli placed the Miocene-Pliocene boundary at the top of the *G. altispira/G. crassaformis* zone, the upper limit of which corresponds to the extinction of *G. altispira* (Pl. 1). In the Indonesian section, *G. fistulosus* makes its first appearance shortly before the extinction of *G. altispira*, as is found in our study. Therefore, Bolli's top of the Miocene correlates with a horizon between the Kaena and Mammoth events in our cores, about 2.9 m.y. ago. Near the middle Miocene-upper Miocene boundary of Bolli, there is a marked coiling change in *Pulleniatina* from sinistral to dextral. This change shortly follows the last occurrence of *G. nepenthes* and is associated with an interval dominated by dextrally coiled forms of *G. tumida*. On these bases, we correlate his upper Miocene-middle Miocene boundary with our Datum VIII, the top of the Gilbert "a" event about 3.7 m.y. ago.

Ericson and others (1963) used the first abundant occurrence of *G. truncatulinoides* as an important criterion for recognizing the Pliocene-Pleistocene boundary in the North Atlantic Ocean. In a North Atlantic deep-sea core, Berggren and others (1967) compared the phylogeny of *Globorotalia truncatulinoides* with paleomagnetic stratigraphy. The evolutionary transition from *Globorotalia tosaensis* to *G. truncatulinoides* was found to occur within the Olduvai normal event, and the Pliocene-Pleistocene boundary, as determined by the first evolutionary appearance of *G. truncatulinoides* (Banner and Blow, 1965) was estimated to be about 1.85 m.y. old, thereby verifying the usefulness of the *G. truncatulinoides* criterion of Ericson and others (1963) as a marker for the Pliocene-Pleistocene boundary. Glass and others (1967), working on a number of cores previously studied by Ericson and others (1963) also observed the first evolutionary appearance of *G. truncatulinoides* near the base of the Olduvai event.

In spite of the fact that we are working in essentially the same area as Parker (1967), our species ranges do not always coincide with hers. Since *S. seminulina* is one of the most solution-resistant species, the extinction of this species is taken as the primary basis for correlating our data with hers. As a further check, two samples from core LSDH 78P (472–4 and 506–8 cm) across Parker's Miocene-Pliocene boundary were studied to ensure the correlation with Parker.

Examination of the Radiolaria in both of these samples shows that they contain *Pterocanium prismatium*, but lack *Panarium penultimum*. The first appearance of *P. prismatium* takes place between the Gilbert "c" and "b" events while *P. penultimum* as defined here becomes extinct above the Gilbert "c" event. Foraminiferal faunas from these two samples include among others *Globigerina nepenthes*, sinistrally coiled, primitive *Pulleniatina spectabilis*, and sinistrally coiled *Pulleniatina primalis*. Thus, foraminifera and Radiolaria indicate that Parker's Miocene-Pliocene boundary falls somewhere between the Gilbert "c" and "b" events, suggesting an age of about 4.2 m.y. It is noteworthy that Dymond (1966) obtained a K-Ar date of 4.3 ± 0.3 m.y. on an ash layer near a Miocene-Pliocene boundary determined by Martini and Bramlette (1963) in the experimental Mohole core. This paleontological boundary was later confirmed by Parker (1967). However, since our core V24–59 contains *S. dehiscens* throughout, we assume that the Miocene-Pliocene boundary as defined by the first appearance of *S. dehiscens* (Banner and Blow, 1965) must lie below the base of our core which we estimate to have an age of about 4.5 m.y. For the time within the range of *S. dehiscens*, Banner and Blow (1965) established six planktonic foraminiferal zones (N 18 to N 23). The Miocene-Pliocene and Pliocene-Pleistocene boundaries are drawn below the zonal boundaries of N 18/N 19 and N 21/N 22, respectively (Pl. 1). Zone N 19 is defined as the interval between the first appearance of *S. dehiscens* and the last occurrence of *G. altispira;* Zone N 20 is the interval subsequent to the extinction of *G. altispira* and prior to the advent of *G. tosaensis;* Zone N 21 is characterized by the occurrence of *G. tosaensis* prior to the first appearance of *G. truncatulinoides;* Zone N 22 is defined by the occurrence of *G. truncatulinoides* prior to the first appearance of *Globigerina calida* (Parker) and *S. dehiscens excavata* Banner and Blow.

The overlapping ranges of *G. altispira* and *G. tosaensis* in Parker's (1967) cores and in our cores precludes the existence of Banner and Blow's N 20 in the tropical Indo-Pacific regions. Blow (1968) subsequently redefined Zone N 20 of Banner and Blow with the introduction of two new species, *Globorotalia acostaensis pseudopima* and *Globorotalia tosaensis tenuitheca*. The first occurrence of the former defines the lower limit and the first appearance of the latter marks the upper limit of the zone. Owing to the absence or scarcity of many of

Banner and Blow's (1965) index species, direct application of their zonation to our cores is not possible. On the basis of other species, however, the ranges of which have been determined in terms of Banner and Blow's zones, we can estimate the ages of their zonal boundaries, namely, Zone N 22/N 23 boundary at about 700,000 years B.P., Zone N 21/N 22 boundary at about 1.9 m.y. B.P., Zone N 20/N 21 boundary at about 3 m.y. B.P., and Zone N 19/N 20 boundary at about 3.3 m.y. B.P. (Pl. 1).

In working on Scripps core LSDH 78P, Blow (1968, Figs. 38 and 41), tentatively placed the Zone N 18/N 19 boundary at about 500 cm. Our own examination of this core indicates that this horizon correlates with a level between the Gilbert "b" and "c" events at about 4.1 m.y. B.P. If we date Blow's Miocene-Pliocene boundary on the basis of this core, it would have an age of about 4.3 m.y. B.P. However, on the basis of our core V24–59, when Blow's criteria is applied, his Miocene-Pliocene boundary would have an age greater than 4.5 m.y. B.P., as previously stated.

The Radiolaria in our cores plus the magnetic stratigraphy allow us to correlate equatorial Radiolaria with the radiolarian zonation of high southern latitudes (Hays, 1965; Hays and Opdyke, 1967). Riedel (1969) has proposed a radiolarian zonation for the Neogene of the equatorial Pacific. For the time included in our cores he has established the following zones: Zone of *Pterocanium prismatium* defined by the occurrence of *P. prismatium* subsequent to the extinction of *Eucyrtidium elongatum peregrinum;* Zone of *Spongaster pentas* defined by the co-occurrence of *P. prismatium* and *Eucyrtidium elongatum peregrinum;* and a Zone of *Eucyrtidium elongatum peregrinum* defined by the occurrence of *E. elongatum peregrinum* prior to the earliest occurrence of *P. prismatium.* Although there are few species in common between the equatorial Pacific and the Antarctic, it is possible to correlate the Antarctic radiolarian zones of Hays with the equatorial Pacific zones of Riedel using paleomagnetic stratigraphy. Within the Quaternary sediments above Riedel's *Pterocanium prismatium* zone, *Stylatractus* sp. becomes extinct. Since the upper limit of this species is roughly the same in both the Antarctic and the equatorial Pacific, it can be used as a point of correlation. *Clathrocyclas bicornis* in the Antarctic and *Pterocanium prismatium* in the equatorial Pacific have similar upper limits so that Riedel's (1969) zone of *P. prismatium* includes

the Antarctic Φ zone. The *Spongaster pentas* zone roughly corresponds with the Antarctic Ύ zone, and the upper limit of the *Eucyrtidium elongatum peregrinum* zone has roughly the same upper limit as the Antarctic T zone.

Riedel and others (1963) approximately correlated the extinction of *Pterocanium prismatium* with the Pliocene-Pleistocene boundary drawn by Ericson and others (1963) in low-latitude Atlantic cores. Harrison and Funnell (1964) compared the paleomagnetic stratigraphy with the radiolarian stratigraphy in two equatorial Pacific cores and concluded that *Pterocanium prismatium* disappeared coincidentally with the last reversal of the earth's magnetic field (700,000 years B.P.). Our data shows that *Pterocanium prismatium* consistently disappears just above the Olduvai event (1.8 m.y. B.P.) (Figs. 5, 6, 8, and 9). The discrepancy between our results and Harrison and Funnell's is probably due to hiatuses in their considerably shorter cores.

Bolli and others (1968) correlated a Pliocene-Pleistocene boundary based on the extinctions of various nannoplankton species in the long SUBMAREX core from the Caribbean with the mid portions of Swedish Deep-Sea cores 58 and 62 from the equatorial Pacific. The level they correlate with in Swedish cores 58 and 62 is approximately the level of the extinction of nannoplankton species and *Pterocanium prismatium,* which Harrison and Funnell (1964) thought corresponded with the last reversal of the earth's magnetic field (700,000 years B.P.). Bolli and others (1968), therefore, assigned an age to their Pliocene-Pleistocene boundary of 700,000 years, but as we have shown, this level has an age of about 1.8 m.y. B.P. We agree, however, that this is the Pliocene-Pleistocene boundary.

CARBONATE AND CLIMATIC CYCLES

In a detailed investigation of deep-sea cores taken by the ALBATROSS during the Swedish Deep-Sea Expedition (1947–1948) in the eastern equatorial Pacific, Arrhenius (1952) reported cyclic carbonate sedimentation. He suggested a correlation with glacial stages of those stratigraphic intervals richer in calcium carbonate relative to clay content, and with nonglacial stages of those poorer in calcium carbonate. He reasoned that during glacial intervals the trade winds would be more intense than during nonglacial intervals and would stimulate the equatorial current system producing greater upwelling and hence greater productivity at the equatorial divergence. This higher productivity would engender higher calcium carbonate production in the surface waters and a higher rate of sedimentation in this region on the sea floor. The higher rate of accumulation would inhibit solution at the bottom, so increased productivity of carbonate plus reduced solution would work together to produce higher concentrations of carbonate in the sediments during glacial periods. Emiliani (1955) presented oxygen isotope data supporting Arrhenius' (1952) interpretation.

To further pursue the study of the carbonate cycles, we have measured the percentage of calcium carbonate in four equatorial Pacific cores, V24–58, 59, 62, and RC11–209, (Figs. 12 and 13) at a 10 cm or closer interval by the rapid gasometric technique described by Hülsemann (1966). No correction was made for the salt content of the cores.

Variations in carbonate content occur throughout the length of all cores, except V24–62 where the carbonate content drops to and remains near zero below 950 cm. V24–58 and RC11–209 were raised from about the same depth (4400 and 4490 m) and show similar carbonate percentages. The correlation of carbonate peaks and troughs between these two cores is striking (Fig. 12). Not only do the same number of peaks occur during the Brunhes (eight), but the shapes of individual peaks having the same position in the sequence are similar. Cores V24–59 and 62 are from deeper water (4662 and 4834 m, respectively) and show on average a lower carbonate content. The amplitude of the cycles is greater than that of the shallower cores and probably can be attributed best to increased solution during times of reduced productivity which may be due in part to greater water depth as a result of eustatic rise of sea level. The increased solution in the deeper cores is evidenced by the absence of solution-susceptible species of planktonic foraminifers.

In general, the four cores show the following features: (1) Eight carbonate peaks occur within the last 700,000 years (Brunhes) with the last and first being the highest; (2) the last five cycles have periods of about 75,000 years, while the first three Brunhes cycles have periods exceeding 100,000 years; (3) at the base of the Brunhes (700,000 years ago) there is a general drop in the level of carbonate content from an average of 84 percent in the Brunhes for core V24–58 to 77 percent for the interval between the Brunhes and Olduvai; (4) there are eleven peaks between the base

of the Brunhes and the top of the Olduvai event in V24–58 and RC11–209 not spaced as regularly as those of the Brunhes and having an average period of about 100,000 years; (5) both V24–59 and V24–62, which penetrate into the Gilbert, show prominent carbonate troughs at the top of the Gauss and two carbonate peaks within the upper Gauss normal. Below the middle of the Gauss series (above the Kaena event), the carbonate content drops to the lowest level in the core and remains at a low level into the upper Gilbert. In V24–59, the carbonate content rises again below the Gilbert "a" event.

With faunal control it is possible to correlate the carbonate variations of Arrhenius (1952) with those observed by us. It has been shown by Glass and others (1967) and by our study that S. dehiscens shows an abrupt drop in abundance at the Brunhes-Matuyama boundary (our Datum I). This datum has been previously observed in Swedish cores 58 and 62, and at this level a marked drop in carbonate content occurs which can be correlated with a similar decrease at the Brunhes-Matuyama boundary in our cores. Also, the level of disappearance of Pterocanium prismatium has been shown to occur just above the Olduvai event in our study. In addition, this level has been determined in cores 58 and 62 of the Swedish Deep-Sea Expedition (Riedel and others, 1963). It is possible, therefore, to use these two levels to correlate the carbonate variations between the two studies (Fig. 14). In Swedish core 58, the drop in abundance of S. dehiscens occurs at a depth of 410 cm and correlates with the Brunhes-Matuyama boundary. Arrhenius numbered the carbonate peaks and troughs from this point to the top of the core and considered this level (410 cm) to be the base of the glacial Pleistocene. In our study we have numbered the troughs (odd numbers) and peaks (even numbers) in the Brunhes series from youngest to oldest and given them a prefix B. A few prominent troughs are numbered below the Brunhes series preceded by letters indicating the magnetic series in which they occur (Figs. 12 and 13).

Dating Carbonate Cycles

A sample taken at the end of the last carbonate peak in core RC11–209 (8–14 cm) has yielded a C^{14} age of 11,600 ± 500 years B.P. (Thurber, personal commun., 1965). The last reversal occurs in this core at 700 cm indicating an average rate of sedimentation during the

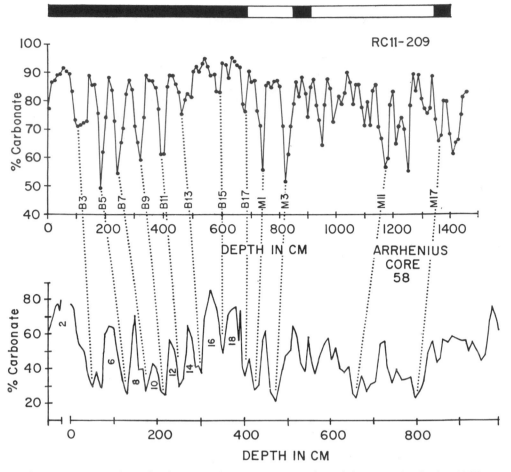

Figure 14.   Comparison of carbonate cycles in RC11–209 and Swedish core 58 (Arrhenius, 1952).

last 700,000 years of 1 cm/1000 years which agrees with the C[14] date. Of the four cores studied in detail, RC11–209 and V24–58 probably give the most reliable age estimates for the carbonate cycles because they have the highest rates of sedimentation (Fig. 15). The first large carbonate peak in the Brunhes series begins about 680,000 years ago and lasts for about 200,000 years. The amplitude of the peaks then decline to a pronounced carbonate low about 250,000 years ago. The carbonate returns to the same level as the early Brunhes high during the last 100,000 years.

Climatic Implications of Carbonate Cycles

The termination of the last glacial stage (Wisconsin) is generally accepted as occurring 11,000 years B.P. (Rubin and Suess, 1955; Ericson and others, 1956). The maximum

extent of the last glacial occurred 7,000 years earlier (Emiliani and Flint, 1963). Veeh (1965) documents a sea stand from 2 to 6 meters above the present level at 120,000 years ago, while Stearns and Thurber (1965) and Broecker and others (1968) have shown additional high stands at 80,000 and 103,000 years ago. Karlstrom (1965) has reported isotopically dated high-sea stands from Alaska at 78,000 to 100,000 years B.P. (Pelukian), 170,000 to 175,000 years B.P. (Kotzebuan) and 210,000 to 224,000 years B.P. (Middletonian).

The duration of our last calcium carbonate high ranges from the extrapolated date of about 80,000 years to 11,600 ± 500 years B.P. by C[14]. The previous calcium carbonate low has extrapolated ages ranging between 90,000 and 130,000 years B.P. These ages are in excellent agreement with previous data for the

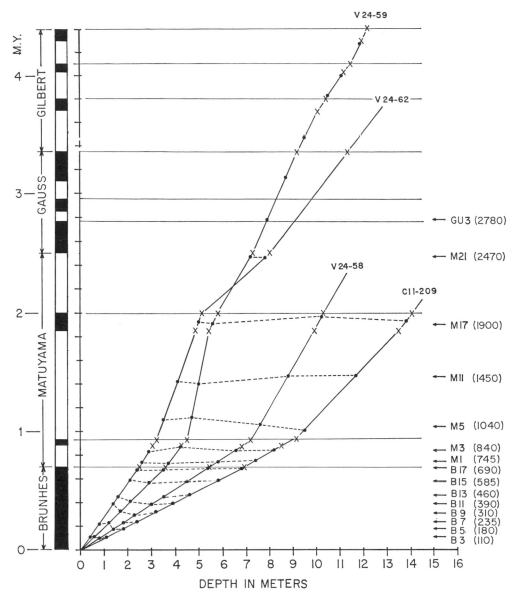

Figure 15. Time *versus* depth plots of magnetic stratigraphy and carbonate minima for four Pliocene-Pleistocene cores. Ages given in thousands of years.

age of the last glacial and last interglacial age. The ages of the high stands of sea level reported from Alaska all correspond well with the estimated ages for our calcium carbonate lows (B3, B5, and B7, Fig. 15). We do not see the fine detail of stadials and interstadials recorded for the Wisconsin (Goldthwait and others, 1965). We believe, therefore, the carbonate peaks within the Brunhes series represent climatic oscillations that may be correlative with glacial stages of varying intensity as Arrhenius (1952) originally postulated. Under this hypothesis, there were eight glacial stages during the last 700,000 years, and the first two and the last were the most severe. Furthermore, it is unlikely that we will find high-sea level stands between 690,000 and 460,000 years B.P. comparable with those that occurred in the upper Brunhes epoch. Prior to 700,000 years ago, the generally lower concentration of calcium carbonate suggests a warmer average climate. This change in carbonate concentration is most pronounced in the deeper cores V24–59 and V24–62. The upward decrease in abundance of the tropical species *S. dehiscens* at the Brunhes/Matuyama boundary is further evidence of this cooling.

Selli (1967), from work in the Po plain of northern Italy, has suggested that the Pleistocene can be divided roughly in half with the upper half (about a million years) representing the glacial era and the lower half, a preglacial Pleistocene. Hays (1965) noted a general cooling in Antarctic cores at his Ψ/X boundary later dated by Opdyke at 700,000 years ago, and Berggren and others (1967) postulated a similar cooling in a North Atlantic core at about one million years ago.

One of the oldest Pleistocene tills in North America (Sherwin Till) underlies the Bishop Tuff dated at 710,000 years B.P. (Dalrymple and others, 1965). Sharp (1968) estimates the Sherwin Till may be as old as 750,000 years. It may well be that the 700,000-year cooling marks the advent of continental glaciation in temperate latitudes as suggested by Hays and Berggren (1970). It is well known, however, that glaciations in higher latitudes precede this date. Fluctuations in calcium carbonate content of varying amplitude continue down through the Matuyama series and may reflect fluctuations in total world climate. We do not wish to imply that carbonate highs below the Brunhes series represent major advances of ice in temperate latitudes. A major change in level of calcium carbonate occurs in V24–59 and V24–62 in the upper Gauss normal series. This sharp upward rise in calcium carbonate in these two cores may reflect a Pliocene cooling which may be related to glaciations recorded at about this time in Iceland at 3 m.y. (Rutten and Wensink, 1960), in the Sierra Nevada range of the United States at 2.7 m.y. (Curry, 1966), and a cooling in New Zealand older than 2.5 m.y. (Mathews and Curtis, 1966). Hays and Opdyke (1967) reported a pronounced faunal change in Antarctic cores at about 2.5 to 3 m.y. In Iceland, Rutten and Wensink (1960) reported alternating tills and lava flows above the oldest till (3 m.y., McDougall and Wensink, 1966). The carbonate peaks in our cores above the middle of the Gauss may reflect these older climatic fluctuations.

Although we see no marked change in carbonate content at the Olduvai event, a number of changes do occur in several microfossil groups. Ericson and others (1963) and Glass and others (1967) interpret their foraminiferal data to indicate a cooling trend across the Pliocene-Pleistocene boundary. McIntyre and others (1967) working with Ericson's material, interpreted the coccolith data to indicate a warming trend across the boundary. The diatom data from our cores supports the conclusion of McIntyre and others (1967).

In Antarctic cores there is a lithologic change from clay below to diatom ooze above in the neighborhood of the Olduvai event (Opdyke and others, 1966) which is believed due to an increase in vertical circulation (Hays, 1965). In all probability, increased vertical circulation in the Antarctic would be caused by cooling rather than warming. Oxygen isotope studies by Emiliani and others (1961) revealed a general cooling trend across the Pliocene-Pleistocene boundary at Le Castella in Calabria, southern Italy, from an average summer surface water temperature of 23° C to 25° C in the late Pliocene to about 15° C in the Calabrian. From our data, whatever change in temperature occurred at this time, was not pronounced and must have been small compared with the change at about 3 m.y. and 700,000 years B.P.

Ewing and others (1968) have reported acoustically highly stratified sediments to basement in the areas where our cores show carbonate layering. Where we observe no carbonate layering, the sediments are acoustically transparent. Therefore, the stratified sediments probably represent high and low carbonate layers and apparently have been collecting through most of the Tertiary. This implies that the climatically induced carbonate cycles of the past few hundred thousand years are not peculiar to the Quaternary, but have been going on throughout the Tertiary or, for that matter, throughout all geologic time. It does not follow, however, that the earlier carbonate cycles correspond to glaciations.

Emiliani (1955, 1958, 1964, 1966) has thoroughly documented cyclic variations in O^{18} content of foraminiferal tests preserved in deep-sea sediments from the Atlantic and Caribbean during late Pleistocene time. He concludes that these variations are the result of fluctuations in surface-water temperatures responding to glacial and interglacial conditions. Using C^{14} and Pa^{231} to Th^{230} ages, he has suggested that the O^{18} variations have a periodicity of 40,000 to 50,000 years similar to that for variations in the summer radiation received in the northern hemisphere, resulting from changes in the eccentricity of the earth's orbit, changes in the earth's axial tilt, and changes associated with the precession of the earth's axis. From these data, Emiliani concludes that insolation variations are the major cause of the large oscillations in climate which have taken place over the last several hundred thousand years.

Ericson and others (1963, 1964) have constructed a climatic curve based on the abundance of the temperature sensitive species *Globorotalia menardii*. In a well-dated core, V12-122, from the Caribbean, Ericson and Wollin (1968) show six warm intervals in the last 320,000 ± 32,000 years (their U/V boundary).

In a careful redetermination of the Pa^{231} and Th^{230} in this core, Broecker and others (1968) have re-estimated the rate of sedimentation and conclude that it is 2.38 ± 0.10 cm/1000 years which is in good agreement with Sackett's (1965) estimate. Using this rate Broecker and others (1968) obtained an age of 126,000 ± 6,000 years for Ericson's W/X boundary. The only direct measurement of the age of the U/V boundary was by Ku and Broecker (1966) in this same core. The result was 320,000 ± 32,000 years. A reanalysis of this date has yielded an age of 380,000 ± 35,000 (Broecker, personal commun., 1968). This age of about 380,000 years B.P. for the U/V boundary is also supported by extrapolated ages derived from the age of the Brunhes/Matuyama boundary in other cores.

Using these new estimates of the age of the W/X and U/V boundaries, we have replotted the *G. menardii* curve of Ericson and Wollin (1968) against time (Fig. 16).

Core V12-122 can be correlated with Emiliani's (1966) Caribbean core P6304-9 by using coiling direction changes of *Globorotalia truncatulinoides*. Using this criteria, Ericson's U/V boundary would fall at about 1060 cm in this core and Ericson's W/X boundary would fall about 320 cm. This latter level was previously dated by Rosholt and others (1961) at about 100,000 years B.P. Replotting Emiliani's paleotemperature curve for P6304-9 against this new time scale, there is close agreement with the *G. menardii* curve of Ericson and Wollin (1968) (Fig. 16).

Using the C^{14} date of 11,600 ± 500 years at 8-14 cm and the Brunhes-Matuyama boundary (700,000 years B.P.) at 695 cm in RC11-209, we can compare our carbonate curve with the climatic curves of Ericson and Wollin (1968) and Emiliani (1966). Considering the wide spacing of absolute ages on RC11-209, the agreement between the three curves is good (Fig. 16). All three methods yield five temperature minima in the last 400,000 years. The cycles in all three cores have periods of approximately 70,000 to 100,000 years, with the last two cycles being longer than the previous three.

These data indicate that temperature fluctuations were probably synchronous between the Pacific and the Atlantic during the last 400,000 years. It is significant that the cycles obtained by these three methods have similar periodicities (70,000 to 100,000 years) and any theories attempting to explain climatic fluctuations must take this into account.

RATES OF SEDIMENTATION

Rates of sedimentation have been calculated for all cores in this study on which paleomagnetic determinations were made using the Brunhes-Matuyama boundary as a datum. The age of this boundary is taken as 700,000 years (Cox and Dalrymple, 1967a). Along the VEMA 24 track east of V24-58, the cores are dominantly calcareous and the rates of sedimentation range from 7.8 mm/1000 years to 17.5 mm/1000 years, with an average of about 10 mm/1000 years (Table 4). The depths from which these cores were taken range from 4189 to 4490 m. Although core-to-core variations in rates of sedimentation occur, there is no overall trend related to longitude (Fig. 4). West of 140° W. longitude, the water depth steadily increases to 4834 m, and the rates of sedimentation decrease from 5.1 mm/1000 years (V24-59) to 3.6 mm/1000 years (V24-62).

In the CONRAD 11 track the rates of sedimentation steadily increase from RC11-208 (7.6 mm/1000 years) to RC11-210 (16.8 mm/1000 years). From RC11-211, which has a rate in excess of 13.5 mm/1000 years, the rates decrease southward to RC11-213 (5.9 mm/1000 years). The pattern which emerges is that of a steady increase in rate of sedimentation as the equator is approached. This pattern was first suggested by Arrhenius (1952) and is confirmed here. Although the rates of sedimentation are highest under the equator, the sediments are thickest a few degrees north.

Seven of the cores used in this study pass through the Olduvai event and average 44 percent lower sedimentation rates in the Matuyama series than in the Brunhes series (Table 4). This decrease occurs between the Jaramillo and Olduvai events. A similarly lower rate for

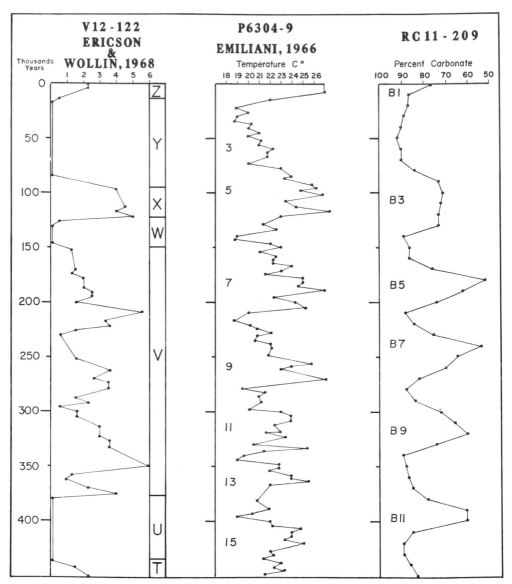

Figure 16. Comparison of temperature curves for Caribbean cores of Ericson and Wollin (1968) and Emiliani (1966) with carbonate curve from equatorial Pacific.

the Matuyama has been observed in Antarctic cores (Opdyke and others, 1966; Hays and Opdyke, 1967).

PRE-PLIOCENE SEDIMENTS

Our age determination of pre-Pliocene sediment cores of the tropical Pacific Ocean is based chiefly on planktonic foraminifers for carbonate deposits and on Radiolaria for siliceous oozes. The term Eocene for calcareous oozes is used in the sense of Bolli (1966b) and Berggren and others (1967), the latter defined its base by the first appearance of *Pseudohastigerina wilcoxensis* (Cushman and Ponton) and the last occurrence of *Globorotalia velascoensis* (Cushman), and its top by the extinction of *Globorotalia centralis* (Cushman) and *Globorotalia cerroazulensis* (Cole). Radio-

TABLE 4. RATES OF SEDIMENTATION IN MILLIMETERS PER THOUSAND YEARS
Interval*

	Brunhes	Base Jaramillo Base Olduvai	Base Olduvai Top Gauss	Top Gauss Top Gilbert
V24–51	11.6	..	..	..
V24–53	8.1	..	..	..
V24–54	11.2	..	..	..
V24–55	6.7	..	..	..
V24–56	17.5	..	..	..
V24–57	12.7	..	..	..
V24–58	7.8	2.6	..	..
V24–59	5.2	1.1	3.2	2.3
V24–60	6.2	2.5	4.8	..
V24–62	3.6	1.7	5.6	3.8
RC11–208	7.8	3.5	..	..
RC11–209	9.9	3.7	..	..
RC11–210	16.5	..	..	..
RC11–211	>13.5	..	..	..
RC11–213	5.9	3.2	..	..

* Dates used in calculating the rates are as follows: Brunhes-Matuyama boundary, 0.7 m.y.; base of Jaramillo, 0.927 m.y.; base Olduvai, 2 m.y.; Gauss-Matuyama boundary, 2.5 m.y.; Gauss-Gilbert boundary, 3.36 m.y.

laria-rich sediments containing abundant representatives of such species as *Sethamphora mongolfieri* (Ehrenberg), *Calocyclas turris* (Ehrenberg) and *Dictyophimus babylonis* (Clark and Campbell) are considered Eocene.

Our interpretation of the Oligocene follows the recommendation made at the 4th International Congress of the Committee of the Mediterranean Neogene (Bologna, Italy, 1967) to place its upper boundary at the base of the stratotype Aquitanian which can be identified by the first evolutionary appearance of the planktonic-foraminiferal genus *Globigerinoides* in areas outside of the stratotype locality. This boundary lies near the base of the *Globorotalia kugleri* zone of Bolli (1957) and between Zones N 3 and N 4 of Banner and Blow (1965). For siliceous sediments, the top of the Oligocene is drawn at the upper limit of *Trigonactura? angusta* Riedel and below the first appearance of *Lychnocanium bipes* Riedel. This boundary should fall below the *Globorotalia kugleri* zone.

The sediments penetrated by long piston cores in the area of this study increase in age from the equator northward and less distinctly southward (Fig. 1). The ages shown in Figure 1 are the age of the sediment at the bottom of the core, but frequently the cores have a similar age at the top. This is particularly true of those cores that penetrate Oligocene and Eocene sediments. Cores penetrating Miocene sediments that are not Miocene at the top have a short interval of late Pleistocene overlying

Miocene or an unfossiliferous interval overlying fossil-rich Miocene sediments. The boundary between Plio-Pleistocene and pre-Pliocene sediments is the Clipperton fracture zone (Fig. 2). Cores penetrating Quaternary and Pliocene sediments occur together along the equator and contain complete or nearly complete sequences. Miocene sediments on the other hand crop out or nearly crop out north of the Clipperton fracture zone (Fig. 1). Oligocene and Eocene sediments are encountered progressively farther north. The middle and upper Miocene and Eocene sediments are predominantly siliceous while lower Miocene and Oligocene cores contain calcium carbonate as well, as has been previously reported for the Oligocene (Riedel and Funnell, 1964). Of the 42 cores raised by Lamont vessels from the siliceous zone of Figure 2, 37 are pre-Pliocene. In the calcareous ooze zone of 123 cores, all except 8 are Pliocene or younger. The occurrence of pre-Pliocene sediments in the calcareous-ooze belt usually can be related to steep slopes, but this is not so for the siliceous-ooze belt.

The near outcropping of Miocene and older sediments north of the calcareous-ooze belt suggests that some bottom process has removed much or all of the younger sediment or that conditions have significantly changed, resulting in very little sediment accumulation since Miocene time. It is hard to visualize, even under highly unproductive surface waters, virtually no sediment accumulation during the last several million years. It is equally difficult to imagine no sediment accumulation under surface waters as productive as those that produced the highly fossiliferous Eocene through Miocene oozes. We believe, therefore, that there has been a reduction in the productivity of the surface waters overlying the radiolarian-ooze belt since Miocene time and that the amount of sediment accumulating may have been further limited by transportation away from this region by bottom currents.

Arrhenius (1952) has attributed the Holocene biogenous sediments accumulating beneath the equator to high productivity associated with the equatorial divergence. He presented evidence that the rate of sedimentation in this region is at a maximum beneath the equator and decreases toward the north and south. Our rates of sedimentation support this viewpoint, but it fails to explain the apparently wider zone of high productivity that existed in middle Tertiary time.

One possibility is that the belt of high productivity was no wider in middle Tertiary than it is now, but that ocean-floor spreading (Hess, 1962; Dietz, 1961) caused the sea floor to gradually move west-northwestward moving the sediments that accumulated beneath the equator in the middle Tertiary out from under the zone of high productivity. Supporting this possibility is the fact that the crest of the thick sediment belt (Ewing and others, 1968) is offset northward from the equator a few degrees, and, therefore, does not lie under the zone of maximum productivity along the equator, as might be expected.

A second possibility is that since the Isthmus of Panama was open during early and middle Tertiary time (not closing completely until Pliocene, Lloyd, 1963), the addition of Atlantic water through this channel stimulated the circulation of the eastern equatorial Pacific and increased and broadened the belt of high productivity. Such a flow of Atlantic water into the Pacific also might increase the rate of equatorial Atlantic and Caribbean circulation. No radiolarian oozes have been recovered yet from recent sediment deposits in the equatorial Atlantic or Caribbean. Eocene radiolarian oozes, however, have been recovered from the western tropical Atlantic (RC8–2, Saito and others, 1966), Caribbean (RC9–55, 56, 58 and 59, Ewing and others, 1967), and the island of Barbados (Beckman, 1954). This suggests an Eocene belt of high productivity stretching from the equatorial Atlantic through the Caribbean into the Pacific.

CONCLUSIONS

Paleomagnetic measurements were made on samples from 15 sediment cores raised from the equatorial Pacific Ocean. Four of these cores were given intensive paleontological study.

From these data we can draw the following conclusions about geologic events within the last 4.5 m.y.:

(1) The magnetic stratigraphy confirms the previously established sequences of reversals including the Kaena reversed polarity event in the Gauss normal epoch and three normal events in Gilbert reversed polarity epoch. We see no evidence of a third large normal event within Matuyama reversed polarity epoch which could be interpreted as the Gilsa normal event.

(2) The diatoms and silicoflagellates are useful stratigraphic indicators. Of special interest are a silicoflagellate (*Mesocena elliptica*) which brackets Jaramillo event and diatoms *Rhizosolenia praebergonii*, *Fragilariopsis antarctica forma nova*, and *Thalassiosira convexa*, the first two of which have upper limits just above Olduvai event and the last in the lower Matuyama.

(3) Nine datum levels are established based upon extinctions and coiling changes of foraminiferal species. These are an upward reduction in abundance of *Spaeroidinella dehiscens* at base of Brunhes series, first appearance of *Pulleniatina finalis* halfway between Olduvai and Jaramillo events, extinction of *Globigerinoides fistulosus* at top of Olduvai event, first appearance of *Globorotalia truncatulinoides* near base of Olduvai event, extinctions of species of *Sphaeroidinellopsis* at top of Mammoth event, extinction of *Globorotalia margaritae* at base of Gauss, marked coiling change of *Pulleniatina* just above Gilbert "a" event, extinction of *Globigerina nepenthes* and *Pulleniatina spectabilis* at top of Gilbert "a" event, and the first appearance of *P. spectabilis* near the top of Gilbert "c" event.

(4) Radiolaria that become extinct in this interval are *Stylatractus* sp. in the middle of Brunhes epoch, *Pterocanium prismatium* just above Olduvai event, and *Eucyrtidium elongatum peregrinum* just above Kaena event. The evolutionary transition from *Panarium penultimum* to *Penartus tetrathalamus* occurs just above Gilbert "c" event.

(5) Two zones of major paleontological change occur first in the middle of Gauss normal epoch and second near Olduvai event.

(6) Seven equatorial foraminiferal species, two radiolarian species, and two diatom species become extinct within a few tens of centimeters of reversals.

(7) The establishment of the true chronostratigraphic relationships of the microfossil species by comparison with the paleomagnetic record allows us to date zonations of previous workers.

(8) Eight distinct fluctuations in the percentage of calcium carbonate are present in Brunhes series. The last carbonate high correlates with the Wisconsin glaciation (80,000 years B.P. to 11,500 years B.P.). It is probable that the preceding seven carbonate highs also correspond to glacial stages. The carbonate cycles during the last 400,000 years correlate with previously constructed temperature curves from the Caribbean.

(9) The rates of sedimentation in this area range from 3.6 mm/1000 years for low carbonate cores to a maximum of 17.5 mm/1000 years for high carbonate cores.

(10) Outcropping pre-Pliocene sediments show increasing age westward from the crest of the East Pacific Rise and northward from the equator. The former trend is attributed to ocean-floor spreading, the latter to a wider belt of high productivity in pre-Pliocene time caused by the introduction of Atlantic water into the Pacific due to the absence of the Isthmus of Panama.

REFERENCES CITED

Akers, W. H., 1965, Pliocene-Pleistocene boundary, northern Gulf of Mexico: Science, v. 149, p. 741–742

Arrhenius, G., 1952, Sediment cores from the East Pacific: Swedish Deep-Sea Exped. (1947–1948) Repts., v. 5, fasc. 1, 89 p.

—— 1963, Pelagic sediments *in* The Sea: New York, Interscience, v. 3, p. 655–727.

Bandy, O. L., 1963, Miocene-Pliocene boundary in the Philippines as related to late Tertiary, stratigraphy of deep-sea sediments: Science, v. 142, p. 1290–1292.

—— 1967, Problems of Tertiary foraminiferal and radiolarian zonation, circum-Pacific area: Tertiary correlations and climatic changes in the Pacific, Symposium no. 25, Sendai (Japan), Sasaki, p. 95–102.

Bandy, O. L., and Wade, M. E., 1967, Miocene-Pliocene-Pleistocene boundaries in deep-water environments: Progress in oceanography, Pergamon Press, v. 4, p. 51–66.

Banner, F. T., and Blow, W. H., 1965, Progress in the planktonic foraminiferal biostratigraphy of the Neogene: Nature, v. 208, p. 1164–1166.

—— 1967, The origin, evolution and taxonomy of the foraminiferal genus *Pulleniatina* Cushman, 1927: Micropaleontology, v. 13, p. 133–162.

Beckman, J. P., 1954, Foraminiferen der oceanic formation, Barbados: Ecologae Geol. Helvetiae, v. 46, p. 301–407.

Berger, W. H., 1968, Planktonic foraminifera: a selective solution and paleoclimatic interpretation: Deep-Sea Research, v. 15, p. 31–43.

Berggren, W. A., Olsson, R. K., and Reyment, R. A., 1967, Origin and development of the foraminiferal genus *Pseudohastigerina* Banner and Blow, 1959: Micropaleontology, v. 13, p. 265–288.

Berggren, W. A., Phillips, J. D., Bertels, A., and Wall, D., 1967, Late Pliocene-Pleistocene stratigraphy in deep-sea cores from the south-central North Atlantic: Nature, v. 216, p. 253–254.

Black, D. I., 1967, Cosmic ray effects and faunal extinctions at geomagnetic field reversals: Earth and Planetary Sci. Letters, v. 3, p. 225–236.

Blow, W. H., 1968, Late middle Eocene to Recent planktonic foraminiferal biostratigraphy: Archives des Sciences de Geneve (in press).

Bolli, H. M., 1957, Planktonic foraminifera from the Oligocene-Miocene Cipero and Lengua Formations of Trinidad, B.W.I.: U.S. Natl. Mus. Bull., v. 215, p. 97–123.

—— 1966a, The planktonic foraminifera in Well Bodjonegoro-I of Java: Ecologae Geol. Helvetiae, v. 59, p. 449–465.

—— 1966b, Zonation of Cretaceous to Pliocene marine sediments based on planktonic foraminifera: Asoc. Venezolana Geol., Mineria y Petroleo Bol., v. 9, p. 3–32.

Bolli, H. M., and Bermúdez, P. J., 1965, Zonation based on planktonic foraminifera of middle Miocene to Pliocene warm-water sediments: Asoc. Venezolana Geol., Mineria y Petroleo. Bol., v. 8, p. 121–149.

Bolli, H. M., Boudreau, J. E., Emiliani, C., Hay, W. W., Hurley, R. J., and Jones, J. I., 1968, Biostratigraphy and paleotemperatures of a section cored on the Nicaragua Rise, Caribbean Sea: Geol. Soc. America Bull., v. 79, no. 4, p. 459–470.

Bradshaw, J. S., 1959, Ecology of living planktonic foraminifera in the north and equatorial Pacific Ocean: Cushman Found. Foram. Research Contr., v. 10, p. 25–64.

Broecker, W. S., Thurber, D. L., Goddard, J., Ku, T., Mathews, R. K., and Mesolella, K. J., 1968, Milankovitch hypothesis supported by precise dating of coral reefs and deep-sea sediments: Science, v. 159, p. 297–300.

Bullard, E. C., 1955, The stability of a homopolar dynamo: Cambridge Philos. Soc. Proc., v. 51, p. 744–760.

Burckle, L. H., Ewing, J., Saito, T., and Leydon, R., 1967, Tertiary sediment from the East Pacific Rise: Science, v. 157, p. 537–540.

Cox, A., and Dalrymple, G. B., 1967a, Statistical analysis of geomagnetic reversal data and the precision of potassium-argon dating: Jour. Geophys. Research, v. 72, no. 10, p. 2603–2614.

—— 1967b, Geomagnetic polarity epochs: Nunivak Islands, Alaska: Earth and Planetary Sci. Letters, v. 3, p. 173–177.

Curry, R. R., 1966, Glaciation about 3,000,000 years ago in Sierra Nevada: Science, v. 154, p. 770–771.

Dalrymple, G. B., Cox, A., and Doell, R. R., 1965, Potassium-argon age and paleomagnetism of the Bishop Tuff, California: Geol. Soc. America Bull., v. 76, no. 6, p. 665–674.

Dietz, R. S., 1961, Continent and ocean basin evolution by spreading of the sea floor: Nature, v. 190, p. 854–857.

Dymond, J. R., 1966, Potassium-argon geochronology of deep-sea sediments: Science, v. 152, p. 1239–1241.

Emiliani, C., 1955, Pleistocene temperatures: Jour. Geology, v. 63, p. 538–578.

—— 1958, Paleotemperature analysis of core 280 and Pleistocene correlations: Jour. Geology, v. 66, p. 264–275.

—— 1964, Paleotemperature analysis of the Caribbean cores A254-Br-C and Cp-28: Geol. Soc. America Bull., v. 75, no. 2, p. 129–144.

—— 1966, Paleotemperature analysis of the Caribbean cores P6304-8 and P6304-9 and a generalized temperature curve for the last 425,000 years: Jour. Geology, v. 74, no. 6, p. 109–126.

Emiliani, C., and Flint, R. F., 1963, The Pleistocene Record, in Hill, M. N. Editor, The Sea: New York, Interscience, v. 3, p. 888–927.

Emiliani, C., Mayeda, T., and Selli, R., 1961, Paleotemperature analysis of the Plio-Pleistocene section at Le Castella, Calabria, Southern Italy, Geol. Soc. America Bull., v. 72, p. 679–688.

Ericson, D. B., Broecker, W. S., Kulp, J. L., and Wollin, G., 1956, Late-Pleistocene climates and deep-sea sediments: Science, v. 124, p. 385–389.

Ericson, D. B., Wollin, G., and Ewing, M., 1963, Pliocene-Pleistocene boundary in deep-sea sediments: Science, v. 139, p. 727–737.

Ericson, D. B., Ewing, M., and Wollin, G., 1964, The Pleistocene epoch in deep-sea sediments: Science, v. 146, p. 723–732.

Ericson, D. B., and Wollin, G., 1968, Pleistocene climates and chronology in deep-sea sediments: Science, v. 162, p. 1227–1234.

Ewing, J., Talwani, M., Ewing, M., and Edgar, T., 1967, Sediments of the Caribbean: Studies Tropical Oceanography, v. 5, p. 88–102.

Ewing, J., Ewing, M., Aitken, T., and Ludwig, W. J., 1968, North Pacific sediment layers measured by seismic profiling, in Knopoff, L., Drake, C. L., and Hart, P. J., Editors, The Crust and upper mantle of the Pacific area: Am. Geophys. Union, Geophys. Mon. 12, p. 147–186.

Foster, J. H., 1966, A paleomagnetic spinner magnetometer using a fluxgate gradiometer: Earth and Planetary Sci. Letters, v. 1, p. 463–466.

Glass, B., Ericson, D. B., Heezen, B. C., Opdyke, N. D., and Glass, J. A., 1967, Geomagnetic reversals and Pleistocene chronology: Nature, v. 216, p. 437–442.

Goldthwait, R. P., Dreimanis, A., Forsyth, J. L., Karrow, P. F., and White, G. W., 1965, Pleistocene deposits of the Erie lake, in The Quaternary of the United States: Princeton, Princeton Univ. Press, p. 85–97.

Harrison, C.G.A., 1968, Evolutionary processes and reversals of the Earth's magnetic field: Nature, v. 217, p. 46–47.

Harrison, C.G.A., and Funnell, B. M., 1964, Relationship of paleomagnetic reversals and micropaleontology in two late Cenozoic cores from the Pacific Ocean: Nature, v. 204, p. 566.

Harrison, C.G.A., and Somayajulu, B.L.K., 1966, Behavior of the earth's magnetic field during a reversal: Nature, v. 212, p. 1193–1195.

Hays, J. D., 1965, Radiolaria and late Tertiary and Quaternary history of Antarctic seas, in Biology of the Antarctic seas II: Antarctic Research ser. 5, Am. Geophys. Union, p. 125–184.

Hays, J. D., and Berggren, W. A., 1970, Quaternary boundaries and micropaleontology of marine bottom sediments: Cambridge University Press (in press).

Hays, J. D., and Opdyke, N. D., 1967, Antarctic Radiolaria, magnetic reversals and climatic change: Science, v. 158, p. 1001–1011.

Hess, H. H., 1962, History of ocean basins, in Petrologic studies, A Volume to Honor A. F. Buddington: Geol. Soc. America, p. 599–620.

Hülsemann, J., 1966, On the routine analysis of carbonates in unconsolidated sediments: Jour. Sed. Petrology, v. 36, p. 622–625.

Ingle, J. C., 1967, Foraminiferal biofacies variation and the Miocene-Pliocene boundary in Southern California: Bulls. Am. Paleontology, v. 52, no. 236, p. 217–394.

Karlstrom, T.N.Y., 1965, Isotope-dated interglacial marine transgressions and the Cook Inlet glacial chronology: 8th INQUA Congress, Denver, Colorado, Abstracts, p. 258.

Kolbe, R. W., 1954, Diatoms from equatorial Pacific cores: Swedish Deep-Sea Expedition Repts., v. VI, no. 1, p. 3–49.

Ku, T. L., 1966, Uranium series disequilibrium in deep-sea sediments: Ph.D. thesis, Columbia University, New York, New York (available on microfilm).

Ku, T. L., and Broecker, W. S., 1966, Atlantic deep-sea stratigraphy: Extension of absolute chronology to 320,000 years: Science, v. 151, p. 448–450.

Lloyd, J. J., 1963, Tectonic history of the south-central American orogen: The backbone of the Americas—tectonic history from pole to pole, a symposium: Am. Assoc. Petroleum Geologists Mem., no. 2, p. 88–100.

Martini, E., and Bramlette, M. N., 1963, Calcareous nannoplankton from the experimental Mohole drilling: Jour. Paleontology, v. 37, p. 845–856.

Mathews, W. H., and Curtis, G. H., 1966, Date of the Pliocene-Pleistocene boundary in New Zealand: Nature, v. 212, p. 979–980.

McDougall, I., and Wensink, H., 1966, Paleomagnetism and geochronology of the Pliocene-Pleistocene lavas in Iceland: Earth and Planetary Sci. Letters, v. 1, p. 232–236.

McIntyre, A., Bé, A.W.H., and Preikstas, R., 1967, Coccoliths and the Plio-Pleistocene boundary: Progress in Oceanography, New York, Pergamon Press, v. 4, p. 3–25.

Menard, H. W., 1967, Extension of Northeastern Pacific fracture zones: Science, v. 155, p. 72–74.

Muhina, V. V., 1963, Biostratigraphic analyses of bottom sediments from Station 3802 in the equatorial zone of the Pacific Ocean (in Russian): Okeanologiya, v. 3, no. 5, p. 861–869.

Ninkovich, D., Opdyke, N. D., Heezen, B. C., and Foster, J. H., 1966, Paleomagnetic stratigraphy, rates of deposition and tephrachronology in North Pacific deep-sea sediments: Earth and Planetary Sci. Letters, v. 1, p. 476–492.

Opdyke, N. D., 1969, The Jaramillo event as detected in oceanic cores: Nato Conference on Internal Constitution of Earth and Planets (in press).

Opdyke, N. D., Glass, B., Hays, J. D., and Foster, J., 1966, Paleomagnetic study of Antarctic deep-sea cores: Science, v. 154, p. 349–357.

Parker, F. L., 1967, Late Tertiary biostratigraphy (planktonic Foraminifera) of tropical Indo-Pacific deep-sea cores: Bulls. Am. Paleontology, v. 52, no. 235, p. 115–203.

Pitman, W. C., and Heirtzler, J. R., 1966, Magnetic anomalies over the Pacific-Antarctic ridge: Science, v. 154, p. 1164.

Revelle, R. R., 1944, Scientific results of Cruise VII of the CARNEGIE during 1928–1929, Marine bottom samples collected in the Pacific Ocean: Carnegie Inst. Washington Pub. 556, Oceanography-II, pt. 1, 193 p.

Revelle, R. R., Bramlette, M. N., Arrhenius G., and Goldberg, E. D., 1955, Pelagic sediments of the Pacific: p. 221–236 in Poldervaart, Arie, Editor, Crust of the Earth: Geol. Soc. America Spec. Paper 62, 762 p.

Riedel, W. R., 1957, Radiolaria: A preliminary stratigraphy: Swedish Deep-Sea Expedition Repts., v. 6, fasc. 3, p. 61–96.

—— 1959, Oligocene and lower Miocene Radiolaria in tropical Pacific sediments: Micropaleontology, .. 5, p. 285–302.

—— 1967, Radiolarian evidence consistent with spreading of the Pacific floor: Science, v. 157, p. 540–542.

—— 1969, Neogene Radiolarian zones: Cambridge University Press (in press).

Riedel, W. R., and Funnell, B. M., 1964, Tertiary sediment cores and microfossils from the Pacific Ocean floor: Geol. Soc. London Quart. Jour., v. 120, p. 305–368.

Riedel, W. R., Parker, F. L., and Bramlette, M. N., 1963, "Pliocene-Pleistocene" boundary in deep-sea sediments: Science, v. 140, p. 1238–1240.

Riedel, W. R., and Sanfilippo, A., 1969, Initial core description: JOIDES Deep-Sea Drilling Project Initial Core Description (in press).

Rosholt, J. N., Emiliani, C., Geiss, J., Koczy, F. F., and Wangersky, P. J., 1961, Absolute dating of deep-sea cores by the Pa^{231}/Th^{230} method: Jour. Geology, v. 69, no. 2, p. 162–185.

Rubin, M., and Suess, H. E., 1955, U.S. Geological Survey natural radiocarbon dates II: Science, v. 121, p. 481–488.

Ruddiman, W. F., and Heezen, B. C., 1967, Differential solution of planktonic Foraminifera: Deep-Sea Research, v. 14, p. 801–808.

Rutten, M. G., and Wensink, H., 1960, Paleomagnetic dating, glaciations and the chronology of the Plio-Pleistocene in Iceland: Internat. Geol. Congr. 21st Session, Norway, part 4, p. 62–70.

Sackett, W. M., 1965, Deposition rates by the Protactinium method, in Symposium on marine geochemistry, Narragansett Marine Laboratory, Rhode Island Univ. Occas. Pub., no. 3, p. 29–40.

Saito, T., Ewing, M., and Burckle, L. H., 1966, Tertiary sediment from the mid-Atlantic ridge: Science, v. 151, p. 1075–1079.

Selli, R., 1967, The Pliocene-Pleistocene boundary in Italian marine sections and its relationship to continental stratigraphies: Progress in Oceanography, v. 4, p. 67–86.

Sharp, R. P., 1968, Sherwin Till-Bishop Tuff geological relationships, Sierra Nevada, California: Geol. Soc. America Bull., v. 79, no. 3, p. 351–363.

Stearns, C. E., and Thurber, D. L., 1965, Th-230 and U-234 dates of late Pleistocene marine fossils from the Mediterranean and Moroccan littorals: Quaternaria, v. VII, p. 29–42.

Tongiorgi, F., and Tongiorgi, M., 1964, Age of the Miocene-Pliocene limit in Italy: Nature, v. 201, p. 365–367.

Uffen, R. J., 1963, Influence of the earth's core on the origin and evolution of life: Nature, v. 198, p. 143.

Veeh, H. H., 1965, Th^{230}/U^{238} and U^{234}/U^{238} ages of elevated Pleistocene coral reefs and their geological implications: Ph.D. thesis, Scripps Inst. Oceanography, La Jolla, California (available by interlibrary loan).

Waddington, C. J., 1967, Paleomagnetic field reversals and cosmic radiation: Science, v. 158, p. 913–915.

Watkins, N. D., and Goodell, H. G., 1967, Geomagnetic polarity change and faunal extinction in the southern ocean: Science, v. 156, p. 1083–1087.

Wray, J. L., and Ellis, C. H., 1965, Discoaster extinction in neritic sediments northern Gulf of Mexico: Am. Assoc. Petroleum Geologists Bull., v. 49, p. 98–99.

11

A Cenozoic time-scale — some implications for regional geology and paleobiogeography

WILLIAM A. BERGGREN

Recent data generally substantiate the most recent incarnation of the Cenozoic time-scale (Berggren, 1969c). Newly obtained dates in the type section of the Chattian (Upper Oligocene) support a previous suggestion that a hiatus may separate the top of the stratotype Chattian from the base of the stratotype Aquitanian. The *Orbulina* Datum is placed at 16 my and the *Globigerina nepenthes* at 13.5 my. The junction of Eurasia and Africa in the early Miocene (ca. 18 my ago) and of Europe and Africa in the middle Miocene (ca. 15 my ago) markedly affected the Tethyan paleogeography and, concomitantly, the biogeographic distribution of larger and smaller Foraminifera, as well as various land mammal groups, including hominoids.

[*Editor's Note:* The article itself has not been reproduced here.]

Figure 4 (over). Cenozoic radiometric time-scale, chronostratigraphy, planktonic foraminiferal zonation and datum planes (levels).

GEOMAGNETIC TIME SCALE	RADIO-METRIC TIME SCALE IN m.y.	GEOMAGNETIC POLARITY HISTORY	EPOCH/SERIES	NORTH AMERICAN MAMMALIAN STAGES	WEST COAST (CALIFORNIA) MARINE STAGES	NEW ZEALAND MARINE STAGES	δO¹⁸ - PALEOTEMPERATURE CURVE - NEW ZEALAND (DEVEREUX, 1967)	EUROPEAN STAGES	CENOZOIC PLANKTONIC FORAMINIFERAL ZONES (BANNER & BLOW, 1965; BLOW, 1969; BLOW and BERGGREN, unpubl)		CENOZOIC PLANKTONIC FORAMINIFERAL DATUM PLANES

$δO^{18}$ - PALEOTEMPERATURE CURVE - NEW ZEALAND (DEVEREUX, 1967) 20° 15° 10°

(Chart content — Cenozoic planktonic foraminiferal datum planes:)

- N23
- N22 — G. truncatulinoides P-R-Z — (1.85) Globorotalia truncatulinoides Datum
- N21 — G. tosaensis tenuitheca Cons-R-Z — (3) Globoquadrina altispira - Sphaeroidinellopsis seminulina (extinction) Datum
- N20 — G. multicamerata - P. obliquiloculata — (3.7) Globigerina nepenthes (extinction) Datum
- N19 — S. dehiscens S. dehiscens/G. altispira P-R-Z — (5) Sphaeroidinella dehiscens Datum
- N18 — G. tumida - S. subdehiscens paenedehiscens P-R-Z
- N17 — G. tumida plesiotumida Cons-R-Z — (6.5) Pulleniatina Datum
- N16 — G. acostaensis G. merotumida P-R-Z — (10) Globorotalia acostaensis Datum
- N15 — G. continuosa Cons-R-Z — (10.5) Candeina Datum
- N14 — G. nepenthes / G. siakensis Conc-R-Z — (13.5) Globigerina nepenthes Datum; (14) Cassigerinella (extinction) Datum
- N13 — S. subdehiscens - G. druryi P-R-Z
- N12 — G. fohsi P-R-Z — (15) Globorotalia fohsi s.l. Datum
- N11 — G. praefohsi Cons-R-Z
- N10 — G. peripheroacuta Cons-R-Z — (16) Orbulina Datum
- N9 — G. suturalis - G. peripheroronda P-R-Z — (17) G. stainforthi (extinction) Datum
- N8 — G. sicanus - P-R-Z - G. insueta — (17.5) G. dissimilis (extinction) Datum
- N7 — G. insueta G. quadrilobatus trilobus P-R-Z
- N6 — G. insueta / G. dissimilis Conc-R-Z — (19) G. insueta Datum
- N5 — G. dehiscens praedehiscens G. dehiscens P-R-Z
- N4 — G. quadrilobatus primordius / G. kugleri Conc-R-Z — (22.5) Globigerinoides Datum
- N3 / P22 — Globigerina angulisuturalis P-R-Z
- N2 / P21 — G. angulisuturalis / G. opima opima Conc-R-Z — (26) Globorotalia opima s.s. (extinction) Datum
- N1 / P20 — G. ampliapertura P-R-Z — (30) Globigerina angulisuturalis Datum
- P19 — G. sellii / P. barbadoensis Conc-R-Z — (32) Pseudohastigerina (extinction) Datum
- P18 — G. tapuriensis Conc-R-Z
- P17 — G. gortanii P-R-Z G. centralis — (38) Hantkenina (extinction) Datum
- P16 — C. inflata T-R-Z
- P15 — G. mexicana P-R-Z — (41) Globigerapsis mexicana (extinction) Datum
- P14 — T. rohri - G. howei P-R-Z — (43) Globorotalia lehneri - Truncorotaloides rohri (extinction) Datum
- P13 — O. beckmanni T-R-Z
- P12 — G. lehneri P-R-Z
- P11 — G. kugleri P-R-Z — (47.5) Globorotalia lehneri - Truncorotaloides topilensis Datum
- P10 — H. aragonensis P-R-Z — (49) Hantkenina Datum
- P9 — A. densa P-R-Z
- P8 — G. aragonensis P-R-Z
- P7 — G. formosa P-R-Z — (52) Globorotalia aragonensis Datum
- P6 b — G. subbotinae P. wilcoxensis Conc-R-Z
- P6 a — G. velascoensis / G. subbotinae Conc-R-Z — (53.5) Pseudohastigerina Datum
- P5 — G. velascoensis P-R-Z — (56) Globorotalia pseudomenardii (extinction) Datum
- P4 — G. pseudomenardii T-R-Z — (58) Globorotalia pseudomenardii Datum
- P3 — G. pusilla pusilla - G. angulata Conc-R-Z — (60) Globorotalia angulata Datum
- P2 — G. uncinata - G. spiralis Conc-R-Z — (61.5) Globoconusa daubjergensis (extinction) Datum
- P1 — G. trinidadensis G. inconstans pseudobulloides trinidadensis triloculinoides eobulloides — (65) Globigerina pseudobulloides (extinction) Datum
- G. mayaroensis T-R-Z
- G. gansseri P-R-Z

North American Mammalian Stages: IRVINGTONIAN, BLANCAN, HEMPHILLIAN, CLARENDONIAN, BARSTOVIAN, HEMINGFORDIAN, ARIKAREEAN, ORELLAN, CHADRONIAN, DUCHESNIAN, UINTAN, BRIDGERIAN, WASATCHIAN, CLARKFORKIAN, TIFFANIAN, TORREJONIAN, PUERCAN/DRAGONIAN

West Coast (California) Marine Stages: WHEELERIAN, VENTURIAN, REPETTIAN, DELMONTIAN, MOHNIAN, LUISIAN, RELIZIAN, SAUCESIAN, ZEMORRIAN, REFUGIAN, NARIZIAN, ULATISIAN, PENUTIAN, BULITIAN, YNEZIAN

New Zealand Marine Stages: CASTLECLIFFIAN, OPOITIAN, KAPITEAN, TONGAPORUTUAN, WAIAUAN, LILLBURNIAN, CLIFDENIAN, ALTONIAN, AWAMOAN, HUTCHINSONIAN, OTAIAN, WAITAKIAN, DUNTROONIAN, WHAINGAROAN, RUNANGAN, KAIATAN, BORTONIAN, PORANGAN, HERETAUNGAN, MANGAORAPAN, WAIPAWAN, TEURIAN

European Stages: CALABRIAN, ASTIAN, PIACENZIAN, ZANCLIAN, TORTONIAN/MESSINIAN, SERRAVALLIAN, LANGHIAN, BURDIGALIAN, AQUITANIAN, CHATTIAN, RUPELIAN, LATTORFIAN, BARTONIAN/PRIABONIAN, LUTETIAN, YPRESIAN, THANETIAN, DANIAN, MAESTRICHTIAN

Epoch/Series: PLEISTOCENE, PLIOCENE, MIOCENE (LATE, MIDDLE, EARLY), OLIGOCENE (LATE, EARLY), EOCENE (LATE, MIDDLE, EARLY), PALEOCENE, CRETACEOUS

12

Copyright © 1974 by Elsevier Scientific Publishing Co.

Reprinted from *Earth and Planetary Sci. Letters* **22**:300–306 (1974)

THE EXTENSION OF THE MAGNETIC TIME SCALE IN SEDIMENTS
OF THE CENTRAL PACIFIC OCEAN*

N.D. OPDYKE

Lamont-Doherty Geological Observatory of Columbia University, Palisades, N.Y. (USA)

L.H. BURCKLE

*Hunter College of C.U.N.Y., New York, N.Y. (USA) and Lamont-Doherty Geological Observatory
of Columbia University, Palisades, N.Y. (USA)*

and

A. TODD

Lamont-Doherty Geological Observatory of Columbia University, Palisades, N.Y. (USA)

Received February 22, 1974
Revised version received June 4, 1974

The extension of the magnetic reversal record back to the early Miocene is presented. This record is pieced together with the aid of microfloral analysis from three low sedimentation rate siliceous deep sea cores from the Equatorial Pacific.

Nineteen Magnetic Epochs are now recognized from the earliest Miocene to the Present. By correlating the micropaleontological data in our cores with selected foraminiferal datums from DSDP Leg IX we correlate these datums with the following magnetic epochs: the *Pulleniatina* Datum occurs in the lower part of Epoch 5, the *G. acostaensis* Datum occurs in Epoch 11, the *G. nepenthes* Datum occurs in Epoch 12, the *Orbulina* Datum at the Epoch 15/16 boundary and the *G. dissimilis* Datum in the lower part of Epoch 16. The Early/Middle Miocene boundary (*Orbulina* Datum) is tentatively placed at the top of Epoch 16.

1. Introduction

The use of magnetic stratigraphy, allied with the study of microfaunal and floral elements, in studies of sediments from the Central Pacific area has proved to be of great utility in the past [1]. Cores from the region have previously been utilized in studies extending the magnetic stratigraphy and its correlation to faunal and floral elements back in time to the Middle Miocene [2–4] and have helped to elucidate the age of the Miocene/Pliocene boundary (5 m.y. B.P. – T. Saito, personal communication). The purpose of the present study is to attempt to extend the magnetic stratigraphy back in time by correlating the magnetic reversal pattern to the diatom flora contained in the cores.

* Lamont-Doherty Geological Observatory Contribution No. 2106.

In order to do this, four cores were selected for study in addition to RC12-65, which has a known magnetic and floral record back to the Middle Miocene. The cores were selected on the basis of diatom flora, which is known to be as old as earliest Miocene in the oldest core studied. The core locations and water depths are given in Table 1. Unconformities which allow the piston cores to reach older horizons are present in the four cores used in this study. On the basis of the diatom

TABLE 1

Core locations

Core	Lat.	Long.	Water depth (m)
RC12-65	04°39'N	144°58'W	4868
RC11-207	07°24.5'N	139°56'W	5081
RC12-62	06°44'N	141°22.5'W	5073
RC13-24	04°56.8'N	175°04'W	5316
RC13-22	00°10.1'S	171°14'W	5316

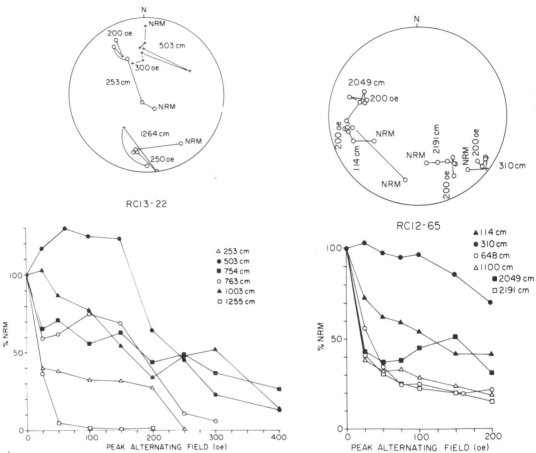

Fig. 1. Alternating field demagnetization curves of samples from various depths in RC12-65 and RC13-22. The change in direction is plotted on a Wulf stereographic projection with arbitrary coordinates.

flora, it is believed that major unconformities are not present in the bottom parts of the cores.

All cores were taken using the Ewing piston corer, and internal orientation of the cores was maintained using the technique described by Foster and Opdyke [2]. The cores were sampled for study at 10-cm intervals. Alternating field (a.f.) partial demagnetization was carried out on selected specimens from each core using methods and equipment similar to that described by McElhinny [5]. On the basis of a.f. demagnetization curves, fields for uniform partial demagnetization of the cores were chosen.

2. Magnetic stability

The a.f. demagnetization curves and resultant changes in direction for RC12-65 and RC13-22 are shown in Fig. 1. In a majority of the samples studied from all cores, including RC12-65, a large, very soft component of magnetization is present. This effect is most serious in RC13-24 where the median destructive field, MDF (the alternating field necessary to reduce the NRM to 50% of its initial value), for all samples is less than 25 oe. Individual samples from all the other cores also have MDF's of 25 oe or less,

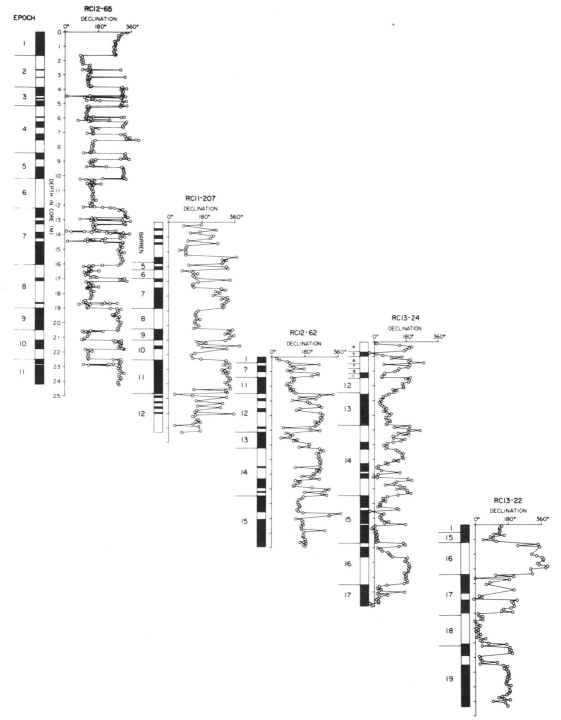

Fig. 2. The change in declination with depth is shown for all the cores used is this study. The change in declination is arbitrary since the cores are not oriented with respect to true north. Normally and reversely magnetized sections of the cores are shown as black and white bars respectively.

173

and this includes RC12-65 originally studied by Foster and Opdyke [2]. On the other hand, core RC13-22, although possessing some samples with low coercivities. has others with MDF's exceeding 200 oe. Because of these large unstable components, partial a.f. demagnetization was carried out in peak alternating fields of 50 oe. These fields were chosen because higher fields often introduced undesirable secondary ARM during the a.f. demagnetization process due to the large unstable component. The fact that a coherent reversal pattern emerges after treatment, for example in RC12-65, is adequate justification of the use of low fields on those sediments. It should be pointed out, however, that the change of direction in individual specimens may not reflect field behavior and the interpretation of any individual sample direction should be treated with caution unless supported by samples from adjacent specimens.

One of the difficulties in working with cores so near the equator is that inclination data are usually an unreliable index of polarity except in certain cases [6]. Therefore, the problem that must be resolved is that although changes in declination can often be clearly seen, the polarity is unresolved because the absolute orientation of the core is not known. In some cores, a small section of Brunhes age sediment is present above the unconformity and in these cases changes of polarity can be followed down the core. This is the case in RC12-62. This is a safe procedure, except in cores where excessive twisting of the core barrel has occurred as the coring device enters the sediment [1]. These areas can sometimes be seen as a spiralling of the declination down the core.

Another way in which the polarity can sometimes be determined is from the behavior of the magnetic intensity on a.f. demagnetization. Upon a.f. demagnetization it is often observed that in reversed specimens with good coercivities the remanent intensity will rise in low fields due to the removal of a secondary component in the direction of the present field which is opposed to the primary direction of magnetization. A good example of this type of behavior is seen in the demagnetization curve from RC13-22 in the specimen taken from 503 cm (Fig. 1).

A third method employed in this study is the use of fossil detritus, in this case diatoms, to correlate between a core with good magnetic stratigraphy (RC12-65) to cores with a less clear magnetic record,

in this case RC11-207 and RC12-62. This is most important and after such a correlation is affected, a coherent study must emerge.

3. Paleontological basis for correlation of magnetics

Although all the cores used in this study are very diatomaceous, a few key selected species serve to demonstrate the between-core correlation. The *Coscinodiscus praeyabei-yabei* complex first appears in the early part of Magnetic Epoch 14 and disappears during Magnetic Epoch 8. The last appearance has been clearly demonstrated in cores RC11-207 and RC12-65. Its first appearance can be seen in RC12-62 and RC13-24 (Fig. 2).

Overlapping the range of *C. praeyabei/C. yabei* is a previously undescribed species, here designated *Coscinodiscus* sp. As seen in Fig. 2, this species first appears in the early part of Magnetic Epoch 15 in core RC12-62 and RC13-24. Its first appearance in the early part of this normal epoch clearly establishes the correlation between these two cores.

Additional evidence for this comes from a third species, *Cestodiscus peplum*. It disappears in the early part of Magnetic Epoch 15 in three cores (RC12-62, RC13-24 and RC13-22). The reader will note that its range overlaps with that of *Coscinodiscus* sp. in core RC-12-62 but not in core RC13-24. This is likely due to differences in sedimentation rates between these two cores. Berger and Heath [7] have pointed out that lower sedimentation rate cores exhibit a greater degree of both upward and downward mixing. We suggest this as the major reason for the difference in range of *C. peplum* between RC12-62 and RC13-24.

In core RC13-22 Brunhes Epoch sediments sit directly on top of Magnetic Epoch 15 sediments. From the core top to 50 cm depth, diatoms typical of the Brunhes are found. Below 50 cm, *C. peplum* is present but not *Coscinodiscus* sp. On this basis, we suggest, therefore, that this is the basal part of Magnetic Epoch 15. Corroborating evidence for this comes from a short ranging species, *Anellus californicus,* which is found mainly in the upper part of Magnetic Epoch 16 in both RC13-22 and RC13-24. Floral evidence suggests that the basal portion of RC13-22 is in the earliest Miocene.

The changes in declination in cores RC12-65, RC11-207, RC12-62, RC13-24 and RC13-22 are show

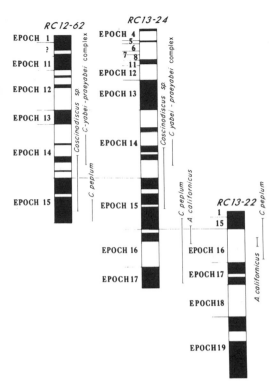

Fig. 3. Correlation of RC12-62, RC13-24 and RC13-22 based on the ranges of selected diatom species.

in Fig. 2. Fig. 3 shows the correlation of the five cores used in this study with the resulting magnetic stratigraphy. Foster and Opdyke [2] established the magnetic epochs through Magnetic Epoch 11 using the numbering system proposed by Hays and Opdyke [8]. It can be seen in Fig. 3 that this numbering system has been carried back in time to Magnetic Epoch 19.

Clearly, an attempt should be made to date the magnetic epochs and events delineated in this study. This is difficult for several reasons, the most obvious being that no way exists to get a direct date on these cores.

The cores used in this study are not calcareous. However, it is possible to draw correlations with foraminiferal datums by comparing the siliceous fossils in our cores with samples from Leg IX of DSDP [9] which contain both diatom and foraminiferal remains. Such comparisons permit us to place, within limits,

certain foraminiferal datums with respect to the magnetic reversal stratigraphy. The ongoing work of Berggren [10, 11] in dating these datum levels makes it possible to give an approximate age to magnetic reversals.

The correlation method described above indicates that the *Pulleniatina* Datum occurs in the early part of Magnetic Epoch 5 at approximately 6 m.y. B.P. This has also been pointed out by T. Saito (personal communication). The *G. acostaensis* Datum occurs in the late part of Magnetic Epoch 11 at approximately 10.5 m.y. B.P. [11]. Prior to that, the *G. nepenthes* Datum occurs in Magnetic Epoch 12 or the later part of Epoch 13. Still earlier, the *G. dissimilis* extinction datum, at 17.5 m.y., is found in the early part of Magnetic Epoch 16.

4. The Early/Middle Miocene boundary

Berggren [11, 19] places the Early/Middle Miocene boundary at the base of Zone N 9 of Blow [12] coincidental with the first appearance of *Orbulina* (the *Orbulina* Datum). In Site 77 of DSDP Leg IX the first appearance of *Orbulina* (The *Orbulina* Datum) occurs in the upper part of Barrel 4. Our examination of this section and sections above and below show the presence of such stratigraphically significant diatoms as *Macrora stella, Annellus californicus, C. lewisianus, C. praepaleaceus* and *Cestodiscus peplum.* Of special significance is the presence of *A. californicus.* Previous studies by many workers [13—15] have pointed out that this is an extremely short ranging indicator species for the Early and Middle Miocene. Our studies indicate that *A. californicus* first appears in Barrel 26 section 2 and disappears in Barrel 24 section 2 of Site 77. In other words, the upper part of its range is approximately coincident with the first appearance of the *Orbulina* Datum.

In core RC13-24, *A. californicus* first appears in Magnetic Epoch 16 and disappears in the lower part of Magnetic Epoch 15. A similar pattern of first and last appearance of this species can be seen in core RC13-22 although the section has been truncated by an unconformity.

On the basis of this line of evidence, we conclude that the *Orbulina* Datum (or base of the Middle Miocene) must occur at or close to the top of Magnetic

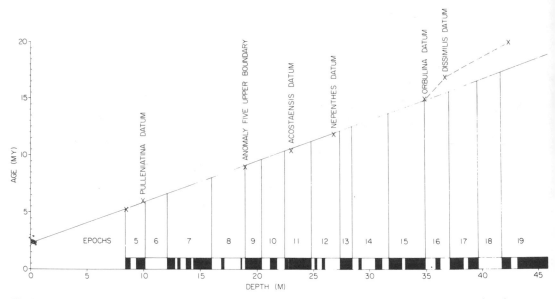

Fig. 4. A plot of age against accumulated sediment with the magnetic epochs as given in Fig. 2. The ages of the various datums are discussed in the text. The first X is the Magnetic Epoch 5 upper boundary and the last X is the age of the anomaly 6 upper bound-. ary. The abrupt change in slope in Magnetic Epoch 16 is shown by the dashed lines and is presumed to give the most correct age for the bottom part of the sequence.

Epoch 16. Indeed, for the sake of convenience we provisionally place the Early/Middle Miocene boundary at the Magnetic Epoch 15/16 boundary with an age of 15 m.y.

Fig. 4 shows an attempt to obtain a provisional time scale for this magnetic stratigraphy. The abscissa of the figure represents the combined stratigraphy for the cumulative length of the cores and the correlation of the cores was done as previously described. The relative position and duration of events within this magnetic sequence is difficult to ascertain with certainty. Events are shown on Fig. 4 if they are represented in at least two cores by more than a single point. In some cases no overlap between cores is available and in this part of the sequence events are inserted if we believe that the data quality warrants it. The authors realize that this is a subjective procedure but in most cases the authors have tried to be conservative in interpretation. The resulting magnetic stratigraphy is then plotted against depth. The ages of the foraminifera datums as given by Berggren [11] are placed in the magnetic stratigraphy as previously described. Addi-

tional points in time are obtained by the estimated age of the Gilbert/Magnetic Epoch 5 boundary at 5.1 m.y. B.P. [8]. Another apparent age is provided by the correlation of the top of Magnetic Epoch 7 to the top of anomaly 5 (9 m.y. [16]) as suggested by Ryan (personal communication). The oldest portion of the time scale shown must be close to the Miocene/Oligocene boundary. We, therefore, think that it is reasonable to correlate Magnetic Epoch 19 with the magnetic anomaly 6 as suggested by Theyer and Hammond [17]. The oldest point on Fig. 4 is provided by this correlation. It can be seen from Fig. 4 that the data fall on a straight line from Magnetic Epoch 5 to the *Orbulina* Datum which would seem to indicate that for this portion of the time scale the accumulate rates of sedimentation were relatively constant, and also that the Berggren time scale contains no large errors in the Late and Middle Miocene. An indirect check on the stratigraphy and dating of this portion of the magnetic stratigraphy is provided by two dates on volcanic ash done on the experimenta Mohole by Dymond [18]. Burckle has been able to correlate the sediment adjacent to the two dated ashes (ages of 11.2 and 12.2 m.y.) to Magnetic Epoch 12 and 13 of this study. Although we cannot place these two dated ashes precisely in our stratigraphy, the dates that have been obtained are clearly close to that predicted in Fig. 4.

It can be seen by an inspection of Fig. 4 that between the *Orbulina* Datum and the *G. dissimilis* Datum a distinct break in slope occurs. If the dates of these two datums are correct this would indicate that Magnetic Epoch 16 should be longer than represented in the cores of this study. Indeed, Theyer and Hammond [17] appear to have a more complete and longer Magnetic Epoch 16 than is represented in this study.

We realize that considerable error may result from the technique employed here due to the indirect way in which the dates on the magnetic sequence have been obtained and to a varying rate of sedimentation in the cores studied. The end result, however, makes a reasonably coherent picture, placing the bottom part of the sequence in the lowermost Miocene.

At the present time, a one-to-one correlation with the sea-floor spreading polarity sequence of Pitman and Heirtzler cannot be made. We are hopeful, however, that future JOIDES drilling in the time span between 10 and 20 m.y. may provide a correlation of the magnetic and fossil record of the sediments with the more complete magnetic record from the sea floor.

Acknowledgements

We thank W. Lowrie and D. Kent for critically reviewing the manuscript and offering many helpful suggestions. Discussions with R. Goll, R. Larson, T. Saito and J.D. Hays of Lamont-Doherty and W. Berggren of Woods Hole Oceanographic Institution were also useful and are much appreciated. Funds to obtain the cores used in this study were provided by the U.S. Navy (N00014-67-A-0108-0004) and the National Science Foundation (Joint Oceanographic Institutions for Deep Earth Sampling, JOIDES). Funds for core analysis were provided by grants from the National Science Foundation (GX 28671, GA 35463X and GA 30569X). Finally, we wish to acknowledge the Deep Sea Drilling Project for making Leg IX samples available to us.

References

1 J.D. Hays, T. Saito, N.D. Opdyke and L.H. Burckle, Bull. Geol. Soc. Am. 80 (1969) 1481.
2 J.H. Foster and N.D. Opdyke, J. Geophys. Res. 75 (1970) 4465.
3 L.H. Burckle, in: 1st Symposium on Recent and Fossil Marine Diatoms, ed., R. Seimonsen (Verlag von J. Cramer, Bremerhaven, 1972) 217.
4 R.M. Goll, Micropaleontology 18 (1973) 443.
5 M.W. McElhinny, Geophys. J. R. Astron. Soc. 10 (1966) 369.
6 N.D. Opdyke and K.W. Henry, Earth Planet. Sci. Lett. 6 (1969) 139.
7 W.H. Berger and G.R. Heath, J. Mar. Res. 26 (1968) 134.
8 J.D. Hays and N.D. Opdyke, Science 158 (1967) 1001.
9 J.D. Hays, H.E. Cook, D.G. Jenkins, F.M. Cook, J.T. Fuller, R.M. Goll, E.D. Milow and W.N. Orr, Initial Reports of the Deep Sea Drilling Project, Leg IX (Govt. Printing Office, Washington, 1971) 3.
10 W.A. Berggren, Hungarian Geol. Soc. Bull. 101 (1971) 162.
11 W.A. Berggren, Lethaia 5 (1972) 193.
12 W.H. Blow, 1st International Conference Planktonic Microfossils Proc. 1 (1969) 199.
13 G.D. Hanna, Acad. Sci. XX (1932) 161.
14 T.H. Reinhold, Ned. Kolon. Geol. Mijnbouwk. Genoots. Verh., Geol. Ser. 12 (1937) 43.
15 L.H. Burckle, Pacific Geol. (1973) in press.
16 W.C. Pitman III and J.R. Heirtzler, Science 154 (1966) 1164.
17 F. Theyer and S.R. Hammond, Earth Planet. Sci. Lett. 22(1974) 307.
18 J.R. Dymond, Science 152 (1966) 1239.
19 W.A. Berggren, in: Symposium on Messinian Events in the Mediterranean, Utrecht, 1973, K. Ned. Akad. Wet., Publ. (1974).

13

PALEOMAGNETIC POLARITY SEQUENCE AND RADIOLARIAN ZONES, BRUNHES TO POLARITY EPOCH 20

F. THEYER and S.R. HAMMOND

Hawaii Institute of Geophysics, University of Hawaii, Honolulu, Hawaii (USA)

Superposition of paleomagnetic polarity logs of seven chronologically overlapping piston cores from the central equatorial Pacific, using the established tropical radiolarian zonation as a stratigraphic reference, produced a nearly continuous correlation of magnetic and radiolarian events ranging from late Pleistocene to earliest Miocene. Twenty magnetic polarity epochs, and possibly as many as 30 polarity events, occur during this time span. Epoch 16 (reverse polarity) appears to be the longest interval (~ 14.8–17.6 m.y. B.P.) among these Neogene magnetostratigraphic unit. The middle/late Miocene boundary is shown to fall within latest Epoch 11 (normal) and its approximate age is between 10.5 and 11 m.y. B.P. The early/middle Miocene boundary occurs within the top of Epoch 16 at a suggested age of about 15 m.y. B.P.

[*Editor's Note:* The article itself has not been reproduced here.]

Part III

EXTENSION OF POLARITY SCALE TO THE CLASSICAL GEOLOGICAL REGIONS: UPLIFTED MARINE SECTIONS

Editor's Comments
on Papers 14 Through 17

Until the last decade almost all knowledge of marine Cenozoic stratigraphy and paleontology had been derived from about a century of investigations of marine sedimentary sections uplifted on land. Because much of the early geological work had been carried out on European Cenozoic marine sections, especially in Italy and France, these became the global standards of reference for chronostratigraphic correlations. Thus through historical precedence, certain key sections have been assigned to represent the type stages of the world to which all other chronostratigraphic schemes are correlated. The selection of these sections was not based on global quality. Indeed, several other areas of the world have much more representative marine sequences, which are less complicated and often more continuous than those in the Mediterranean marked by isolation from open oceanic areas. For instance, the tectonically active circum-Pacific region generally contains better exposed, fresher sections, exhibiting higher sedimentation rates, fewer unconformities, and richer planktonic microfossil assemblages. In New Zealand excellent marine sections at temperate latitudes have been studied extensively since the early 1940s and have served as a standard of reference for temperate biostratigraphic sequences elsewhere in the world. Likewise, marine sections in Japan and California have been well examined and formed an important reference for later work elsewhere. Furthermore, in lying adjacent to major oceanic boundary currents, these sequences have provided vital information on late Cenozoic oscillations of tropical and temperate to subarctic water masses.

During the late 1960s, the emphasis of investigation on land-based marine sections suddenly switched to that of deep-sea sediments, including sections of latest Cenozoic age obtained by piston coring and those of greater age obtained by the Deep Sea Drilling Project. Hundreds of workers began to carry out a truly diverse range of investigations of deep-sea sediments. These included studies of all the preserved microfossil groups, of the sediments themselves, and of the geological, geophysical, and paleoceanographic implications resulting from these investigations. This progress has demanded an ability to correlate between deep-sea sequences and the classical marine sequences on land, especially the type sections of Europe. Before magnetostratigraphy had been developed, nearly all stratigraphic approaches used in correlation of marine sediments, including the use of various planktonic microfossil groups, had first been applied to problems involving land-based marine sections and later extended to the study of deep-sea sequences. This was not the case with magnetostratigraphy, which, like oxygen isotopic stratigraphy, was first used on deep-sea sequences.

Marine sections exposed on land are very familiar to most geologists as these sections are often conspicuously exposed in road cuts and river gorges. The sediments contained in these sequences are almost always of shallower facies than their deep-sea counterparts, are generally coarser, and may contain rich assemblages of macrofossils such as the mollusca and solitary corals. These shallow marine sections normally contain persistent planktonic foraminiferal and calcareous nannofossil sequences but are poor in siliceous microfossils (radiolaria and diatoms). Shallow marine sections suffer from two important disadvantages: They are weathered and do not provide continuously outcropping sequences, although the latter problem is often partly compensated by high sedimentation rates, allowing detailed stratigraphic analyses.

During the late 1960s, at the time when magnetostratigraphy had become accepted as a powerful tool for correlating deep-sea sediments, it became apparent to a few groups working independently at different universities throughout the world, that the application of magnetostratigraphy should be extended, if possible, to these land-based marine sections for several purposes: to establish an urgently needed chronology for the geological history derived from their study; to correlate these sections with those in the deep-sea and with the type marine sections in Europe; and to capitalize on the much higher sedimentation rates of these sections to provide potentially higher resolution for study of short-term polarity events.

Early attention on land-based magnetostratigraphy had actually been focused on *nonmarine* red beds because of their comparatively strong remanent magnetization. However, such sequences contain few fossils and hence have provided only restricted paleoenvironmental information as well as being difficult to correlate over long distances. New studies were necessary on marine sections containing abundant microfossils for intercalibration with the polarity time scale.

With these problems in mind, field programs were to be organized in the late 1960s in New Zealand and Japan. James Kennett had recently joined with Norman Watkins at Florida State University and was anxious to attempt magnetostratigraphy on New Zealand Late Cenozoic sections he had worked on for his Ph.D. dissertation in the early 1960s. The paleomagnetic field work involved the collection by Watkins and Kennett of paleomagnetic and biostratigraphic samples in about 18 sections ranging in age from the early Late Miocene through the Quaternary. Of these, a particularly important Pliocene to early Quaternary section at Mangaopari Stream in the southern North Island was studied in collaboration with Dr. Paul Vella, who had been major professor for James Kennett a few years earlier. This sequence was of particular interest because oxygen isotopic and detailed planktonic foraminiferal trends had already been established, and the sequence was known to contain the Pliocene-Pleistocene boundary as recognized in New Zealand (Devereux et al., 1970). The oxygen isotopic record was one of the first to be developed anywhere for the Pliocene and contained evidence for the build-up of Northern Hemisphere ice caps. This event needed to be dated. A magnetostratigraphy was successfully developed for this section, defining a polarity record from the Matuyama through Gauss (Kennett et al., Paper 14). The Pliocene-Pleistocene boundary, as generally accepted in New

Zealand, was found to be closely associated with the most conspicuous normal event in the middle of the Matuyama (called the Gilsa by Kennett et al., Paper 14) and hence coincident with the Pliocene-Pleistocene boundary as recognized in deep-sea sediments (Berggren et al., 1967). The principal significance of this paper, however, was its demonstration that magnetostratigraphy could be applied to land-based marine sections at least in favorable circumstances and hence be of assistance in establishing the chronology of late Cenozoic climatic history. The work was still confined to the Late Pliocene and Early Pleistocene. A chronology was still required for the New Zealand Early Pliocene and Late Miocene sequences. Several sections were collected by Watkins and Kennett for this purpose, and two were discovered to exhibit a suitably high paleomagnetic intensity to provide an acceptable magnetostratigraphy. These sections were at Mangapoike and Blind Rivers, and the results are reproduced in Paper 15. Independently, a group from Victoria University of Wellington, headed by Drs. Paul Vella and D. A. Christoffel, began work on a similar problem in a section at Hinakura in southern North Island (Lienert et al., 1972). Two of the principal objectives of these studies were to establish the age of the Miocene-Pliocene boundary as recognized in New Zealand and to date conspicuous Late Cenozoic paleoclimatic oscillations recorded in these sections. Both studies showed that the Miocene-Pliocene boundary placed near the base of the Gilbert Reversed Epoch; at 4.6 m.y. in the case of Lienert et al. (1972) and at about 4.3 m.y. by Kennett and Watkins (Paper 15). These interpretations immediately created an international correlation problem because the type (International) Miocene-Pliocene boundary in Italy was considered to lie between 4.9 and 5.1 m.y. (Berggren, Paper 11), and hence the difference in ages required an explanation. Ryan et al. (1974) reinterpreted the paleomagnetic stratigraphy of these sections by assuming quite different sedimentation rates throughout the sequences and thus stretching and squeezing the magnetostratigraphy of Kennett and Watkins. Although the ages applied to the New Zealand Late Miocene to Early Pliocene sequences by Ryan et al. (1974) are acceptable, their modifications of the New Zealand paleomagnetic stratigraphy are unacceptable for biostratigraphic reasons (Kennett, 1977). As pointed out by Hornibrook (1977), it is more likely that the ages assigned to the Late Miocene polarity changes are in error in the Mangapoike River section because of a previously unrecognized unconformity close to the Miocene-Pliocene boundary. Furthermore, a 5.2-m.y. K/Ar date has since been obtained for earliest

Pliocene sediments of the Chatham Islands, which supports an age for the New Zealand Miocene-Pliocene boundary closer to 5 m.y. (Hornibrook, 1977). Thus at this time the exact age of the Miocene-Pliocene boundary as applied to New Zealand sequences is still in doubt, although it apparently falls within the interval from Epoch 5 through earliest Gilbert Epoch. This question may be difficult to resolve because several additional sections sampled by Watkins and Kennett, which apparently represent continuous sedimentation over the boundary, exhibit very low intensities of magnetization, and it has been found using various demagnetization techniques that the remanent magnetism cannot be differentiated from a normal polarity chemical overprint, acquired during the last 0.69 m.y.

At the same time as the South Pacific field work was being carried out by Watkins and Kennett, similar investigations had begun on late Cenozoic sequences in Japan by a group from Tohoku University, Sendai, headed by Dr. H. Nakagawa and including Dr. N. Niitsuma. Their work commenced with the well-exposed and thick ($\sim$ 450 m) marine section at Boso Peninsula, southern Honshu (Nakagawa et al., 1969). These sediments seem to be ideal for magnetostratigraphic work because of generally good exposures and suitably high intensity of the natural remanent magnetism. In a short time the Tohoku group produced a long magnetostratigraphic record interpreted to range from Epoch 7 through the Brunhes Epoch in terms of the polarity time scale of Cox et al. (1964). Later more detailed work by Niitsuma (1976) required some reinterpretation especially in that part of the sequence originally assigned to Epochs 5 through 7. This was because of previously unrecognized complexities resulting from secondary paleomagnetic overprinting of the natural remanent magnetism. The chronological framework established on the late Cenozoic sequences at Boso Peninsula provided ages of the Pliocene-Pleistocene and Miocene-Pliocene boundaries as recognized in Japan, and a framework for correlation with other sections (Nakagawa et al., 1975).

The problem that secondary paleomagnetic overprinting can cause in magnetostratigraphic interpretations of the Japanese sections was clearly demonstrated by Dr. Dennis Kent, a geophysist, and Dr. Lloyd Burckle, a paleontologist from Lamont-Doherty Geological Observatory, New York. These workers had traveled to Japan in the early 1970s, under the support of a U.S.-Japanese cooperative scientific program, to conduct combined magnetostratigraphic and biostratigraphic studies of critical marine sequences at Oga Peninsula, northwest Japan. Biostratigraphic subdivisions that had been previously established in this sequence by Japanese workers are used as a standard Neogene time-stratigraphic framework in Japan, and it was hoped that a chronology could be established for the sequence. Unfortunately, the study (Kent, 1973) revealed only one true reversal of the earth's field. Most other parts of the sequence were dominated by normal polarity interpreted as a viscous remanent magnetization (VRM) acquired during the last 0.69 m.y. Thus the magnetostratigraphy is of only very limited assistance for applying a chronology to this sequence.

As magnetostratigraphy was being developed in sequences in Japan and New Zealand, it became apparent that similar work was needed to provide a chronology for the type sections of Europe. This was clearly critical in the assignment of correct stratigraphic ages in the sequences distant from the European types. The Tohoku group promptly began field work on many of the Italian-type sections, which are so central to the stratigraphic subdivision and correlation of the Neogene. The first such analyses were carried out on the type sections of the Pliocene-Pleistocene boundary at Calabria, southern Italy, and polarity records were produced for the type sections at Le Castella (Nakagawa et al., 1971) and Santa Maria Catanzaro (Nakagawa et al., 1975). These workers favored an interpretation of the magnetostratigraphy as assigning the Pliocene-Pleistocene boundary to be close to the major normal event in the Matuyama (termed "Gilsa" by Nakagawa et al., 1971), thus providing an age of 1.8 m.y. B.P. These results were brought into question by work shortly carried out afterward by Watkins and Kennett. The results of these workers (Watkins et al., Paper 16) clearly provided strong evidence for a normal polarity overprint acquired during the last 0.69 m.y., produced by postdepositional chemical precipitation of a magnetically unstable superparamagnetic material. The different results produced by two groups on the same sections are further discussed by Nakagawa et al. (1977) who suggest that they may have been caused by different laboratory techniques. Nevertheless, the results are in conflict, and further work is required to ascertain the validity of age assignments, which are of particular importance because they are concerned with the type sections.

These differences in results between different laboratories are not a reflection of any inherent weakness in the paleomagnetic approach when applied to marine sections as a whole, but the realization that certain sections may not be favorable for producing an acceptable magnetostratigraphy. For instance, several other late Cenozoic sections studied by Nakagawa and his col-

185

leagues in Italy and described in a series of papers (Nakagawa et al., 1971, 1974, 1975, 1977) do exhibit magnetostratigraphic patterns similar to the established polarity time scale of Cox (Paper 1). Of particular importance are those at Santerno and Tortono, northern Italy. The stratigraphy established for these sections has already been used for correlation of the type European stages with those in deep-sea cores and in Japan (Ryan et al., 1974; Nakagawa et al., 1974, 1977). As in New Zealand and Japan, not all sections are found to be suitable for magnetostratigraphic investigations, but those that do produce apparently valid magnetostratigraphic records are very important for international correlation. Nevertheless, caution should be exercised in uncritical acceptance of the magnetostratigraphy from land-based marine sections because several factors may complicate this record, especially normal polarity overprinting resulting from postdepositional chemical changes in the sediment.

As has been emphasized during the previous discussions, the study of weakly magnetized sediments such as carbonates has not been easy. This problem has been much alleviated by the development during the 1970s of more rapid and sensitive magnetometers such as the digital spinners and the cryogenic instruments. A direct result of this improved technology is a classical group of papers published in 1977 on the geomagnetic polarity history of Middle Cretaceous to Paleocene calcareous marine sediments at Gubbio, in the Umbrian Apennines of Italy [Arthur and Fischer, 1977; Premoli Silva, 1977; Lowrie and Alvarez, 1977; Roggenthen and Napoleone, 1977; Alvarez et al., 1977 (Paper 17)]. The results of these papers are of particular importance because they are the first to furnish a long, continuous sequence of magnetic polarity zones in sediment of Middle Cretaceous to Paleocene age, including the Cretaceous-Tertiary (Maastrichtian-Paleocene) boundary and because they have been intercalibrated with planktonic foraminiferal zones. In almost all of the documented Late Cretaceous-Early Tertiary sections of the world, the boundary between these periods is represented by a hiatus. The Gubbio section is almost unique in apparently containing an unbroken sequence across the Cretaceous-Tertiary boundary (Luterbacher and Premoli Silva, 1964). The results of the Gubbio study are important because the magnetostratigraphy closely matches the polarity sequence inferred from marine magnetic anomaly profiles. Furthermore, the polarity sequence exhibits a long normal zone through the Late Cretaceous, ending close to the Santonian-Campanian boundary with the onset of the Gubbio A-Reversed Zone. This is again followed by a relatively long normal zone continuing to the early Maastrichtian. It is perhaps of significance that more rapid polarity changes, providing polarity zones of shorter duration, commence in the latest Cretaceous Maastrichtian Stage and continue into the Paleocene and younger. Thus the latest Cretaceous is represented by a major change in the tempo of change of the earth's magnetic polarity, which has continued through the Cenoozoic. The very sharp, faunally defined Cretaceous-Tertiary boundary falls close to the top of polarity zone Gubbio G-, which corresponds in the marine magnetic profiles to the reversed segment immediately preceding anomaly 29. This is in agreement with the work of Sclater et al. (1974) who placed the Cretaceous-Tertiary boundary between anomalies 29 and 30. The long normal intervals near the end of the Cretaceous Period correspond with the Cretaceous Quiet Zone of the marine magnetic anomaly profiles.

REFERENCES

Arthur, M. A., and A. G. Fischer. 1977. Upper Cretaceous—Paleocene magnetic stratigraphy at Gubbio, Italy. I. Lithostratigraphy and sedimentology. *Geol. Soc. America Bull.* **88**:367–371.

Berggren, W. A., J. D. Phillips, A. Bertels, and D. Wall. 1967. Late Pliocene-Pleistocene stratigraphy in deep-sea cores from the South-Central North Atlantic. *Nature* **216**:253–255.

Cox, A., R. R. Doell, and G. B. Dalrymple. 1964. Reversals of the earth's magnetic field. *Science* **144**:1537–1543.

Devereux, I., C. H. Hendy, and P. Vella. 1970. Pliocene and Early Pleistocene sea temperature fluctuations, Mangaopari Stream, New Zealand. *Earth and Planetary Sci. Letters* **8**:163–168.

Hornibrook, N. de B. 1977. Dating of Pliocene volcanic rocks. *New Zealand Jour. Geology and Geophysics* **20**:466–467.

Kennett, J. P. 1977. Late Neogene paleoceanography of the South Pacific (Abstract). In *Proceedings of the First International Congress on Pacific Neogene Stratigraphy.* Tokyo: Kaiyo Shuppan Co., pp. 81–82.

Kent, D. V. 1973. Paleomagentism of some Neogene sedimentary rocks on Oga Peninsula, Japan. *Jour. Geomagnetism and Geoelectricity* **25**:87–103.

Lienert, B. R., D. A. Christoffel, and P. Vella. 1972. Geomagnetic dates on a New Zealand Upper Miocene-Pliocene section. *Earth and Planetary Sci. Letters* **16**:195–199.

Lowrie, W., and W. Alvarez. 1977. Upper Cretaceous-Paleocene magnetic stratigraphy at Gubbio, Italy. III. Upper Cretaceous magnetic stratigraphy. *Geol. Soc. America Bull.* **88**:374–377.

Luterbacher, H. P., and I. Premoli Silva. 1964. Biostratigrafia del limite Cretaceo-Terziario nell' Appennino Centrale. *Riv. Italiana Paleontologia e Stratigrafia* **70**:67–128.

Nakagawa, H., N. Kitamura, Y. Takayanagi, T. Sakai, M. Oda, K. Asano, N. Niitsuma, T. Takayama, Y. Matoba, and H. Kitazato. 1977. Magnetostratigraphic correlation of Neogene and Pleistocene between the Japanese Islands, Central Pacific and Mediterranean Regions. *Proceedings of the First International Congress on Pacific Neogene Stratigraphy.* Tokyo: Kaiyo Shuppan Co., pp. 285–310.

Nakagawa, H., N. Niitsuma, and C. Elmi. 1971. Pliocene and Pleistocene magnetic stratigraphy in Le Castella area, Southern Italy—A preliminary report. *Quaternary Research* 1:360–368.

Nakagawa, H., N. Niitsuma, and I. Hayasaka. 1969. Late Cenozoic geomagnetic chronology of the Boso Peninsula. *Jour. Geol. Soc. Japan* 75:267–280.

Nakagawa, H., N. Niitsuma, K. Kimura, and T. Sakai. 1975. Magnetic stratigraphy of Late Cenozoic stages in Italy and their correlatives in Japan. In *Late Neogene Epoch Boundaries,* edited by T. Saito and L. H. Burckle. New York: Micropaleontology Press, pp. 64–70.

Nakagawa, H., N. Niitsuma, N. Kitamura, Y. Matoba, T. Takayama, and K. Asano. 1974. Preliminary results on magnetostratigraphy of Neogene stage stratotype sections in Italy. *Riv. Italiana Paleontologia e Stratigrafia* 80:615–630.

Niitsuma, N. 1976. Magnetic stratigraphy in the Boso Peninsula. *Jour. Geol. Soc. Japan* 82:163–181.

Premoli Silva, I. 1977. Upper Cretaceous-Paleocene magnetic stratigraphy at Gubbio, Italy. II. Biostratigraphy. *Geol. Soc. America Bull.* 88:371–374.

Roggenthen, W. M., and G. Napoleone. 1977. Upper Cretaceous-Paleocene magnetic stratigraphy at Gubbio, Italy. IV. Upper Maastrichtian-Paleocene magnetic stratigraphy. *Geol. Soc. America Bull.* 88:378–382.

Ryan, W. B. F., M. B. Cita, M. Dreyfus Rawson, L. H. Burckle, and T. Saito. 1974. A paleomagnetic assignment of Neogene stage boundaries and the development of isochronous datum planes between the Mediterranean, the Pacific and Indian Oceans in order to investigate the response of the World Ocean to the Mediterranean "salinity crisis." *Riv. Italiana Paleontologia e Stratigrafia* 80:631–688.

Sclater, J. G., R. D. Jarrard, B. McGowran, and S. Gartner. 1974. Comparison of the magnetic and biostratigraphic time scales since the Late Cretaceous. In *Initial Reports of the Deep Sea Drilling Project,* Vol. XXII, edited by C. C. Van der Borch et al. Washington, D.C.: U.S. Government Printing Office, pp. 381–386.

ADDITIONAL READINGS

Baag, C.-G., and C. E. Helsley. 1974. Evidence for penecontemporaneous magnetization of the Moenkopi formation. *Jour. Geophys. Research* 79:3308–3320.

Berggren, W. A., D. P. McKenzie, J. G. Sclater, and J. E. van Hinte. 1975. World-wide correlation of Mesozoic magnetic anomalies and its implications: Discussion and reply. *Geol. Soc. America Bull.* 86:267–272.

Brock, A., and R. L. Hay. 1976. The Olduvai event at Olduvai Gorge. *Earth and Planetary Sci. Letters* 29:126–130.

Brock, A., and G. L. Isaac. 1974. Paleomagnetic stratigraphy and chronology of hominid-bearing sediments east of Lake Rudolf, Kenya. *Nature* **247**:344–348.

Bucha, V. 1970. Geomagnetic reversals in Quaternary revealed from a palaeomagnetic investigation of sedimentary rocks. *Jour. Geomagnetism and Geoelectricity* **22**:253–271.

Couvering, J. A. van, and W. A. Berggren. 1977. Biostratigraphical basis of the Neogene time scale. In *Concepts and Methods of Biostratigraphy*, edited by E. G. Kauffman and J. E. Hazel. Stroudsburg, Pa.: Dowden, Hutchinson & Ross, pp. 283–306.

Hillhouse, J., and A. Cox. 1976. Brunhes-Matuyama polarity transition. *Earth and Planetary Sci. Letters* **29**:51–64.

Irving, E., and R. W. Couillard. 1973. Cretaceous normal polarity interval. *Nature Phys. Sci.* **244**:10–11.

Irving, E., and G. Pullaiah. 1976. Reversals of the geomagnetic field, magnetostratigraphy and relative magnitude of paleosecular variation in the Phanerozoic. *Earth-Sci. Rev.* **12**:35–64.

Johnson, A. H., and A. E. M. Nairn. 1972. Jurassic palaeomagnetism. *Nature* **240**:551–552.

Johnson, A. H., A. E. M. Nairn, and D. N. Peterson. 1972. Mesozoic reversal stratigraphy. *Nature Phys. Sci.* **237**:9–10.

Khramov, A. N. 1957. Paleomagnetism—The basis of a new method of correlation and subdivision of sedimentary strata. *Acad. Sci. USSR Doklady, Earth Sci. Sec. Proc.* **112**:129–132.

Kukla, G. J. 1977. Pleistocene land-sea correlation. I. Europe. *Earth-Sci. Rev.* **13**:307–374.

Larson, R. L., and C. E. Helsley. 1975. Mesozoic reversal sequence. *Rev. Geophysics and Space Physics* **13**:174–176.

Larson, R. L., and T. W. C., Hilde. 1975. A revised time scale of magnetic reversals for the Early Cretaceous and Late Jurassic. *Jour. Geophys. Research* **80**:2586–2594.

Larson, R. L., and W. C. Pitman, III. 1972. World-wide correlation of Mesozoic magnetic anomalies and its implications. *Geol. Soc. America Bull.* **83**:3645–3662.

Robinson, E., and J. L. Lamb. 1970. Preliminary palaeomagnetic data from the Plio-Pleistocene of Jamaica. *Nature* **227**:1236–1237.

Vella, P. 1975. The boundaries of the Pliocene in New Zealand. In *Late Neogene Epoch Boundaries*, edited by T. Saito and L. H. Burckle. New York: Micropaleontology Press, pp. 85–93.

Verosub, K. L. 1975. Paleomagnetic excursions as magnetostratigraphic horizons: A cautionary note. *Science* **190**:48–50.

14

Reprinted from *Science* 171:276–279 (1971)

PALEOMAGNETIC CHRONOLOGY OF PLIOCENE-EARLY PLEISTOCENE CLIMATES AND THE PLIO-PLEISTOCENE BOUNDARY IN NEW ZEALAND

J. P. Kennett, N. D. Watkins, and P. Vella

Geomagnetic polarity reversals for the last 5 million years (*1*) have been used to provide ages of and correlations between many deep-sea sedimentary cores (*2*) but have only rarely been applied to studies of Cenozoic marine sedimentary sequences exposed on land (*3*). We have extended the method to relatively shallow-water Pliocene to Early Pleistocene marine sediments exposed on New Zealand.

The remarkably complete Tertiary and Quaternary stratigraphic and paleontological record in New Zealand has enabled clear definition of relative paleotemperature oscillations resulting from periodic regional migrations of subantarctic water to the south and warmer subtropical water to the north (*4, 5*). These paleotemperature oscillations have been difficult to correlate in other parts of the world, however, because of the provincial character of New Zealand marine faunal assemblages and the local controls on paleooceanographic conditions.

The paleomagnetic stratigraphy was determined in the nearly continuous Pliocene to Early Pleistocene sequence at Mangaopari Stream–Makara River (Fig. 1), at the southern end of the North Island of New Zealand. Fresh sediments exposed in the stream beds and adjacent road cuts total 650 m of foraminifer-rich marine mudstone, sandstone, and limestone of upper bathyal to middle neritic depths. The section is of considerable importance because of prior biostratigraphic, paleoclimatic, and oxygen isotope studies (*6, 7*) (Fig. 2). Objectives of this study include determination of the stratigraphically defined Pliocene-Pleistocene boundary by correlation with the base of the marine Calabrian stage of southern Italy; the evaluation of relations between this stratigraphically defined boundary and marked climatic cool-

ings, including that which defines the currently accepted Plio-Pleistocene boundary determined primarily by the appearance of cool-water Mollusca (*4, 6, 8*). Finer definition of the geomagnetic polarity history during the Matuyama epoch (0.69 to 2.43 million years ago) (*1*) may also result from detailed studies of such a section.

Three cores, 2.5 cm in diameter, were drilled with a gasoline-powered portable drill from each of 61 sites throughout the section (Figs. 1 and 2), as well as in two slightly younger (Marahauan) stratigraphic horizons in a nearby section (*9*). All cores were oriented in geographic coordinate, while still attached to outcrop, to an accuracy of ± 3°. At each site vertical separation between the three cores was no greater than 30 cm, but horizontal separation was up to 3 m. In the laboratory each core was sliced into specimens 2.2 cm long for paleomagnetic analysis. The direction and intensity of the remanent magnetism in each specimen were measured with a spinner magnetometer, operating at 5 cycle/sec (*10*), before and after application of alternating magnetic fields (*11*) of 50 and 75 oersteds.

Intensities of magnetization are of the order of 10^{-7} electromagnetic unit per gram after the viscous components have been removed by the demagnetization treatments. In the log of the paleomagnetic polarities (Fig. 2) the latitude of the virtual geomagnetic pole (*12*) is used to indicate the polarity. Northern latitudes indicate normal polarity, and southern latitudes indicate reversed polarity. The few results with latitudes between 10°N and 10°S are not assigned to any polarity. These are clearly transitional and do not represent major polarity changes.

A section from lowermost Gauss to middle Matuyama geomagnetic polar-

ity epochs is recorded (Fig. 2). This is closely similar to the established geomagnetic polarity scale of Cox (*1*) (Fig. 2). Age assignments, made by simple comparison with the established scale (Fig. 2), are corroborated by the sedimentation rate diagram, which indicates a linear relation between age and sediment thickness (Fig. 3). The sedimentation rate was nearly constant at 40 cm per 1000 years. A uniform rate might be expected for such a sedimentary section, which consists for the most part of rather uniform blue-gray massive mudstone deposited mainly in upper bathyal depths. It is unlikely that such a linear relationship (Fig. 3) would result fortuitously from a combination of a variable sedimentation rate and misidentification of magnetic boundaries. Such rapid rates of sedimentation fall within ranges more typical for flysch and molasse troughs (*13, 14*), although similar rates have been suggested for some nonturbidite terrigenous sediments (*14*).

The variation of the virtual geomagnetic latitude within periods of constant polarity (Fig. 2) may reflect the lack of cancellation of geomagnetic secular variation, which might be expected in sediments accumulating at a rate that is one to two orders of magnitude faster than in deep-sea sediments (*15*), or it may be in part the result of imperfections in recording of the ambient geomagnetic field resulting from burrowing activities of benthonic faunas (*16*).

The geomagnetic stratigraphy, when compared with previous oxygen isotope and foraminiferal paleotemperature curves determined on the same sequences (Fig. 2) by Devereux et al (*6*), indicates pronounced cold phases in the New Zealand Middle Pliocene (Waipipian stage), the Late Pliocene (upper Mangapanian), and at the beginning of the currently accepted

Pleistocene [lower Hautawan (6)]. Additional cold intervals indicated by cold-water megafaunas (7) coincide with the two Hautawan limestone members of the Pukenui Limestone (Fig. 2).

In terms of magnetic stratigraphy and chronology the first major Pliocene cooling (2.50 to 2.30 million years ago; Middle Pliocene) spans the Gauss-Matuyama boundary; the late Pliocene cooling (2.13 to 2.10 million years ago) coincides approximately with the lower part of the Olduvai event; and the Early Pleistocene cooling (1.98 to 1.88 million years ago) approximates the upper part of the Olduvai event (Fig. 2). At least one and possibly two cool intervals (1.77 to about 1.61 million years ago) shown by Vella and Nicol (7) coincide with the Gilsá event. According to the data (Fig. 2), sea temperatures during the intermediate warmer oscillations were not significantly warmer than at present-day similar latitudes (6). Uniform warm-water conditions existed during most of Gauss time, except for a slight cooling coinciding approximately with the end of the Kaena (?) event (2.80 million years ago).

The position and age of the climatically defined Plio-Pleistocene boundary (based on the first severe cooling) and the stratigraphically defined boundary (based on correlation with the Calabrian stage in southern Italy) have been examined by many workers (11–20). Our data are relevant to the interpretation of both definitions, and we first discuss the position of the climatically determined boundary.

The currently accepted Plio-Pleistocene boundary in New Zealand is climatically determined (4) and is at the boundary between the Waitotaran (Mangapanian) and Hautawan stages (18). A duration of 2 to 5 million years for the Pleistocene, suggested by a panel of five New Zealand scientists, was based on relative rates of evolutionary change and on sedimentation within the Cenozoic (19). Isotopic age determinations made by two independent groups (21, 22) show that the first recognized cooling is at 2.5 million years ago, in good agreement with our paleomagnetically dated first cooling in the Mangaopari Stream section (2.50 million years ago). The magnetic stratigraphy shows, however, that this cooling took place in the middle of the Pliocene, and not at the Plio-Pleistocene boundary.

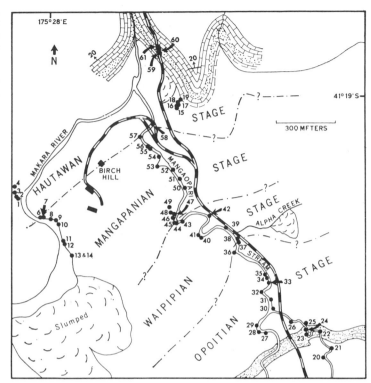

Fig. 1. Map of the Mangaopari Stream–Makara River section showing the location of paleomagnetic sample sites. Opoitian (Lower Pliocene) sandstone and Hautawan (Lower Pleistocene) limestones and sandstones are shown.

At Timaru (South Island of New Zealand), a basalt sheet with reversed polarity (20) overlying sedimentary rocks of earliest Pleistocene and Pliocene age (23) was dated at 2.5 ± 0.4 million years (21). This cooling is almost certainly that which spans the Gauss-Matuyama boundary. In the Wanganui Basin, the base of the Hautawan shell bed, which represents the currently accepted Plio-Pleistocene boundary, has been dated by extrapolation to be a maximum of 2.15 million years old. Beu (24) correlated this shell bed with the base of the Pukenui Limestone in southern Wairarapa, which we date as 1.72 million years (Fig. 2), coinciding with the Gilsá event. Either these stratigraphic units are diachronous or the radiometric age is too old by about 0.5 million years. The cooling events dated in Timaru and Wanganui are unlikely to be synchronous. Basalts thought to be at the Plio-Pleistocene boundary on the west coast of New Zealand (22) provided maximum ages ranging from 2.55 ± 0.1 million years to 2.3 ± 0.2 million years

for the beginning of the "Pleistocene." It was concluded that the age data and limited paleomagnetic results indicated a first climatic cooling before the geomagnetic field changed from normal (Gauss) to reversed (Matuyama). Our data confirm this interpretation but indicate that the cooling is not basal Pleistocene (Hautawan) in age, as suggested from previous floral analyses (25), but rather is the Middle Pliocene Waipipian cooling.

Evidence for possible contemporaneous Pliocene cooling elsewhere includes pronounced faunal change in antarctic cores at about 2.5 to 3 million years (26), glaciations recorded in the Sierra Nevada of the United States at 2.7 million years (27), tillites in Iceland at 3 million years (28, 29), and calcium carbonate peaks reflecting cool temperatures in the upper Gauss of equatorial Pacific cores (30). The climatic cooling we report in the lower and middle Matuyama (Fig. 2) therefore clearly occurred elsewhere (29, 30).

Middle Pliocene cooling is not the

earliest major Late Cenozoic cooling reported in New Zealand. Cold-water faunas and evidence of a fall in sea level occur in the uppermost Miocene Kapitean stage (*31, 5*). The age of the Kapitean stage and supposed Miocene-Pliocene boundary in New Zealand could not be determined in our study because of a disconformity at the base of the Opoitian stage in the section (*32*).

There may be a relationship between paleoclimatic and geomagnetic polarity changes (Fig. 2). Cooling trends commonly began at, or shortly after, geomagnetic polarity changes [for example, at the top of Kaena (2.80 million years); base of Olduvai (2.13 million years); upper part of Olduvai (1.98 million years); and base of Gilsá (1.79 million years)] or cooling maxima coincided with geomagnetic reversals [for example, at the Gauss-Matuyama boundary (2.43 million years)]. Ex-

ceptions are the Mammoth (?) event and base of Kaena (?) event, which do not coincide with any climatic changes, and the top of the lower part of the Olduvai event (2.11 million years), which occurred immediately before a warming trend. Therefore, six of eight definite polarity changes are close to, or at, climatic changes. This may be interpreted to be either coincidence of two unrelated variables or to be support for an indirect connection between

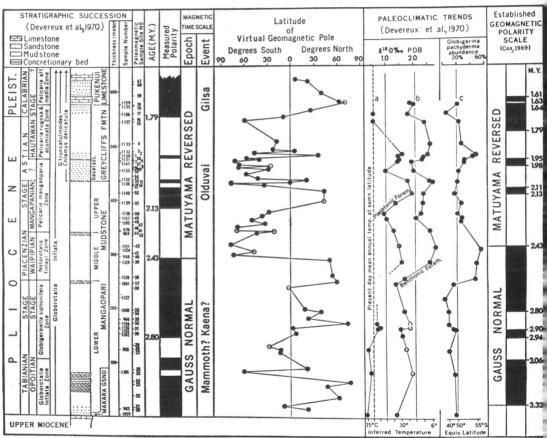

Fig. 2. Paleomagnetic stratigraphy, stratigraphic succession, and paleoclimatic trends (*6, 7*) at Mangaopari Stream–Makara River New Zealand. Sample positions, ranges of some key fossils, local stratigraphic units, and possible European correlations are shown (*M.Y.*, million years). Magnetic polarity in the section (left polarity column) is interpreted from the latitude of the virtual magnetic pole (*12*); black is normal polarity and clear is reversed polarity, which corresponds to the latitude of the virtual geomagnetic pole being higher than 10°N and 10°S, respectively. Latitudes between 10°N and 10°S are not assigned to polarity changes for reasons including the inherent uncertainty of the position of the spin axis and axial dipole in terms of present geographic coordinates. The established geomagnetic polarity scale (*1*) is given at the far right for comparison. The ages of the polarity column are substantiated by Fig. 3. Note that the data points at each site result from the mean remanent magnetism from three separate cores after demagnetization: solid circles have a resultant vector, corresponding to a mean direction, which, for three specimens, is nonrandom at the 95 percent confidence level, whereas the few open circles represent a 5 percent possibility of randomness. For nonrandomness at the 95 percent level when there are three cores per site, the resultant vector must be 2.62, when a unit vector is assigned to each core (*39*). Two sites from a nearby section (*9*), which is stratigraphically slightly younger than the highest sample in this section, are both of reversed polarity (*40*). The paleoclimatic curves are, from left to right: (*a* and *b*) oxygen isotope ratios of planktonic and benthonic foraminiferal tests; (*c*) abundance of *Globigerina pachyderma* as a percentage of all planktonic foraminiferal tests Note that the position of the Plio-Pleistocene boundary and the base of the Calabrian have been modified from that previously determined by Devereux *et al.* (*6*).

polarity changes, climatic variations, and faunal changes (*33*).

We now examine the evidence for the position of the stratigraphically defined Plio-Pleistocene boundary. In the first attempt at determining this boundary in New Zealand, Fleming (*34*) correlated the base of the Hautawan with that of the Calabrian on paleoclimatic evidence. A further attempt was based on the biostratigraphic succession at Mangaopari Stream (*6*), where the boundary was placed at the first appearance of *Globorotalia truncatulinoides* (at the base of the Hautawan stage) corresponding to the upper Olduvai event as shown in Fig. 2. The first definite appearance of *G. truncatulinoides* (*6*), however, is at a level coinciding with the base of the Gilsá event and near the middle of the Hautawan stage. Rare, atypical specimens, previously reported below this level as *G. truncatulinoides*, have been reexamined and are referrable to the *G. crassaformis* complex. The first appearance of *G. truncatulinoides* in tropical deep-sea cores occurs within the major normal event in the Matuyama epoch (*35*), which has often been called the Olduvai but is now regarded as the Gilsá (*1*, *36*), the base of which is dated at 1.79 million years (*1*). The first appearance of *G. truncatulinoides* gives correlation with the type Calabrian (*37*), at the base of the type Italian Pleistocene. The age of the base of the Gilsá (1.79 million years) is close to the age of 1.8 million years for the Plio-Pleistocene boundary in the Calabrian type section based on extrapolation of sedimentation rates (*38*).

We therefore place the stratigraphically defined Plio-Pleistocene boundary at the base of the Gilsá and first definite appearance of *G. truncatulinoides*. This level occurs near the middle of the Hautawan stage, the base of which is the generally accepted position for the Plio-Pleistocene boundary in New Zealand. The coincidence of the first appearance of *G. truncatulinoides* with the base of the Gilsá in our section strongly supports the coincidence of the Gilsá event with at least part of the Calabrian stage. As both Italy and New Zealand are in temperate latitudes, it is likely that there was little or no lag between the two areas in the first appearance of *G. truncatulinoides*.

We conclude that marked cooling occurred much earlier than the strati-

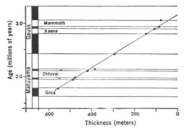

Fig. 3. Sedimentation rate diagram. The paleomagnetic polarity boundaries (Fig. 2) in the Mangaopari Stream–Makara River section are plotted against the assigned ages resulting from the known geomagnetic polarity history (*1*). In the ordinate, polarity epochs are in vertical print; polarity events are in smaller horizontal print; normal polarity is in black; and reversed polarity is clear. The slope of the line indicates a linear sedimentation rate of 40 cm per 1000 years for the section.

graphically defined beginning of the Pleistocene, which is at the base of the Gilsá geomagnetic polarity event. Correlations of a climatically defined Plio-Pleistocene boundary, when the first marked climatic cooling is used, are unlikely to be valid. The Pliocene epoch would be eliminated if the first marked late Cenozoic climatic cooling is taken as the beginning of the Pleistocene, because the first known major cooling took place in Late Miocene time (*5*, *31*). We have shown that paleomagnetic dating of marine sediments now exposed on land can assist in correlation of Plio-Pleistocene strata and climatic changes. The extension of this method to marine sediments exposed on land will hopefully contribute significantly to definition of the geomagnetic polarity history.

Huangaroa River, 0.5 km downstream from Hautotara Bridge. The stratigraphic thickness between this location and the highest sample (sample 61) in the Mangaopari Stream section is still unknown. Sample 62 is from marine fossiliferous mudstone in the east side of the river and the other (sample 63) is from slightly younger conformable lignitic mudstone on the west side of the river.

References and Notes

1. A. Cox, *Science* **163**, 237 (1969). Unless otherwise stated, all age assignments of the known geomagnetic history are after this most recent compilation.
2. N. D. Opdyke, B. Glass, J. D. Hays, J. Foster, *Science* **154**, 349 (1966); N. D. Watkins and H. G. Goodell, *Earth Planet. Sci. Lett.* **2**, 123 (1967).
3. H. Nakagawa, N. Niitsuma, I. Hayasaka, *J. Geol. Soc. Japan* **75**, 267 (1969).
4. C. A. Fleming, *Trans. Roy. Soc. N.Z.* **74**, 207 (1944).
5. J. P. Kennett, *Int. Union Geol. Sci. Comm. Mediter. Neogene Stratigr. G. Geol. Annoli Mus. Geol. Bologna* **35**, 143 (1969).
6. I. Devereux, C. H. Hendy, P. Vella, *Earth Planet. Sci. Lett.* **8**, 163 (1970).
7. P. Vella and W. H. Briggs, Jr., *N.Z. J. Geol. Geophys.*, in press; P. Vella and E. R. Nicol, *ibid.*, in press.
8. A. U. E. Boreham, *ibid.* **6**, 3 (1963).
9. Two additional sites were drilled in slightly younger (Marahauan) sediments 2 km north of the Mangaopari Stream section in the
10. J. H. Foster, *Earth Planet. Sci. Lett.* **1**, 463 (1966).
11. M. W. McElhinny, *Geophys. J. Roy. Astron. Soc.* **10**, 369 (1966).
12. The virtual geomagnetic pole is that pole resulting from a geocentric axial dipole that causes the observed declination and inclination of remanent magnetism at the sample site.
13. D. C. Krause, *N.Z. Dep. Sci. Ind. Res. Bull.*, No. 183 (1967).
14. A. G. Fisher, *Geol. Soc. Amer. Bull.* **80**, 549 (1969).
15. For example, see N. D. Watkins and H. G. Goodell, *Science* **156**, 1083 (1967).
16. N. D. Watkins, *Earth Planet. Sci. Lett.* **4**, 341 (1968).
17. For summary, see I. McDougall and J. J. Stipp, *Nature* **219**, 51 (1968) and (*18–20*).
18. I. McDougall and J. J. Stipp, *Nature* **219**, 51 (1968).
19. D. F. Squires, *N.Z. J. Geol. Geophys.* **3**, 137 (1960).
20. D. S. Coombs and T. Hatherton, *Nature* **184**, 883 (1959).
21. W. H. Mathews and G. H. Curtis, *ibid.* **212**, 979 (1966).
22. J. J. Stipp, J. M. A. Chappell and I. McDougall, *Amer. J. Sci.* **265**, 462 (1967).
23. H. S. Gair, *N.Z. J. Geol. Geophys.* **4**, 89 (1961).
24. A. G. Beu, *ibid.* **12**, 643 (1969).
25. R. A. Couper and D. R. McQueen, *N.Z. J. Sci. Technol. Sect. B* **35**, 398 (1954).
26. J. D. Hays and N. D. Opdyke, *Science* **158**, 1001 (1967).
27. R. R. Curry, *ibid.* **154**, 770 (1966).
28. M. G. Rutten and H. Wensink, *Int. Geol. Congr. Rep. Session, Norden 21st, Part 4*, 1960, 62 (1960).
29. I. McDougall and H. Wensink, *Earth Planet. Sci. Lett.* **1**, 232 (1966).
30. J. D. Hays, T. Saito, N. D. Opdyke, L. H. Burckle, *Geol. Soc. Amer. Bull.* **80**, 1481 (1969).
31. J. P. Kennett, *N.Z. J. Geol. Geophys.* **10**, 1051 (1967).
32. In addition to lithological evidence for disconformity, well-developed *Globorotalia inflata* occurring immediately above the disconformity suggest that much of the lower Opitian is missing. Typical *G. inflata* has not been observed in the lower Opitian of several other New Zealand sections examined by one of us (J.P.K.) but seems to first appear higher in the Opitian.
33. J. P. Kennett and N. D. Watkins, *Nature* **227**, 930 (1970); J. R. Heirtzler, *J. Geomagn. Geoelec.* **22**, 197 (1970).
34. C. A. Fleming, *N.Z. Geol. Surv. Bull.*, No. 52 (1953).
35. W. A. Berggren, J. D. Phillips, A. Bertels, D. Wall, *Nature* **216**, 253 (1967); J. D. Phillips, W. A. Berggren, A. Bertels, D. Wall, *Earth Planet. Sci. Lett.* **4**, 118 (1968); B. Glass, D. B. Ericson, B. C. Heezen, N. D. Opdyke, J. A. Glass, *Nature* **216**, 437 (1967).
36. N. D. Watkins and A. Abdel-Monem, *Geol. Soc. Amer. Bull.*, in press.
37. D. D. Bayliss, *Lethaia* **2**, 133 (1969).
38. R. Selli, *Progr. Oceanogr.* **4**, 67 (1967).
39. S. A. Vincenz and J. M. Bruckshaw, *Proc. Cambridge Phil. Soc.* **56**, 21 (1960).
40. These two sites are important because they show that the normal polarity at the top of the Mangaopari section, identified as the Gilsá event (1.79 to 1.61 million years ago), is not the Brunhes geomagnetic polarity epoch.
41. Supported by NSF grant GA 13093 (Florida State University) and NSF grant GA 27092 (University of Rhode Island). R. Raymond drafted the figures and D. Cassidy assisted in photographic work.

15

Copyright © 1974 by the Geological Society of America

Reprinted from *Geol. Soc. America Bull.* 85:1385–1398 (1974)

Late Miocene–Early Pliocene Paleomagnetic Stratigraphy, Paleoclimatology, and Biostratigraphy in New Zealand

J. P. KENNETT
N. D. WATKINS } *Graduate School of Oceanography, University of Rhode Island, Kingston, Rhode Island 02881*

ABSTRACT

The paleomagnetic stratigraphy, biostratigraphy, and paleoclimatology have been studied in two marine sections of late Miocene to early Pliocene age in New Zealand. A total of over 850 separately oriented cores were collected from 270 sites. The Blind River section (41°43′ S.) is now adjacent to the southernmost subtropical (temperate) water mass, but planktonic foraminifera indicate that the area was covered by subantarctic water during much of late Miocene and early Pliocene time. The Mangapoike River section (38°55′ S.) records temperature oscillations mainly within the subtropical water mass during late Miocene–early Pliocene age, with perhaps one subantarctic interval during latest Miocene time.

The Miocene-Pliocene boundary in New Zealand has consistently been placed at the first evolutionary appearance of *Globorotalia puncticulata* at the boundary between the late Miocene Kapitean Stage and the early Pliocene Opoitian Stage. This boundary lies within sediments deposited during the Gilbert Reversed Epoch between the Nunivak Event (base at 4.14 m.y. B.P.) and the Gilbert C Event (top at 4.33 m.y. B.P.) in both sections. Thus, the Miocene-Pliocene boundary, as recognized in New Zealand, is dated as 4.3 ± 0.1 m.y., which appears to be slightly younger than the type (International) Miocene-Pliocene boundary in Italy (4.9 to 5.1 m.y.). Biostratigraphic ranges of planktonic foraminifera between New Zealand and the Mediterranean differ in detail, perhaps due to different paleooceanographic histories. A major cooling episode during the early Gilbert Reversed Epoch is recorded at Blind River and Mangapoike River. This cooling is more pronounced in the southern section examined, where it is represented by the occurrence of a central subantarctic planktonic foraminiferal assemblage. In the northern section, cooling was also pronounced, although of shorter duration, represented by a probably northern subantarctic assemblage. The Miocene-Pliocene boundary in Europe has still only been dated indirectly by means of non-Mediterranean sections. Interpretation of late Cenozoic paleomagnetic data from Mediterranean deep-sea cores collected from *Glomar Challenger* is rejected. *Key words: geochronology, foraminifera, Miocene-Pliocene boundary, Antarctic glaciation, Messinian Stage.*

INTRODUCTION

The well-exposed and continuous marine Cenozoic sequences in New Zealand have provided an excellent means to study perate planktonic foraminiferal and careous nannofossil zonation and tem ate paleoclimatology (Hornibrook, 19 Jenkins, 1967a; Edwards, 1971; Kenn 1967; Hornibrook and Edwards, 19

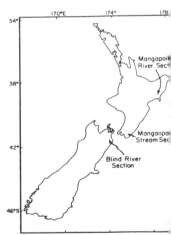

Figure 1. Location of two New Zealand section amined in this study and Mangaopari Stream se previously studied by Kennett and others, 1971.

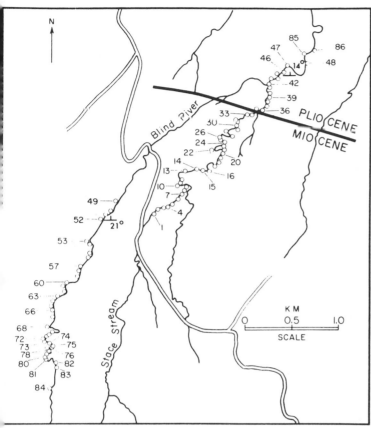

Figure 2. Map of Blind River–Stace Stream section, New Zealand, showing location of paleomagnetic sample tes, numbered corresponding to original collecting sequence.

The two sections examined (Figs. 1 through 3) are the Blind River–Stace Stream section, Marlborough, in the northern part of the South Island (lat. 41°44' S., long. 174°03' E.), and the Mangapoike River section, Hawkes Bay, in central North Island (lat. 38°55' S., long. 177°32' to 177°35' E.).

BIOSTRATIGRAPHIC POSITION OF NEW ZEALAND MIOCENE-PLIOCENE BOUNDARY

The Miocene-Pliocene boundary in New Zealand has been consistently placed at the boundary between the late Miocene Kapitean Stage and the early Pliocene Opoitian Stage (Finlay and Marwick, 1947; Fleming, 1959; Hornibrook, 1958; Jenkins, 1967a). This boundary is most clearly marked by the first appearance of nonkeeled *Globorotalia* that have been variously called *G. inflata* (d'Orbigny) (Finlay and Marwick, 1947; Hornibrook, 1958; Jenkins, 1967a); *G. crassaformis* (Galloway & Wissler; see Kennett, 1966a; 1967) and *G. puncticulata* (d'Orbigny) (Kennett, 1973; Kennett and Watkins, 1972; Hornibrook and Edwards, 1971; Collen and Vella, 1973). Kennett (1966b) described the evolution of *G. puncticulata* under the name of *G. crassaformis* and later recognized the true identity of the species (Kennett and Watkins, 1972; Kennett, 1973). The Miocene-Pliocene boundary in New Zealand is hence placed at the first evolutionary appearance of *G. puncticulata*, which is a clearly defined datum throughout New Zealand marine sections and within site 207 of the Deep Sea Drilling Project (Kennett, 1973).

Early populations of *G. puncticulata* are more compact, less inflated, and have a smaller aperture than later, more advanced populations. *Globorotalia sphericomiozea* Walters is considered to be an early form of *G. puncticulata*, and we refer these forms to *G. puncticulata sphericomiozea* Walters. Within the type Kapitean Stage at Kapitea Creek, Westland, *G. puncticulata sphericomiozea* is represented as only rare end-members in populations of its ancestral form, *G. conomiozea* (Kennett, 1966a; 1966b). The first appearance of *G. puncticulata sphericomiozea* occurs at the base of the Opoitian Stage, where the *G. puncticulata sphericomiozea* morphotype is more abundant than the *G. conomiozea* morphotype (Kennett, 1966a, 1966b).

BLIND RIVER AND MANGAPOIKE GEOLOGY OF THE SECTIONS

The Blind River section consists of 1,200 m of late Miocene to early Pliocene marine

Devereux and others, 1970). Because of temperature-related faunal provincialism, correlations are difficult with the type European sections and with zonal schemes in tropical-warm, subtropical areas. Paleomagnetic stratigraphy in marine sediments, which utilizes global isochrons (polarity changes), greatly enhances long-range correlations. Paleomagnetic stratigraphic methods have recently been applied to marine late Cenozoic sequences in Japan (Nakagawa and others, 1969), New Zealand (Kennett and others, 1971; Lienert and others, 1972) and the type Pliocene-Pleistocene boundary section in Italy (Nakagawa and others, 1971). In New Zealand, the paleomagnetic stratigraphy of the section at Mangaopari Stream (Kennett and others, 1971) facilitated identification of the base of the Gilsa polarity event at $t = .79$ m.y., which is correlated with the

Pliocene-Pleistocene boundary (Berggren and others, 1967). This in turn has enabled biostratigraphic and paleoclimatic correlations to be made with sequences in various deep-sea areas.

We here describe the paleomagnetic stratigraphy for two late Miocene to early middle Pliocene sections in New Zealand, one in the South Island and the other in the North Island (Fig. 1). Our main purpose is to establish the age of the long-adopted Miocene-Pliocene boundary in New Zealand and paleoclimatic-paleo-oceanographic changes that occurred in late Miocene and Pliocene times. The contrasting climatic zones represented are useful paleomagnetically, because any magnetic overprint due to precipitation of magnetic minerals (Watkins and Kennett, 1973) will not be the same in both sections.

mudstone, siltstone, and minor sandstone (Fig. 4), which is moderately well exposed in the stream beds of Blind River and Stace Stream (Kennett, 1965, 1966a). The streams run almost normal to the regional strike of the strata. The dip decreases upwards from 25 to 10°. The base of the section consists of coarse-grained sandstone and conglomerate unconformably overlying severely deformed, indurated graywacke and argillite of Permian-Jurassic age (Kennett, 1965).

The section is now adjacent to the southernmost part of the subtropical water mass, ~250 km north of the subtropical convergence that separates the subantarctic and subtropical water masses. East coast waters of northern South Island, at least as far north as the Blind River section, are cooled by a north-flowing current with intermixed subantarctic water (Garner, 1959; Heath, 1972a). Furthermore, present-day winds of the coastal Marlborough area, where Blind River is located (Fig. 1), cause upwelling of cool water; this further lowers the surface-water temperatures on the coast (Heath, 1972b). These local conditions may also have enhanced cooler planktonic foraminiferal faunas than normally would occur at these latitudes during the late Cenozoic.

The North Island section (Fig. 1) is well exposed along the Mangapoike River and adjacent road (Fig. 3). The section is summarized in Table 1 and consists mostly of marine sandstones and argillaceous siltstones and mudstones with a total thickness of 850 m (Fig. 5). The benthonic and planktonic foraminifera have been described by B. A. McInnes (unpub. data).

The river and road run almost normal to the regional strike of the strata, which dip about 25° to the northwest throughout the section (Fig. 3). The lower and the upper parts (samples 150–193 and 97–149) were collected from river outcrops. The middle part of the section (samples 1 to 96) was collected from road outcrops that are high above the river.

MIOCENE-PLIOCENE BOUNDARY IN EACH SECTION

In the Blind River section, the evolution of *Globorotalia miozea conoidea* Walters (see *G. miozea* Finlay of Kennett, 1966a) to *G. conomiozea* Kennett to *G. puncticulata sphericomiozea* Walters (see *G. crassaformis* of Kennett, 1966a) is clearly recorded. The boundary between Miocene and Pliocene times is placed at the level where the *G. conomiozea* morphotype becomes dominated by the *G. puncticulata sphericomiozea* morphotype (Fig. 4).

In the Mangapoike River section, the

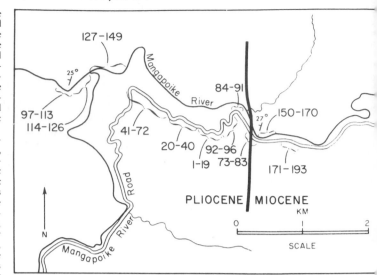

Figure 3. Map of Mangapoike River section, New Zealand, showing general location of paleomagnetic sampling sites in numbers according to original sampling sequence. Sampling density was too high to show exact location of each sample.

evolution is also recorded of *G. miozea conoidea* to *G. conomiozea* to *G. puncticulata sphericomiozea* to *G. puncticulata puncticulata*. The boundary between Miocene and Pliocene times and the Kapitean and Opoitian Stages is placed at the level where *G. puncticulata sphericomiozea* first appears (Fig. 5). The Mangapoike River section occurs within the type area of the Opoitian Stage. No type section has been designated for the Opoitian Stage, but N. Hornibrook (oral commun.) considers that the Mangapoike River section is the most satisfactory one for this purpose. It is important to note that the base of the Opoitian Stage has generally been considered to coincide with the base of the Opoiti Series of Ongley (1928, 1930), which is at the disconformity marked by a sandstone with basal limestone (Table 1, Fig. 5; Finlay, 1939; Fleming, 1959). This disconformity is above the first appearance of *G. puncticulata sphericomiozea* (McInnes, 1965) and hence is above the horizon accepted as the Miocene-Pliocene boundary. The first appearances of *G. puncticulata puncticulata* and *Globorotalia crassaformis* are in the sandstone immediately above the disconformity.

METHODS USED

In the field, three cores 2.5 cm in diam and averaging 12 cm in length were taken, by using a gasoline-powered portable drill, at each of 86 sites in Blind River and 193

sites in Mangapoike River. In most cases the outcrop was cleaned off to a depth ~30 cm before drilling. All cores were oriented in geographic coordinates to a precision of ±3°. Sites were chosen where possible to maintain a vertical stratigraphic interval of about 3 m. Samples as much as kg were taken for micropaleontologic study at selected intervals.

In the laboratory, a 2.2-cm-length segment was sliced from each core for paleomagnetic analysis. The natural remanent magnetism was measured with a 5-Hz spinner magnetometer. Unstable components of some of the cores were examined

TABLE 1. SEQUENCE OF STRATA
AT MANGAPOIKE RIVER*

Series		Meters
Ormond	Massive yellow calcarenite	~30
	Disconformity	
Opoiti	Shelly calcarenite and quartz sandstone grading down to siltstone	
	Argillaceous siltstone and quartz sandstone	4?
	Massive calcareous quartz sandstone with calcarenite at base	8?
	Disconformity	
Mapiri	Mudstone, siltstone, and sandstone	3?

* Data from Ongley (1928, 1930).

y demagnetizing with incremental alternating magnetic fields as high as 400 Oe. ubsequently with other samples, the alternating field method was abandoned in vor of incremental thermal demagnetization, with final heating to 600°C, using a nitrogen atmosphere to obviate oxidation during the heating cycles.

All paleomagnetic data were analyzed using conventional statistical methods (Fisher, 1953). For each site, the latitude of the virtual geomagnetic poles (the surface expression of the geocentric dipole that would cause the observed direction of remanent magnetism at the site) is used as the index of polarity and is considered to be normal for latitudes above 45° N., reversed with latitudes higher than 45° S., and intermediate for latitudes of less than 45°.

Micropaleontologic samples were washed over a 0.063-mm sieve to concentrate the foraminifera; after drying, they were divided with a 0.124-mm sieve. Samples of ~300 specimens of the fraction >0.124 mm were separated with a modified Otto microsplitter. In these samples, specimens of individual species were counted. Coiling of *Globigerina pachyderma* was determined by counting as many as 200 specimens in the same size fraction.

Our purpose is to define global polarity changes — not any shorter period excursions or secular variation components of the ancient geomagnetic field. Therefore, we have employed a three-point running average of the virtual geomagnetic poles, applying unit vector per site.

RESULTS AND DISCUSSION

Magnetic Properties

An overprint of normal polarity, or large unstable component, was found in most, but not all, samples from both sections. The intensity of magnetization (J) of the natural remanent magnetization is of the order of 10^{-5} to 10^{-6} emu g^{-1}. Treatment of unstable components in 21 specimens from 7 selected sites in each section by successively higher alternating magnetic fields at 50-Oe intervals was not consistently effective, because the scatter of directions within sites did not always decrease with the treatment. In contrast, thermal demagnetization at 250°C consistently resulted in decrease in the scatter of directions and, in several sites, resulted in a clear indication of the removal of a single component normal polarity overprint. Simultaneously, there was a reduction of J by 50 to 75 percent in most samples. Above 300°C, an increase in J by an order of magnitude and loss of in-site coherence of the remanent magnetism indicate the formation of a different magnetic phase in the minerals present. Dehydration of an iron oxyhydroxide, such as goethite, is probably the cause. For our experimental purpose (in order to measure the polarity during the deposition of the sediments), thermal demagnetization at 250°C was established as the optimum cleaning treatment. Significantly, no highly unstable magnetic phase was produced by the heating process and, therefore, no superparamagnetic component (which could in-

BLIND RIVER

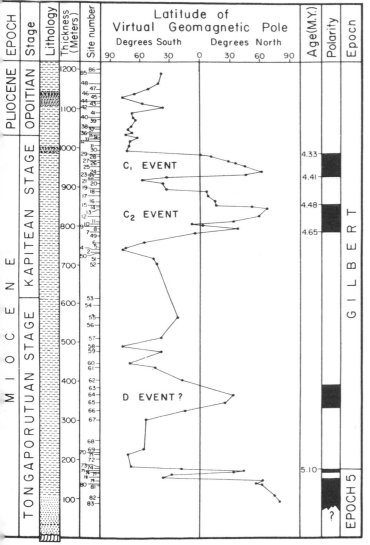

Figure 4. Paleomagnetic stratigraphy and stratigraphic succession of Blind River section, New Zealand. Sample tions and New Zealand stages shown. Sample numbers correspond to site locations (Fig. 2). Magnetic polarity in ion interpreted from latitude of virtual magnetic pole; black is normal polarity and clear is reversed polarity. s of polarity boundaries after Opdyke (1972). Sediment lithologic symbols: dots = sandstone; fine dashes = tone; coarser dashes = mudstone.

dicate the presence of a finely precipitated secondary component) was detected (Watkins and others, 1974). This does not prove that coarse precipitated components are absent, but as we shall later show, between-section comparisons render this an unlikely possibility.

Micropaleontology

Dominant species throughout the Blind River section are sinistral *Globigerina pachyderma* (Ehrenberg) and *Globigerina bulloides* (d'Orbigny) (Fig. 6). *Globigerina woodi* Jenkins is important in the Tongap-

orutuan Stage (early late Miocene) b insignificant in the latest Miocene Kapite Stage and early Pliocene Opoitian Sta Other consistent species that occur in re tively low frequencies are the *Globorota miozea conoidea. G. conomiozea — puncticulata sphericomiozea* comple

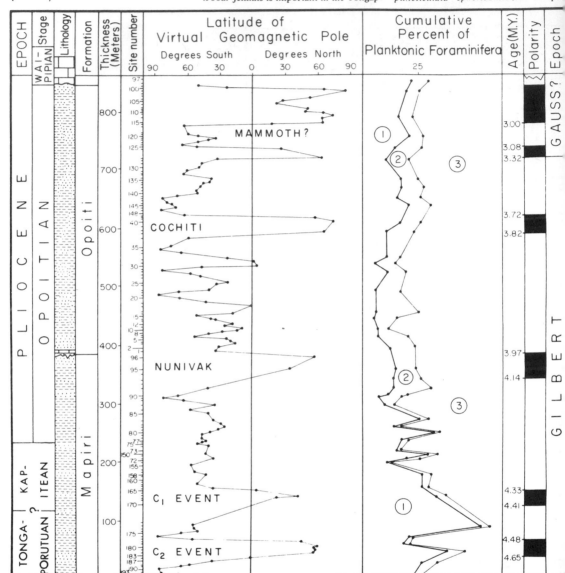

Figure 5. Paleomagnetic stratigraphy, stratigraphic succession, and planktonic foraminiferal trends at Mangapoike River section. Sample positions, local formations, New Zealand stages shown. Sample numbers correspond to site locations in Figure 3. Magnetic polarity in section as in Figure 4. Ages of polarity boundaries after Cox (19 and Opdyke (1972). Frequency oscillations in planktonic foraminiferal species: 1 = sinistral *Globigerina pachyderma*, 2 = dextral *Globigerina pachyderma*, and 3 = o species. Sediment lithologic symbols: dots = sand; dashes = mudstone and siltstone; bricks = limestone.

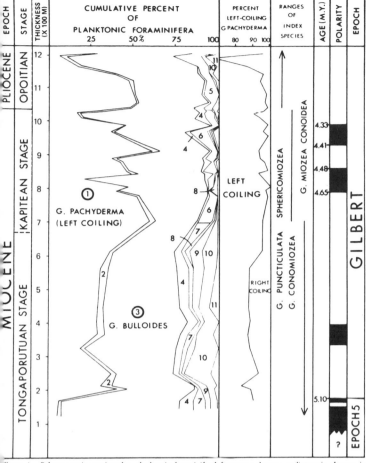

Figure 6. Paleomagnetic stratigraphy, planktonic foraminiferal frequency changes, coiling ratio changes in *Globigerina pachyderma*, and ranges of some key planktonic foraminiferal species in Blind River section. New Zealand late Miocene–early Pliocene stages shown. Age and polarity from Figure 4. Key to planktonic foraminifera: 1 = sinistral *Globigerina pachyderma*, 2 = dextral *Globigerina pachyderma*, 3 = *Globigerina bulloides*, 4 = *Globigerina quinqueloba*, 5 = *Globorotalia puncticulata sphericomiozea*, 6 = *Globorotalia conomiozea*, 7 = *Globorotalia miozea conoidea*, 8 = *Globorotalia scitula*, 9 = *Globigerinita glutinata*, 10 = *Globigerina woodi*, and 11 = other species.

Globigerina quinqueloba Natland, dextral *G. pachyderma*, *Globorotalia scitula* (Brady), and *Globigerinita glutinata* (Egger). Sinistral forms of *G. pachyderma* greatly predominate over dextral forms and constitute over 70 percent of the population of this species. *Orbulina universa* d'Orbigny and a few other species occur rarely. Oscillations in frequencies of several planktonic foraminifera record paleoceanographic changes discussed later.

In the more northern Mangapoike River section (Fig. 7), species diversity is slightly higher than in the Blind River section. Sinistral *G. pachyderma* is much less abundant,

while other species are more important, reflecting warmer conditions. The most abundant species is *G. bulloides*. The *G. puncticulata* group is relatively abundant in early Pliocene time. In the late Miocene and earliest Pliocene, *G. glutinata* and *Globigerina woodi* are abundant at several levels. Other consistent but less common forms are *Globigerina falconensis* Blow, *G. quinqueloba*, and dextral *G. pachyderma*. *G. pachyderma* is predominantly left coiling in most of the section and is predominantly right coiling in part of the early Pliocene only.

The framework of the biostratigraphic

subdivision in both sections is the evolutionary sequence of forms from *Globorotalia miozea conoidea* in late Miocene time to *G. puncticulata* in early Pliocene (Figs. 6, 7). In the Blind River section, specimens that make up the bioseries have more thickly encrusted tests than those at Mangapoike River. Thus, at Blind River, *G. miozea conoidea* is the more typical thickened form, whereas at Mangapoike River, forms that resemble *G. miotumida* Jenkins are more common. Gibson (1967) has pointed out that these two forms represent the same species, differing only in the degree of test encrustation. *G. conomiozea* populations in late Miocene time at Mangapoike River and other northern North Island sections are less strongly conical than typical forms occurring in more southern, cooler water late Miocene sequences. As a result, populations of *G. conomiozea* at Mangapoike River more closely resemble *G. miozea conoidea* but differ in having an average of four and one-half chambers in the final whorl, compared with five chambers in *G. miozea conoidea*. We suggest that a cline existed in *G. conomiozea* with a more conical southern phenotype grading northward into a less conical phenotype. The absence of typical *G. conomiozea* in the late Miocene of the Mangapoike River section is not considered to reflect a latest Miocene hiatus that has removed that part of the evolutionary bioseries which occurs in the Kapitean Stage, but the absence does reflect that the morphotypes which make up the bioseries differ slightly with latitude. Frequency oscillations in *G. pachyderma* at the Mangapoike River section suggest that the late Miocene is continuous. McInnes (1965) has indeed already described the evolutionary development of the bioseries from a keeled form with four and one-fourth chambers in the final whorl (which he termed *G. miozea* [spherical variety]) to a nonkeeled, four chambered form (which he termed *G. inflata*). The earliest form in the lineage described by McInnes is considered by us to represent a low conical *G. conomiozea*, while the latest described form is *G. puncticulata puncticulata*. Intermediate nonkeeled, four-chambered forms are *G. puncticulata sphericomiozea*.

Paleomagnetic Stratigraphy

The latitude of the virtual geomagnetic pole for each site is presented in stratigraphic order in Figures 4 and 5. The sections are discussed separately.

Blind River Section. Late Opoitian strata with *Globorotalia inflata* have mostly normal polarity and have been assigned to the Gauss geomagnetic epoch at two southern North Island sections (Kennett and others, 1971; Lienert and others, 1972). This suggested age range has been confirmed by a fission-track date (P. Vella, oral commun.), which provides unambiguous evidence that the original detrital remanent magnetism in the sections was preserved. As the Blind River section has reversed polarity (Fig. 4) and the top of the section is early Opoitian in age, it is almost certain on stratigraphic grounds alone that the section represents the Gilbert Reversed Epoch. There is no doubt that the section is neither dominantly Gauss Epoch or Epoch 5; these both have mainly normal polarity magnetization. The two short normal polarity sequences in the upper part of the Kapitean State (Fig. 4) are best matched in character by the split Gilbert C Event. The normal section at the base is accordingly assigned to Epoch 5.

Corresponding ages applied to the section indicate a decrease in sedimentation rate (Fig. 8) from about 130 cm per 1,000 yr in the Tongaporutuan–Early Kapitean, to about 60 cm per 1,000 yr in the late Kapitean and Opoitian (latest Miocene–earliest Pliocene). Such rapid sedimentation rates can be expected in the marginal sedimentary basins of New Zealand (Kennett and others, 1971). The change in rate is consistent with lithological changes. The more rapidly deposited lower part contains mostly massive mudstone and turbidites, which are deposited in a continental slope environment. The less rapidly deposited upper part contains mudstone and bedded siltstone and sandstone; this is characteristic of the continental shelf and upper continental slope environment.

A brief interval of normal polarity in the early part of the Gilbert Epoch, which is labeled Event "d" (Fig. 4), is represented by only two sampling sites. This may be an unusually short polarity event that can be distinguished only where the sedimentation has been rapid, but it requires verification by corresponding work in other sections.

Mangapoike River Section. As at Blind River, the polarity of sediments at Mangapoike is mostly reversed. Comparison of the biostratigraphy with that at Blind River and the paleomagnetic data suggests that the sequence ranges from early Gilbert to middle Gauss. It is proposed that the split Gilbert C (C1 and C2 events) can be recognized in the Kapitean Stage and latest Tongaporutuan Stage, which form the lower part of the section. The Nunivak and Cochiti Events (Cox, 1969) are also well defined — the upper boundary of the Nunivak corresponds to the disconformity that separates the Mapiri and Opoiti Formations.

A nearly linear sedimentation rate (Fig.

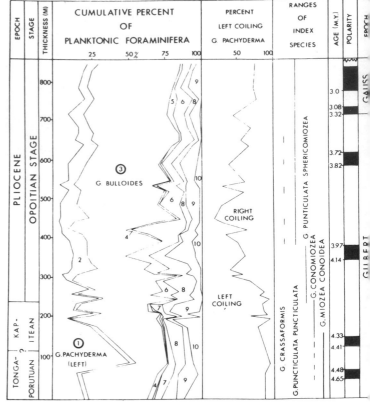

Figure 7. Paleomagnetic stratigraphy, planktonic foraminiferal frequency changes, coiling ratio changes *Globigerina pachyderma*, and ranges of some key planktonic foraminiferal species in Mangapoike River section New Zealand late Miocene and early Pliocene stages shown. Age and polarity from Figure 5. Key to plankto foraminifera: 1 = sinistral *Globigerina pachyderma*, 2 = dextral *Globigerina pachyderma*, 3 = *Globigerina bu loides*, 4 = *Globigerina falconensis*, 5 = *Globigerina quinqueloba*, 6 = *Globorotalia puncticulata*, 7 = *Globorota conomiozea*, 8 = *Globigerinita glutinata*, 9 = *Globigerina woodi*, 10 = other species.

9) supports these identifications of events within the Gilbert Epoch and also indicates that the Mapiri-Opoiti disconformity represents only a brief time interval, as independently suggested by the fossils. The sedimentation rate is 64 cm per 1,000 yr, similar to that at Mangaopari Stream (Kennett and others, 1971) and the upper part of the Blind River section (Fig. 8). The upper part of the Mangapoike River section is primarily normal in polarity, with shorter intervals of reversed polarity. Our preferred interpretation (which we stress is not the only possible interpretation) is that a highly condensed sequence of early Gauss Epoch age (Fig. 5) is represented. The implicit reduction in the sedimentation rate during the latest Gilbert and early Gauss (Fig. 9) is consistent with a marked change in sedimentary facies to shallow-water calcarenite and sandstone (Table 1). The

sedimentary facies further changed duri the Waipipian Age with deposition of mo massive calcarenite (Fig. 5). Assignment the upper part of the Mangapoike Riv section to Gauss Normal Epoch is su ported by correlation with the Mangaop Stream section (Kennett and others, 197 in which the late Opoitian is also of Gat age.

Compilation of Two Sections. Ma gapoike River and Blind River pr vide two overlapping polarity sequenc that represent the Gilbert Epoch. The G bert C1 event and C2 event are clea defined in the Kapitean Stage and possit the latest Tongaporutuan Stage (Figs. 4, which provide a valuable datum for int national correlations. Such between-secti similarity is convincing support for the v idity of the method.

The predominantly reversed polarity

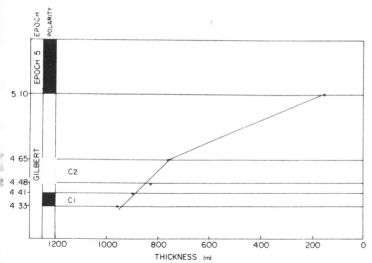

Figure 8. Sedimentation rate diagram for Blind River section. Paleomagnetic polarity boundaries (Fig. 4) plotted against assigned ages resulting from known geomagnetic polarity history (Cox, 1969; Opdyke, 1972). In the ordinate, polarity epochs are in vertical print; polarity events are in horizontal print; normal polarity is in black, and reversed polarity is clear. Slope of line indicates sedimentation rates of 130 cm per 1,000 yr during earliest part of Ibert Epoch and ~60 cm per 1,000 yr during Gilbert C events.

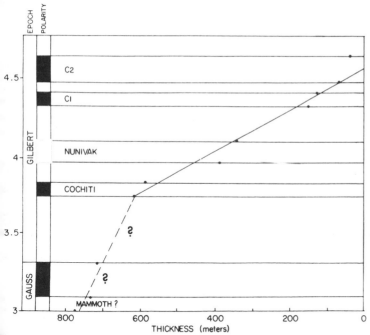

Figure 9. Sedimentation rate diagram for Mangapoike River section. Paleomagnetic polarity boundaries (Fig. 5) plotted against assigned ages resulting from known geomagnetic polarity history (Cox, 1969; Opdyke, 1972). Ordinate as in Figure 8. Slope of line indicates approximately linear sedimentation rate of 64 cm per 1,000 yr for early Ibert Epoch. Our preferred interpretation of magnetic stratigraphy in upper part of section suggests much reduced sedimentation rates of 22 cm per 1,000 yr.

both sections is a convincing demonstration that the results (Figs. 4, 5) represent the original polarity of the sediment. It can neither represent the Gauss Epoch nor Epoch 5. The reversed polarity can in no way reflect an unstable component acquired during the last 0.7 m.y., which is a period of normal polarity. The great similarity of the data between the sections, superimposed on the contrasting paleoenvironments within and between sections, is strong support for minimal effects by precipitation of magnetic minerals after deposition (Watkins and others, 1974). The fact that the polarity data of Lienert and others (1972) for the Hinakura section is mostly normal (despite assignment to the reversed Gilbert Epoch) suggests, in contrast, incomplete removal of an unstable normal polarity component. Similar results for the Pliocene-Pleistocene of Italy (Nakagawa and others, 1971), in which very dominantly normal polarity sections are assigned (presumably because of stratigraphic requirements) to the reversed Matuyama Epoch, probably reflect the same problem.

The paleomagnetic ages we have applied to the New Zealand late Miocene to early Pliocene biostratigraphic sequence and those ages of Kennett and others (1971) for the late Pliocene–early Pleistocene have been used by Kennett and Vella (1974) in a study of DSDP site 284 in the Tasman Sea west of New Zealand. Paleomagnetic ages for 5 distinct biostratigraphic horizons (base of *G. conomiozea*, t = 4.7 m.y.; base of *G. puncticulata*, t = 4.3 m.y.; peak of cooling in Waipipian Stage, t = 2.4 m.y.; base of *G. truncatulinoides*, t = 1.8 m.y.; and cooling in late Hautawan Stage, t = 1.7 m.y.), inferred for the site 284 sequence, fall near a straight line on an age-versus-depth plot. This supports the paleomagnetic age assignments for the New Zealand late Miocene to early Pleistocene sequence (Kennett and Vella, 1974).

Relation of New Zealand Miocene-Pliocene Boundary to Paleomagnetic Stratigraphy

The first appearance of *Globorotalia puncticulata sphericomiozea*, which marks the traditionally accepted Miocene-Pliocene boundary in New Zealand, occurs between the Gilbert C Event (t = 4.65 to 4.33 m.y.) and the Nunivak Event (t = 4.14 to 3.97 m.y.) in both sections. The New Zealand Miocene boundary is therefore 4.3 ± 0.1 m.y. old. Another potentially valuable datum shown by this study is the first appearance at or close to the upper part of the Nunivak Event (t = 3.97 ± 0.1 m.y.) of both *G. puncticulata puncticulata* and *G. crassaformis*. Because *G. puncticulata puncticulata* in New Zealand appears in evolutionary succession from *G. punc-*

ticulata sphericomiozea, this datum should be valuable for correlation and dating in the southwest Pacific.

Figure 10 shows the paleomagnetic stratigraphy and derived ages for the Blind River, Mangapoike, and Mangaopari sections that overlap to form a continuous sequence representing the paleomagnetic polarity history from near the top of Epoch 5 ($t = 5.10$ m.y.) to the Gilsa Event ($t = 1.61$ to 1.79 m.y.) near the middle to the Matuyama Reversed Epoch. In this diagram, the Miocene-Pliocene and the Pliocene-Pleistocene boundaries are taken as datum levels and *sedimentation rates have not been adjusted to force correlation with the established geomagnetic scale* (Cox, 1969). The diagram enables ages to be assigned to New Zealand stages as follows: top of the Tongaporutuan Stage at about 4.75 ± 0.1 m.y.; Kapitean Stage between 4.75 ± 0.1 and 4.3 ± 0.1 m.y.; Opoitian Stage between 4.3 ± 0.1 m.y. and 2.6 ± 0.1 m.y.; Waipipian Stage between 2.6 ± 0.1 and 2.3 ± 0.1 m.y.; Mangapanian Stage between 2.3 ± 0.1 and 1.9 ± 0.1 m.y., and the base of the Hautawan Stage at 1.9 ± 0.1 m.y. The Opoitian Stage is much longer (1.7 ± 0.1 m.y.) than the other dated stages, which range between 0.45 ± 0.1 and 0.3 ± 0.1 m.y. B.P.

Paleoclimatic History

Because the Blind River section is close to the northern limit of subantarctic waters, it might be expected to record oscillations of the southern subtropical (temperate) and subantarctic water masses. Contrary to this expectation, examination of the planktonic foraminifera in the Blind River section reveals that, throughout the entire late Miocene to earliest Pliocene, this area was under subantarctic water. The evidence for this consists of relatively low diversity of assemblages (maximum of 10 species observed at any level) and the great dominance of sinistral *G. pachyderma* and *G. bulloides* (Fig. 6). In the Mangapoike River section, greater diversity of the assemblages, higher frequencies of *G. bulloides* and several other species at the expense of sinistral *G. pachyderma,* and higher percentages of dextral forms within populations of *G. pachyderma* (Fig. 7) indicate that subtropical waters prevailed — except for a brief interval during the latest Miocene when subantarctic waters penetrated as far north as central North Island. Planktonic foraminiferal changes throughout most of this section reflect oscillations of water temperatures within the subtropical water mass.

In the Blind River section, fluctuations in the relative abundances of sinistral *G.*

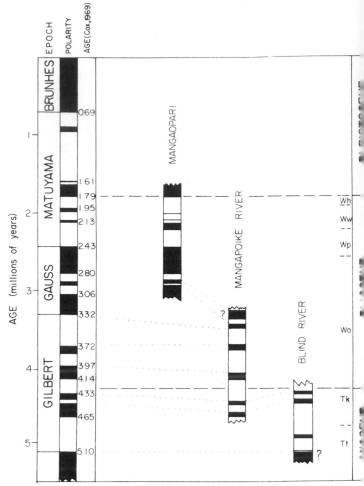

Figure 10. Paleomagnetic stratigraphy of two sections described in this paper and Mangaopari Stream sec (Kennett and others, 1971). Miocene-Pliocene boundary (a datum based on evolutionary change) and Plioc Pleistocene boundaries form datum levels with respect to which magnetic polarity results are plotted. Note sedimentation rates for each section have not been adjusted to force correlation with established geomagnetic s shown on left (Cox, 1969; Opdyke, 1972), but are taken from Figures 8 and 9. Broad departures from li sedimentation rates can be inferred from lack of perfect correspondence of known polarity scale and polarity da shown. Symbols on right represent New Zealand stages and are as follows: Tt = Tongaporutuan Stage, Tk = K tean Stage, Wo = Opoitian Stage, Wp = Waipipian Stage, Ww = Mangapanian Stage, and Wh = Hautawan Sta

pachyderma and of *G. bulloides* (Fig. 11) are reciprocal and reflect oscillations in water temperature. Higher frequencies of *G. pachyderma* record colder episodes and higher frequencies of *G. bulloides* record warmer episodes. A trend toward cooler conditions at the end of the Miocene is indicated by progressively more sinistral *G. pachyderma* populations, a significant increase in abundance of sinistral *G.*

pachyderma, a corresponding decrease importance of other species, a gradual crease in faunal diversity (Fig. 6), and pr ably a decrease in abundance of *G. w* (Fig. 11). Extremely cold conditions vailed during Kapitean and earliest O tian times, in contrast to the warmer co tions of the preceding Tongaporutuan later Opoitian times. Planktonic forami eral assemblages of the Kapitean Stag

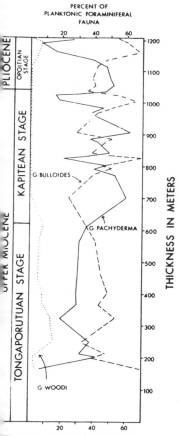

PERCENT OF
PLANKTONIC FORAMINIFERAL
FAUNA

Figure 11. Frequency fluctuations of temperature-
sensitive planktonic foraminiferal species in Blind River
section. Note reciprocal oscillations of sinistral
Globigerina pachyderma, which is a cooler species, and
Globigerina bulloides, which is a warmer species.

...ind River have features in common with
...olocene faunas in central subantarctic
...aters 800 km south of New Zealand
...ennett 1968a, 1969). This presumably
...flects substantial northward movement of
...bantarctic water in late Miocene times.

In the Mangapoike River section, the
...test Tongaporutuan to earliest Opoitian
...oling is also marked by high frequencies
...sinistral G. *pachyderma*. Percentages of
...her species, such as G. *miozea conoidea*,
...conomiozea, G. *glutinata*, and G.
...oodi, are not substantially reduced
...g. 7).

Warming later in the early Pliocene
...pper Opoitian Stage) is indicated by a re-
...ction in sinistral G. *pachyderma*,

---

increase in dextral G. *pachyderma*, increase
in other species compared with G.
pachyderma and G. *bulloides*, and a substantial increase in the numbers of dextral
forms within G. *pachyderma* populations
(Fig. 7). A reversal in these trends in the late
Opoitian indicates a return to cool conditions, probably immediately preceding the
cooling known to have occurred in the succeeding Waipipian Stage (Kennett and
others, 1971).

Substantial cooling associated with the
Kapitean Stage has previously been indicated by Kennett (1967, 1968b) and more
tentatively by Jenkins (1967b), based
mainly on coiling direction changes in G.
pachyderma. It has also been previously
pointed out that planktonic foraminiferal
assemblages are depleted in species
throughout New Zealand during the Kapitean, consistent with cool conditions (Jenkins, 1971; Kennett, 1968b; Collen and
Vella, 1973). The use of a single parameter
for only one species (such as sense of coiling) as a paleoclimatic index should be regarded with caution, because the environmental tolerance of a species, or particular
forms of a species, may change with time
(Keany and Kennett, 1972). The present
work, however, substantiates the previous
interpretations of an extremely cool episode
during the latest Miocene. We also now
show that this climatic episode occurs
within the period 4.7 m.y. and 4.2 m.y.
B.P., close to the time of the Gilbert C
Event. The late Miocene cooling is the oldest well-documented late Cenozoic cooling
event recorded in New Zealand and is associated with shallowing sedimentary facies
considered by Kennett (1967, 1968b) to
reflect a glacio-eustatic lowering of sea
level.

The cooling coincides with a cooling inferred from high carbonate content below
the Cochiti Event in the equatorial Pacific
(Hays and others, 1969). Evidence of possibly synchronous late Miocene cooling has
also been reported in California (Ingle,
1967, 1973). Widespread ice-rafting in the
Southern Ocean is known to have occurred
during the late Miocene–early Pliocene
(since about 4 to 5 m.y. B.P.; Goodell and
others, 1968; Kennett, 1972; Bandy and
others, 1971). In Eltanin core 34–5 from
northern Antarctic waters, the first appearance of ice-rafted quartz is associated with
the Gilbert C Event ($t = 4.65$ to 4.33 m.y.
B.P.) and may possibly represent a buildup
of Antarctic ice (Kennett and Brunner,
1973). Deep-sea drilling in areas adjacent
to Antarctica has demonstrated that glaciation has been active on Antarctica since at
least the beginning of the Neogene about 25
m.y. ago (Hayes and others, 1973).
Hence, the late Miocene cooling records an

---

increase in glaciation of Antarctica, rather
than initiation of late Cenozoic glaciation.
It is not known whether these earlier late
Cenozoic coolings corresponded with
major advance of ice in temperate latitudes,
although Bandy and others (1969) reported
major marine coolings associated with
nonmarine tillites about 13 or 14 m.y. B.P.
in southeastern Alaska.

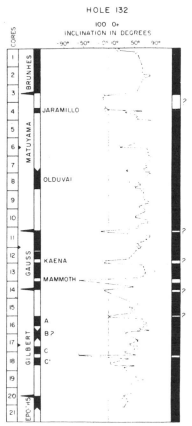

Figure 12. Paleomagnetic results for cores from
DSDP site 132 (Mediterranean Sea), showing published
polarity interpretation (Ryan and Flood, 1973, fig. 3) on
left and interpretation favored in this paper, on right.
Columns from left to right are DSDP core numbers; interpreted polarity epochs; published interpreted polarity
(black = normal, clear = reversed); data for each
specimen (inclination only = direction of remanent
magnetism after treatment in alternating magnetic field
of 100 Oe; positive inclination = no sign; negative inclination as indicated: note that present-day inclination
at site is positive 55°); interpreted polarity, using best
nonsubjective index of positive inclination = normal polarity; negative inclination = reversed polarity. Note
that when inclination is <15°, interpreted polarity indicated by a question mark, since virtual geomagnetic
latitude would be only transitional, and not definitely
indicative of true polarity reversal.

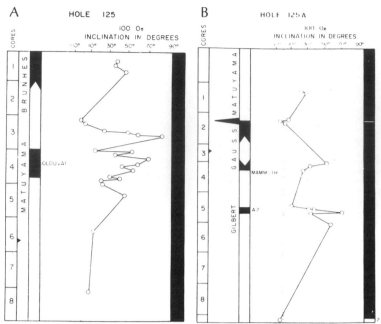

Figure 13. Paleomagnetic results for cores from DSDP holes 125 (diagram A) and 125A (diagram B), showing published polarity interpretations (Ryan and Flood, 1973, figs. 2, 3) on left of each hole and interpretation favored in this paper on right of each hole. For explanation of other symbols, see Figure 12.

Correlation of Miocene-Pliocene Boundary between New Zealand and Europe

The type Miocene-Pliocene boundary occurs between the late Miocene Messinian Stage and the early Pliocene Tabianian (equals Zanclean) Stage. Berggren (1973) has summarized the evidence for the age of this type boundary and placed it between 4.9 and 5.1 m.y. B.P., which is between the base of the Gilbert C Event ($t = 4.55$ m.y.) and the top of Epoch 5 ($t = 5.18$ m.y., according to Talwani and others, 1971). Little direct dating is available for the type sections in Europe; the date suggested by Berggren is derived by mainly biostratigraphic correlation with paleomagnetically and radiometrically dated cores outside of the Mediterranean area (Hays and others, 1969; Saito and others, 1974; Gill and McDougall, 1973). An approximate 5-m.y. date for the type Miocene-Pliocene boundary (Van Couvering and Miller, 1971; Berggren, 1973) is supported by two radiometric dates from the late Miocene–early Pliocene sequence. A K-Ar date of about 4.7 m.y. has been determined for the Orciatico trachyte, which metamorphosed early Pliocene clays in Tuscany (Tongiorgi, 1970) and a K-Ar date

of about 6 m.y. for late Miocene green tuffs in Ischia (Civetta and others, 1970).

Paleomagnetic stratigraphy is yet to be determined on Italian sections containing the Miocene-Pliocene boundary. The paleomagnetic stratigraphy of deep-sea drill sites (Leg 13) in the Mediterranean near Italy (Ryan and Flood, 1973; Ryan, 1973; Cita and Ryan, 1973) has assumed importance since Cita (1973) and Berggren (1973) have used this data in support of a 5-m.y. date for the boundary. For these reasons, it is necessary to comment in detail on the results of Ryan and Flood (1973), Ryan (1973), and Cita and Ryan (1973).

We reject the interpretations of those authors, because the data presented bear little resemblance to the inferred polarity sequence. Ryan (1973) admitted that his interpretation technique was subjective and depended heavily on a satisfactory comparison with the micropaleontologic results from the same core, as well as correlation with micropaleontologic and paleomagnetic results from Atlantic and Indian Ocean piston cores. We consider that his interpretation depended almost entirely on circular argument, which is quite unacceptable to us.

Figures 12 and 13 (A and B) show the

inclinations derived by Ryan (1973), following the indicated demagnetizatio treatment, and show the different interpretations which are given to the data. Ryan (1973, p. 1382) lists nine criteria that he has employed to derive the polarity log shown on the left-hand side of Figures 1 and 13 (A and B). While some of these a conventional (depending on the direction of remanent magnetism after magnetic clearing), others are not and are applied inconsistently. Production of the polarity logs Figures 12 and 13 (A and B) did not in fa require any paleomagnetic data. If su treatment of paleomagnetic data is san tioned, a major potential use of this metho to delineate global late Cenozoic m cropalentologic isochrons will not b realized — the reinforcement of possibly i correct paleontologic correlations will b more probable. The credibility of th paleomagnetic method requires rigorous objective use. It appears that th paleomagnetic data for holes 125, 125 and 132 (Figs. 12, 13A, 13B) reflect either dominant normal polarity overprint (due precipitation of magnetic minerals durii the last 0.7 m.y.), which cannot be remove by alternating magnetic field demagnetiz tion, or a dominant normal polarity form by the results of mixing during drilling. Th great range of Eh and pH environmen during the past 6 m.y. in the Mediterrane (Hsü and others, 1973) would certain have enhanced prospects of the former po sibility (Watkins and others, 1974). Th biostratigraphy indicates without ar doubt that the sediments were not all d posited during a period of normal polarit

Although no assistance is provided by th Mediterranean paleomagnetic data, a da of about 5 m.y. for the Miocene-Plioce boundary may well be correct, because the biostratigraphic correlations with dat sequences outside of the Mediterranea These employ both planktonic foraminife and calcareous nannofossils and are u likely to be greatly in error. The traditio ally used Miocene-Pliocene boundary New Zealand, based on the first appearan of *Globorotalia puncticulata* at the base the Opoitian Stage, would therefore b about 0.6 m.y. younger, although st within the early part of the Gilbert Epoc This would mean that the true Miocene Pliocene boundary in New Zealand shou be placed near the boundary between th Kapitean and Tongaporutuan Stages an hence, that the profound cooling associate with the Kapitean Stage and formerly r garded as late Miocene must therefore b assigned to the early Pliocene.

Consequently, it is appropriate to revie the biostratigraphic evidence for th Miocene-Pliocene boundary, as outlined b

Cita (1973) and Berggren (1973), to evaluate its validity in New Zealand. Unfortunately, the nannofossil biostratigraphy in the late Cenozoic of New Zealand has not yet been documented in sufficient detail. Planktonic foraminifera that are potentially useful for correlation of late Miocene–early Pliocene sequences include *Globorotalia tumida* (Brady), *Sphaeroidinella dehiscens* (Parker and Jones), *Globorotalia margaritae* (Bolli), *Globorotalia puncticulata*, and *Globoquadrina dehiscens* (Chapman, Parr and Collins) (Cita, 1972; Berggren, 1973).

The first appearances of *G. tumida* and *S. dehiscens* have been considered useful in identifying the Miocene-Pliocene boundary. Both *G. tumida* and *S. dehiscens* are tropical species, are absent in the New Zealand and Mediterranean late Cenozoic sequences, and are of no assistance in correlation between temperate and tropical sections.

In most Mediterranean sections, the first appearance of *G. puncticulata* is in the early Pliocene well above the Miocene-Pliocene boundary (Cita, 1973; Cati and others, 1968; Bertolino and others, 1968). This is not recorded as an evolutionary first appearance as in New Zealand, and, thus, the early Pliocene appearance of *G. puncticulata* in the Mediterranean sequence appears to have resulted from migration after its evolution in the South Pacific, and perhaps elsewhere from *G. conomiozea* (Kennett, 1966a).

In sections from Sicily, D'Onofrio (1964) and Colalongo (1970) have reported *G. puncticulata* in the latest Messinian Stage (latest Miocene). M. B. Cita (written commun., August 31, 1973) does not consider these to be genuine occurrences, but instead she considers that these have resulted from contamination from the overlying Trubi Marls of early Pliocene age. Different states of preservation of the late Messinian foraminifera and *G. puncticulata* found together in the same samples support this opinion. Furthermore, the late Messinian environment of deposition was brackish, shallow water (containing *Cyprideis pannonica* and *Ammonia beccarii tepida*) and hence probably unfavorable for planktonic foraminifera. It should also be noted that forms figured as *G. conomiozea* from the early Messinian of Sicily (Catalano and Provieri, 1971) are similar to the type New Zealand forms[1]. If reports of *G. puncticulata* and *G. conomiozea* from the Messinian Stage are confirmed, the Kapitean age of New Zealand correlates on bio-

stratigraphic evidence with at least part of the Messinian Stage of Europe, and a conflict would therefore exist between the paleomagnetic age of the New Zealand Miocene-Pliocene boundary and the 5 m.y. age given for the Mediterranean Miocene-Pliocene boundary (Van Couvering and Miller, 1971; Berggren, 1973).

The first appearance of *G. margaritae* also coincides closely with the Mediterranean Miocene-Pliocene boundary (Cita, 1973). This species is rare in New Zealand latitudes because of cooler conditions and first appears slightly above the evolutionary appearance of *G. puncticulata* (Site 207, 207A, Kennett, 1973). Its rarity makes it an unreliable index for international correlation.

The uncertainty of biostratigraphic correlation in sequences with different paleooceanographic histories is further exemplified by varying ranges of *Globoquadrina dehiscens*. Berggren (1973) reported the extinction of this species as a reliable datum for distinguishing the Miocene-Pliocene boundary. In New Zealand, the last appearances of *G. dehiscens* define a diachronous surface decreasing in age from south to north, because the ecologic limit of the species moved progressively further north in response to progressive cooling during the late Miocene (Kennett, 1968b). This species has not been reported from the Kapitean Stage, presumably because of the intense cooling but occurs in the Opoitian of Kaawa Creek in the northwest of the North Island (Jenkins, 1971, p. 165). Jenkins considered that these occurrences may be due to reworking, but the species could have lived at this latitude during the early Pliocene (upper Opoitian) warm interval (Collen and Vella, 1973). No early Pliocene occurrences were detected in deep-sea drilled sites from similar latitudes by Kennett (1973) and Kennett and Vella (1974). However, we have observed *G. dehiscens* in association with *G. margaritae* and other Pliocene species in the early Pliocene of a core (E35–12) from the Great Australian Bight (Watkins and Kennett, 1972). Srinivasan and Srivastava (1974) have also reported *G. dehiscens* in definite early Pliocene of tropical facies at Car Nicobar Islands, Bay of Bengal. The extinction of *G. dehiscens* evidently was controlled mainly by paleo-oceanographic cooling and is not a reliable datum for identifying the Miocene-Pliocene boundary.

It appears, therefore, that biostratigraphic ranges of the few potentially useful planktonic foraminiferal species are different in the Mediterranean than in the cooler New Zealand region. This explains the difficulty in correlating between New Zealand and the Mediterranean, which has

been acknowledged by previous authors (Hornibrook, 1958; Jenkins, 1971). As emphasized by Vella (1974), fission track, radiometric, and paleomagnetic dating provide the most promising means for exact correlations. Ideally, these techniques should be applied assiduously and vigorously to the type Messinian and type Pliocene and neighboring sections in Italy, if the present ambiguities in the dating of time-stratigraphic boundaries are to be resolved. Unfortunately, serious difficulty may exist in applying the paleomagnetic method to the classic Italian section, because postdepositional precipitation of magnetic minerals appears to be enhanced by local paleoenvironments (Watkins and others, 1974).

CONCLUSIONS

Two overlapping late Miocene–early Pliocene marine sections in New Zealand are dominantly confined to the Gilbert Epoch. Within and between sections, comparisons of the paleomagnetic data, as well as results of thermal demagnetization treatments of specimens, provide strong evidence that postdepositional chemical remanent magnetism has not destroyed the original magnetic signature. When combined with previous work, the data provide a total paleomagnetic polarity sequence for the period 5 to 1.70 m.y. B.P. in New Zealand.

The traditional position of the New Zealand Miocene-Pliocene boundary, based on the first appearance of *G. puncticulata*, is dated as 4.3 ± 0.1 m.y. old.

Paleomagnetic ages applied to late Cenozoic biostratigraphic units in Mediterranean deep-sea drilled sites are rejected, because the published paleomagnetic polarity variation was not based on the data involved.

The Miocene-Pliocene boundary in Italy has been indirectly dated at about 5.0 m.y. by biostratigraphic correlation with paleomagnetically dated sequences outside of the Mediterranean area. If this is correct, the New Zealand Miocene-Pliocene boundary is 0.6 to 0.8 m.y. younger than the type boundary. Thus, the Kapitean Stage of New Zealand perhaps should be placed in the lower Pliocene instead of the upper Miocene.

Ranges of critical late Miocene–early Pliocene planktonic foraminifera differ somewhat between New Zealand and the Mediterranean area. Precise biostratigraphic correlation is not possible because of different paleo-oceanographic histories, and conflicting correlations result from differences in ranges of planktonic foraminifera in the two areas.

---

[1] Forms referred to *G. conomiozea* by Colalongo (1970) from the early Messinian do not closely resemble *conomiozea* from the type in New Zealand.

A late Miocene to early Pliocene severe climatic cooling is recorded by the planktonic foraminifera in the New Zealand Kapitean Stage and lower Opoitian Stage. Subantarctic waters covered much of the New Zealand east coast region. The climax of the cooling is paleomagnetically dated as between 4.7 ± 0.1 and 4.2 ± 0.1 m.y. B.P. within the early Gilbert Reversed Epoch. This was preceded by progressive cooling during the late Tongaporutuan Stage (same as earliest Gilbert) and was followed by substantially warmer conditions during most of the Opoitian Stage (same as latest Gilbert and Gauss).

ACKNOWLEDGMENTS

We thank N. de B. Hornibrook (New Zealand Geological Survey) for advice on sampling the Mangapoike River section; Alan Carter (University of New South Wales, Sydney) for assistance in the sampling of this section; P. Vella (Victoria University of Wellington) for his critical suggestions on the manuscript; and Charlotte Brunner for valuable technical assistance.

This research was supported by National Science Foundation Grant GA-27092 (Geophysics Division.)

REFERENCES CITED

Bandy, O. L., Butler, E. A., and Wright, R. C., 1969, Alaskan upper Miocene marine glacial deposits and the *Turborotalia pachyderma* datum plane: Science, v. 166, p. 607–609.

Bandy, O. L., Casey, R. E., and Wright, R. C., 1971, Late Neogene planktonic zonation, magnetic reversals, and radiometric dates, Antarctic to the tropics: Am. Geophys. Union Antarctic Research Ser., v. 15, p. 1–26.

Berggren, W. A., 1973, The Pliocene time scale: Calibration of planktonic foraminiferal and calcareous nannoplankton zones: Nature, v. 243, p. 391–397.

Berggren, W. A., Phillips, J. D., Bertels, A., and Wall, D., 1967, Late Pliocene-Pleistocene stratigraphy in deep-sea cores from the south-central North Atlantic: Nature, v. 216, p. 253–254.

Bertolino, V., and others, 1968, Proposal for a biostratigraphy of the Neogene in Italy based on planktonic foraminifera: Gior. Geologia, v. 35, p. 23–30.

Catalano, R., and Sprovieri, R., 1971, Biostratigrafia di alcune serie Saheliane (Messiniano Inferiore) in Sicilia, *in* Farinacci, A., ed., Proceedings of the II planktonic conference, Rome, 1970: Rome, Edizioni Tecnoscienza, p. 211–249.

Cati, F., and others, 1968, Biostratigrafia del Neogene mediterraneo basata sui foraminiferi planktonici: Italy Soc. Geol. Bull., v. 87, p. 491–503.

Cita, M. B., 1972, The Miocene/Pliocene boundary: History and definition [abs.]: Internat. Geol. Cong., 24th, Montreal 1972, p. 536.

——1973, Pliocene biostratigraphy and chronostratigraphy, *in* Ryan, W.B.F., Hsü, K. J., and others, eds., 1973, Initial reports of the Deep Sea Drilling Project, Vol. 13: Washington, D.C., U.S. Govt. Printing Office, p. 1343–1379.

Cita, M. B., and Ryan, W.B.F., 1973, Timescale and general synthesis, *in* Ryan, W.B.F., Hsü, K. J., and others, eds., 1973, Initial reports of the Deep Sea Drilling Project, Vol. 13: Washington, D.C., U.S. Govt. Printing Office, p. 1405–1415.

Civetta, L., Gasparini, P., and Adams, J.A.S., 1970, Geochronology and geochemical trends of volcanic rocks from Campania, S. Italy: Eclogae Geol. Helvetiae, v. 63, p. 57–68.

Colalongo, M. L., 1970, Appunti biostratigrafici sul Messiniano: Gior. Geologia, v. 36, p. 515–542.

Collen, J. D., and Vella, P., 1973, Pliocene planktonic foraminifera, southern North Island, New Zealand: Jour. Foram. Research, v. 3, p. 13–29.

Cox, A., 1969, Geomagnetic reversals: Science, v. 163, p. 237–245.

Devereux, I., Hendy, C. H., and Vella, P., 1970, Pliocene and early Pleistocene sea temperature fluctuations, Mangaopari Stream, New Zealand: Earth and Planetary Sci. Letters, v. 8, p. 163–168.

D'Onofrio, S., 1964, I foraminiferi del neostratotipo del Messiniano: Gior. Geologia, v. 32, p. 409–461.

Edwards, A. R., 1971, A calcareous nannoplankton zonation of the New Zealand Paleogene, *in* Farinacci, A., ed., Proceedings of the II Planktonic Conference, Rome, 1970: Rome, Edizioni Tecnoscienza, p. 381–419.

Finlay, H. J., 1939, New Zealand foraminifera: Key species in stratigraphy: Royal Soc. New Zealand Trans., v. 68, p. 530.

Finlay, H. J., and Marwick, J., 1947, New divisions of the New Zealand Upper Cretaceous and Tertiary: New Zealand Jour. Sci. Technology, v. 28, p. 235–236.

Fisher, R., 1953, Dispersion on a sphere: Royal Soc. Proc., v. A. 217, p. 295–305.

Fleming, C. A., ed., 1959, Lexique stratigraphique international 6 Océanic (4), New Zealand: Paris, Centre nationale de la Recherche, scientifique, 527 p.

Garner, D. M., 1959, The sub-tropical convergence in New Zealand surface waters: New Zealand Jour. Geology and Geophysics, v. 2, p. 315–337.

Gibson, G. W., 1967, The foraminifera and stratigraphy of the Tongaporutuan Stage in the Taranaki coastal and six other sections. Part 1. Systematics and distribution: Royal Soc. New Zealand, Trans. Geol., v. 5, no. 1, p. 1–70.

Gill, J. B., and McDougall, I., 1973, Biostratigraphic and geological significance of Miocene-Pliocene volcanism in Fiji: Nature, v. 241, p. 176–180.

Goodell, H. G., Watkins, N. D., Mather, T. T., and Koster, S., 1968, The Antarctic glacial history recorded in sediments of the Southern Ocean: Palaeogeography, Palaeoclimatology, Palaeoecology, v. 5, p. 41–62.

Hayes, D. E., and others, 1973, Leg 28 deep-se drilling in the southern ocean: Geotimes, 18, p. 19–24.

Hays, J. D., Saito, T., Opdyke, N. D., and Burckle, L. H., 1969, Pliocene-Pleistocen sediments of the equatorial Pacific; The paleomagnetic, biostratigraphic, and c matic record: Geol. Soc. America Bull., 80, p. 1481–1514.

Heath, R. A., 1972a, The Southland Curren New Zealand Jour. Marine and Freshwat Research, v. 6, no. 4, p. 497–533.

——1972b, Oceanic upwelling produced l northerly winds on the north Canterbu coast, New Zealand: New Zealand Jou Marine and Freshwater Research, v. 6, n 3, p. 343–351.

Hornibrook, N. de B., 1958, New Zealand U per Cretaceous and Tertiary foraminifer zones and some overseas correlations: M cropaleontology Press, Am. Mus. Na History, v. 4, p. 25–38.

Hornibrook, N. de B., and Edwards, A. R., 197 Integrated planktonic foraminiferal ar calcareous nannoplankton datum levels the New Zealand Cenozoic, *in* Farinacc A., ed., Proceedings of the II planktor conference, Rome, 1970: Rome, Edizio Tecnoscienza, p. 649–657.

Hsü, K. J., Ryan, W.B.F., and Cita, M. B., 197 Late Miocene dessication of the Mediterr nean: Nature, v. 242, p. 240–244.

Ingle, J. C., Jr., 1967, Foraminiferal biofaci variation and the Mio-Pliocene boundary southern California: Bull. Am. Paleonto ogy, v. 52, p. 236.

——1973, Neogene foraminifera from t northeastern Pacific Ocean, Leg 18, Dec Sea Drilling Project, *in* Kulm, L. D., v Huene, R., and others, eds., 1973, Init reports of the Deep Sea Drilling Proje Vol. 18: Washington, D.C., U.S. Go Printing Office, p. 517–567.

Jenkins, D. G., 1967a, Planktonic foraminifer zones and new taxa from the low Miocene to the Pleistocene of New Ze land: New Zealand Jour. Geology ar Geophysics, v. 10, p. 1064–1078.

——1967b, Recent distribution, origin, and co ing ratio changes in *Globorotal. pachyderma* (Ehrenberg): Micropaleonto ogy, v. 13, p. 195–203.

——1971, New Zealand Cenozoic planktor foraminifera: New Zealand Geol. Surve Bull., v. 42, p. 1–278.

Keany, J., and Kennett, J. P., 197 Pliocene–early Pleistocene paleoclima history recorded in Antarctic-subantarct deep-sea cores: Deep-Sea Research, v. 19 529–548.

Kennett, J. P., 1965, Faunal succession in tw upper Miocene–lower Pliocene section Marlborough, New Zealand: Royal So New Zealand Trans., v. 3, p. 197–213.

——1966a, The *Globorotalia crassaform* bioseries in north Westland and Mar borough, New Zealand: Micropaleonto ogy, v. 12, p. 235–245.

——1966b, Stratigraphy and fauna of the ty section and neighbouring sections of t Kapitean Stage, Greymouth, N.Z.: Roy Soc. New Zealand Trans., v. 4, p. 1–77.

Kennett, J. P., 1967, Recognition and correlation of the Kapitean Stage (upper Miocene, New Zealand): New Zealand Jour. Geology and Geophysics, v. 10, p. 143–156.

——1968a, Latitudinal variation in Globigerina pachyderma (Ehrenberg) in surface sediments of the southwest Pacific Ocean: Micropaleontology, v. 14, p. 305–318.

——1968b, Paleo-oceanographic aspects of the foraminiferal zonation in the upper Miocene–lower Pliocene of New Zealand: Gior. Geologia, v. 35, p. 143–156.

——1969, Distribution of planktonic foraminifera in surface sediments to the southeast of New Zealand, in Bronnimann, P., and Renz, H. H., eds., Proceedings first international conference on planktonic microfossils: v. 2, p. 307–322.

——1972, The climatic and glacial record in Cenozoic sediments of the Southern Ocean, in Bakker, E. M. van Zinderen, ed., Paleoecology of Africa and Antarctica: S.C.A.R., v. 6, p. 59–78.

——1973, Middle and late Cenozoic planktonic foraminiferal biostratigraphy of the southwest Pacific — Deep Sea Drilling Project Leg 21, in Burns, R. E., and Andrews, J. E., eds., Initial reports of the Deep Sea Drilling Project, Vol. 21: Washington, D.C., U.S. Govt. Printing Office, p. 575–639.

Kennett, J. P., and Brunner, C. A., 1973, Antarctic late Cenozoic glaciation: Evidence for initiation of ice rafting and inferred increased bottom-water activity: Geol. Soc. America Bull., v. 84, p. 2043–2052.

Kennett, J. P., and Vella, P., 1974, Late Cenozoic planktonic foraminifera and paleo-oceanography at Deep Sea Drilling Project Site 284 in the cool subtropical South Pacific, in Kennett, J. P., and Houtz, R. E., eds., Initial reports of the Deep Sea Drilling Project, Vol. 29: Washington, D.C., U.S. Govt. Printing Office (in press).

Kennett, J. P., and Watkins, N. D., 1972, The biostratigraphic, climatic, and paleomagnetic records of late Miocene to early Pleistocene sediments in New Zealand [abs.]: Internat. Geol. Cong., 24th, Montreal 1972. p. 538.

Kennett, J. P., Watkins, N. D., and Vella, P., 1971, Paleomagnetic chronology of Pliocene–early Pleistocene climates and the Plio-Pleistocene boundary in New Zealand: Science, v. 171, p. 276–279.

Lienert, B. R., Christoffel, D. A., and Vella, P., 1972, Geomagnetic dates on a New Zealand upper Miocene–Pliocene section: Earth and Planetary Sci. Letters, v. 16, p. 195–199.

McInnes, B. A., 1964, A detailed examination of microfaunas from a Miocene-Pliocene (Taranakian-Opoitian-Waitotaran) section on the Mangapoike River, Northern Hawkes Bay, New Zealand: British Petroleum Shell and Todd Petroleum Development Ltd., Paleontological Note no. 32, p. 10.

——1965, Globorotalia miozea Finlay as an ancestor of Globorotalia inflata (d'Orbigny): New Zealand Jour. Geology and Geophysics, v. 8, p. 104–108.

Nakagawa, H., Niitsuma, N., and Hayasaka, I., 1969, Late Cenozoic geomagnetic chronology of the Boso Peninsula: Japan Jour. Geol. Soc., v. 75, p. 267–280.

Nakagawa, H., Niitsuma, N., and Elmi, C., 1971, Pliocene and Pleistocene magnetic stratigraphy in Le Castella area, southern Italy — A preliminary report: Quaternary Research, v. 1, p. 360–368.

Ongley, M., 1928, Wairoa subdivision: New Zealand Geol. Survey Ann. Rept., 22d, p. 6–8.

——1930, Wairoa subdivision: New Zealand Geol. Survey Ann. Rept., 24th, p. 7–10.

Opdyke, N. O., 1972, Paleomagnetism of deep sea cores: Rev. Geophysics and Space Physics, v. 10, p. 213–249.

Ryan, W.B.F., 1973, Paleomagnetic stratigraphy, in Ryan, W.B.F., Hsü, K. J., and others, eds., 1973, Initial reports of the Deep Sea Drilling Project, Vol. 13: Washington, D.C., U.S. Govt. Printing Office, p. 1380–1387.

Ryan, W.B.F., and Flood, J. D., 1973, Preliminary paleomagnetic measurements on sediments from the Ionian (Site 125) and Tyrrhenian (Site 132) Basins of the Mediterranean Sea, in Ryan, W.B.F., Hsü, K. J., and others, eds., 1973, Initial reports of the Deep Sea Drilling Project, Vol. 13: Washington, D.C., U.S. Govt. Printing Office, p. 599–603.

Saito, T., Burckle, L. H., and Hays, J. D., 1974, Late Neogene epoch boundaries in deep-sea sediments, in Saito, T., and Burckle, L. H., eds., Late Neogene epoch boundaries: Micropaleontology Press, Am. Mus. Nat. History (in press).

Srinivasan, M. S., and Srivastava, S. S., 1974, Sphaeroidinella dehiscens datum and the Miocene-Pliocene boundary: Am. Assoc. Petroleum Geologists Bull., v. 58, p. 304–311.

Talwani, M., Windisch, C. C., and Langseth, M. G., Jr., 1971, Reykjanes ridge crest: A detailed geophysical study: Jour. Geophys. Research, v. 76, p. 473–517.

Tongiorgi, E., 1970, in Selli, R., ed., Report on absolute age: Gior. Geologia: v. 35, no. 1, p. 55.

Van Couvering, J. A., and Miller, J. A., 1971, Late Miocene marine and non-marine time scales in Europe: Nature, v. 230, p. 559–563.

Vella, P., 1974, The boundaries of the Pliocene in New Zealand, in Saito, T., and Burckle, L. H., eds., Late Neogene Epoch boundaries: Micropaleontology Press, Am. Mus. Nat. History (in press).

Watkins, N. D., and Kennett, J. P., 1972, Regional sedimentary disconformities and upper Cenozoic changes in bottom water velocities between Australasia and Antarctica: Am. Geophys. Union Antarctic Research Ser., v. 19, p. 273–293.

——1973, The paleomagnetism of the Pliocene-Pleistocene boundary type sections on Italy and New Zealand [abs.]: EOS (Am. Geophys. Union Trans.), v. 54, p. 1075.

Watkins, N. D., Kester, D. R., and Kennett, J. P., 1974, Palaeomagnetism of the type Pliocene-Pleistocene boundary at Santa Maria de Catenzaro, Italy, and the problem of post-depositional precipitation of magnetic minerals: Earth and Planetary Sci. Letters (in press).

MANUSCRIPT RECEIVED BY THE SOCIETY JULY 30, 1973
REVISED MANUSCRIPT RECEIVED MARCH 7, 1974

16

PALEOMAGNETISM OF THE TYPE PLIOCENE/PLEISTOCENE BOUNDARY SECTION AT SANTA MARIA DE CATANZARO, ITALY, AND THE PROBLEM OF POST-DEPOSITIONAL PRECIPITATION OF MAGNETIC MINERALS

N.D. WATKINS, D.R. KESTER and J.P. KENNETT

Graduate School of Oceanography, University of Rhode Island, Kingston, R.I. (USA)

In order to assist in global correlation of the Pliocene/Pleistocene boundary, a paleomagnetic survey has been made of the world type section for the boundary, in Calabria, Italy. A total of 116 specimens from 36 sites were collected from the section, which as sampled is about 30 m thick. All sites possess stable remanent magnetism and northern hemisphere virtual geomagnetic poles, consistent with deposition during a period of normal polarity. If this period was during the Gilsa event, a minimum sedimentation rate of 17 cm per 1000 years would be required. Since a normal polarity chemical overprint, acquired during the last 0.69 m.y. could also explain the data, thermomagnetic analyses were made of selected samples. The results show the presence of highly unstable superparamagnetic material, which is most likely a product of post-depositional chemical precipitation. This observation enhances the possibility that any longer-duration precipitation could have created coarse and thus magnetically more stable components, the effects of which could not be readily distinguished from the original remanent magnetism.

The difficulties of distinguishing between original depositional remanent magnetism and post-depositional chemical remanent magnetism, in outcropping marine sediments, as well as the large range of possible natural causes of the latter, is summarized in the form of a discussion of feasible Eh and pH changes occurring between original deposition at upper bathyal depths and final sampling above sea level. It is considered that, in the absence of experimental means to distinguish the roles of original and chemically overprinted paleomagnetic signals, similarity of magnetic polarity stratigraphy between sections of sediments representing different paleoenvironments and sedimentation rates is a necessary if insufficient requirement for diagnosis of real geomagnetic behavior during deposition, as opposed to post depositional effects.

[*Editor's Note:* The article itself has not been reproduced here.]

Upper Cretaceous–Paleocene magnetic stratigraphy at Gubbio, Italy V. Type section for the Late Cretaceous–Paleocene geomagnetic reversal time scale

WALTER ALVAREZ *Lamont-Doherty Geological Observatory of Columbia University, Palisades, New York 10964*
MICHAEL A. ARTHUR ⎱ *Department of Geological and Geophysical Sciences, Princeton University, Princeton, New Jersey 08540*
ALFRED G. FISCHER ⎰
WILLIAM LOWRIE *Institut für Geophysik, ETH - Hönggerberg, CH-8049, Zürich, Switzerland*
GIOVANNI NAPOLEONE *Istituto di Geologia e Paleontologia, via Lamarmora, 4, Florence, Italy*
ISABELLA PREMOLI SILVA *Istituto di Paleontologia, University of Milan, Italy*
WILLIAM M. ROGGENTHEN *Department of Geological and Geophysical Sciences, Princeton University, Princeton, New Jersey 08540*

ABSTRACT

A biostratigraphically complete and well-exposed sequence of Upper Cretaceous–Paleocene pelagic limestones at Gubbio, Italy, has provided a record of geomagnetic polarity reversals that closely matches the sequence inferred from marine magnetic anomalies. Abundant foraminifera permit accurate dating of the sequence. Because of these favorable circumstances, the Gubbio locality is formally proposed as the magnetostratigraphic type section for the Upper Cretaceous and Paleocene.

INTRODUCTION

The realization that the Earth's magnetic field has undergone repeated reversals in polarity was in itself a major advance in geophysics. In addition, as the effects of geomagnetic reversals are globally synchronous, they provide geologists with a powerful new correlation tool for the study of Earth history. Geomagnetic polarity is imprinted on various rocks in the form of remanent magnetism, and the history of the geomagnetic field may be deciphered from two independent records: (1) marine magnetic anomalies resulting from magnetic imprints on laterally accreted oceanic crust and (2) the magnetic stratigraphy of vertically accreted sequences of sedimentary and volcanic rocks. In order for magnetic stratigraphy to reach its full potential, these two records need to be precisely correlated with each other and with the biostratigraphic time scale, and to be calibrated in terms of absolute age by correlation with radiometrically dated rocks.

In the four preceding papers, we have given the results of lithostratigraphic, paleontologic, and paleomagnetic studies of what appears to be a continuously deposited sequence of Upper Cretaceous and Paleocene sediments at Gubbio, Italy. In this paper, we tie these three kinds of investigation together in order to obtain accurate paleontological dating for the reversal sequence at Gubbio and, by correlation, for the marine magnetic anomaly pattern. Recognizing the need for a firm stratigraphic basis for inferred magnetic history, we propose the Gubbio section as a magnetostratigraphic type section for the Upper Cretaceous and Paleocene.

DATING SEA-FLOOR ANOMALIES

A number of pioneers (Cox and others, 1963a, 1963b; McDougall and Tarling, 1964; Doell and Dalrymple, 1966; McDougall and Chamalaun, 1966) working on late Neogene volcanic sequences, established in the 1960s the reality of correlatable *magnetic polarity zones*, and inferred that these had resulted from periodic reversals of the geomagnetic dipole. Vine and Matthews (1963) recognized that the anomaly stripes on magnetic maps of ocean floors resulted from lateral accretion of oceanic crust in this reversing geomagnetic field, and that the sea floor potentially reveals the history of these reversals. On this basis, Heirtzler and others (1968) proposed the first comprehensive polarity chronology extending back to the Late Cretaceous. Helsley and Steiner (1969), Larson and

Pitman (1972), Larson (1974), and a number of other workers have delineated four major episodes in geomagnetic history since the breakup of Pangaea in the early Mesozoic (Table 1). These major episodes, now established in various ocean basins, have led to refinement of paleogeography and to partial understanding of the spreading history of oceanic crust.

A precise history of the oceans requires dating of the crustal strips which generate the individual anomalies, and this has not yet been achieved to the desired accuracy. Direct methods useful in dating these crustal strips are: (1) determining the radiometric age of basaltic basement rocks recovered beneath known anomalies in DSDP holes, (2) dating the oldest overlying sediments by paleontological means, and (3) interpolating between and extrapolating beyond well-dated anomalies, assuming a constant spreading rate. Each of these approaches has limitations. Samples are few, and many of them are not suitable for radiometric dating. The oceanic basement is highly complex, with superimposed extrusive lavas and intrusive dikes and sills whose ages may differ from the country rock. Furthermore, alteration is common. Consequently, single radiometric dates may be misleading. Paleontologic dates are too young if the oldest fossiliferous beds were preceded by unfossiliferous deposits or are separated

TABLE 1. MAJOR EPISODES IN GEOMAGNETIC FIELD BEHAVIOR

Episode	Marine magnetic anomaly pattern	Geomagnetic field behavior
4	Cretaceous-Cenozoic anomaly sequence	Cretaceous-Cenozoic reversal sequence (Late Cretaceous to present)
3	Cretaceous Quiet Zone	Cretaceous Long Normal polarity chron (Early to Late Cretaceous)
2	Keathley, Japanese, Hawaiian, and Phoenix magnetic lineations	M-sequence reversals (Late Jurassic to Early Cretaceous)
1	Jurassic Quiet Zone	Field behavior uncertain

from the underlying basement by a depositional hiatus, or if the hole bottoms in a sill instead of in true basement. Finally, variations in spreading rates have been demonstrated over time intervals on the order of 10^7 yr (Sclater and others, 1974; Larson and Pitman, 1972; Hays and Pitman, 1973; Lowrie and Alvarez, this issue, Fig. 2). The fundamental problem in determining the ages of geomagnetic reversals by dating basalt or overlying sediments beneath known anomalies is that the age determination is made on a tiny fraction of the geologically complex volume of rock that contributes to the magnetic record of the ancient field polarity. More accurate dates for the anomalies of the last several million years have resulted from correlations with radiometrically dated magnetic stratigraphy (for example, Doell and Dalrymple, 1966; Watkins and others, 1975), and in theory this should be possible for older anomalies as well.

MAGNETIC STRATIGRAPHY

All sedimentary and volcanic rocks are deposited or emplaced in the geomagnetic field, and those with suitable mineralogy record the field direction. Thus a history of field reversals may be revealed in volcanic sequences in which the course of events is double checked by superposition and by radiometric dating. Such work has been carried back about 7 m.y. (Watkins and others, 1975), but beyond that time, the method becomes difficult because of the irregularity of volcanic events and the increasing uncertainties in radiometric age determinations. Furthermore, results in volcanic sequences are not easily tied into the paleontologic scale. Accordingly, there has been much interest in developing a magnetic stratigraphy in sedimentary rocks.

Early attention was focused mainly on red beds because of their comparatively strong remanent magnetization, but such sequences are rarely well dated by fossils. The discovery that marine biogenic oozes yield reliable magnetic results (Opdyke and others, 1966; Opdyke, 1972) stimulated interest in indurated pelagic and hemipelagic limestones and marls, which have the virtues of more continuous sedimentation and abundant datable fossils. However, the study of these weakly magnetized rocks remained unfeasible in many cases until the development of more rapid and sensitive magnetometers, such as the digital spinners and cryogenic instruments used in this investigation.

The globally synchronous reversals of the magnetic field offer the prospect of time markers for the precise correlation of di-

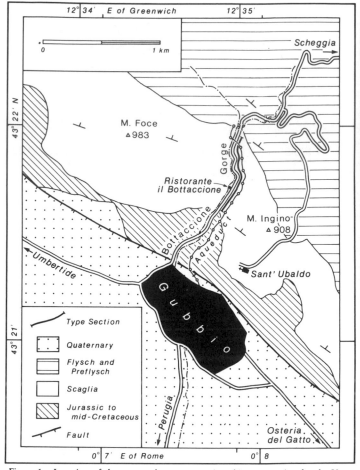

Figure 1. Location of the proposed magnetostratigraphic type section for the Upper Cretaceous and Paleocene. The geology is simplified after Barnaba (1958).

verse events on the Earth's surface. A detailed network of correlations between (1) the sea-floor anomalies, which provide the closest approach to a continuous record of geomagnetic field polarity; (2) the radiometrically dated volcanic rock sequences on land; and (3) sedimentary sections which contain the standard biostratigraphic zones is essential to further progress along these lines. A number of studies of magnetic stratigraphy in sedimentary sections have been published recently (Opdyke and others, 1974; Theyer and Hammond, 1974; Shive and Frerichs, 1974). Previous reversal time scales for the Late Cretaceous

and Paleocene have been pieced together from a number of partial sections (van Hinte, 1972, 1976; Watkins, 1975; Keating and others, 1975; Thierstein, 1977). The Gubbio section provides the first complete biostratigraphic and magnetic record from a single locality.

THE GUBBIO SECTION

In the course of the sedimentological, paleontological, and paleomagnetic investigations described in the four preceding papers, we have delineated the sequence of magnetic polarity zones in the Gubbio section and have tied this sequence directly to the standard planktonic foraminiferal zonation.

The advantages offered by the Gubbio section for this kind of study are (1) nearly complete exposure in fresh roadcuts; (2) relatively constant rate and type of sedimentation, yielding limestones which provide both a magnetic record and the full sequence of known planktic foraminiferal zones — this stratigraphic continuity apparently extends across the Maastrichtian-Paleocene boundary, which in much of the world is marked by a hiatus; (3) abundant foraminifera; (4) remanent magnetization that is stable and well clustered after AF or thermal removal of secondary components, and which, although weak, can be measured routinely on digital spinner magnetometers or cryogenic instruments; (5) good access; and (6) a well-indurated lithology which is easily sampled with a portable drill.

The principal disadvantage is that the foraminifera cannot usually be separated from the matrix and must be identified in thin section, while nannoplankton are not well preserved, and other fossils are rare. A second disadvantage is the presence of some small faults, most of which have displacements of less than 1 m, but a few of which may have somewhat larger offsets. However, the continuity of faunal zones and the close match between the reversal sequence at Gubbio and the marine magnetic anomaly sequence (Lowrie and Alvarez, this issue, Fig. 2) confirm the absence of major structural disturbances.

PROPOSED MAGNETOSTRATIGRAPHIC TYPE SECTION

Inasmuch as magnetic stratigraphy will have to be patched together from many scattered sections, as has biostratigraphy, it is important to avoid procedural pitfalls that could lead to the kind of ambiguity and fruitless argument that have troubled classical stratigraphy. We thus, in general, adhere to the recommendations of the Subcommission of Magnetostratigraphic Nomenclature (1973). The most essential of their recommendations is the conceptual separation of concrete, measurable and mappable rock units from inferred units of time. We feel that it is equally important that magnetostratigraphic and chronostratigraphic entities or concepts be tied to type sections, where they may subsequently be tested, emended, and elaborated.

Because of the apparently complete reversal sequence recorded in the Scaglia Rossa at Gubbio, we propose that this be

taken as the type section for the Late Cretaceous–Paleocene geomagnetic reversal sequence. In accordance with the procedure for establishing a type section given in the Code of Stratigraphic Nomenclature (Am. Commission on Stratigraphic Nomenclature, 1961), we note that the section is located at 43°21.6–22.0′N, 12°34.9′E of Greenwich (0°7.8′E of Rome, Monte Mario) on Italian topographic map sheet 116-III-SE "Gubbio" (1:25,000). The outcrops are on the southeast side of the Gubbio-Scheggia road, between kilometre posts 10 and 11 (km 41-42 in the old numbering shown on the topographic map), below the mediaeval aqueduct known as the "Bottaccione." A location map is given in Figure 1.

The physical magnetostratigraphic units defined at Gubbio are sequences of beds characterized by a common polarity. These are polarity zones; we have recognized 23 in the Gubbio section. For purposes of reference they require identifying labels. The lowest of the zones, which appears to represent the Long Normal episode of the Cretaceous, is here identified as the "Gubbio Long Normal Zone." The succeeding zones are designated by letters with the prefix "Gubbio," in ascending alphabetical order; normal zones are distinguished by "+" and reversed zones by "−." In accordance with the hierarchy of units recommended by the Subcommission on Magnetostratigraphic Nomenclature (1973), we have chosen to regard some of the thinnest polarity units as polarity subzones within thicker polarity zones. Thus subzones Gubbio D1+, D2−, and D3+ make up zone Gubbio D+. This procedure is analogous to standard practice in lithostratigraphy, where the section must be completely covered by a sequence of formations, while some (but not necessarily all) of the formations may be divided into members.

The zonal designations are shown in our Figure 2; in paper III, Figures 1 and 2; and in paper IV, Figure 3. The positions of magnetic zonal boundaries relative to foraminiferal zones and to various proposed absolute chronologies are also given in Figure 2. Metre numbers have been painted on the outcrop, and more permanent markers for polarity-zone boundaries have been cemented into the rock.

We infer that each polarity zone recognized in the Gubbio section was formed during a period of corresponding normal or reversed geomagnetic polarity. These abstract units of Earth history we term "polarity chrons" in adaptation to Hedberg (1976, p. 69), rather than "polarity intervals" as suggested by the Subcommission (1973), inasmuch as "chron" refers clearly

to time, whereas "interval" is ambiguous, as pointed out to us by Hedberg. These units of time may be designated with the same labels as their corresponding zones in the rock record; thus one may refer to the "Gubbio G− Polarity Chron," a unit of time that included the very latest Cretaceous and the very earliest Paleocene.

Correlations between the sequence of polarity zones at Gubbio and the sequence of marine magnetic anomalies have been presented in the preceding papers. Lowrie and Alvarez (paper III) assumed the Gubbio Long Normal Zone and the "Cretaceous Quiet Zone" of the sea floor to be common expressions of the Cretaceous Long Normal polarity chron and found an excellent match of successive longer and shorter polarity zones at Gubbio with the sequence of marine magnetic anomalies. In this match, the Cretaceous-Tertiary boundary falls into polarity zone Gubbio G− and between anomalies 29 and 30. Independently, Roggenthen and Napoleone (paper IV) correlated the long zone Gubbio M− with the broad negative anomaly between 26 and 27 and used the paleontologic data available from DSDP–JOIDES (paper II, Table 1) as tie points (paper IV, Fig. 4). Matching polarity zones with sea-floor anomalies upward and downward yielded a correlation for the top of the Cretaceous identical to that arrived at by Lowrie and Alvarez (paper III) and identified anomaly 26 as belonging to the late Paleocene *pseudomenardii* zone. These correlations agree with the age assignments of marine magnetic anomalies by Sclater and others (1974).

PALEONTOLOGICAL CALIBRATION

In Figure 2, the stratigraphic positions of the reversal boundaries in the Gubbio section are compared with the stage and foraminiferal zone boundaries determined by Premoli Silva (this issue). The following observations can be made:

1. Although the base of the Gubbio (Cretaceous) Long Normal Zone is not dated by this study, it is at least as old as Cenomanian.

2. The Gubbio Long Normal Zone ends very close to the Santonian-Campanian boundary, with the onset of the Gubbio A− Zone.

3. The Campanian-Maastrichtian boundary falls near the top of the relatively long zone Gubbio B+.

4. All Late Cretaceous reversed polarity zones except Gubbio A− are of Maastrichtian age.

5. In contrast with the Campanian, the Paleocene and Maastrichtian are generally

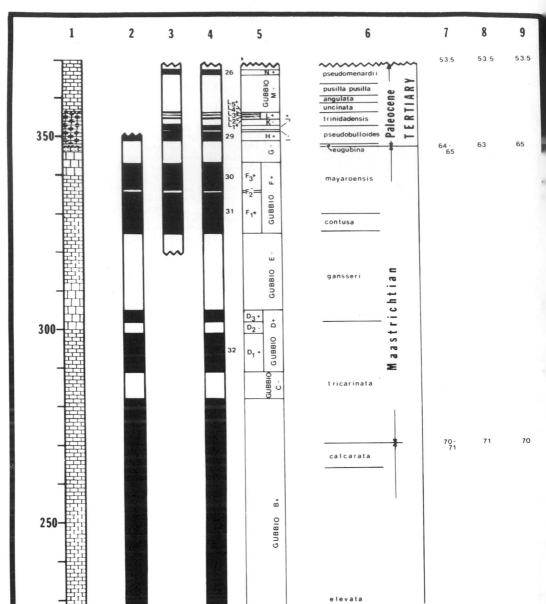

Figure 2. Results of stratigraphic studies of the Gubbio section. Column 1: measured section in metres with lithology (Arthur a Fischer, paper I). Column 2: magnetic results of Lowrie and Alvarez (paper III); black is normal polarity, white is reversed. Column magnetic results of Roggenthen and Napoleone (paper IV). Column 4: combined magnetic results (numbers to the right of the colu show correlations with the marine magnetic anomaly sequence). Column 5: designation of polarity zones; "+" indicates normal polar "−" indicates reversed polarity; zones Gubbio D+, F+, and L+ have been divided into subzones. Column 6: planktic foraminiferal zo from Premoli Silva (paper II), with chronostratigraphic units on the right. Columns 7 through 9: absolute age calibration accordin three proposed time scales. 7: Obradovich and Cobban (1975). 8: Bukry and Douglas (unpub. data; 1974, personal commun.). 9: Hinte (1972). Paleocene-Eocene boundary age from Berggren (1972) in all cases. Exact metre levels of boundaries between magn polarity zones and biostratigraphic zones in the measured section are reported in the preceding papers.

212

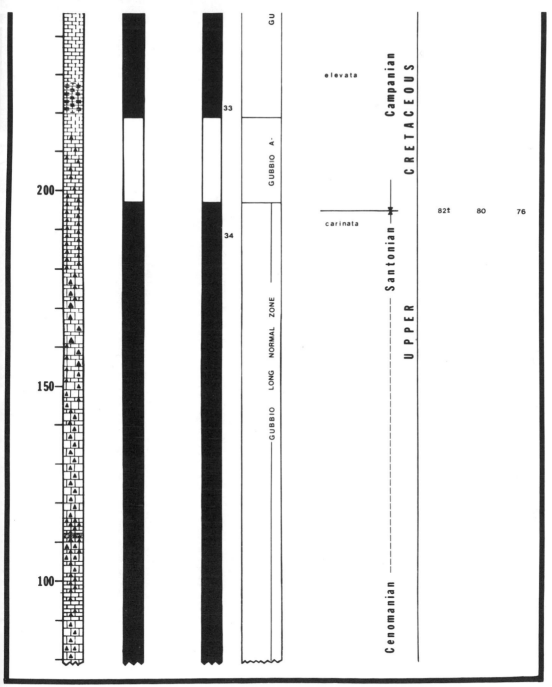

Figure 2. (*Continued*).

213

characterized by polarity zones that represent short time intervals.

6. The Cretaceous-Tertiary boundary falls nearly at the top of polarity zone Gubbio G−.

7. The section studied does not reach the Paleocene-Eocene boundary, which must occur slightly above the Gubbio N+ zone.

CALIBRATION BY ABSOLUTE AGE

Calibration of geomagnetic polarity history in terms of years remains somewhat uncertain because of inaccuracies in radiometric age determinations and in correlations between radiometrically dated rocks and those rocks that contain fossils. The present state of knowledge is illustrated by the dates assigned to various biostratigraphic boundaries in three contemporary time scales, as shown in Figure 2 (cols. 7 through 9).

ACKNOWLEDGMENTS

Individual acknowledgments are given in the preceding sections. In addition, we thank Roger L. Larson, Richard A. Schweickert, Hollis D. Hedberg, and Norman D. Watkins for reviewing this manuscript. Contribution no. 2379, Lamont-Doherty Geological Observatory of Columbia University. Contribution no. 139, Geophysical Institute, ETH — Zürich.

Financial support was provided by National Science Foundation Grants DES 73-06663, DES 74-22214, and EAR 76-03272; by Princeton University's Tuttle Bequest; by the Consiglio Nazionale delle Ricerche (Comitate per le Scienze Geologiche e Minerarie, Milano); and by a NATO-CNR Fellowship.

COMBINED REFERENCES CITED

Adams, P. R., and Frerichs, W. E., 1973, Foraminifera from the type area of the Hilliard Formation, in Greater Green River Basin Symposium Guidebook: Casper, Wyo., Wyoming Geol. Assoc., p. 193–199.

Alvarez, W., and Lowrie, W., 1974, Paleomagnetismo della Scaglia Rossa umbra e rotazione della penisola italiana: Soc. Geol. Italiana Boll., v. 93, p. 883–891.

American Commission on Stratigraphic Nomenclature, 1961, Code of stratigraphic nomenclature: Am. Assoc. Petroleum Geologists Bull., v. 45, p. 645–665.

Arthur, M. A., and Roggenthen, W. M., 1976, Primary bedding rhythms in pelagic calcareous sediments [abs.]: Soc. Econ. Paleontologists and Mineralogists and Am. Assoc. Petroleum Geologists, Ann. Mtg., New Orleans.

Barnaba, P. F., 1958, Geologia dei Monti di Gubbio: Soc. Geol. Italiana Boll., v. 77, p. 39–70.

Berggren, W., 1972, A Cenozoic time-scale — Some implications for regional geology and paleobiogeography: Lethaia, v. 5, p. 195–215.

Bolli, H. M., 1966, Zonation of Cretaceous to Pliocene marine sediments based on planktonic foraminifera: Assoc. Venozolana Min. Pet., v. 9, p. 3.

Bortolotti, V., Passerini, P., Sagri, M., and Sestini, G., 1970, The miogeosynclinal sequences, in Sestini, G., ed., Development of the northern Apennines geosyncline: Sed. Geology, v. 4, p. 341–444.

Channell, J.E.T., and Tarling, D. H., 1975, Palaeomagnetism and the rotation of Italy: Earth and Planetary Sci. Letters, v. 25, p. 177–188.

Cobban, W. A., and Reeside, J. B., Jr., 1952, Correlation of the Cretaceous formations of the western interior of the United States: Geol. Soc. America Bull., v. 63, p. 1011–1044.

Cox, A., Doell, R. R., and Dalrymple, G. B., 1963a, Geomagnetic polarity epochs and Pleistocene geochronometry: Nature, v. 198, p. 1049–1051.

——1963b, Geomagnetic polarity epochs: Sierra Nevada II: Science, v. 142, p. 382–385.

Doell, R. R., and Dalrymple, G. B., 1966, Geomagnetic polarity epochs: A new polarity event and the age of the Brunhes-Matuyama boundary: Science, v. 152, p. 1060–1061.

Fischer, A. G., 1969, Geological time-distance rates: The Bubnoff Unit: Geol. Soc. America Bull., v. 80, p. 549–552.

Fischer, A. G., and Arthur, M. A., 1977, Secular variation in the pelagic realm, in Cook, H. E., and Enos, P., eds., Deep marine carbonate environments: Soc. Econ. Paleontologists and Mineralogists Spec. Pub. No. 25, (in press).

Frerichs, W. E., and Adams, P. R., 1973, Correlation of the Hilliard Formation with the Niobrara Formation, in Greater Green River Basin Symposium Guidebook: Casper, Wyo., Wyoming Geol. Assoc., p. 187–192.

Goree, W. S., and Fuller, M. D., 1977, Magnetometers using RF driven, weak line (SQUID) sensors and their applications in rock magnetism and paleomagnetism: Rev. Geophys. and Space Physics (in press).

Guinasso, N. L., Jr., and Schink, D. R., 1975, Quantitative estimates of biological mixing rates in abyssal sediments: Jour. Geophys. Research, v. 80, no. 21, p. 3032–3043.

Hays, J. D., and Pitman, W. C., III, 1973, Lithospheric plate motion, sea level changes and climatic and ecological consequences: Nature, v. 246, p. 18–22.

Hedberg, H. D., ed., 1976, International stratigraphic guide: New York, John Wiley & Sons, Inc., 200 p.

Heirtzler, J. R., Dickson, G. O., Herron, E. M., Pitman, W. C., III, and Le Pichon, X., 1968, Marine magnetic anomalies, geomagnetic field reversals and motions of the ocean floor and continents: Jour. Geophys. Research, v. 73, p. 2119–2136.

Helsley, C. E., and Steiner, M. B., 1969, Evidence for long intervals of normal polarity during

the Cretaceous period: Earth and Planetary Sci. Letters, v. 5, p. 325–332.

Hinte, J. E. van, 1972, The Cretaceous time scale and planktonic foraminiferal zones: Koninkl. Nederlandse Akad. Wetensch. Proc., ser. B, v. 75 (1), p. 1–8.

——1976, A Cretaceous time scale: Am. Assoc. Petroleum Geologists Bull., v. 60, p. 498–516.

Keating, B., Helsley, C. E., and Pessagno, E. A., Jr., 1975, Late Cretaceous reversal sequence: Geology, v. 3, p. 73–76.

Kent, D. V., 1973, Post-depositional remanent magnetization in deep sea sediment: Nature, v. 246, p. 32–34.

Klootwijk, C. T., and Van den Berg, J., 1975, The rotation of Italy: Preliminary palaeomagnetic data from the Umbrian sequence, northern Apennines: Earth and Planetary Sci. Letters, v. 25, p. 263–273.

Ladd, J. W., 1974, South Atlantic sea-floor spreading and Caribbean tectonics [Ph.D thesis]: New York, Columbia Univ., 251 p.

Larson, R. L., 1974, An updated time scale of magnetic reversals for the late Mesozoic [abs.]: EOS (Am. Geophys. Union Trans.), v. 55, p. 236.

Larson, R. L., and Pitman, W. C., III, 1972, World-wide correlation of Mesozoic magnetic anomalies, and its implication: Geol. Soc. America Bull., v. 83, p. 3645–3662.

Løvlie, R., 1974, Post-depositional remanent magnetization in a re-deposited deep sea sediment: Earth and Planetary Sci. Letters, v. 21, p. 315–320.

Løvlie, R., Lowrie, W., and Jacobs, M., 1971, Magnetic properties and mineralogy of four deep sea cores: Earth and Planetary Sci. Letters, v. 15, p. 157–168.

Lowrie, W., and Alvarez, W., 1974, Rotation of the Italian Peninsula: Nature, v. 251, p. 285–288.

——1975, Paleomagnetic evidence for rotation of the Italian Peninsula: Jour. Geophys. Research, v. 80, p. 1579–1592.

Luterbacher, H. P., 1964, Studies on some Globoratalia from the Paleocene and lower Eocene of the central Apennines: Ecloga Geol. Helvetiae, v. 57, p. 631–730.

Luterbacher, H. P., and Premoli Silva, I., 1962, Note préliminaire sur une revision du profil de Gubbio, Italie: Riv. Italiana Paleontologia e Stratigrafia, v. 68, p. 253–288.

——1964, Biostratigrafia del limite Cretaceo-Terziario nell' Appennino centrale: Riv. Italiana Paleontologia e Stratigrafia, v. 70, p. 67–128.

Mauritsch, H. J., and Turner, P., 1975, The identification of magnetite in limestones using the low-temperature transition: Earth and Planetary Sci. Letters, v. 24, p. 414–418.

McDougall, I., and Chamalaun, F. H., 1966, Geomagnetic polarity scale of time: Nature, v. 212, p. 1415–1418.

McDougall, I., and Tarling, D. H., 1964, Dating geomagnetic polarity zones: Nature, v. 202, p. 171–172.

Mohler, H. P., 1966, Stratigraphische Untersuchungen in den Giswiler Klippen (Préalpes Médianes) und ihrer helvetisch-ultrahelvetischen Unterlage: Beitr. Geologie

Karte Schweiz, no. 129, XV-84 p.

Molyneux, L., 1972, A complete results magnetometer for measuring the remanent magnetization of rocks: Geophys. Jour., v. 24, p. 429–435.

Monechi, S., and Pirini Radrizzani, C., 1975, Nannoplankton from Scaglia Umbra Formation (Gubbio) at Cretaceous-Tertiary boundary: Riv. Italiana Paleontologia e Stratigrafia, v. 81, p. 45–87.

Obradovich, J. D., and Cobban, W. A., 1975, A time-scale for the Late Cretaceous of the Western Interior of North America, in Geol Assoc. Canada Spec. Paper no. 13, p. 31–54.

Opdyke, N. D., 1972, Paleomagnetism of deep-sea cores: Rev. Geophysics and Space Physics, v. 10, p. 213–249.

Opdyke, N. D., and Foster, J. H., 1970, Paleomagnetism of cores from the North Pacific: Geol. Soc. America Mem. 126, p. 83–119.

Opdyke, N. D., Glass, B. P., Hays, J. D., and Foster, J., 1966, Paleomagnetic study of Antarctic deep-sea cores: Science, v. 154, p. 349–357.

Opdyke, N. D., Burckle, L. H., and Todd, A., 1974, The extension of the magnetic time scale in sediments of the central Pacific Ocean: Earth and Planetary Sci. Letters, v. 22, p. 300–306.

Pessagno, E. A., 1967, Upper Cretaceous planktonic foraminifera from the western Gulf Coastal Plain: Paleontographica Americana, v. 5, p. 245–445.

Pitman, W. C., III, Larson, R. L., and Herron, E. M., 1974, Magnetic lineations of the oceans: Geol. Soc. America Map and Chart Ser. MC-6.

Porthault, B., 1974, Le Crétacé supérieur de la "Fosse Vocontienne" et des régions limitrophes (France Sud-Est) [thèse, 257]: Lyon, Univ. Claude Bernard, 342 p.

Postuma, J. A., 1971, Manual of planktonic foraminifera: Amsterdam, Elsevier Publ. Co., 397 p.

Premoli Silva, I., and Bolli, H. M., 1973, Late Cretaceous to Eocene planktonic foraminifera and stratigraphy of Leg 15 sites in the Caribbean Sea, in Initial reports of Deep Sea Drilling Project, Vol. 15: Washington

D.C., U.S. Govt. Printing Office, p. 499–547.

Premoli Silva, I., Napoleone, G., and Fischer, A. G., 1974, Risultati preliminari sulla stratigrafia paleomagnetica della Scaglia cretaceo-paleocenica della sezione di Gubbio (Appennino centrale): Soc. Geol. Italiana Boll., v. 93, p. 647–659.

Raff, A. D., 1966, Boundaries of an area of very long magnetic anomalies in the northeast Pacific: Jour. Geophys. Research, v. 71, p. 2631–2636.

Renz, O., 1936, Stratigraphische und mikropaleontologische Untersuchung der Scaglia (Obere Kreide-Tertiar) im zentralen Apennin: Ecologae Geol. Helvetiae, v. 29, p. 1–149.

——1951, Ricerche stratigrafiche e micropaleontologiche sulla Scaglia (Cretaceo superiore-Terziario) dell'Appennino centrale): Mem. Descr. Carta Geologica d'Italia, v. 29, p. 1–173 [Italian trans. of Renz, 1936].

Scientific Staff, DSDP Leg 37, 1974, Leg 37 — The volcanic layer: Geotimes, v. 19, p. 16–18.

——DSDP Leg 39, 1975, Leg 39 examines facies changes in South Atlantic: Geotimes, v. 20, p. 26–28.

Sclater, J. G., and Fisher, R. L., 1974, Evolution of the east-central Indian Ocean, with emphasis on the tectonic setting of the Ninetyeast Ridge: Geol. Soc. America Bull., v. 85, p. 683–702.

Sclater, J. G., Jarrard, R. D., McGowran, B., and Gartner, S., 1974, Comparison of the magnetic and biostratigraphic time scales since the Late Cretaceous, in Initial reports of the Deep Sea Drilling Project, Vol. 22: Washington, D.C., U.S. Govt. Printing Office, p. 381–386.

Shive, P. N., and Frerichs, W. E., 1974, Paleomagnetism of the Niobrara Formation in Wyoming, Colorado, and Kansas: Jour. Geophys. Research, v. 79, p. 3001–3008.

Simpson, E.S.W., Schlich, R., and others, 1974, Initial reports of the Deep Sea Drilling Project, Vol. 25: Washington, D.C., U.S. Govt. Printing Office, 884 p.

Subcommission on Magnetostratigraphic Nomenclature, 1973, Magnetic polarity

time scale: Geotimes, v. 18, p. 21–22.

Symons, D.T.A., and Stupavsky, M., 1974, A rational paleomagnetic stability index: Jour. Geophys. Research, v. 79, p. 1718–1720.

Theyer, F., and Hammond, S. R., 1974, Paleomagnetic polarity sequence and radiolarian zones, Brunhes to Polarity Epoch 20: Earth and Planetary Sci. Letters, v. 22, p. 307–319.

Thierstein, H. R., 1977, Biostratigraphy of marine Mesozoic sediments by calcareous nannoplankton: 3rd Planktonic Conference Proc. (in press).

Van den Berg, J., Klootwijk, C. T., and Wonders, T., 1975, Implications for the rotational movement of Italy from current palaeomagnetic research in the Umbrian Sequence, northern Apennines, in Progress in Geodynamics: Amsterdam, Royal Netherlands Acad. Arts and Sciences, p. 165–175.

Van der Voo, R., and French, R. B., 1974, Apparent polar wandering for the Atlantic-bordering continents: Late Carboniferous to Eocene: Earth Sci. Rev., v. 10, p. 99–119.

Vine, F. J., and Matthews, D. H., 1963, Magnetic anomalies over oceanic ridges: Nature, v. 199, p. 947–949.

Watkins, N. D., 1975, Correlating stratigraphic zones and magnetic polarities: Geotimes, v. 20, p. 26–27.

Watkins, N. D., and Baksi, A. K., 1974, Magnetostratigraphy and oroclinal folding of the Columbia River, Steens and Owyhee Basalts in Oregon, Washington and Idaho: Am. Jour. Sci., v. 274, p. 148–189.

Watkins, N. D., McDougall, I., Kristjansson, L., Saemundsson, K., Johannesson, H., and Walker, G.P.L., 1975, Geomagnetic time scale: Detection and dating of Epoch 5, Epoch 6, and "Anomaly 5" in Iceland [abs.]: EOS (Am. Geophys. Union Trans.), v. 56, p. 974–975.

Part IV

MAGNETOSTRATIGRAPHY OF NONMARINE SEDIMENTS: REDBEDS AND MAMMALS

Editor's Comments
on Papers 18 and 19

18   HELSLEY and STEINER
Paleomagnetism of the Lower Triassic Moenkopi Formation

19   OPDYKE et al.
The Paleomagnetism and Magnetic Polarity Stratigraphy of the Mammal-Bearing Section of Anza Borrego State Park, California

Cenozoic chronostratigraphy was founded on dated sequences of fossils; largely mollusca and foraminifera in the case of marine sequences and mammalian fossils for terrestrial sequences. The correlation of terrestrial and marine strata has always been very difficult because of their major environmental differences, which allow almost no overlap in the character of fossil assemblages. Furthermore, intercalibration is made even more difficult because of the scarcity of interdigitation of terrestrial with marine sediments that have deep enough facies to contain the planktonic microfossils so important in long-distance correlation. In addition, terrestrial sequences often contain poor fossil assemblages allowing only general biostratigraphic subdivision and related low resolution in correlation.

Until recently stratigraphic correlation in terrestrial sequences was approaching a level of ultimate resolution and, like the Quaternary marine record in the 1960s, was in need of a breakthrough to stimulate the field and to significantly enhance more detailed subdivision and correlation. The conspicuous success of magnetostratigraphy of deep-sea sediments assisted in prompting stratigraphers to use the approach on fresh-water and subaerial-formed sedimentary deposits. However, as is often the case in science, the first studies of magnetostratigraphy of terrestrial sediments had been performed by workers even long before the approach became commonly used in the study of marine cores. Russian workers had examined Pliocene-Quaternary terrestrial sequences in western Turkmenia (Khramov, 1958) well before coherent polarity stratigraphies were developed in North America

218

beginning in the early 1960s. Other work on terrestrial sequences followed soon after the development of the Late Cenozoic polarity time scale in the middle 1960s by Cox, McDougall, and their associates. These included studies that dealt with Paleozoic magnetostratigraphy by Irving (1966); Mesozoic magnetostratigraphy of red beds by Picard (1964), Helsley (1969), Burek (1970), and Helsley and Steiner (1969, 1974); and a wide diversity of other Late Cenozoic sedimentary facies including loess (Bucha et al., 1969), fresh-water clastics (Van Montfrans and Hospers, 1969; Van Montfrans, 1971) and more recently of mammal-rich sequences (Johnson et al., 1975; Opdyke et al., 1977 and Lindsay et al., 1978). As summarized by Johnson et al. (1975), if mammal-bearing terrestrial deposits could provide genuine paleomagentic polarity records, then on an intercontinental scale, absolute time correlations would be possible with the marine record. Such a development would then make possible the combination of all the relevant marine and continental data into a comprehensive chronology.

Because of space limitations here, summaries are included of only two general groups of studies on terrestrial magnetostratigraphy: The largely nonfossiliferous red-bed sequences of Mesozoic age and the rich mammalian sequences of Late Cenozoic age.

Early work on the Triassic sedimentary sequences of the western United States by Runcorn (1956) and later workers demonstrated the presence of both normal and reversed polarities, although no attempt was made to describe the magnetostratigraphy, probably because the polarity time scale had not yet been produced. The development of a polarity record for Mesozoic strata essentially began in the 1960s after the early versions of late Cenozoic polarity time scale began to appear. One of the earliest studies was that of Picard (1964) on the Chugwater Formation of Wyoming in which he defined seven polarity zones and used them for correlation. This work was followed by more intensive studies by McMahon and Strangeway (1967, 1968a, b) and Helsley (1969).

Using high-density sampling, Helsley (1969) examined the paleomagnetism of the Early Triassic, largely unfossiliferous, Moenkopi Formation of the Colorado Plateau and recognized clear evidence for mixed polarity with at least 11 reversals. These were tentatively used to carry out West Coast North American correlations including the sections previously examined by Picard (1964). The primary objective of this initial study was not so much to develop the approach as a stratigraphic tool but to provide a framework for interpretation of the magnetic anomaly patterns observed at sea. A further important aspect of the work was to discriminate remanent from secondary magnetization resulting from subaerial weathering. Related to the latter problem, Helsley found that magnetization is relatively stable in these rocks and that thermal demagnetization is effective in removing any secondary component that may have been present.

The early work of Helsley (1969) was later extended to younger parts of the Moenkopi Formation by Helsley and Steiner (Paper 18). They confirmed the mixed polarity character of the Early Triassic and perhaps the earliest part of the Middle Triassic. While Helsley and colleagues were carrying out their Mesozoic work in North America, similar studies were being independently carried out by Burek (1970) on the Triassic Upper Buntsandstein of southwest Germany, which consists of clastic sediments with discontinuities, some of which are associated with soil horizons. Clearly, the next logical step in this work was the correlation between the North American and European sequences. Unfortunately this has not yet been possible using magnetostratigraphy because in the absence of fossils, there is no way of separating magnetic polarity changes from each other. Thus detailed intercontinental correlation of Early Mesozoic sedimentary sequences awaits further stratigraphic developments.

Both Burek (1964; 1970) and Helsley and Steiner (Paper 18) have emphasized the importance of differences between the Early Triassic mixed polarity record and long periods of constant polarities in both underlying Permian and overlying later Mesozoic (Late Triassic to Cretaceous) sediments. The Permian largely exhibits reversed polarity (Irving and Parry, 1963; Irving, 1964). It remains unclear how close to the Paleozoic-Mesozoic boundary the character of the polarity change takes place, but the potential for application of magnetostratigraphy in the definition of the Paleozoic-Mesozoic boundary was recognized by Burek (1970). Following the Early Triassic polarity reversal sequence, much of the remainder of the Mesozoic before the latest Cretaceous largely exhibits constant polarity separated by shorter intervals of more frequent polarity change. Specifically, the polarity history of the Late Triassic and Jurassic is uncertain but appears to be constant based on the marine magnetic anomaly (sea-floor spreading) pattern in the Jurassic, which is represented by a Quiet Zone (Burek, 1970; Alvarez et al., Paper 17). This was in turn followed by a shorter sequence of reversals (M-sequence) during the Late Jurassic to Early Cretaceous and then by dominantly normal polarity through almost all of the Cretaceous (Helsley and Steiner, 1969), reflected in the Cretaceous Quiet Zone of the oceans. A fundamental dif-

ference in the polarity history occurred during the latest Cretaceous through the entire Cenozoic, which is represented by a
rapidly oscillating reversal sequence.

One discovery of possible substantial importance that has
emerged from these various studies, which is clearly summarized
by Helsley and Steiner (Paper 18), is that major geologic eras, defined on the basis of major changes in the fossil record, may end
with a low frequency of reversals and begin with a high reversal
frequency. They speculated that the major changes in life forms
that take place at these times may have resulted from evolutionary
adaptation during the long periods of constant polarity in which the
magnetic field is used in the regulation of some necessary or vital
function; and then this changes to the disadvantage of some forms.
Regardless of the mechanism involved, it is clear that major crises
in the earth's biota at the Paleozoic-Mesozoic boundary and at the
Mesozoic-Cenozoic boundary are associated with major changes
in the tempo of polarity changes of the earth's magnetic field.

A second highly significant development related to magnetostratigraphy of terrestrial sediments has been concerned with the
dating of Late Cenozoic mammalian sequences, largely in North
America. However, the first magnetostratigraphic studies on rocks
associated with mammalian fossils were not carried out on sediments but on volcanic rocks (welded tuffs, ignimbrites, and basalt
lava flows) in east Africa by Gromme and Hay (1967; 1971). Paleomagnetism and radiometric dating were carried out on these volcanic sequences in the well-known Olduvai Gorge area. The volcanic rocks are interbedded with sediments containing rich mammalian assemblages including primitive primates considered to be
directly ancestral to man. The ages derived are thus important
anthropologically. Resulting radiometric dates established an age
of this sequence as Middle Matuyama and the Olduvai Normal
Event was recognized, providing an age of 1.8 m.y. B.P. for part
of the sequence.

The study of the magnetostratigraphy of late Cenozoic terrestrial sections containing important mammal fossils began in the
early 1970s by a group that included Dr. Neil Opdyke of the Lamont-
Doherty Geological Observatory and two vertebrate fossil experts, Dr. E. H. Lindsay of the Universtiy of Arizona and N. M.
Johnson of Dartmouth College, New Hampshire. The primary objective of these studies was to correlate the already well-known
Pliocene-Pleistocene mammalian sequence in the western United
States with the late Cenozoic polarity record. If successful, such
an investigation would enable the numerous mammalian events

to be dated; the position of the Pliocene-Pleistocene boundary as applied in marine sequences to be intercalibrated with the mammalian sequence; and the timing to be established of significant migrations of mammals into North America from Asia via the Bering land bridge and South America via the Panama Isthmus. This in turn would provide much needed information on the history of intercontinental connections among Asia, North America, and South America.

To investigate these problems, the group began their work on the Late Cenozoic mammal-rich bearing sequences in the San Pedro Valley, Arizona (Johnson et al., 1975), and the Anza-Borrego State Park, southern California (Opdyke et al., Paper 19). As with deep-sea cores and marine sequences on land, it was discovered that not all terrestrial sections are suitable for the establishment of a magnetostratigraphy. However, these workers found by trial and error that those sections most useful for establishing paleomagnetic correlations need to have sufficient time represented; to be dominated by fine sediments; be strongly magnetized; be largely free of secondary mineralization or weathering; and devoid of a history of lightning strikes (Lindsay et al., 1976).

The first study at San Pedro Valley, Arizona (Johnson et al., 1975), identified a sequence from the upper Gilbert Epoch through to the Brunhes Epoch. Four mammal datum planes were paleomagnetically dated, including the first appearance of the hare *Lepus* at 1.9 m.y. marking the appearance of a definitive Irvingtonian mammal assemblage. The significance of this assemblage is that it contains Eurasian elements reflecting migration via the Bering Sea land bridge, which developed during low stands of sea level. This biogeographic event was thus dated by Johnson et al. (1975) as occurring shortly before the Olduvai Event, surprisingly close to the Pliocene-Pleistocene boundary as defined in the marine sedimentary record.

A similar study was carried out in the Anza-Borrego State Park (Opdyke et al., Paper 19) on a sequence extending from the Late Gilbert Epoch (below Cochiti Event) to the Matuyama Epoch. The polarity record enabled eight faunal events to be dated, including appearances and/or extinctions of a variety of mammals, including horses, bovids such as *Euceratherium*, and carnivores including the saber-toothed cat *Smilodon*. The appearance of *Smilodon* and *Euceratherium* are significant because these forms, like the hare *Lepus*, are known only from the Irvingtonian and the younger Rancholabrean faunas in North America and not before. Their appearance is almost certainly linked with the formation of the Bering land bridge, which is thus inferred to have been exposed as a migration route at about 1.9 m.y. ago. The initial appearance of certain other well-known Eurasian (old-world) forms such as *Mammuthus* is not yet well dated and requires additional paleomagnetic stratigraphy.

A natural extension of the studies by this group (Lindsay et al. 1967), was into the type areas of the Late Cenozoic mammalian stages to establish their paleomagnetic ages and relations with the San Pedro and Anza-Borrego sections. It was discovered, however, that sediments in the type sequences of the Texas Panhandle area are weakly magnetized and only partially stable in alternating fields. Second, the sequences are too thin to represent much time for clear recognition of the magnetic polarity zones. Third, the type sections are marked by a large secondary magnetization. Nevertheless, the establishment of a magnetostratigraphy in several of the sections enabled tentative placement within the polarity time scale and the tentative dating of mammalian events. The total range in ages of the assemblages is from Epoch 5 to the Brunhes Epoch. Most of the mammalian changes are considered to represent biogeographic changes, the earlier ones resulting from migration from South America (including *Glyptotherium*) and the later ones, of migration from Eurasia. It seems significant that the appearance of two South American immigrants (*Paramylodon* and *Glyptotherium*) have been dated by Lindsay et al. (1976) at about 2.6 m.y. associated with the Late Blancan Fauna. Marine evidence suggests that the Panama Isthmus had already developed 0.5 to 1.0 m.y. before this (Keigwin, 1978) suggesting several possibilities: that the age suggested by Lindsay et al. (1976) for this migration is in the early Gauss Epoch rather than the Late Gauss; that there was a delay in northward migration of the South American forms well after the Panama Isthmus developed enough to prevent oceanic migration of plankton but not enough for land-mammal migrations; that other environmental factors delayed northward migration of these forms even after the intercontinental connection was established; or that the dating of the development of the Panamanian land bridge is still in error. Examination of the original paleomagnetic logs of Lindsay et al. (1976) suggests that the age assignments may be incorrect in the thin sequence at Cita Canyon, a particularly important section for the development of these ideas.

The dating of late Cenozoic terrestrial sediments containing mammalian fossil remains was recently extended to Pakistan (Opdyke et al., 1977; Keller et al., 1977). A magnetostratigraphy has been established on fluvial molasse derived from the Himalayas.

The derivation of sediments from various parts of the Himalayas as been adjusted at certain times during the Late Cenozoic. These changes have been dated by the magnetostratigraphy, as have several of the mammalian events. The earliest record of the one-toed horse *Equus* in this region is dated at 2.5 m.y. while the three-toed horse *Hipparion* persisted until 1.8 m.y. ago.

The strength of paleomagnetic stratigraphy, as applied to problems of correlation, has been particularly well illustrated in a recent study involving latest Cretaceous and Early Tertiary terrestrial and marine sediments. The study involves the dating and correlation of the dramatic biotic change that has long been known to have occurred at the Cretaceous/Tertiary boundary. As discussed in Part II, this boundary, as recognized in the marine sequence at Gubbio, Italy, has been placed high in the reversed magnetozone between anomaly 29 and 30 (Alvarez et al., Paper 17). The paleomagnetic age of the Cretaceous/Tertiary boundary, as defined by vertebrate fossils, has now been determined in nonmarine sedimentary deposits of the western interior of North America by Butler et al. (1977) and Lindsay et al. (1978). These workers have found that the extinction of the dinosaurs in the San Juan Basin, New Mexico, occurred slightly later in the polarity time scale but, nevertheless, close to the marine faunal change in the northern Italian section. The extinction of the dinosaurs took place within magnetic anomaly 29, and critical early Paleocene mammals first occur within magnetic anomaly 28. The age difference between the Cretaceous/Tertiary boundary as defined in marine and terrestrial sequences was estimated by Butler et al. (1977) to be between 1.5 and 0.5 m.y. Considering the vast amount of time that has lapsed since this event, the degree of synchroneity of the terrestrial and marine faunal changes is remarkable; on the other hand, the diachronism of the event, even though slight, seems to be of critical importance in the development of theories to explain the faunal extinctions.

REFERENCES

Bucha, V., J. Horacek, A. Koci, and J. Kukla. 1969. Die paleomagnetische messungen in dem Lossen. In *Periglanzialzone-Loss und Paleolithikum in der Tschechoslowakei*, edited by J. Demek. Brno: 8th INQUA Congress, pp. 123–131.

Burek, P. J. 1964. Korrelation revers magnetisierter Gesteinfolgen in Oberen Bundstandstein SW-Deutschlands. *Geol. Jahrb.* **84**:591–616.

Burek, P. J. 1970. Magnetic reversals: Their application to stratigraphic problems. *Am. Assoc. Petroleum Geologists Bull.* **54**:1120–1139.

Butler, R. F., E. H. Lindsay, L. L. Jacobs, and N. M. Johnson. 1977. Magnetostratigraphy of the Cretaceous-Tertiary boundary in the San Juan Basin, New Mexico. *Nature* **267**:318–323.

Grommé, C. S., and R. L. Hay. 1967. Geomagnetic polarity epochs: New data from Olduvai Gorge, Tanganyika. *Earth and Planetary Sci. Letters* **2**:111–115.

Grommé, C. S., and R. L. Hay. 1971. Geomagnetic polarity epochs: Age and duration of the Olduvai normal polarity event. *Earth and Planetary Sci. Letters* **10**:179–185.

Helsley, C. E. 1969. Magnetic reversal stratigraphy of the Lower Triassic Moenkopi Formation of Western Colorado. *Geol. Soc. America Bull.* **80**:2431–2450.

Helsley, C. E., and M. B. Steiner. 1969. Evidence for long intervals of normal polarity during the Cretaceous period. *Earth and Planetary Sci. Letters* **5**:325–332.

Irving, E. 1964. *Paleomagnetism.* New York: John Wiley & Sons, 399 pp.

Irving, E. 1966. Paleomagnetism of some carboniferous rocks from New South Wales and its relation to geological events. *Jour. Geophys. Research* **71**:6025–6051.

Irving, E., and L. G. Parry. 1963. The magnetism of some Permian rocks from New South Wales. *Royal Astron. Soc. Geophys. Jour.* **7**:395–411.

Johnson, N. M., N. D. Opdyke, and E. H. Lindsay. 1975. Magnetic polarity stratigraphy of Pliocene-Pleistocene terrestrial deposits and vertebrate faunas, San Pedro Valley, Arizona. *Geol. Soc. America Bull.* **86**:5–12.

Keigwin, L. D. 1978. Pliocene closing of the Isthmus of Panama, based on biostratigraphic evidence from nearby Pacific Ocean and Caribbean Sea Cores. *Geology* **6**:630–634.

Keller, H. M., R. A. K. Tahirkheli, M. A. Mirza, G. D. Johnson, N. M. Johnson, and N. D. Opdyke. 1977. Magnetic polarity stratigraphy of the Upper Siwalik Deposits, Pabbi Hills, Pakistan. *Earth and Planetary Sci. Letters* **36**:187–201.

Khramov, A. N. 1958. *Paleomagnetism and Stratigraphic Correlation,* A. J. Lojkline, trans. Canberra: Australian National University, 204 pp.

Lindsay, E. H., L. L. Jacobs, and R. F. Butler. 1978. Biostratigraphy and magnetostratigraphy of Paleocene terrestrial deposits, San Juan Basin, New Mexico. *Geology* **6**:425–429.

Lindsay, E. H., N. M. Johnson, and N. D. Opdyke. 1976. Preliminary correlation of North American land mammal ages and geomagnetic chronology. In *Studies on Cenozoic Paleontology and Stratigraphy,* edited by G. R. Smith and N. E. Friedlands. Ann Arbor: University of Michigan Press, pp. 111–119.

McMahon, B. E., and D. W. Strangway. 1967. Kiaman magnetic interval in the Western United States. *Science* **155**:1012–1013.

McMahon, B. E., and D. W. Strangway. 1968a. Investigation of Kiaman magnetic division of Colorado red beds. *Royal Astron. Soc. Geophys. Jour.* **15**:265–285.

McMahon, B. E., and D. W. Strangway. 1968b. Stratigraphic implication of paleomagnetic data from Upper Paleozoic-Lower Triassic redbeds of Colorado. *Geol. Soc. America Bull.* **79**:417–428.

Montfrans, H. M. van. 1971. *Paleomagnetic Dating in the North Sea Basin.* Rotterdam: Princo N.V., 113 pp.

Montfrans, H. M. van, and J. Hospers. 1969. A preliminary report on the stratigraphical position of the Matuyama-Brunhes Geomagnetic field reversal in the Quaternary sediments of the Netherlands. *Geologie en Mijnbouw* **48**:565–572.

Opdyke, N. D., N. Johnson, G. Johnson, E. Lindsay, and R. A. K. Tahirkheli. 1977. Paleomagnetism of the Upper Siwalik sediments of Pakistan (Abstract). *EOS* **58**:379.

Picard, M. D. 1964. Paleomagnetic correlation of units within Chugwater (Triassic) Formation, West-Central Wyoming. *Am. Assoc. Petroleum Geologists Bull.* **48**:269–291.

Runcorn, S. K. 1956. Paleomagnetic survey in Arizona and Utah: Preliminary results. *Geol. Soc. America Bull.* **67**:301–316.

ADDITIONAL READINGS

Baag, C.-G., and C. E. Helsley. 1974. Evidence for penecontemporaneous magnetization of the Moenkopi Formation. *Jour. Geophys. Research* **79**:3308–3320.

Berggren, W. A., D. P. McKenzie, J. G. Sclater, and J. E. van Hinte. 1975. World-wide correlation of Mesozoic magnetic anomalies and its implications: Discussion and reply. *Geol. Soc. America Bull.* **86**:267–272.

Brock, A., and R. L. Hay. 1976. The Olduvai event at Olduvai Gorge. *Earth and Planetary Sci. Letters* **29**:126–130.

Brock, A., and G. L. Isaac. 1974. Paleomagnetic stratigraphy and chronology of hominid-bearing sediments east of Lake Rudolf, Kenya. *Nature* **247**:344–348.

Bucha, V. 1970. Geomagnetic reversals in Quaternary revealed from a palaeomagnetic investigation of sedimentary rocks. *Jour. Geomagnetism and Geoelectricity* **22**:253–271.

Hillhouse, J., and A. Cox. 1976. Brunhes-Matuyama polarity transition. *Earth and Planetary Sci. Letters* **29**:51–64.

Irving, E., and R. W. Couillard. 1973. Cretaceous normal polarity interval. *Nature Phys. Sci.* **244**:10–11.

Irving, E., and G. Pullaiah. 1976. Reversals of the geomagnetic field, magnetostratigraphy and relative magnitude of paleosecular variation in the Phanerozoic. *Earth-Sci. Rev.* **12**:35–64.

Johnson, A. H., and A. E. M. Nairn. 1972. Jurassic palaeomagnetism. *Nature* **240**:551–552.

Johnson, A. H., A. E. M. Nairn, and D. N. Peterson. 1972. Mesozoic reversal stratigraphy. *Nature Phys. Sci.* **237**:9–10

Khramov, A. N. 1957. Paleomagnetism—The basis of a new method of correlation and subdivision of sedimentary strata. *Acad. Sci. USSR Doklady, Earth Sci. Sec. Proc.* **112**:129–132.

Larson, R. L., and C. E. Helsley. 1975. Mesozoic reversal sequence. *Rev. Geophysics and Space Physics* **13**:174–176.

Larson, R. L., and T. W. C. Hilde. 1975. A revised time scale of magnetic reversals for the Early Cretaceous and Late Jurassic. *Jour. Geophys. Research* **80**:2586–2594.

Larson, R. L., and W. C. Pitman, III. 1972. World-wide correlation of Mesozoic magnetic anomalies, and its implications. *Geol. Soc. America Bull.* **83**:3645–3662.

McElhinny, M. W., and P. J. Burek. 1971. Mesozoic palaeomagnetic stratigraphy. *Nature* **232**:98–102.

Shuey, R. T., F. H. Brown, and M. K. Croes. 1974. Magnetostratigraphy of the Shungura Formation, Southwestern Ethiopia: Fine structure of the Lower Matuyama polarity epoch. *Earth and Planetary Sci. Letters* **23**:249–260.

Steiner, M. B., and C. E. Helsley. 1974. Reproducible anomalous Upper Triassic magnetization. *Geology* **2**:195–198.

Steiner, M. B., and C. E. Helsley. 1975. Late Jurassic magnetic polarity sequence. *Earth and Planetary Sci. Letters* **27**:108–112.

Storetvedt, K. M. 1970. On remagnetization problems in palaeomagnetism: Further considerations. *Earth and Planetary Sci. Letters* **9**: 407–415.

Thompson, R. 1977. Stratigraphic consequences of palaeomagnetic studies of Pleistocene and Recent sediments. *Jour. Geol. Soc.* **133**:51–59.

Valencio, D. A. 1972. Intercontinental correlation of Late Palaeozoic South American rocks on the basis of their magnetic remanences. *Acad. Brasileira Ciênc. Anais* **44**:357–364.

18

Copyright © 1974 by the Geological Society of America

Reprinted from *Geol. Soc. America Bull.* **85**:457–464 (1974)

Paleomagnetism of the Lower Triassic Moenkopi Formation

C. E. HELSLEY } *Institute for Geological Sciences, University of Texas at Dallas, P.O. Box 30365, Dallas, Texas 75230*
M. B. STEINER }

ABSTRACT

Paleomagnetic studies on the upper half of the Moenkopi Formation have identified three reversed and three normal polarity zones; this brings the total number of polarity zones observed in this formation to ten. The variable thickness and presumed duration of these polarity zones when combined with smaller variations internal to individual polarity zones suggest that this reversal sequence should be very useful as a regional correlation tool.

Several zones of anomalous magnetization have been recognized within the formation. These anomalous zones constitute less than 5 percent of the total sample population but are very important, because they may assist in the recognition of individual polarity zones.

The pole position derived from this study is at 103° E., 55° N. (dp 1.6°, dm 3.2°) and, due to the large number of samples and the length of time represented, probably represents an excellent average for the Lower Triassic of North America.

INTRODUCTION

Observations of both normal and reversed magnetic polarity have been made on rocks of all ages, yet for rocks older than the late Cenozoic, the time sequence of these reversals is virtually unknown. Two long periods of constant polarity have been identified, one in the Permian (Irving and Parry, 1963) and one in the Cretaceous (Helsley and Steiner, 1969). Other than these periods of constant polarity, no definitive polarity sequences are recognized generally, although several attempts have been made to use locally developed reversal sequences as an aid in stratigraphic correlation (Khramov, 1958; Picard, 1964; Burek, 1967; Helsley, 1969). The purpose of this paper is to extend our knowledge of the polarity-reversal sequence for Lower Triassic rocks of the Colorado Plateau so that the polarity reversal sequence can be used in the future in regional correlation of these otherwise unfossiliferous rocks.

PREVIOUS WORK

Studies of Triassic rocks from the Colorado Plateau (Runcorn, 1956; Kint-zinger, 1957; Collinson and Runcorn, 1960) have shown that both normal and reversed polarities are characteristically present, although no attempt was made to define the polarity sequence. Helsley (1969) has defined a polarity-reversal sequence for the lower half of the Lower Triassic Moenkopi Formation and has done reconnaissance sampling of the upper portion of the Moenkopi Formation. He showed that these rocks are remarkably stable and that thermal demagnetization was effective in removing any secondary component that may have been present. The magnetization of the Moenkopi Formation has been shown to have been acquired penecontemporaneously with deposition (Baag and Helsley, 1974).

GEOLOGIC SETTING

The Moenkopi Formation crops out throughout the Colorado Plateau region. Its regional stratigraphy and history are discussed by McKee (1954) and more recently by Blakey (1973). In the western portions of its outcrop region, the Moenkopi Formation is characterized by a marine sequence including gray shales, carbonates, and evaporites in which red to maroon shales are interbedded. To the east, the formation becomes more clastic, and in eastern Utah and western Colorado, it consists of brown and red-brown sandstone, siltstone, and shale with minor amounts of gypsum and a few locally developed conglomerate lenses.

The region sampled for these studies is in the thickest and most continuously deposited portion of the clastic facies of the Moenkopi Formation south of Gateway, Colorado (Fig. 1), in an area in which the geology has been described by Shoemaker (1955). Bedding within the Moenkopi Formation is generally even and continuous, which greatly facilitates local lithologic correlations. In the area of this study, the formation consists of about 240 m of interbedded reddish-brown sandstone, siltstone, and shale with some interbedded conglomerate. The formation has been subdivided into four members on the basis of lithology by Shoemaker and Newman (1959). These members provide a means of local correlation between the outcrops sampled along the Dolores River and those sampled along Salt Creek (Fig. 1). These members are (from bottom to top): the Tenderfoot Member, a sequence of massively bedded bimodal sandstone, regionally equivalent to the Hoskinnini Member (Stewart, 1959); the Ali Baba Member, a sequence of interbedded sandstone, siltstone, and conglomerate; the Sewemup Member, a sequence of fine-grained sandstone, siltstone, and shale; and the Pariott Member, a sequence of multicolored siltstone, sandstone, and conglomerate. The lower boundary of the Sewemup Member is a distinct lithologic boundary in this area and has been used as a basis for stratigraphic correlation between adjacent sampling sections.

The age of the Moenkopi Formation of this region is well established as Lower Triassic (Dane, 1935; Shoemaker and Newman, 1959); however, the range of time represented by the formation is poorly known. On the basis of regional correlation with other strata of Permian and Triassic age, Shoemaker and Newman (1959) conclude that the lowest member of the Moenkopi may be either Permian or Triassic and that it is likely that the boundary lies within this member. The uppermost units of the Moenkopi Formation are considered by the same authors to be perhaps as young as Middle Triassic. Thus, one can only conclude that the rocks sampled for this study were deposited during an unknown period of time between uppermost Permian to lowermost Middle Triassic. Inasmuch as no major diastems have been noticed in this section, one can assume that sedimentation was essentially continuous.

Helsley's 1969 study systematically sampled the lower half of the formation up to the lower part of the Sewemup Member along the base of the cliffs bordering the Dolores River 15 km south of Gateway, Colorado. The localities from which these samples came are shown by the dots in Figure 1. Samples from the upper half of the formation (for the present study) were collected near the junction of Salt Creek with the Dolores River, and the sites are shown in Figure 1 by the asterisks. Sampling for the present study was begun about 33 m below the Sewemup–Ali Baba boundary and overlaps the previous work (Helsley, 1969) by approximately 75 m. Stratigraphic

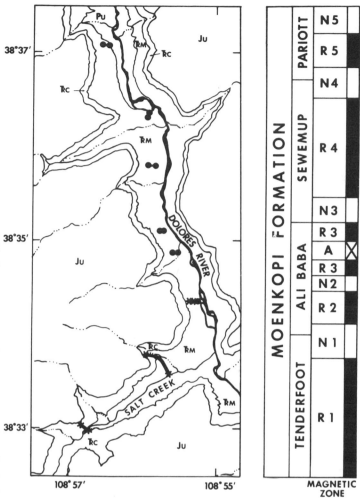

Figure 1. Location map showing geology of the collecting sites in central western Colorado and the general reversal sequence as related to the four members of the Moenkopi Formation. Pu = Permian Cutler Fm., Trm = Triassic Moenkopi Fm., Trc = Triassic Chinle Fm., Ju = undivided Jurassic Fms. Single and double dots represent Helsley's (1969) localities; asterisks indicate localities of the present study. The long continuous sample sequence in Salt Creek is indicated by broad line starred on either end.

sampling at 35-cm intervals was carried out up to the base of the overlying Chinle Formation, for a total section of 150 m. Cores were collected by means of a hand-held gasoline powered drill (Helsley, 1967) and were oriented by magnetic compass.

Three kilometers farther up Salt Creek, the basal portions of the uppermost member of the Moenkopi Formation are exposed. This unit, the Pariott Member, is present in only a few places in the western Colorado–eastern Utah area. The Pariott

has tentatively been correlated with exposures of the Moenkopi Formation in Arizona and may be of Middle (?) Triassic age (Shoemaker and Newman, 1959). At the locality in Salt Creek, 35 m of the Pariott Member and 8 m of the unconformably overlying Chinle Formation were sampled.

NATURAL REMANENT MAGNETIZATION

The field cores were cut into 2.5-cm-long cylinders and measured on a PAR SM-1

spinner magnetometer. Thermal demagnetization was done in a noninductively wound, twelve-sample capacity furnace. Temperature was monitored at four levels within the furnace by platinum thermocouples referenced to an ice-water mixture and read on a digital voltmeter. Care was taken to ensure that the samples cooled in a zero field region ($\pm$ 10 gammas). After cooling from each demagnetization step, the samples were measured on the PAR magnetometer.

The natural remanent magnetization (NRM) of all samples of the upper part of the formation is summarized in Figure 2. For purposes of discussion and identification, the formation has been subdivided into polarity zones beginning at the base of the formation (Fig. 1). Sampling of this second profile of the Moenkopi Formation began in the reversed zone R3 (Helsley, 1969) and overlaps the previous study from zone $R3_A$ to the lower part of zone R4 (Figs. 1 and 4).

The detailed study of the upper part of the Moenkopi Formation reveals one thick reversed zone followed by three thinner polarity zones. Sampling was continued across the Moenkopi boundary into the overlying Chinle Formation. The uppermost Moenkopi samples measured in this study are of normal polarity (N5), and the overlying unconformable basal units of the Chinle Formation are reversed.

The NRM of the reversed groups has a mean declination of 159° and inclination of -2° and the normal intervals, 348° and 24°. The two NRM means are approximately 160° apart. The NRM directions of several portions of the upper Moenkopi Formation show large dispersion, particularly in zones $R3_B$, R5, and N5. All parts of the section showing highly scattered NRM directions have abundant coarse sandstone units, and thus the greater scatter can be correlated directly to the presence of coarse material.

R4 has a much smaller degree of scattering of NRM directions than any of the other reversed sequences. This consistent direction persists for a long stratigraphic interval (approximately 60 m) and the low degree of scatter is probably related to lithology, for this reversed zone occurs in a section of finer grained sediments in comparison to the lithology of the other reversed zones. The extreme scatter characterizing the NRM of the A and R3 zones ceases abruptly near the base of N3 and corresponds to the lithologic boundary between the coarse sands of the Ali Baba Member and the siltstones and fine sands of the Sewemup Member.

The continuation of work on the Moenkopi Formation brought to light a correlation error in Helsley's original collection. In the 1969 paper, four normal

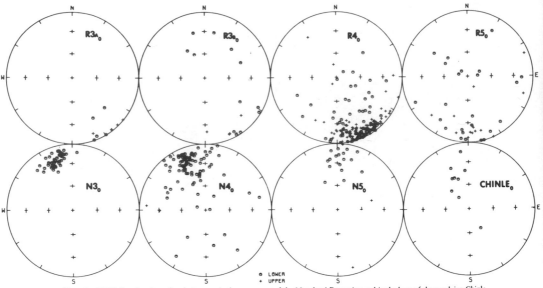

Figure 2. NRM directions in each polarity zone in the upper part of the Moenkopi Formation and in the base of the overlying Chinle.

and five reversed zones were described. It has since become apparent that the section containing polarity change N3 to R4 was sampled twice; thus N4 and R5 of Helsley (1969) are the same as N3 and R4. The lower part of the Moenkopi therefore contains three normal and four reversed zones. To verify the new correlation, an additional 12 m of section above Helsley's 1969 section were sampled, thus extending that section well into the R4 zone and permitting an unambiguous lithologic correlation to be made between Helsley's previous sections and the present study.

DEMAGNETIZATION STUDIES

Helsley (1969) demagnetized a portion of the lower Moenkopi samples in alternating fields of up to 1400 oersteds. It was found that the coercive force of the secondary component was approximately the same as that of the NRM, and thus AC demagnetization was not an effective cleaning procedure for samples from the lower part of the Moenkopi Formation. Thermal demagnetization proved to be quite effective in removing the secondary components responsible for the scatter in the samples from the lower part of the Moenkopi Formation, and thus all samples from the upper Moenkopi section were thermally demagnetized in two steps—550°C and 620°C. Some samples were also heated to 660°C. In general, the direction of magnetization does not change when the samples are heated to

temperatures in excess of 550°C. These results are shown in Figures 3 and 4. The mean of the reversed samples after demagnetization is 160°, −13°; the mean for the normal samples is 342°, 17°. The two means are 176° apart, indicating that a stable magnetization has been isolated. Table 1 shows the mean direction of magnetization of each polarity zone after thermal demagnetization.

Thermal demagnetization significantly reduced the scatter of this sample population. The result is a highly consistent set of data showing well-defined polarity zones. The internal consistency of the data allows the identification of several small magnetization anomalies within the section. The "anomalous" zones constitute only about 5 percent of the samples, and this suggests that anomalous behavior should be a minor part of most sample sets. The anomalies observed are an "excursion" of directions in the R4 zone, an anomalous interval of magnetization within the R3 zone (identified as zone A), and an undulating variation of magnetization direction throughout N3-R4-N4 zones.

In the R4 zone, a group of samples, about 15 m above the base of the interval (Fig. 4), seem to show a transitional movement away from and back to the typical reversed direction. An intensity low precedes the variation in direction, a feature also observed in the succeeding reversal of polarity (R4-N4). The systematic variation

of direction and intensity of magnetization observed in this interval suggest that this anomaly may be a short term excursion of the Earth's magnetic field. Its observation in a second sampling of the lower part of the R4 zone several kilometers away further substantiates its validity.

The zone referred to as PF by Helsley (1969) is again present in this section and is hereafter referred to as zone A. Zone A is present in the position expected from the previously sampled section (approximately 5 km away) and encompasses approximately the same stratigraphic thickness (Fig. 4). The NRM directions (Fig. 5) from this collection of zone A are more scattered than were the magnetization directions of the samples collected previously, and the intensity is low compared to the mean intensity of the rest of the upper Moenkopi samples (3.43 versus 7.18 x 10^{-6} emu/cc). The pole position obtained from the most consistent part of the zone is four degrees from the zone A pole obtained by Helsley (1969; see Fig. 11 below). Zone A is therefore reproducible at neighboring sections and may well be a real feature and not just a local anomaly.

In Figure 4, an undulatory behavior of the magnetization with respect to stratigraphic position is quite noticeable, particularly in zone R4. This behavior is most obvious in the part of the section consisting of N3, R4, and N4 and is restricted to samples from the Sewemup Member, a unit of fine-grained

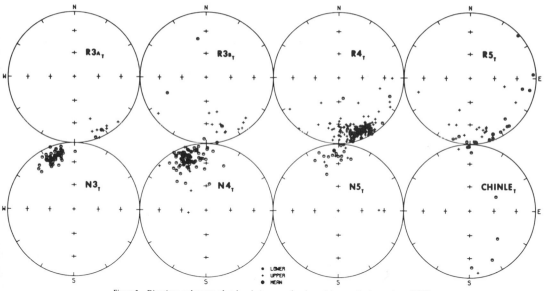

Figure 3. Directions and means of each polarity zone after thermal demagnetization to about 550°C.

lithology. In order to test the reality of this behavior, the original sample population was subdivided into two groups, and, to make these groups independent, every other sample was assigned to alternate groups. Running averages, over ten samples, were calculated on each of the two sample groups. The striking similarity of the resulting curves (Fig. 6) shows that the undulatory behavior is indeed a real feature of the remanence and not just an artifact produced by a few samples. In Figure 6 the polarity of the normal zones has been inverted so that undulation of the directions could be observed more readily. Since the polarity boundaries cannot be identified in the

running average data (Fig. 6), nor can they be seen in a plot of individual sample directions converted to one polarity, one is led to the conclusion that the variation is independent of polarity reversals. This suggests that the anomalous (secular ?) field is independent of the main dipolar field.

Poles have been calculated for every 3-m stratigraphic interval for all of the upper part of the formation and are shown in Figure 7, a through g. These poles, particularly those from the N3-R4-N4 section (7, b through f), have directions that ·range from a Lower Permian pole position (Gose and Helsley, 1972) to a Lower Jurassic position (Steiner and Helsley,

1972). This is not one continuous change but seems to oscillate back and forth. Helsley's previous collection of the Moenkopi Formation was also analyzed in the same manner for that part of the section corresponding to R4. Although his was only the lower half of R4, the poles produced exactly the same pattern as those from this sampling of zone R4 (compare Fig. 7, b through d with 7h).

To demonstrate the range of the variation, six groups of samples, one from each of the more or less constant directions observed in Figure 6, were analyzed for a pole position. The result was either a "Permian" pole, a "Lower Triassic" pole, or an "uppermost Triassic" pole, depending on where one sampled the section. Therefore, there is a large probability that a pole position derived from a short stratigraphic section would not have been a valid representation of the pole for the entire formation. If one wishes to remove this long-term variation, one must sample the section more or less uniformly (and presumably over a large stratigraphic (and presumably time) interval.

INTENSITY

The mean NRM intensity of magnetization for the upper part of the Moenkopi Formation is 7.8×10^{-6} emu per cc, and intensities range from 0.5 to 35.0×10^{-6} emu per cc. The intensity after thermal demagnetization to the tightest sample grouping (550°C for 85 percent of the samples and 630°C for the remainder) is 6.20×10^{-6} emu per cc. The intensity after demagnetization for this part of the formation is approxi-

TABLE 1. ANALYSES OF DIRECTIONS OF EACH POLARITY INTERVAL

	D	I	N	R	K	α_{95}	Long	Lat	dp	dm
					All Data					
R3$_A$	153.2	-9.9	9	8.79	39.0	8.3	113.7	48.6	4.3	8.4
R3$_B$	151.6	-21.9	16	13.69	6.5	15.7	107.1	58.3	8.8	16.6
N3	338.3	16.7	44	43.29	60.5	2.8	109.6	53.4	1.5	2.9
R4	160.0	-13.3	171	161.36	17.6	2.6	105.8	53.5	1.4	2.7
N4	340.9	18.1	80	75.05	16.0	4.1	106.4	56.1	2.2	4.2
R5	158.3	-9.3	35	27.97	4.8	12.4	106.8	50.9	6.3	12.5
N5	355.9	10.5	19	17.83	15.4	8.8	78.5	56.6	4.5	9.0
CHINLE	176.3	-11.5	9	7.60	5.7	23.6	77.8	57.1	12.1	23.9
					Selected Data					
R3$_A$	153.2	-9.9	9	8.79	39.0	8.3	113.7	48.6	4.3	8.4
R3$_B$	156.9	-20.2	9	8.66	23.4	10.9	113.6	55.1	6.0	11.4
N3	338.8	16.4	43	42.36	65.3	2.7	108.9	54.3	1.4	2.8
R4	160.3	-13.0	156	151.21	32.3	2.0	105.2	53.5	1.0	2.0
N4	341.7	16.9	72	69.73	31.3	3.0	104.6	55.9	1.6	3.1
R5	165.5	-8.1	23	21.92	20.4	6.9	95.5	53.2	3.5	6.9
N5	353.5	17.5	14	13.57	30.3	7.3	83.9	59.9	3.9	7.6
CHINLE	175.2	-3.8	5	4.96	94.4	7.9	77.3	53.3	4.0	7.9

Longitude is east; latitude is north.
N = Number of samples or zones.
R = Resultant vector.
K = Precision parameter.
α_{95} = Semiangle of 95% confidence cone.
dp, dm = Semiaxis of error ellipse.

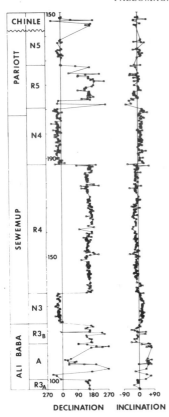

Figure 4. Stratigraphic plot of sample declinations and inclinations after about 550°C demagnetization for the entire Moenkopi Formation. The stratigraphic sequence on the right is from Helsley (1969).

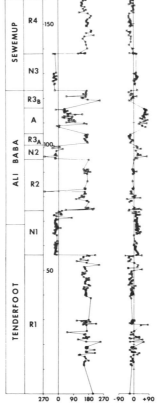

netization as are those of the reversed polarities. And, as in the lower part of the Moenkopi Formation, the normal and reversed intensities are not the same before demagnetization. The mean normal intensity is higher by 0.87 x 10⁻⁶ emu per cc, but is only 0.36 x 10⁻⁶ emu per cc higher after demagnetization.

The mean intensities of each polarity zone are shown in Figure 9. Helsley (1969) noted an increase in intensity upward in the lower part of the Moenkopi Formation. In contrast, there seems to be a decrease upward in the upper part of the Moenkopi Formation. The combination of these data result in the pattern seen in Figure 9 for the entire Moenkopi Formation . This variation parallels the variation of lithology and indicates that the finer grained lithologies of the Sewemup and Ali Baba Members have relatively high intensities, but that the coarser sediments of the Pariott, Ali Baba, and Tenderfoot Members have low intensities.

POLES

No standard method of analysis has been adopted for computing a pole position for a lengthy stratigraphic interval with a large number of samples. One may analyze all samples together; but for the sake of comparison, it is obvious that such a pole will represent a tremendously greater time span than, for example, a pole from one or several lava flows. Thus, it is felt that it is more meaningful to compute poles by intervals throughout a long stratigraphic section. It is reasonable to examine intervals of different polarity or intervals of an arbitrary stratigraphic length.

Polarity interval analysis is advantageous in that changes in polarity represent natural divisions of the magnetic record. However, there is the disadvantage of computing pole positions from large and nonuniform units of time per pole. A more uniform analysis can be obtained by analyzing the observations in terms of a fixed stratigraphic interval whose length is determined by the shortest polarity interval observed, provided the density of sampling is high enough. The shortest polarity interval recorded in this study is 6 m. For this reason, the observations were analyzed in 6-m intervals. This interval per pole (15 samples per pole) proves to be sufficient to average out lithologic variation and samples of uncharacteristic magnetization. The precision of the pole as reflected in the oval of confidence (dp, dm) is quite acceptable for the 6-m groups, averaging about 6° to 8° for the long axis (dm).

However, the 6-m interval tends to average out much of the long period variation noted above, and so a shorter interval of 3 m was also used. Analyzing by 3-m intervals with the sampling density used in this study gives an average of about eight samples per pole calculation. Analysis by 3-m intervals (Fig. 7, a through g) reflects the

mately the same, though slightly higher, than that for the lower part of the Moenkopi Formation (4.56 x 10⁻⁶ emu per cc; Helsley, 1969). Figure 8 is a histogram of sample intensities, grouped according to normal or

reversed polarity, both before and after thermal demagnetization. As was seen in the lower part of the Moenkopi Formation, the intensity of the normal polarities is virtually the same before and after thermal demag-

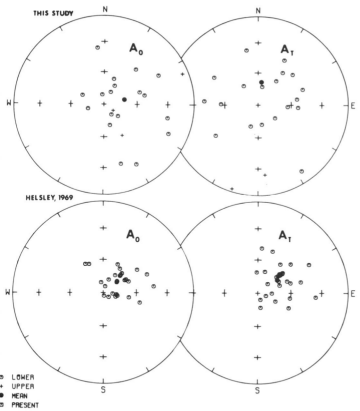

Figure 5. Anomalous zone NRM (A₀) directions and directions after about 600°C (A_T) from both the present study and Helsley's 1969 study.

For the analysis of the upper part of the Moenkopi Formation, however, the question of the best analysis is primarily academic, as even the NRM pole position is within 4° of the variously computed demagnetized poles. The agreement between the NRM and demagnetized poles is at first surprising, considering the amount of instability shown in Figure 2. The agreement results from the method of analysis, which was to invert any data point that was reversed before adding it to the accumulating mean. For the points in the streak between the reversed direction and the present Earth's field, this meant that the directions more than 90° from the mean normal direction were inverted while those closer than 90° to the normal direction were left unchanged. The result was to artificially balance the streaked points on either side of the true normal direction. A good agreement between the NRM and demagnetized data can always be expected for a set of samples collected from a mixed polarity sequence whose unstable samples are more or less evenly distributed along the instability great circle, provided some stable samples are present to determine the boundary between the data to be left unchanged and that to be inverted.

The pole position (6-m interval, selected)

long-term variations of the pole seen in N3-R4-N4 and may be the best way of analyzing this aspect of the data.

All analyses are complicated by the problem that a few samples have directions far removed from the means of their respective groups. These can perturb the mean of a small group significantly. In a large collection, these abnormal directions could quite reasonably stem from orientation errors in the field, from errors in recording the dip and strike of the core in the field, or from sampling a lithologic unit unsuitable for paleomagnetic study (for example, a coarse sand lens). As experimental and natural errors cannot be ruled out, it does not seem unreasonable to exclude the very abnormal samples from any group. Thus, one can calculate the above four analyses of poles with all data present (unselected) or with abnormal samples removed (selected). Demagnetized data are

preferred over NRM data according to theory, but an NRM pole analysis was also computed. (Due to the presence of a secondary magnetization in the NRM data, the distinction between added secondary directions and aberrant directions is not possible, and no selection was made.)

The results of these different types of analyses in both the selected and unselected forms are presented in Table 2. An over-all pole was calculated from the group means for each type of analysis. Figure 10 shows that all methods give almost exactly the same pole. If one considers the question of statistical accuracy of the pole position (Table 2), the analysis of 6-m intervals seems to give the best combination of a small circle of confidence (α_{95}) and a high precision parameter (k). The 6-m interval poles average out most of the long-term variations observed and are used as the estimate of the pole for this sequence of sediments.

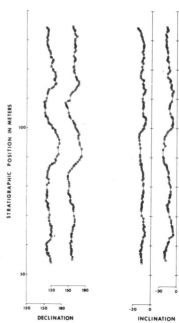

Figure 6. Running averages over ten samples for two independent samplings of the N3-R4-N4 interval obtained by assigning every other sample to alternate groups. Directions used are from samples demagnetized to approximately 550°C.

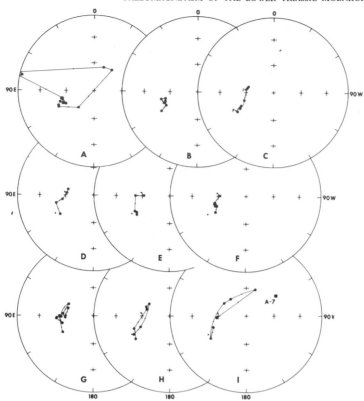

other time-equivalent units. They are the reversed zones R1 (including R1S) and R4, each of which is about 60 m long. Unfortunately, R1 is incomplete because its base is an unconformity. Polarity zone R4 covers a lengthy stratigraphic interval between thinner zones of normal and reversed polarities. In addition to its greater thickness (and presumably longer time duration), R4 has recorded a systematic undulatory variation throughout its extent and a small excursion in its lower part. These features can be correlated for several kilometers at least, and thus the R4 zone seems to present a very useful tool for correlation within the Lower Triassic.

The Moenkopi Formation and, if its age designations are correct, the Lower Triassic and possibly the lowest part of the Middle Triassic are characterized by at least nine polarity changes of the geomagnetic field. In view of the facts that the short polarity zones of the Cenozoic follow a long constant polarity period in the Cretaceous (Helsley and Steiner, 1969), and the short Triassic polarity zones follow the long constant

Figure 7. a through g. Three-meter-interval poles from the upper Moenkopi Formation. h. Three-meter-interval poles for the lower half of R4 in Helsley's 1969 collection. i. Pole for each sample of the excursion in R4 and the pole for anomalous zone A. Gose and Helsley's (1972) Permian pole (P) and Steiner and Helsley's (1972) Lower Jurassic pole (J) are plotted for reference as small diamonds.

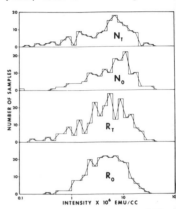

Figure 8. Intensity histograms of normal (N) and reversed (R) NRM (O) and demagnetized (T) data.

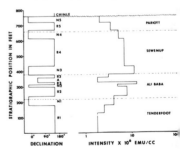

Figure 9. Average intensities of each polarity interval (indicated by a schematic declination plot) plotted according to stratigraphic position. Members of the Moenkopi Formation are indicated on the right.

for the upper Moenkopi (103° E., 55° N.; dp 1.6°, dm 3.2°) compares well with the one from the lower Moenkopi (98° E., 57° N.; dp 2.5°, dm 4.9°) as would be expected. The slightly more northerly position of the pole from the lower part of the formation is probably artificial, as the normal samples contributing to it were, for the most part, not demagnetized. The upper Moenkopi Formation pole is compared to the lower Moenkopi Formation pole (Helsley, 1969) and other published Lower Triassic poles in Figure 11. The upper and lower Moenkopi Formation poles are very close together and, in general, agree well with the other observations of Triassic pole positions (summarized by Helsley, 1969, Table 8).

DISCUSSION OF RESULTS

In the course of sampling the higher part of the Moenkopi Formation, a certain

amount of the lower part of the formation was resampled. This dual sampling is interesting because it allows correlation of part of the polarity sequence over a distance of several kilometers. The overlap of the sections include the zones R3$_A$, A, R3$_B$, N3, and the lower part of R4. The same reversal sequence is present in both, and the stratigraphic lengths of the polarity zones are approximately the same. Moreover, the long-term variation of the pole observed in R4, the "anomalous" zone A, and the "excursion" in the lower part of R4 are present in both sections and suggest that small-scale irregularities may be very useful in separating one polarity zone from another when they are used in regional correlation.

Two polarity zones are strikingly different from the others in the Moenkopi Formation and may be useful in regional correlation within the Moenkopi Formation and within

TABLE 2. ANALYSES OF UPPER MOENKOPI FORMATION SAMPLE DIRECTIONS

	D	I	N	R	K	a_{95}	Long	Lat	dp	dm
All samples NRM	160.0	-12.3	387	322.56	6.0	3.2	105.6	53.0	1.7	3.3
All samples demag. (un)	160.9	-14.7	374	346.78	13.7	2.0	105.0	54.5	1.1	2.1
All samples demag. (sel)	161.0	-14.3	326	315.38	30.6	1.4	104.6	54.4	0.7	1.5
By polarity zones (un)	161.2	-14.3	7	6.94	93.6	6.3	104.4	54.4	3.3	6.4
By polarity zones (sel)	161.4	-14.7	7	6.95	110.7	5.8	104.2	54.7	3.0	5.9
By 6-m intervals (un)	162.1	-14.8	22	21.78	96.9	3.2	103.2	55.0	1.7	3.2
By 6-m intervals (sel)	162.0	-14.5	22	21.79	101.9	3.1	103.3	54.8	1.6	3.2
By 3-m intervals (un)	162.1	-14.5	43	41.81	35.2	3.7	103.1	54.8	2.0	3.8
By 3-m intervals (sel)	162.4	-14.5	43	42.40	70.0	2.6	102.6	54.9	1.4	2.7
A zone (un)		8.3	69.3	21	15.67	3.8	19.1	90.1	74.4	27.8	32.6
A zone (sel) A-7		53.0	68.4	14	12.49	8.6	14.4	56.6	51.3	20.4	24.2
A zone 1969		46.7	66.9	28	26.59	19.2	6.4	52.9	55.3	8.7	10.6

See Table 1 for explanation of heading abbreviations.
Un = unselected (all samples are in analysis).
Sel = selected (deviant samples removed).
Demag. = thermal demagnetization.

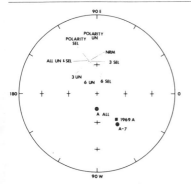

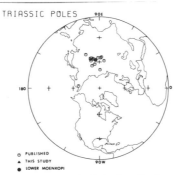

TRIASSIC POLES

○ PUBLISHED
▲ THIS STUDY
● LOWER MOENKOPI

Figure 10. Poles (dots at center of spoke) computed by various analyses (polarity zones, 3-m intervals, and 6-m intervals) for the entire Moenkopi Formation. SEL = selected data, UN = all data used. Anomalous zone poles for selected (A-7) and all data and the 1969 study are also shown.

Figure 11. Upper Moenkopi pole position shown with the lower Moenkopi pole (Helsley, 1969) and published Lower Triassic poles for North America (summarized in Table 8 of Helsley, 1969).

polarity of the Upper Paleozoic Reversed Interval, it is tempting to speculate that major geologic eras, defined on the basis of major changes in the fossil record, may end with a very low frequency of reversals and begin with a high reversal frequency. The major changes in life forms that take place at these times may be the result of evolutionary adaptation during the long period of constant polarity in which the magnetic field is used in the regulation of some necessary or vital function. Such organisms would be at a severe disadvantage when a period of frequent reversals once again began; for at each reversal, the intensity of the field has been observed to decrease to a low or near-zero value. Such a speculation is supported by recent experiments in which profound biologic changes were observed during embryonic development when the organism was grown in a region of near-zero field (Russell, 1972).

ACKNOWLEDGMENTS

Assistance in field collections was provided by W. A. Gose, Wray Curtis, Joan Helsley, and Anne Helsley; their help is greatly appreciated. This work was supported by National Science Foundation Grants GA–2205 and GA–15999.

REFERENCES CITED

Baag, C., and Helsley, C., 1974, Evidence for penecontemporaneous magnetization of the Moenkopi Formation: Jour. Geophys. Research (in press).

Blakey, R. C., 1973, Stratigraphy and origin of the Moenkopi Formation (Triassic) of southeastern Utah: Mtn. Geologist, v. 10, p. 1–17.

Burek, P. J., 1967, Korrelation revers magnetisierter Gesteinsfolgen im Oberen Buntsandstein SW-Deutschlands (Correlation of reversely magnetized rock sequences in the Upper Buntsandstein, SW-Germany): Geol. Jahrb., v. 84, p. 591–616.

Collinson, D. W., and Runcorn, S. K., 1960, Polar wandering and continental drift: Evidence of paleomagnetic observations in the United States: Geol. Soc. America Bull., v. 71, p. 915–958.

Dane, C. H., 1935, Geology of the Salt Valley anticline and adjacent areas, Grand County, Utah: U.S. Geol. Survey Bull. 863, 184 p.

Gose, W. A., and Helsley, C. E., 1972,

Paleomagnetic and rock-magnetic studies of the Permian Cutler and Elephant Canyon Formations in Utah: Jour. Geophys. Research, v. 77, p. 1534–1548.

Helsley, C. E., 1967, Advantages of field-drilling of samples for paleomagnetic studies, in Collinson, D. W., Creer, K. M., and Runcorn, S. K., eds., Methods in paleomagnetism: New York, Elsevier Pub. Co., p. 26–30.

——1969, Magnetic reversal stratigraphy of the Lower Triassic Moenkopi Formation of western Colorado: Geol. Soc. America Bull., v. 80, p. 2431–2450.

Helsley, C. E., and Steiner, M. B., 1969, Evidence for long intervals of normal polarity during the Cretaceous period: Earth and Planetary Sci. Letters, v. 5, p. 325–332.

Irving, E., and Parry, L. G., 1963, The magnetism of some Permian rocks from New South Wales: Royal Astron. Soc. Geophys. Jour., v. 7, p. 395–411.

Khramov, A. N., 1958, Paleomagnetism and stratigraphic correlation: Leningrad, Gostoptechizdat; [English trans. by A. J. Lojkline] Canberra, Australian National Univ. (1960), 204 p.

Kintzinger, P. R., 1957, Paleomagnetic survey of Triassic rocks from Arizona: Geol. Soc. America Bull., v. 68, p. 931–932.

McKee, E. D., 1954, Stratigraphy and history of the Moenkopi Formation of Triassic age: Geol. Soc. America Mem. 61, 133 p.

Picard, M. D., 1964, Paleomagnetic correlation of units within Chugwater (Triassic) Formation, west-central Wyoming: Am. Assoc. Petroleum Geologists Bull., v. 48, p. 269–291.

Runcorn, S. K., 1956, Paleomagnetic survey in Arizona and Utah: Preliminary results: Geol. Soc. America Bull., v. 67, p. 301–316.

Russell, John R., III, 1972, Baseline studies on the effects of low magnetic fields (< 50 gammas) on developmental processes in Fundulus heteroclitus [abs.]: Alabama Acad. Sci. Jour. (April).

Shoemaker, E. M., 1955, Geology of the Juanita Arch quadrangle, Colorado: U.S. Geol. Survey Geol. Quad. Map 81.

Shoemaker, E. M., and Newman, W. L., 1959, Moenkopi Formation (Triassic? and Triassic) in salt anticline region, Colorado and Utah: Am. Assoc. Petroleum Geologists Bull., v. 43, p. 1835–1851.

Steiner, M. B., and Helsley, C. E., 1972, Jurassic polar movement relative to North America: Jour. Geophys. Research, v. 77, p. 4981–4993.

Stewart, J. H., 1959, Stratigraphic relations of Hoskinnini Member (Triassic?) or Moenkopi Formation on Colorado Plateau: Am. Assoc. Petroleum Geologists Bull., v. 43, p. 1852–1868.

Manuscript Received by the Society January 4, 1972

Revised Manuscript Received September 13, 1973

Contribution no. 205 Institute of Geological Sciences, University of Texas at Dallas, Dallas, Texas

19

Reprinted from *Quaternary Research* 7:316–329 (1977)

The Paleomagnetism and Magnetic Polarity Stratigraphy of the Mammal-Bearing Section of Anza Borrego State Park, California

N. D. Opdyke, E. H. Lindsay, N. M. Johnson, and T. Downs

We sampled 150 sites in fine-grained Plio–Pleistocene sediments of the Palm Springs and Imperial Formations. Sampling was confined to 3000 m of stratigraphically continuous section containing abundant vertebrate fossil remains of the Vallecito Creek, Arroyo Seco, and Layer Cake local faunas of the Irvingtonian and Blancan Land Mammal Ages. The magnetic stability of these sediments was sufficient to delineate the magnetic stratigraphy, which ranges from below the Cochiti event at the base, to the Matuyama reversed magnetic epoch at the top of the section. Eight faunal events are placed relative to the magnetic polarity sequence. They are cf. *Pliohippus* extinction, *Geomys* appearance, cf. *Equus* appearance, *Tremarctos* appearance, *Hypolagus* extinction, cf. *Odocoileus* appearance, *Smilodon* and *? Euceratherium* appearance. The latter two faunal events characterize Irvingtonian Land Mammal Age. The transition from the Blancan to Irvingtonian Land Mammal Age occurs in the lower Matuyama magnetic epoch close to the Pliocene–Pleistocene boundary. The appearance of European migrants during the lower Matuyama magnetic epoch indicates that the Bering land bridge was exposed for animal migration between Europe and North America at this time.

INTRODUCTION

A thick section (4600 m) of poorly indurated Plio-Pleistocene sands, muds, silts, and clays are exposed in homoclinally dipping beds within the Anza Borrego State Park, California (Fig. 1). These beds are richly fossiliferous, containing vertebrate fossils that have Blancan and Irvingtonian affinities. The purposes of this study are to correlate these fossils with the known magnetic time scale and with the previously studied vertebrate fossil localities in the San Pedro Valley of Southern Arizona. These detailed biostratigraphic data may help to elucidate the faunal change from Blancan to Irvingtonian Land Mammal Ages.

GEOLOGIC SETTING

The section chosen for study includes the entire Palm Springs Formation (Woodring, 1932), which averages 3000 m in thickness and is composed predominantly of

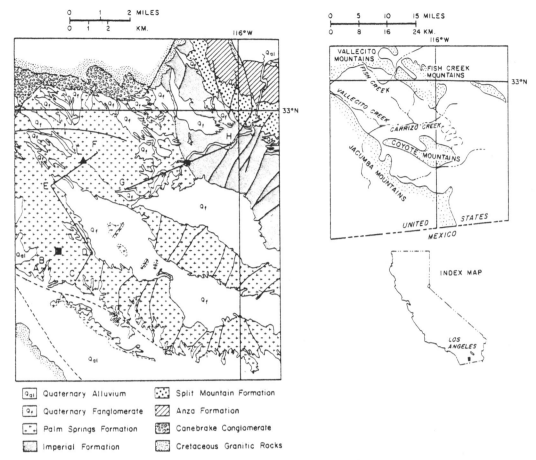

FIG. 1. Geological map of the region under study adapted from Woodard (1974) indicating schematically the sampled sections. The three large filled symbols denote the locations of three faunal zones of Downs and White (1968): Zone 7 (circle) occurs at the contact of the Palm Springs and Imperial Formations at the corner of sections 3, 4, 9, and 10, T. 14S, R. 8E. Zone 29 (triangle) intersects vertebrate locality 1711, which is located adjacent to Fish Creek in the NE 1/4 of section 12, T. 14S, R. 7E. Zone 56 (square) intersects a prominent topographic ridge at the ''view of Badlands'' in the east half of section 23, T. 14S, R. 7E.

terrestrial sediments in a sequence of inter-bedded sandstone, claystone, arkosic sand-stone, pebble conglomerate, and occasional freshwater marl. Littoral marine environments are indicated by the presence of an impoverished marine invertebrate fauna which occurs sporadically in the lower part of the formation (Downs and Woodard, 1961: Woodard, 1974). The presence of detrital foraminifera of Cretaceous age (Merriam and Bandy, 1965) indicates that the sediments of the Palm Springs Formation were being supplied partly by the ancestral Colorado River, along with locally derived coarse detritus from the rising

Peninsular Range along the western margin of the basin. Our sampled section included only the upper part of the Imperial Formation which has yielded the earliest vertebrate fossils known from the region. The Imperial Formation (Tarbet, 1951) contains a middle member of conspicuously rhythmically bedded gray silty mudstone and very fine quartz sand. The upper part of the Imperial Formation consists of alternating siltstone and sandstone with intercalated biostromal limestones and calc-arenite beds (Woodard, 1974).

Our magnetic polarity section (Figs. 1 and 7) is tied into the faunal zones of Downs

237

and White (1968) for biostratigraphic and geographic control. Baselines for the faunal zones of Downs and White are zones 29 and 56. Zone 29 is located at the Gauss–Gilbert boundary (2255 m) in our section. Zone 56 crosses Arroyo Hueso (NE ¼ sect. 25, T. 14 S, R.7E) at 450 m in our section. Another reference point is the contact of the Palm Springs and Imperial Formations (Downs and White zone 7), at 3600 m in our section.

SAMPLING

We sampled 150 sites during 3 visits to the area in 1973 and 1974. Three samples were collected per site using the methods and techniques described by Johnson and others (1975). The sites were located on aerial photographs and the stratigraphic position and total thickness were calculated from the aerial photographs.

MAGNETIC ANALYSIS

The samples were cut into 2.5 cm cubes and measured on a high-sensitivity slow-speed Digico magnetometer, as described by Molyneux (1971). Alternating field demagnetization curves were done on 31 samples from sites spread throughout the sampled section. Alternating field demagnetization was carried out in steps in peak fields of up to 500 Oe using apparatus and techniques similar to those described by McElhinny (1966). Figure 2 illustrates the behavior of eight representative samples, four of whose NRM directions are toward the north with positive inclinations and four which are directed toward the south with both positive and negative inclinations. All four of these latter samples were, however, probably originally magnetized in a reversed field. It can be seen for three of the four reversely magnetized samples that stable end points were not achieved until peak fields on the order of 300 Oe were applied. In the case of one sample (101B) a negative inclination was not reached even in peak fields as high as

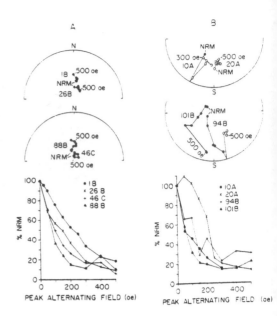

FIG. 2. Alternating field demagnetization curves and changes of directions on demagnetization of (a) four normally and (b) four reversely magnetized samples from the Anza Borrego section.

500 Oe. Because a secondary vector of considerable stability is present, therefore, all sites were routinely demagnetized in peak fields of 300 Oe. After the demagnetization step, all sites were measured and the mean directions of magnetization were determined. Figure 3 shows the mean site directions of magnetization of all sites that are not statistically random before and after a.f. demagnetization in peak fields of 300 Oe. The mean directions of magnetization after demagnetization may be classified into three groups. One category of sites are northerly directed with positive inclinations which are not significantly different from the earth's dipole field. A second category are sites which yield southerly directions with negative and some positive inclinations which are not grouped close to the earth's dipole field. A smaller set of samples exhibit westerly declinations and mostly positive inclinations. This latter category is classed as an intermediate direction of magnetization. It is clear that on the whole the reversed directions are essentially a strung distribution toward the present field and are

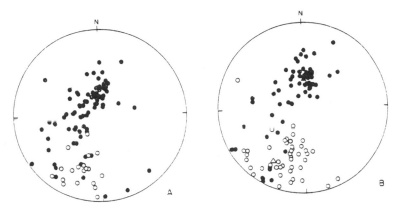

Fɪɢ. 3. (a) Site mean directions of NRM for statistically significant sites. (b) Mean site directions for statistically significant sites after a.f. demagnetization of 300 Oe peak field. The plot is a stereographic projection. The filled circles are directions with north-seeking components directed below the horizontal, and the open circles represent inclinations above the horizontal. The mean direction of magnetization for the normally magnetized sites in 3b is represented by the filled triangle and the reversed sites by the open triangle.

most probably a result of a secondary direction of magnetization that remains after a.f. demagnetization in peak fields as high as 300 Oe. The secondary component was clearly acquired after folding of the beds (strike = 300°, dip = 20°SW) in a normally directed field, probably during the Brunhes normal polarity epoch. This would account for the too shallow reversed directions and the strung distribution of the site mean directions toward the west.

MAGNETIC MINERALOGY

The magnetic fractions from six of the sites were obtained using methods and techniques described by Løvlie *et al.* (1971).

TABLE 1

CURIE TEMPERATURE (θ), ISOTHERMAL REMANENT MAGNETIZATION (J_{IRM}) AND REMANENT COERCIVITY (H_{cr}) DATA FOR SELECTED SAMPLES

Site	θ (°C)	J_{IRM} ($\times 10^{-3}$ emu g)	H_{cr} (Oe)
AB20	633	1.112	308
AB37	558	7.015	322
AB47	563	0.996	564
AB51	598	1.812	245
AB80	560		
AB101	656	0.200	458

The Curie points of the magnetic separates were determined using the Curie point balance described by Kent and Lowrie (1974). The Curie points are given in Table 1 and range from 558 to 656°C. Representative Curie temperature curves are shown in Fig. 4. The Curie temperatures for three of the sites give results indicating that the dominant magnetic mineral is magnetite or titanomagnetite. Three of the sites, however, have Curie temperatures above 600°C, which are higher values than can be expected from the magnetites and titanomagnetites. We believe that the most probable mineral is a stabilized form of maghemite similar to that described by Harrison and Peterson (1965) from deep-sea cores in the Indian Ocean.

Isothermal remanent magnetization (IRM) acquisition curves and remanent coercivities (H_{cr}) were determined for five sites. The values for IRM and H_{cr} are given in Table 1, and the curves from which these values were determined are shown in Fig. 5. In general, the samples acquire 80% or more of the final saturation value in fields as low as 1000 Oe. Some samples do, however, continue to increase in moment up to the maximum fields available (8000 Oe). This asymptotic behavior may be due to a small **239** fraction of highly coercive material, perhaps haematite. Several samples show pronounced kinks in the back field required to determine the H_{cr}. This is particularly true of sample 47A (Fig. 5). It is clear in this case that a mixture of material with at least two coercivities is present, one of which is much higher than the other. This could be due to a mixture of grain sizes and/or minerals. It would appear that one of the principal results of the magnetic mineralogy study is the great variability of these properties. This is understandable, as at least two provenances are postulated for these sediments, as well as the long time involved in their deposition. It is also true that both marine and nonmarine environments of deposition are present in this sediment sequence.

Studies of viscous remanent magnetization (VRM) were carried out on several sites (Fig. 6), and it is clear that VRM can account for a significant fraction of the natural remanent magnetization (NRM) in certain cases (50% in the sample given).

MAGNETIC STRATIGRAPHY

The data from the preceding section (Fig. 3) show that a significant portion of a secondary component remains after a.f. demagnetization in peak fields of up to 300 Oe. It is possible, however, to unambiguously determine the magnetic polarity in most instances. In Fig. 7 we plot the magnetic

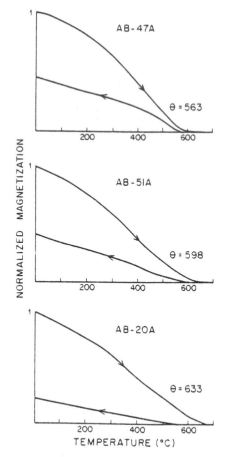

Fig. 4. Curie temperature curves for samples from three different sites.

declination for all sites which have mean directions of magnetization which are statistically significant (Watson, 1956). If only these sites are used, the broad outlines

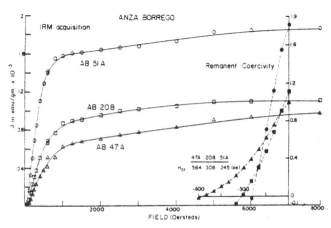

240 Fig. 5. IRM acquisition curves and remanent coercivity determinations for three samples from Anza Borrego.

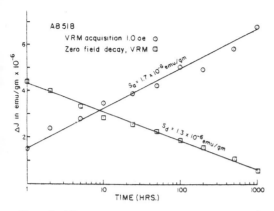

FIG. 6. VRM acquisition and decay curves for normally magnetized sample AB51B from the uppermost part of the section.

have normal polarity. Below these three sites, 434 m of reversely magnetized sediment is present. A normal magnetozone occurs between 427 and 503 m. Below this occurs 488 m of reversed sediment, which is followed by a thick normal zone of sediment which extends from 991 m to 2248 m. This normal magnetozone is punctuated by two relatively short reversed sections between 1314 and 1379 m and 1494 to 1692 m. A long series of reversed sites occurs below 2248 to 3414 m and a clearly defined normal section extends from 3414 to 3616 m. Below this magnetozone three reversely magnetized sites occur.

of the magnetic stratigraphy are immediately apparent because the site density per unit time is high enough to delineate the polarity structure of the sedimentary column. However, for a full evaluation of the magnetic stratigraphy in Fig. 7, all the information is used in the final interpretation. The sites which are statistically significant are called Class I sites. Class II sites have only two samples per site but the directions are concordant. These cases usually arise when one of the samples from the site has been lost during the sample preparation procedure. In these cases there is no question about the polarity of the sites. Class III sites are those sites which have statistically random directions of magnetization but where two of the three samples are directed in, say, a reversed direction, whereas the third is widely divergent in some intermediate direction. Class IV sites are those which have strung distributions but where the directions change upon a.f. demagnetization in a systematic way so that the polarity information has been obtained. This kind of hierarchy has the advantage of presenting some idea as to the total reliability of the data. The magnetic stratigraphy on the left of the figure is derived from all the magnetic data.

The data yield the following results. The upper three sites representing 27 m

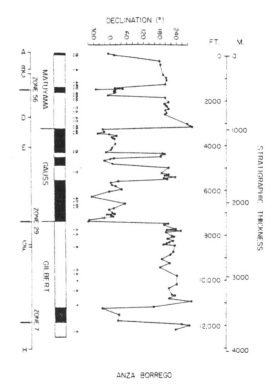

ANZA BORREGO

FIG. 7. Magnetic stratigraphy of the Anza Borrego section. Plotted points of the value of the declination are from those sites which give statistically valid mean directions of magnetization (Class I sites). Class II sites, filled triangle; Class III sites, open triangle; Class IV sites, open circle; Class V sites, cross. See text for explanation of classes. Beside each symbol is plotted the interpreted polarity of the site, either (+) normal or (−) reversed. The final interpretation is given on the left side of the figure and is represented by the normal (black) and reversed (white) polarity diagram.

Four sites were sampled below 3800 m, but the sampling density was not great enough to allow the construction of the magnetic stratigraphy beyond this level.

CORRELATION WITH THE REVERSAL TIME SCALE

The question remains as to which part of the reversal time scale this series of reversals of the earth's magnetic field belongs. The only independent means of correlation available at the present time is the vertebrate fauna. The Layer Cake and Arroyo Seco fauna have been placed in the Blancan Mammal Age, while the younger Vallicito Creek fauna has Irvingtonian affinities (Downs and White, 1968). Johnson *et al.* (1975) have shown that the oldest Irvingtonian based on the occurrence of *Lepus* (hare) and other small mammals occurs within the Matuyama reversed polarity epoch in the region of the Olduvai event, whilst the Blancan faunas characterize the Gauss normal polarity epoch and range into the lower Matuyama epoch. It seems certain, therefore that the section under study can be interpreted as Matuyama in the upper 991 m to 2248 m and the upper Gilbert reversed magnetic epoch below 2248 m. This interpretation is given considerable weight by the fact that two reversed events are seen in this section, the Kaena and Mammoth, which is correlative to the Gauss magnetic epoch.

It appears that both boundaries of the Gauss magnetic epoch are present, and therefore we can estimate the age of the normal events which lie above and below the sediments of the Gauss magnetic epoch. A simplistic method of doing this would be to extrapolate the rate of sedimentation from the sediments deposited during the entire Gauss magnetic epoch (140 cm/1000 yr). However, it is clear from the relative position of the two reversed events within the sediments of the Gauss epoch that the sedimentary section in the upper Gauss magnetic epoch is not **242** long enough when compared with sedi-

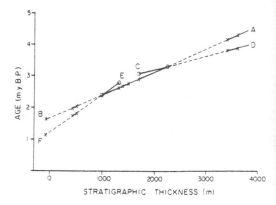

Fig. 8. Plot of rate of sedimentation vs time. Slope AB is the extrapolation for the entire Gauss. Slope CD is the extrapolation upward using the sedimentation rate of the upper Gauss, and EF is an extrapolation downward using the sedimentation rate of the lower Gauss.

ments of the lower Gauss. Therefore a change in the rate of sedimentation occurs at about 3 my. A more meaningful method would be to use the sedimentation rate derived from sediments deposited during upper Gauss time (between the Matuyama/Gauss boundary and the upper reversed, or Kaena, event) for extrapolation into the Matuyama; and to use the sedimentation rate for lower Gauss time sediments deposited between the lower reversed, or Mammoth, event and the Gauss/Gilbert boundary for extrapolation into the Gilbert. The results of both of these analyses are shown in Fig. 8. By using the averaged sedimentation rate during the whole Gauss magnetic epoch, the normal, or Cochiti, event in the Gilbert epoch can be assigned an age older than 4 my (4.16 to 4.31 my). If only the rate for sediments deposited during the lower Gauss magnetic epoch is used, the Cochiti event appears younger than 4 my (3.83 to 3.91 my), which is closer to its age estimated from other studies (3.72 to 3.82 my; Opdyke, 1972). By applying the same rationale, to evaluate the age of the event in the lower Matuyama it can be seen that by extrapolation of the average rate for the total Gauss epoch, the age of this event would be older than 2 my (2.00 to 2.05 my),

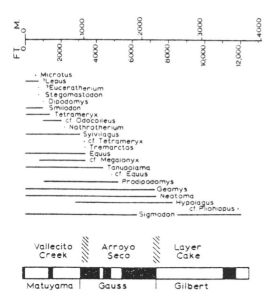

FIG. 9. Stratigraphic range of Anza Borrego faunas plotted against the magnetic stratigraphy.

VERTEBRATE PALEONTOLOGY

A comprehensive faunal list and biostratigraphic division of the mammal fauna from the Anza Borrego State Park was published by Downs and White (1968). Additional fossils have been collected since 1968, and subsequent taxonomic revisions require a reevaluation of the stratigraphic

close to the radiometrically determined age of the Réunion event (2.02 my; Mac-Dougall and Watkins, 1973). However, if the age is determined using only the extrapolation of sedimentary rates from the upper Gauss, the age of the event would be younger than 2 my (1.75 to 1.83 my), which is close to the measured age of the Olduvai event. On the evidence available, the event observed above the Gauss could be either the Olduvai or the Réunion event and the normal polarity zone at the top of the section could be either the basal Jaramillo or Olduvai event, depending on which interpretation is chosen. There are no conclusive means to differentiate between the two interpretations based on magnetic polarity and sedimentation rate.

limits of the fauna. For this study, a check of taxonomic identifications was made by Downs, White, and Zakrzewski. Stratigraphic ranges were checked by Downs and Lindsay. Additions and changes in the faunal list include the addition of the cricetid rodent *Synaptomys* (lemming) (Zakrzewski, 1974); the bat previously identified as *?Myotis* (brown bat) is now identified as a new genus *Anzanycteris* (White, 1969); and *?Mustela* (weasel) is now assigned to the mustelid subfamily *Galictinae*. The faunal division of 1968 is unchanged, although stratigraphic ranges of a few taxa have been extended.

As defined by Downs and White (1968), the Layer Cake fauna is probably early Blancan, the Arroyo Seco fauna is late Blancan, and the Vallecito Creek fauna is early Irvingtonian based on the small mammal fauna. The stratigraphic position with respect to the paleomagnetic transitions of the faunas are shown in Fig. 9. Downs and White (1968) noted that definite boundaries between Layer Cake, Arroyo Seco, and Vallecito Creek faunas could not be delineated. By coincidence, the stratigraphic interval they assigned to the Layer Cake fauna corresponds very closely with the magnetic polarity we assign to the Gilbert magnetic epoch. Similarly, the sediments containing the Arroyo Seco fauna were mostly deposited during the Gauss magnetic epoch, and likewise the Vallecito Creek fauna is found in sediments whose magnetic polarity we assign to the Matuyama magnetic epoch.

The stratigraphic ranges of 20 representative genera within the Anza Borrego sequence are shown in Fig. 9 and discussed below. These genera were selected because of their chronologic significance and their representation in other vertebrate faunas of comparable age. Ranges of other Anza Borrego mammal taxa can be plotted relative to these genera, using the ranges provided in Fig. 2 of Downs and White (1968). The 20 selected genera may be used to define 8 faunal events in the **243**

Anza Borrego stratigraphic sequence. These faunal events, which we name faunal datum planes (Fig. 10), mark the lowest or highest stratigraphic occurrence of a faunal taxon within a continuous stratified sequence correlated to the paleomagnetic time scale. In effect, they correspond to a biostratigraphic boundary, i.e., the limit of a teilzone. As used herein, the faunal datum planes are tied to the magnetic polarity sequence in the Anza Borrego deposits. The faunal datum planes established in this paper apply only to the Anza Borrego sequence. The same faunal events may (but do not necessarily have to) occur synchronously in other stratigraphic sequences in North America. Datum planes provide more precise chronologic resolution than previous faunal divisions (Downs and White, 1968; and Tedford, 1970, Fig. 3 and p. 683). In the discussion that follows the lowest stratigraphic occurrence and the highest stratigraphic occurrence are designated LSD and HSD, respectively.

The cf. *Pliohippus* LSD (~3.8 my) marks the highest (and only) stratigraphic occurrence of the horse *Pliohippus* in the Anza Borrego section (*Pliohippus* is considered a Clarendonian and Hemphillian genus). Its record in the Layer Cake fauna indicates the lowest vertebrate fossil level in the Anza Borrego section may be Hemphillian Land Mammal Age. A characteristic Blancan (and younger) genus, the rodent *Sigmodon* (cotton rat), occurs stratigraphically lower than cf. *Pliohippus*.

The *Geomys* LSD (~3.3 my) marks the lowest stratigraphic occurrence of the gopher *Geomys* in the Anza Borrego section. The cricetid rodent *Neotoma* (pack rat) appears in the sequence slightly lower than *Geomys*. Both genera, *Geomys* and *Neotoma*, occur in Blancan and younger faunas in North America.

The cf. *Equus* LSD (~2.9 my) marks the lowest stratigraphic level that yields fossils of the modern horse *Equus* in the Anza

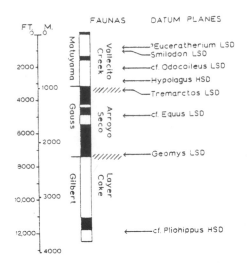

FIG. 10. Vertebrate fossil datum planes from the Anza Borrego section.

Borrego section. *Equus* is also characteristic of Blancan and younger faunas in North America.

The *Tremarctos* LSD (~2.5 my) marks the lowest stratigraphic occurrence of the bear *Tremarctos* in the Anza Borrego section. The rabbit *Sylvilagus*, unquestionable *Equus*, and the pronghorn cf. *Tetrameryx* also appear at the stratigraphic level of the *Tremarctos* LSD. *Tremarctos* is rare in North American faunas; this probably represents its earliest known occurrence in North America.

The *Hypolagus* HSD (~2.3 my) marks the highest stratigraphic occurrence of the rabbit *Hypolagus* in the Anza Borrego section. *Hypolagus* is a common rabbit in Hemphillian and older faunas of North America. It is replaced in Pleistocene and Recent faunas by the cotton tail *Sylvilagus* and the hare *Lepus*.

The cf. *Odocoileus* LSD (~2.1 my) marks the lowest occurrence of a true cervid, e.g., *Odocoileus* or *Cervus*, in the Anza Borrego section. Probably the earliest occurrence of *Odocoileus* in North America is in the Fox Canyon fauna of Kansas. This datum event in the Anza Borrego sequence also approximates the occurrence of the ground sloth *Nothrotherium*, and the highest occurrence of the stilt-legged camel, cf. *Titanotylopus*, in the sequence.

The *Smilodon* LSD (~1.9 or 1.6 my) marks the lowest occurrence of the well-known saber-toothed cat *Smilodon*, which is known only from Irvingtonian and Rancholabrean faunas in North America. The kangaroo rat *Dipodomys* replaces the ancestral kangaroo rat, *Prodipodomys*, at (approximately) this same stratigraphic level in the Anza Borrego section.

The *?Euceratherium* LSD (~1.9 or 1.6 my) marks the only known occurrence of this bovid in the Anza Borrego section. *Euceratherium*, a Eurasian immigrant, is probably the earliest bovid to inhabit North America. Bovids are not known from Blancan deposits of North America; they are characteristic of Irvingtonian and especially Rancholabrean faunas. The Blancan gomphothere *Stegomastodon* and the hare *Lepus* also occur at this stratigraphic level. The rodent *Microtus* (vole) occurs slightly higher stratigraphically.

The Layer Cake fauna has a duration of about 0.7 my. A single datum event, cf. *Pliohippus* HSD marks the base of this fauna at the base of the Cochiti event in the Gilbert magnetic epoch. The cf. *Pliohippus* HSD should be very near the Hemphillian–Blancan boundary. The Layer Cake fauna correlates approximately with the Fox Canyon fauna of Kansas, based on its magnetic polarity and fauna (Lindsay *et al.*, 1975).

The Arroyo Seco fauna has a duration of about 1.1 my, and is marked by lower (*Geomys*, LSD), middle (*Equus*, LSD), and upper (*Tremarctos*, LSD) faunal datum events. This fauna approximately correlates (through both the fauna and magnetic polarity) with the Benson fauna of southern Arizona, the Red Corral and Cita Canyon faunas of Texas, and the Rexroad, Bender, and Saunders faunas of Kansas (Lindsay *et al.*, 1975).

The Vallecito Creek fauna has a duration of about 0.8 or 1.5 my (depending on the interpretation of the normal magnetic polarity event in the middle of the Vallecito Creek section), and is marked by four faunal datum events. According to magnetic polarity stratigraphy the *Hypolagus* HSD correlates approximately in time with the Borchers fauna of Kansas. The cf. *Odocoileus* LSD also correlates approximately with time of the Borchers fauna of Kansas, or with that of the Mt. Blanco fauna of Texas. The *Smilodon* and *?Euceratherium* LSD correlate approximately with the Mt. Blanco fauna of Texas or the Curtis Ranch fauna of Arizona (Lindsay *et al.*, 1975). These magnetostratigraphic correlations indicate that most of the Vallecito Creek fauna occurs within Blancan Land Mammal Age, but that it also includes the early part of the Irvingtonian Land Mammal Age.

COMPARISON OF SAN PEDRO VALLEY BEDS WITH ANZA BORREGO

A previous study by Johnson and others (1975) has succeeded in placing the Blancan and Irvingtonian faunas of the San Pedro Valley, Arizona, within the framework of magnetic polarity stratigraphy. It would be instructive, therefore, to more closely compare the faunas of these two areas, since the time spans of the sediments involved are similar. Out of 32 genera, 22 (about 50%) are common to the 2 regions. In the San Pedro Valley the fossil-bearing beds begin about the level of the Mammoth event. At that level *Sigmodon*, *Geomys*, and *Neotoma* are initially recorded. In the Anza Borrego section, however, all three of these species occur earlier in time —*Geomys* and *Neotoma* near the base of the Gauss and *Sigmodon* at the base of the Cochiti event. Clearly a better and more complete record of these animals is present in the Anza Borrego area. Their absence in the lower strata of the San Pedro Valley probably results from a sedimentary regime that did not favor fossil preservation. *Equus* appears in both sections at about 3 my. It seems probable that *Equus* evolved within the time span between the *Pliohippus* occurrence at 4 my and the first *Equus* occurrence at about 3 my. This is in conflict with recent K–Ar dates from Idaho (Armstrong *et al.*, 1975), which would place the well-known Hagerman *Equus* fauna at 5 my BP. It seems to us that the Hagerman age assignment is too old by at least 1 my.

The lowest occurrence of *Sylvilagus* is very similar in both sections, appearing in the lowermost Matuyama in the San Pedro Valley and in the uppermost Gauss at Anza Borrego. *Sylvilagus* may turn out to be a very important guide fossil for the late Pliocene/early Pleistocene. *Lepus*, the hare, occurs in both regions and is important in the definition of the Ivringtonian Land Mammal Age. In Arizona, its first occurrence was placed below the Olduvai event along with the last occurrence of *Stegomastodon*. Interestingly enough, the appearance of *Lepus* at Anza Borrego is also associated with the only occurrence of *Stegomastodon*. At Anza Borrego the first appearance of *Lepus* occurs above the first event within the Matuyama. The ambiguous age of this event has previously been discussed. If this event is the Olduvai, then this would imply that *Lepus* appears 300,000 yr later at Anza Borrego than in the San Pedro Valley. If, however, the event is the Réunion event, then *Lepus* would appear almost simultaneously in both regions.

Several other animals occur in the two sections, one being *Odocoileus*, the deer, which appears in sediments of the upper Gauss epoch at McRae Wash in the San Pedro Valley and in sediments of the lower Matuyama epoch at Anza Borrego. *Tanupolama* appears near the Kaena event in both localities. *Dipodomys* (kangaroo rat) appears in both sections within the lower Matuyama epoch.

It is clear that an overall similarity of the ranges of common genera exists between the two areas even though the Anza Borrego section has a much more abundant fauna and higher rate of deposition than the San Pedro section. It is also clear that some important elements of the fauna which are present in the one area are missing from the other. For example, *Nannippus*, a three-toed horse, forms an important element of the San Pedro Valley fauna during the Gauss magnetic epoch, yet it is entirely missing at Anza Borrego. This difference may reflect different environments, or perhaps the geographic center of distribution of *Nannippus* was the high plains and the San Pedro Valley was on the margin of that range. The record of *Nannippus* is later (lower Matuyama epoch) in Texas. Similarly, sloths (e.g., *Megalonyx* and *Nothrotherium*) are present in the Anza Borrego region but absent from the San Pedro Valley during the Gauss epoch.

EURASIAN IMMIGRANTS

The ancestry of most genera of the Anza Borrego fauna is clearly from the New World; however, several genera (e.g., *Euceratherium* and *Odocoileus*) probably have an Old World origin. *Euceratherium*, the shrub ox, and *Odocoileus*, the deer, appear in the Anza Borrego section in the lower Matuyama epoch (1.7–2.0 my). It is reasonable to assume that the Bering land bridge was open to migration during the lower Matuyama magnetic epoch, allowing an exchange of some fauna at that time. *Mammuthus* (elephant), an Irvingtonian guide fossil, may or may not have crossed from Europe at this time. Unfortunately, this animal does not occur in the stratigraphically continuous part of the section under study. *Mammuthus* occurs 25 miles north of the Anza Borrego Desert, near Borrego, in a steeply dipping, truncated and faulted section which cannot be precisely correlated with the continuous section studied here. Samples of sediment from above and below the occurrence of *Mammuthus* near Borrego are reversely magnetized. All that can be said is that

Mammuthus occurs in the Matuyama reversed magnetic epoch. *Microtus*, which occurs in the Anza Borrego section just above *Euceratherium*, has been ascribed a Eurasian origin by Repenning (1967). However, its earliest occurrence in Eurasia is within the Biharian at a level now known to be younger than 0.9 my BP (Berggren and Van Couvering, 1974). Thus, *Microtus* occurs in California approximately 1 my prior to its appearance in Europe. Interestingly enough, *Lepus*, which has been ascribed a Eurasian origin, occurs almost synchronously on both continents, 1.9 my in Europe (Flint, 1971; Zagwijn, 1974) and simultaneously in the San Pedro Valley (Johnson *et al.*, 1975). Therefore, this leaves the continent of origin for *Lepus* in doubt; it does suggest the Bering land bridge was open in the lower Matuyama magnetic epoch (about 2 my BP).

THE BLANCAN–IRVINGTONIAN BOUNDARY

Paleomagnetic and faunal data from Texas (Mt. Blanco), Kansas (Borchers), and California (Anza Borrego) indicate the transition between the Blancan and Irvingtonian faunas (as defined by the small mammals) occurs within the lower Matuyama magnetic epoch (Lindsay *et al.*, 1975). This transition is well marked in the San Pedro Valley (Johnson *et al.*, 1975) and the Anza Borrego Desert. In both stratigraphic sequences the change from one fauna to the other is transitional and deposition is continuous across this transition, with gradual replacement of Blancan genera by Irvingtonian genera. For example, the earliest record of genera more characteristic of Irvingtonian than Blancan in each sequence is *Tetrameryx* (pronghorn) in the Anza Borrego sequence and *Ondatra* (muskrat) in the San Pedro Valley sequence. Both of these genera occur with an assemblage of dominantly Blancan genera. *Tetrameryx* is also recorded from the Blancan Cita Canyon

fauna of Texas (Johnston and Savage, 1955) which has a normal magnetic polarity, and is regarded as being in the Gauss magnetic epoch (Lindsay *et al.*, 1975). Recently, Zakrzewski (1974) reviewed the fossil record of Ondatrini, and suggested the *Pliopotamys meadensis* of the Dixon and Deer Park faunas of Kansas may have given rise to *Ondatra idahoensis* of the Blancan Grand View fauna of Idaho. Hence, *Tetrameryx* and *Ondatra*, though characteristic of Irvingtonian faunas occur in several late Blancan faunas. Other mammals which are more characteristic of Blancan faunas persist into faunas which have an Irvingtonian aspect; this is particularly true of *Stegomastodon*, both at the San Pedro Valley and at Anza Borrego.

A problem in defining the Blancan–Irvingtonian boundary in both the Anza Borrego and in San Pedro Valley sequences is the absence of *Mammuthus*, the diagnostic Irvingtonian guide fossil (Savage, 1951). Faunal evidence in these two sequences indicates that a better characterization of Irvingtonian is needed. Lundelius *et al.* (in press) suggest the beginning of the Irvingtonian be recognized by the appearance of *Mammuthus* and *Lepus* plus the absence of diagnostic Blancan species, e.g., *Sigmodon medius*, *Nannippus phlegon*, *Equus* (*Dolichohippus*) *simplicidens*, and *Borophagus diversidens*.

The Blancan–Irvingtonian boundary is arbitrarily placed at the *Smilodon* LSD in the Anza Borrego sequence, and at the *Lepus* LSD in the San Pedro Valley sequence. The *Smilodon* LSD marks the appearance of *Smilodon* and *Dipodomys* in the Anza Borrego sequence; the *Lepus* LSD marks the appearance of *Lepus* and *Dipodomys* in the San Pedro Valley sequence. All of these genera are characteristic of Irvingtonian faunas. The ?*Euceratherium* LSD that overlies the *Smilodon* LSD in the Anza Borrego sequence marks the appearance of *Euceratherium* and *Lepus*, two other characteristic Irvingtonian genera. The sediments of the Palm Springs

Formation indicate a gradual replacement of the Blancan fauna by Irvingtonian forms through evolution and emigration over a period of time of several hundred thousand years. Therefore, the exact placement of the boundary between the Blancan and Irvingtonian Land Mammal Ages is more academic than real. The boundaries as given above are really a convenience to indicate the point at which the fauna becomes more characteristic of Irvingtonian than Blancan. This boundary occurs at about 1.9 my (below the Olduvai event) in the San Pedro Valley, and about 1.6 to 2.0 my (above the Olduvai or Réunion event) in the Anza Borrego section. On both instances, the change to a dominantly Irvingtonian fauna occurs close in time to the Pliocene–Pleistocene boundary, as defined in marine sediments within the Olduvai event (Berggren *et al.*, 1967).

ACKNOWLEDGMENTS

The writers would like to thank Harley O. Garbani, who was of great assistance during the collection of the samples. We would also like to acknowledge Richard Merriam and G. Davidson Woodard, who provided useful background data. This research was supported by the National Science Foundation through Grants GB 12120 and BMS 72-02014-A03 to the Natural History Museum of Los Angeles County and through Grant DES 74-13860 to Columbia University. Lamont-Doherty Geological Observatory Contribution No. 2462.

REFERENCES

Armstrong, R. L., Leeman, W. P., and Malde, H. E. (1975). K–Ar dating, Quaternary and Neogene volcanic rocks of the Snake River Plain, Idaho. *American Journal of Science* **275**, 225–251.

Berggren, W. A., Phillips, J. D., Bertels, A., and Wall, D. (1967). Late Pliocene–Pleistocene stratigraphy in deep-sea cores from the south central North Atlantic. *Nature (London)* **216**, 253–254.

Berggren, W. A., and Van Couvering, J. A. (1974). The late Neogene. Biostratigraphy, geochronology and paleoclimatology of the last 15 million years in marine and continental sequences. *Paleogeography, Paleoclimatology and Paleoecology* **16**, ix–216.

Downs, T., and White, J. A. (1968). A vertebrate faunal succession in superposed sediments from Late Pliocene to Middle Pleistocene in California. *In* "Proceedings of XXIII International Geological Congress, Section 10, Tertiary–Quaternary Boundary," pp. 41–47. Academica Press, Prague.

Downs, T., and Woodard, G. D. (1961). Middle Pleistocene extension of the Gulf of California into the Imperial Valley (abs.). *Geological Society of America Special Papers* **68**, Number 12, 31.

Flint, R. F. (1971). "Quaternary Geology," p. 750. Wiley, New York.

Harrison, C. G. A., and Peterson, M. N. A. (1965). A magnetic mineral from the Indian Ocean. *American Mineralogist* **50**, 704–712.

Johnston, C. S., and Savage, D. E. (1955). A survey of various late Cenozoic vertebrate faunas of the panhandle of Texas. Part 1. Introduction, description of localities, preliminary faunal lists. *University of California Publications in Geological Sciences* **31**, 27–49.

Johnson, N., Opdyke, N. D., and Lindsay, E. (1975). Magnetic polarity stratigraphy of Pliocene–Pleistocene terrestrial deposits and vertebrate fauna, San Pedro Valley, Arizona. *Geological Society of America Bulletin* **86**, 5–11.

Kent, D. V., and Lowrie, W. (1974). Origin of magnetic instability in sediment cores from the central North Pacific. *Journal of Geophysical Research* **79**, 2987–3000.

Lindsay, E. H., Johnson, N. M., and Opdyke, N. D. (1976). Preliminary correlation of North American land mammal ages and geomagnetic chronology. *In* "Studies on Cenozoic Paleontology and Stratigraphy, Claude W. Hibbard Memorial Vol. 3. University of Michigan Papers on Paleontology No. 12" (G. R. Smith and N. E. Friedland, Eds.), pp. 111–119. Univ. of Michigan Press, Ann Arbor.

Løvlie, R., Lowrie, W., and Jacobs, M. (1971). Magnetic properties and mineralogy of four deep-sea cores. *Earth and Planetary Science Letters* **15**, 157–168.

Lundelius, E. L., Churcher, C. S., Downs, T., Harington, C. R., Lindsay, E. H., Schultz, G., Semken, H. A., and Zakrzewski, R. J. (in press). The North American Quaternary sequence.

McDougall, I., and Watkins, N. D. (1973). Age and duration of the Réunion geomagnetic polarity event. *Earth and Planetary Science Letters* **19**, 443–452.

McElhinny, M. W. (1966). An improved method for demagnetizing rocks in alternating magnetic fields. *Geophysical Journal of the Royal Astronomical Society* **10**, 369–374.

Merriam, R., and Bandy, O. R. (1965). Source of upper Cenozoic sediments in Colorado Delta region. *Journal of Sedimentary Petrology* **35**, 911–916.

Molyneux, L. (1971). A complete result magnetometer for measuring the remanent magnetization of rocks. *Geophysical Journal of the Royal Astronomical Society* **24**, 1–5.

Opdyke, N. D. (1972). Paleomagnetism of deep-sea cores. *Reviews of Geophysics and Space Physics* **10**, 213–249.

Repenning, C. A. (1967). Palearctic–Nearctic mammalian dispersal in the late Cenozoic. *In* "The Bering Land Bridge" (D. M. Hopkins, Ed.), pp. 288–311. Stanford Univ. Press, Stanford.

Savage, D. E. (1951). Late Cenozoic vertebrates of the San Francisco Bay Region. *University of California Publications in Geological Science* **28**, 215–314.

Tarbet, L. A. (1951). Imperial Valley, in possible future petroleum provinces of North America. *American Association of Petroleum Geologists Bulletin* **35**, 260–263.

Tedford, R. H. (1970). Principles and practices of mammalian geochronology in North America. *In* "North American Paleontology Convention, Proceedings F, Chicago, 1969," pp. 666–703.

Watson, G. S. (1956). A test for randomness of directions. *Monthly Notices of the Royal Astronomical Society, Geophysical Supplement* **7**, 160–161.

White, J. A. (1969). Late Cenozoic bats (Subfamily Nyctophilinae) from the Anza–Borrego Desert of California. *In* "Contributions in Mammalogy—A Volume Honoring Professor E. Raymond Hall," pp. 275–282. Kansas University Museum of Natural History Miscellaneous Publication 51.

Woodard, G. D. (1974). Redefinition of Cenozoic stratigraphic column in Split Mountain Gorge, Imperial Valley, California. *American Association of Petroleum Geologists Bulletin* **58**, 521–539.

Woodring, W. P. (1932). Distribution and age of the Tertiary deposits of the Colorado Desert. *Carnegie Institute of Washington Publication* **148**, 1–25.

Zagwijn, W. H. (1974). The Pliocene–Pleistocene boundary in western and southern Europe. *Boreas* **3**, 75–97.

Zakrzewski, R. J. (1974). Fossil Ondatrini from western North America. *Journal of Mammalogy* **55**, 284–292.

Part V

DATING THE CLIMATIC RECORD

Editor's Comments
on Papers 20 Through 24

The late Cenozoic has been one of those rare episodes in earth history marked by the presence of bipolar ice caps and by relatively rapid, high-amplitude global climatic changes associated with immense changes in the volume of these ice caps. These large-scale ice-volume changes have occurred on a quasiperiodic basis, particularly in the Northern Hemisphere. The oscillations in ice-sheet volume have in turn been directly responsible for late Cenozoic sea-level changes of up to nearly 120 m between glacial and interglacial episodes and have also led to conspicuous adjustments in oceanic circulation and related biogeographic changes.

The beginnings of our knowledge of late Cenozoic paleoclimatic history, especially where related to glacial changes, resulted from studies of terrestrial sediments transported and deposited by the vast continental ice sheets of the Northern Hemisphere.

These early studies led to a concept of only four major glacial expansions during the Quaternary, separated by warmer, interglacial "stages." It was also believed that ice sheets in the Northern Hemisphere began to form near the beginning of the Quaternary Period, which even in the early 1960s was generally considered to represent approximately the last 1 m.y. of earth history. Almost nothing was known about the age of the Antarctic ice sheet, although it was generally assumed that this should have undergone a similar history to the Northern Hemisphere ice sheets. These concepts were soon to change drastically.

During the early 1930s, Schott's pioneering efforts led to the discovery that planktonic foraminiferal assemblages in deep-sea cores exhibit changes in their character in response to the late Quaternary paleoclimatic changes associated with the changes in the Northern Hemisphere ice sheets. By this simple observation, Schott (1939) was to establish a foundation for later paleoclimatic studies based on microfossils preserved in deep-sea sediments. In the postwar period, these studies began to expand rapidly, initially as a result of the research of David Ericson and Goesta Wollin at Lamont Geological Observatory (Ericson et al., 1956), and later by other workers at Lamont and other institutions. At about the same time Cesare Emiliani, initially at the University of Chicago and later at the University of Miami, had begun to utilize changes in oxygen isotopic composition of planktonic foraminiferal tests in deep-sea cores, particularly from the Caribbean Sea, to study Quaternary paleoclimatic history (Emiliani, 1955). The paleoclimatic curves developed using this approach clearly established, after only a few years of work, the character of middle and late Quaternary climatic oscillations. However, the sequences needed to be dated. Good age determinations were obtained for the youngest part of the paleoclimatic record by the Carbon 14 method, which is effectively restricted to the dating of sequences younger than about 3×10^4 years; and the Ionium/Thorium method, restricted to sediments younger than about 3×10^5 years. The ages of deep-sea sediments older than this were not clearly known. The development of paleomagnetic stratigraphy of deep-sea sediments has provided, at least in part, this necessary chronological extension to include sequences as old as 5 m.y. Therefore its development has played a vital role in both the dating and correlation of climatic history recorded in cores. Even so, the earliest paleomagnetic studies applied to the problem of paleoclimatic history were not carried out on marine sediments but on land-based sections in Iceland by McDougall and Wensink (1966). As a result of paleomagnetic dating of basaltic lava flows containing interbedded tillites, these workers were able to demonstrate that the onset of Icelandic glaciation occurred about 3 m.y. ago, which in turn provided an important basis for later thinking about the development of Northern Hemisphere glaciations during the late Cenozoic. It seems most appropriate to discuss the development of magnetostratigraphy as applied to climatic history in terms of three major climatic regimes of the earth: the polar, temperate, and tropical areas.

The first workers to use magnetostratigraphy for dating climatic history in marine sequences were Opdyke and his colleagues (Paper 6); Hays and Opdyke (Paper 8) and Hays (1967). All these workers carried out preliminary paleomagnetic dating of ice-rafted debris in Antarctic piston cores and of the history of deposition of siliceous ooze as a reflection of increased biological productivity from increases in oceanic upwelling. These studies were to provide a firm foundation for later, more detailed studies of paleoclimatic and paleoglacial development in the Southern Ocean region, an area perhaps distinguished by the largest amount of such work. This earliest research was based largely on general sedimentological observations such as the stratigraphy of ice-rafted debris and diatom ooze. Slightly later contributions by Goodell et al. (1968) and Goodell and Watkins (1968) followed suit and provided much of the first important information on the history of Antarctic glaciation based on the paleomagnetic dating of glacial-marine sequences collected during U.S.N.S. *Eltanin* expeditions. Of particular importance were their demonstrations that sediment rafting by icebergs began to occur in the Southern Ocean prior to 5 m.y. ago (their oldest dated core); that major Antarctic glaciation thus developed much earlier than that of the Northern Hemisphere; and that the Antarctic ice sheet has been a permanent late Cenozoic feature for at least the last 5 m.y. They further provided some of the first evidence to show that major water-mass boundaries in the Southern Ocean have moved northward several degrees during the late Cenozoic but have not migrated much further south than present-day limits, thus demonstrating the relative warmth of the current interglacial episode. The relative abundance of ice-rafted debris in sediments of Gauss Epoch age (t = 2.43 to 3.32 m.y. B.P.) was interpreted to reflect a climax of late Cenozoic Antarctic glaciation, a conclusion later disputed by Warnke (1968; 1970). He argued that the relative abundance of ice-rafted debris may instead be a reflection of the amount of material available on land for later transportation by icebergs rather than a direct measure of the degree of glacial development.

The next phase of investigations related to dating of paleo-climatic history in the Antarctic was to be carried out using quanti-tatively determined oscillations in microfossil assemblages. By the late 1960s, quantitative and semiquantitative studies of micro-fossils in sedimentary sequences were beginning to take place in most parts of the oceans. The Antarctic region was no excep-tion, and the first intensive investigations were to be carried out at Florida State University, expedited by the accessibility of a large collection of Antarctic piston cores stored there in the National Antarctic Core Repository. The first such studies were by James Kennett (1970) who quantitatively examined planktonic foraminifera in cores as old as the middle Matuyama ($t = 1.2$ m.y. B.P.) and by Kurt Geitzenauer (1969; 1972), who for his Ph.D dissertation quantita-tively examined the calcareous nannofossil sequences in several cores. The latter contribution is also notable in being one of the first studies that quantitatively defined calcareous nannofossil trends for paleoclimatic and biostratigraphic information. Both studies identified Southern Ocean paleoclimatic trends during the middle to late Quaternary and dated these using paleomagnetic stratigraphy previously documented by Goodell and Watkins (1968). Kennett (1970) identified 8 intervals of climatic warming during the last 1.2×10^6 years, which in part could be correlated with paleoclimatic history recorded in tropical regions. Later quan-titative studies were also carried out on similar selected cores for radiolaria by Huddlestun (1971) and for silicoflagellates by Jendrze-jewski and Zarillo (1972). These studies were all limited by core length to ages younger than about 1.3 m.y. These ages encompass the Brunhes and latest Matuyama Epochs, which are still well within the Quaternary. It still remained to extend the dated paleoclimatic record into the earliest Quaternary and the Pliocene. This phase of the work was begun by John Keany and James Kennett (Paper 20). Keany was to carry out these studies for his M.S. thesis while at Florida State University. The major thrust of the work was to ex-tend quantitative planktonic foraminiferal and radiolarian trends into the early Quaternary and late Pliocene (to 2.4 m.y. B.P.) in the Southern Ocean sequence in order to establish a paleoclimatic history. The study revealed a change near the Brunhes-Matuyama boundary ($t = 0.7$ m.y. B.P.) from lower amplitude climatic oscilla-tions during the Matuyama Epoch to higher amplitude oscillations during the Brunhes Epoch. The planktonic foraminiferal evidence also indicates cooler conditions during the Matuyama Epoch, in conflict with the analyses of Bandy et al. (1971) who presented radio-larian evidence to indicate cooler conditions during the Brunhes

Epoch. The conflicting conclusions are yet to be resolved. Magnetostratigraphy developed for these sequences continued to enhance intercore correlations within the Southern Ocean region as well as to assist in the correlation of Southern Ocean sequences with those in warmer parts of the world's oceans. John Keany was later to extend this work as his Ph.D. dissertation at the University of Rhode Island to include quantitative radiolarian studies in sequences of early and middle Pliocene age, with particular emphasis on sequences of Gilbert age ($t = 5.18$ to 3.32 m.y. B.P.). His studies showed a distinct early-middle Pliocene cooling trend over a long interval of 1.6 m.y., in sharp contrast to the more rapid climatic changes typical of the Quaternary (Keany, 1973; 1978).

For many years there has been considerable interest in the nature and history of glacial-marine sediments, those sediments that have been dropped to the ocean floor from melting icebergs. The dating of sequences containing glacial-marine sediments has progressed for several years, especially in the Northern Hemisphere (Bramlette and Bradley, 1942; Conolly and Ewing, 1965). The earlier methods were marked by the usual limitations discussed elsewhere, but magnetostratigraphy has now provided a firm chronological basis for such research. The earliest studies were carried out on Southern Ocean sequences (Opdyke, et al., Paper 6; Goodell et al., 1968), although this was followed soon after by similar work in the North Pacific initially by Conolly and Ewing (1970) and later by Kent et al. (1971). In the north Pacific investigations, the sequences were dated using magnetostratigraphy, and correlated using a combination of paleomagnetic stratigraphy, tephrachronology, and biostratigraphy. Kent and his colleagues (1971) were among the first workers to distinguish clear paleoclimatic fluctuations simply on the basis of the number and geographic range of glacial-marine sedimentary layers. A dramatic paleoglacial change was inferred to have occurred during the early Quaternary about 1.2 m.y. ago, based on an upward increase in the number of glacial-marine layers that remained in higher abundance during the middle and late Quaternary.

Most of the early paleoglacial work, however, simply equated observed increases in ice-rafted debris accumulation rates with increased iceberg activity, which in turn was generally assumed to increase during glacial episodes. In 1974 magnetostratigraphy was first systematically applied to determine the chronology of ice-rafted debris over a wide latitudinal region (Watkins et al., Paper 21). Using limited evidence from only three cores, Watkins and his colleagues developed a model to explain the spatial and temporal distribution of ice-rafted debris (Paper 21) in which diachronous relations were inferred to exist of the relative abundances of ice-rafted debris. They proposed that during interglacial episodes, when the 10°C isotherm is at its southernmost position, the accumulation of ice-rafted debris is greater close to the Antarctic continent because of increased iceberg calving and that the debris decreases rapidly northward. During glacial episodes, the deposition of ice-rafted debris is minimal close to the continent because of ice-shelf growth but extends much further northward, corresponding to the movement of the 0°C isotherm. This model was tested later by the same research group using 9 cores (Keany et al., 1976). Although their later work basically confirmed the simple model of diachronous distribution of ice-rafted debris accumulation rates, they nevertheless discovered distortions in lateral variation of relationships between paleotemperature variation and debris accumulation. Unfortunately, attempts to test the detailed model failed because of a lack of enough precision for the intercore correlations even though several stratigraphic approaches were employed including paleomagnetism, biostratigraphy, and paleoclimatic stratigraphy. Their study clearly demonstrated that the stratigraphic and chronological precision obtained from magnetostratigraphy of deep-sea cores is too low for studies that require determination of highly detailed temporal relations.

TEMPERATE AREAS

Compared with the polar and tropical regions of the world, little work has yet been carried out on the paleomagnetic dating of climatic history in temperate areas. Almost all magnetostratigraphic dating of climatic change in temperate regions has been studies on land-based marine sections in New Zealand by Kennett et al. (Paper 14) and Kennett and Watkins (Paper 15). Aspects of the development of this work are also discussed in Part III. Almost no paleoclimatic work has yet been carried out in Northern Hemisphere sequences for which magnetostratigraphy is available, including the critical Neogene sections in Italy studied by Nakagawa and colleagues (1974; 1975; 1977). These sections contain some of the type reference stages for the world.

In New Zealand the sections dated using magnetostratigraphy by Kennett and colleagues contain planktonic foraminiferal assemblages that record late Cenozoic oscillatory movements of Subantarctic and cool-temperate water masses across the New Zealand region. A trend toward cooler conditions near the end

257

of the Miocene is marked by an increase in inferred cool-water planktonic foraminifera and a decrease in warm-water forms. The climax of this cooling was dated by Kennett and Watkins (Paper 15) to be between about 4.7 and 4.2 m.y. B.P. within the early Gilbert Epoch and the age of the Miocene-Pliocene boundary as used in New Zealand was dated at 4.3 m.y. B.P. This is younger than the generally accepted age of the Miocene-Pliocene boundary in the type section in southern Europe at about 5.1 m.y. B.P. (Berggren, Paper 11). The chronological placement of the New Zealand Miocene-Pliocene boundary was later disputed by Ryan et al. (1974), who consider it to be too young by about 0.5 m.y. Their age adjustments of the Miocene-Pliocene boundary in New Zealand were primarily carried out to obtain a correlation between the New Zealand latest Miocene Kapitean Stage marked by cool-water masses and an inferred glacio-eustatic lowering of sea level (Kennett, 1967) and the Messinian Stage of the Mediterranean basin, well known for its evaporitic sedimentary rocks. If the Kapitean and Messinian Stages are coeval, the discrepancy in ages assigned to these stages needed to be resolved. This has perhaps been resolved by Hornibrook (1977), who has reported a disconformity close to the Miocene-Pliocene boundary in the northern stratigraphic section (Mangapoike River) studied by Kennett and Watkins (Paper 15). It is still unclear how much time is represented by this disconformity.

The work of Kennett and Watkins (Paper 15) and by Kennett and others (Paper 14) also revealed that much of the early and middle Pliocene was marked by somewhat warmer conditions compared with the latest Miocene. This was followed, at about 2.4 m.y. ago, by a major late Pliocene cooling of the oceans indicated by planktonic foraminiferal changes and a positive O^{18} excursion indicating a large build-up in polar ice (Kennett et al., Paper 14; Shackleton and Kennett, 1975). This change is considered to represent the initial development of Northern Hemisphere ice sheets (Shackleton and Kennett, 1975; Shackleton and Opdyke, Paper 24). More rapid paleoclimatic oscillations followed during the remainder of the Pliocene and during the Quaternary in association with hugh fluctuations in ice volume of the Northern Hemisphere ice sheets.

TROPICAL AREAS

Within the tropical areas, almost all paleomagnetic dating of late Cenozoic climatic cycles has been carried out in piston cores from the eastern equatorial Pacific region. This region was discovered by Hays et al. (Paper 10) to be particularly useful because sedimentation rates were low enough to assure rather long late Cenozoic sequences. Furthermore, the East Pacific Rise and the ridge systems around the Galapagos Islands were shallow enough to enable accumulation of calcareous biogenic sediment important in paleoclimatic interpretations. The first paleomagnetic dating of tropical climatic cycles was carried out as part of the classical study by Hays and his colleagues at Lamont Geological Observatory (Paper 10). Their paleoclimatic interpretations, rather than being based on micropaleontological or oxygen isotopic changes, were made using oscillations in the relative abundance of calcium carbonate in the sediment. Interpretations were based on an assumption proposed several years earlier by Arrhenius (1952) that glacial episodes in this region are marked by higher calcareous biogenic content relative to clay and that interglacial episodes are poorer in calcium carbonate. During glacial episodes, winds are inferred to have been stronger, thus stimulating greater oceanic upwelling and creating higher biogenic productivity at the equatorial divergence. This in turn produced higher rates of calcareous biogenic sedimentation. These assumptions were later strongly supported by high correlations between the cycles in calcium carbonate abundance and oxygen isotopic changes (Emiliani, 1955; Hays et al., Paper 10; Shackleton and Opdyke, Paper 22). Hays and his colleagues (Paper 10) discovered that 8 carbonate peaks occurred during the Brunhes Epoch and hence, by implication, 8 major glacial episodes. Like the oxygen isotopic data of Emiliani (1955) over a decade earlier, this observation effectively questioned the classical concept of only a few (4) glacial "stages" within the entire Quaternary, which, furthermore, is 1 m.y. longer in duration than the Brunhes Epoch. In the equatorial Pacific, lower calcium carbonate in that part of the sequence older than the Brunhes Epoch was interpreted by Hays and his colleagues to reflect warmer average climatic conditions. A sedimentological change occurring near the Brunhes-Matuyama boundary was considered to reflect the initiation of continental glaciations at temperate latitudes, although the authors were careful to point out that much older glaciations previously developed at higher latitudes, especially the Antarctic. For instance, a major increase in calcium carbonate content in the upper Gauss Epoch was correctly interpreted by Hays et al. (Paper 10) to represent a major late Pliocene cooling event recorded by others in terrestrial and marine sequences (for summary, see Kennett, 1977). Thus these initial paleoclimatic interpretations of dated tropical sequences were made indirectly from sedimentological evidence. The paleomagnetic dating of oxygen isotopic cycles was to begin somewhat later. This approach is a much more powerful and direct tool for paleoclimatic interpretations because oxygen isotopic changes, when measured in certain deep-sea benthonic foraminifera, represent a direct measure of ice-volume fluctuations in the polar regions and hence of global climatic history.

The important step involving paleomagnetic dating of the late Cenozoic oxygen isotopic curves in tropical sequences has been carried out by the combined efforts of Nicholas Shackleton of Cambridge University, England, and Neil Opdyke of Lamont-Doherty Geological Observatory. During the middle 1970s, Nicholas Shackleton had begun to cooperate closely with several scientists at Lamont-Doherty Geological Observatory. In a series of three important papers (Papers 22, 23, and 24), Opdyke was to provide the magnetostratigraphy and Shackleton the oxygen isotopic history. The work commenced with the late Quaternary (Shackleton and Opdyke, Paper 22) and demonstrated the existence of 22 "stages" representing alternating greater and lesser volumes of Northern Hemisphere ice volumes during the last 870,000 years. Furthermore, close similarities between calcium carbonate and oxygen isotopic curves in the same sequences confirmed the earlier assumptions of Arrhenius (1952) and Hays et al. (Paper 10) in which they based their paleoclimatic interpretations on fluctuations in calcium carbonate content. This work was extended in a second paper (Shackleton and Opdyke, Paper 23) to include all the Quaternary Period in a single core (V28–239), thus providing a record spanning the Brunhes and much of the Matuyama Epochs. The particular value of this study was its demonstration of three distinct phases of paleoglacial change during the late Cenozoic as follows: (1) an upper part, from the Jaramillo Event to the present day (t = 0.9 m.y. to P.D.), in which glacial stages of approximately 1×10^5 year intervals have oscillated with interglacial episodes in rather uniform fashion and at the highest amplitude for the entire late Cenozoic; (2) an intermediate interval between 0.9 and 1.4 m.y. ago (middle Matuyama Epoch) during which the oxygen isotopic oscillations occur at about 4×10^4 year intervals; (3) an older sequence, between about 1.4 and 2.5 m.y. (early Matuyama Epoch), during which lower frequency oxygen isotopic oscillations exhibit an amplitude no greater than in the middle section, reflecting more stable paleoglacial conditions, but still about 2/3 of the latest Quaternary glacial maximum. The third paper (Shackleton and Opdyke, Paper 24) extended the record even further back through much of the Pliocene to include the Gauss and Gilbert Epochs. The most important result of this contribution was that it dated the onset, at about 3.2 m.y. (within the Gauss Epoch), of the quasi-cycle glacial-interglacial fluctuations that so clearly distinguish the latest Cenozoic. The isotopic evidence suggests that no large Northern Hemisphere ice sheets had accumulated before this time. Isotopically the ocean was in an interglacial state with relatively constant isotopic composition. Small oscillations may reflect changes in Antarctic ice-sheet volume or in small Northern Hemisphere glaciations, or may be due to analytical error. At about 3 m.y. ago this changed when large ice sheets began to form in Northern Hemisphere areas and show subsequent large oscillations in volume.

REFERENCES

Arrhenius, G. 1952. Sediment cores from the East Pacific. *Rept. Swedish Deep-Sea Exped.* **5**:89.

Bandy, O. L., R. E. Casey, and R. C. Wright. 1971. Late Neogene planktonic zonation, magnetic reversals, and radiometric dates, Antarctic to the Tropics. In *Biology of the Antarctic Seas II, Antarctic Research Series 15.* Washington, D.C.: American Geophysical Union, 26 pp.

Bramlette, M. N., and W. H. Bradley. 1942. Lithology and geological interpretations: Geology and biology of North Atlantic deep-sea cores. *U.S. Geol. Survey Prof. Paper 196*, 34 pp.

Conolly, J. R., and M. Ewing. 1965. Ice-rafted detritus as a climate indicator in Antarctic deep-sea cores. *Science* **150**:1822–1824.

Conolly, J. R., and M. Ewing. 1970. Ice-rafted detritus in Northwest Pacific deep-sea sediments. *Geol. Soc. America Mem. 126*, pp. 219–231.

Emiliani, C. 1955. Pleistocene temperatures. *Jour. Geology* **63**:538–578.

Ericson, D. B., W. S. Broecker, J. L. Kulp, and G. Wollin. 1956. Late-Pleistocene climates and deep-sea sediments. *Science* **124**:385–389.

Geitzenauer, K. R. 1969. Coccoliths as Late Quaternary paleoclimatic indicators in the Subantarctic Pacific Ocean. *Nature* **223**:170–172.

Geitzenauer, K. R. 1972. The Pleistocene calcareous nannoplankton of the Subantarctic Pacific Ocean. *Deep-Sea Research* **19**:45–60.

Goodell, H. G., and N. D. Watkins. 1968. The paleomagnetic stratigraphy of the Southern Ocean: 20° West to 160° East longitude. *Deep-Sea Research* **15**:89–112.

Goodell, H. G., N. D. Watkins, T. T. Mather, and S. Koster. 1968. The Antarctic glacial history recorded in sediments of the Southern Ocean. *Palaeogeography, Palaeoclimatology, Palaeoecology* **5**: 41–62.

Hays, J. D. 1967. Quaternary sediments of the Antarctic Ocean. In *The Quaternary History of the Ocean Basins*, Progress in Oceanography, Vol. 4. Oxford: Pergamon Press, pp. 117–131.

Hornibrook, N. de B. 1977. Dating of Pliocene volcanic rocks. *New Zealand Jour. Geology and Geophysics* **20**:466–467.

Huddlestun. P. 1971. Pleistocene paleoclimates based on radiolaria from Subantarctic deep-sea cores. *Deep-Sea Research* **18**:1141–1143.

Jendrzejewski, J. P., and G. A. Zarillo. 1972. Late Pleistocene paleotemperature oscillations defined by silicoflagellate changes in a Subantarctic deep-sea core. *Deep-Sea Research* **19**:327–329.

Keany, J. 1973. New radiolarian paleoclimatic index in the Plio-Pleistocene of the Southern Ocean. *Nature* **246**:139–141.

Keany, J. 1978. Paleoclimatic trends in Early and Middle Pliocene deep-sea sediments of the Antarctic. *Marine Micropaleo.* **3**:35–49.

Keany, J., M. Ledbetter, N. Watkins, and T.-C. Huang. 1976. Diachronous deposition of ice-rafted debris in Subantarctic deep-sea sediments. *Geol. Soc. America Bull.* **87**:873–882.

Kennett, J. P. 1967. Recognition and correlation of the Kapitean Stage (Upper Miocene, New Zealand). *New Zealand Jour. Geology and Geophysics* **10**:143–156.

Kennett, J. P. 1970. Pleistocene paleoclimates and foraminiferal biostratigraphy in Subantarctic deep-sea cores. *Deep-Sea Research* **17**:125–140.

Kennett, J. P. 1977. Cenozoic evolution of Antarctic glaciation, the Circum-Antarctic Ocean, and their impact on global paleoceanography. *Jour. Geophys. Research* **82**:3843–3860.

Kent, D., N. D. Opdyke, and M. Ewing. 1971. Climate change in the North Pacific using ice-rafted detritus as a climatic indicator. *Geol. Soc. America Bull.* **82**:2741–2754.

McDougall, I., and H. Wensink. 1966. Paleomagnetism and geochronology of the Pliocene-Pleistocene lavas in Iceland. *Earth and Planetary Sci. Letters* **1**:232–236.

Nakagawa, H., N. Kitamura, Y. Takayanagi, T. Sakai, M. Oda, K. Asano, N. Niitsuma, T. Takayama, Y. Matoba, and H. Kitazato. 1977. Magnetostratigraphic correlation of Neogene and Pleistocene between the Japanese Islands, Central Pacific and Mediterranean regions. In *Proceedings of the First International Congress on Pacific Neogene Stratigraphy*. Tokyo: Kaiyo Shuppan Co, pp. 285–310.

Nakagawa, H., N. Niitsuma, K. Kimura, and T. Sakai. 1975. Magnetic stratigraphy of Late Cenozoic stages in Italy and their correlatives in Japan. In *Late Neogene Epoch Boundaries*, edited by T. Saito and L. H. Burckle. New York: Micropaleontology Press, pp. 64–70.

Nakagawa, H., N. Niitsuma, N. Kitamura, Y. Matoba, T. Takayama, and K. Asano. 1974. Preliminary results on magnetostratigraphy of Neogene Stage strato-type sections in Italy. *Riv. Italiana Paleontologia e Stratigrafia* **80**:615–630.

Ryan, W. B. F., M. B. Cita, M. Dreyfus Rawson, L. H. Burckle, and T. Saito. 1974. A paleomagentic assignment of Neogene Stage boundaries and the development of isochronous datum planes between the Mediterranean, the Pacific and Indian Oceans in order to investigate the response of the World Ocean to the Mediterranean "salinity crisis." *Riv. Italiana Paleontologia e Stratigrafia* **80**:631–688.

Schott, W. 1939. Deep-sea sediments of the Indian Ocean. In *Recent Marine Sediments*, edited by, P. D. Trask. Tulsa: American Association of Petroleum Geologists, pp. 396–408.

Shackleton, N. J., and J. P. Kennett. 1975. Late Cenozoic oxygen and carbon isotopic changes at DSDP Site 284: Implications for glacial history of the Northern Hemisphere and Antarctica. In *Initial Reports of the Deep Sea Drilling Project*, Vol. XXIX, edited by J. P. Kennett et al. Washington, D.C.: U.S. Government Printing Office, pp. 801–807.

Warnke, D. A. 1968. Comments on a paper by H. D. Goodell and N. D. Watkins, "The paleomagnetic stratigraphy of the Southern Ocean 20° West to 160° East longitude." *Deep-Sea Research* 15:723–725.

Warnke, D. A. 1970. Glacial erosion, ice rafting and glacial marine sediments: Antarctica and the Southern Ocean. *Am. Jour. Sci.* 269:276–294.

ADDITIONAL READINGS

Bielak, L. E., and M. Briskin. 1978. Pleistocene biostratigraphy, chronostratigraphy and paleocirculation of the Southeast Pacific central water core RC11–220. *Marine Micropaleo.* 3:51–94.

Briskin, M., and W. A. Berggren. 1975. Pleistocene stratigraphy and quantitative paleo-oceanography of tropical North Atlantic core V16–205. In *Late Neogene Epoch Boundaries*, edited by T. Saito and L. H. Burckle. New York: Micropaleontology Press, pp. 167–198.

Ciesielski, P. F., and F. M. Weaver. 1974. Early Pliocene temperature changes in the Antarctic Seas. *Geology* 2:511–516.

Clark, D. L. 1969. Paleoecology and sedimentation in part of the Arctic Basin. *Arctic* 22:233–245.

Clark, D. L. 1970. Magnetic reversals and sedimentation rates in the Arctic Ocean. *Geol. Soc. America Bull.* 81:3129–3134.

Clark, D. L. 1974. Late Mesozoic and Early Cenozoic sediment cores from the Arctic Ocean. *Geology* 2:41–44.

Donahue, J. G. 1970. Pleistocene diatoms as climatic indicators in North Pacific sediments. *Geol. Soc. America Mem.* 126, pp. 121–138.

Ericson, D. B., and G. Wollin. 1968. Pleistocene climates and chronology in deep-sea sediments. *Science* 162:1227–1234.

Erickson, D. B., and G. Wollin. 1970. Pleistocene climates in the Atlantic and Pacific Oceans: A comparison based on deep-sea sediments. *Science* 167:1483–1485.

Fleck, R. J., J. H. Mercer, A. E. M. Nairn, and D. N. Peterson. 1972. Chronology of Late Pliocene and Early Pleistocene glacial and magnetic events in Southern Argentina. *Earth and Planetary Sci. Letters* 16:15–22.

Hunkins, K., A. W. H. Bé, N. D. Opdyke, and G. Mathieu. 1971. The Late Cenozoic history of the Arctic Ocean. In *The Late Cenozoic Glacial Ages*, edited by K. K. Turekian. New Haven: Yale University Press, pp. 215–237.

Keigwin, L. D., Jr. 1978. Pliocene closing of the Isthmus of Panama, based on biostratigraphic evidence from nearby Pacific Ocean and Caribbean sea cores. *Geology* 6:630–634.

Mörner, N.-A., and J. Lanser. 1975. Paleomagnetism in deep-sea core A179–15. *Earth and Planetary Sci. Letters* 26:121–124.

Saito, T. 1976. Geologic significance of coiling direction in the planktonic foraminifera *Pulleniatina*. *Geology* **4**:305–309.

Saito, T., L. H. Burckle, and J. D. Hays. 1975. Late Miocene to Pleistocene biostratigraphy of equatorial Pacific sediments. In *Late Neogene Epoch Boundaries*, edited by T. Saito and L. H. Burckle. New York: Micropaleontology Press, pp. 226–244.

Shackleton, N. J., and J. P. Kennett. 1975. Paleotemperature history of the Cenozoic and the initiation of Antarctic glaciation: Oxygen and carbon isotope analyses in DSDP Sites 279, 277, and 281. In *Initial Reports of the Deep Sea Drilling Project*, Vol. XXIX, edited by J. P. Kennett et al. Washington, D.C.: U.S. Government Printing Office, pp. 743–755.

Theyer, F. 1973. Reply to N. D. Watkins, J. P. Kennett, and P. Vella, 1973. *Nature Phys. Sci.* **244**:47–48.

Watkins, N. D., J. P. Kennett, and P. Vella. 1973. Palaeomagnetism and the *Globorotalia truncatulinoides* datum in the Tasman Sea and Southern Ocean. *Nature Phys. Sci.* **244**:45–47.

Wensink, H. 1964. Paleomagnetic stratigraphy of younger basalts and intercalated Plio-Pleistocene tillites in Iceland. *Geol. Rundschau* **54**:364–384.

20

Reprinted from *Deep-Sea Research* **19**:529 (1972)

Pliocene–early Pleistocene paleoclimatic history recorded in Antarctic–Subantarctic deep-sea cores

JOHN KEANY* and JAMES P. KENNETT[†]

(*Received* 12 *November* 1971; *in revised form* 29 *March* 1972; *accepted* 12 *April* 1972)

Abstract—Micropaleontological studies have been carried out on seven cores of middle Pliocene to early Pleistocene age (lower to upper Matuyama; $t = 0.7$ to 2.43 m.y. BP) from northern Antarctic and Subantarctic waters south of Australia and New Zealand. All of the cores contain abundant foraminiferal and radiolarian faunas, and have been dated by both micropaleontological and paleomagnetic methods. Fluctuations of a cold left-coiling *Globigerina pachyderma* fauna, with a warmer Subantarctic fauna define ten distinct intervals of warming during the Matuyama Reversed Epoch. This compares with six for the Brunhes Normal Epoch ($t = 0.0$ to 0.7 m.y. BP). Peaks formed by increased frequencies of warm-water forms and right-coiling *G. pachyderma* are of lower amplitude than warmer-water peaks in the Brunhes. It is thus inferred that Matuyama climatic fluctuations in the Southern Ocean are of lower magnitude than those in the Brunhes. Furthermore, the foraminiferal evidence indicates that average temperatures during the Matuyama are in general cooler than during the Brunhes. This conflicts with previous paleotemperature determinations based on radiolarian assemblages which suggest average warmer conditions throughout the entire Matuyama Epoch. The disappearance of present-day subtropical radiolarians, *Saturnulus planetes* and *Pterocanium trilobum*, near the Brunhes–Matuyama boundary in the Southern Ocean has previously been regarded as strong evidence for a deterioration of climatic conditions in the Brunhes. Throughout the Matuyama, however, these forms occur in association with cold (northern Antarctic–Subantarctic) planktonic foraminiferal assemblages, and thus apparently lost their environmental tolerance for northern Antarctic–Subantarctic waters about 0.7 m.y. BP. The opposite is true for the planktonic foraminifer *Globorotalia inflata* which adapted to Subantarctic waters approximately 0.7 m.y. BP. Both of these apparent ecological adaptations coincide very closely with the magnetic reversal at the Brunhes–Matuyama boundary. Middle to late Pliocene (lower Matuyama) climatic oscillations within Subantarctic–northern Antarctic waters correspond rather closely with those established for the New Zealand middle to late Pliocene, but appear to be out of phase with equatorial and North Atlantic regions.

[*Editor's Note:* The article itself has not been reproduced here.]

*Department of Geology, Florida State University, Tallahassee, Fla. 32306, U.S.A.
Graduate School of Oceanography, University of Rhode Island, Kingston, R.I. 02881, U.S.A.

21

Reprinted from *Science* **186**:533–536 (1974)

ANTARCTIC GLACIAL HISTORY FROM ANALYSES OF ICE-RAFTED DEPOSITS IN MARINE SEDIMENTS: NEW MODEL AND INITIAL TESTS

Norman D. Watkins, John Keany, Michael T. Ledbetter, and Ter-Chien Huang

Abstract. *Contrasts between the latitudinal distributions of ice-rafted debris deposited in deep-sea sediments during Pleistocene glacial and interglacial periods are predicted by a new model. The model requires the existence of a restricted zone where rates of deposition of ice-rafted debris are essentially independent of glacial-interglacial cycles. Initial tests and published results show that the concept is valid in the Southern Ocean and that it provides a new means of diagnosing major migrations of climatic zones.*

Deep-sea sedimentary cores are prolific recorders of Earth history, simultaneously providing evidence of past changes in climate, microfauna and flora, sea floor dynamic processes, geomagnetism, atmospheric particulate transport, and other phenomena. In high latitudes, marine sediments also include materials deposited by melting icebergs. Study of the spatial and temporal distribution of such ice-rafted debris (IRD) has long been recognized as a promising means of diagnosing the behavior of the polar ice caps (*1*). The advantage of the method over conventional land-based geological techniques is particularly obvious for Antarctica, where almost all relevant geological evidence (especially that for interglacial periods) remains inaccessible (*2*). Studies of deep-sea sedimentary cores have conclusively shown, for example, that Antarctica was a source of icebergs as early as the Eocene (*3, 4*), whereas similar results from studies on the continent are very difficult to obtain (*2*).

As stressed by Denton *et al.* (*2*), however, analyses of IRD have not provided any substantial advances in understanding the details of past fluctuations of the Antarctic ice cap, and it is the unraveling of this history which has maximum prospective significance. Most published results concerning IRD have simply equated observed increases in concentrations with increased iceberg activity, which is, in turn, almost invariably assumed to occur during glacial periods (5–7). In contrast, Denton et al. (2), Fillon (8), and Anderson (9) have warned that IRD maxima may, in some instances, be associated with interglacial periods. It is now well established that ice has been present on Antarctica throughout at least the Pleistocene (6), so that iceberg production has been continuous. Warnke (10) has summarized the factors that prevent simple interpretation of glacial deposits in deep-sea sediments, stressing the hypothesis that the debris available for incorporation into the parent glaciers may systematically diminish as subglacial erosion continues on the continent, to yield icebergs with little or no debris toward the end of an erosional cycle. Another important restriction is that ice shelves, presumably devoid of debris in most cases, are the source of the majority of icebergs (2). In several studies (6) between-core variations of IRD do not appear to be coherent, while in others (11) correlations of IRD have been made between cores. No consideration has hitherto been given to the possibility of diachronism in IRD horizons.

In Fig. 1 we present a simple model that not only reconciles previous conflicts in the interpretation of occurrences of IRD, but also clarifies the full potential value of IRD studies in selected regions.

Figure 1a is designed to present the following ideas with particular reference to the Southern Hemisphere. Interglacial periods, when the 0°C isotherm is in its southernmost latitudes, are marked by minimal ice-shelf extent, and more frequent calving of icebergs from the coastal glaciers can be expected; this creates an accumulation rate of IRD which is high close to the continent but decreases rapidly northward away from the source. During glacial periods, however, ice shelves will assist in creating a relatively barren zone close to the coasts. The more northerly position of the 0°C isotherm will result in a more northerly zone of melting and deposition. The relative amounts of IRD released during glacials and interglacials will be reflected in the areas under the two curves in Fig. 1a: This will depend on durations,

intensities, and rates of onset and retreat of the respective periods.

The relevance of the model to studies of the Pleistocene IRD record in deep-sea sedimentary cores is illustrated in Fig. 1b, which is derived from Fig. 1a, but expanded into a time-dependent series under the assumption of oscillating glacial and interglacial periods: the four hypothetical curves are thus equivalent to the records which sediment cores could ideally be expected to provide for the relative latitudes indicated. Three main points emerge from inspection of Fig. 1b: (i) the amplitude of the IRD signal will vary systematically and simply with lat-

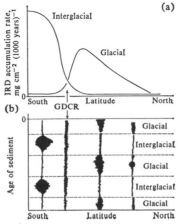

(b) **Fig. 1.** Model explaining the spatial and temporal distribution of ice-rafted debris (IRD) in deep-sea sediments of the Southern Ocean during glacial and interglacial periods. (a) During interglacial periods, when the 0°C isotherm is at its southernmost position, the IRD accumulation rate is higher close to the continent (to the south) because of increased iceberg calving, and decreases rapidly northward. During glacial periods IRD deposition is minimized close to the continent by ice-shelf growth, but extends much farther northward, corresponding to the movement of the 0°C isotherm. (b) Adaptation of the model in (a) to predict IRD accumulation rates during alternating glacial and interglacial periods as a function of latitude in four sites. The IRD deposition maxima will be during interglacials in southernmost latitudes and during glacial periods in more northern latitudes. The glacial debris conjugate region (GDCR) is the zone where the two curves in (a) intersect, so that the glacial debris accumulation rate is essentially independent of glacial-interglacial cycles. The position of the GDCR would migrate substantially in time if the intensities and rates of initiation and termination of glacial periods were variable. See text for further discussion.

itude, but the variation between cores could appear to be complex if glacial and interglacial periods are not independently defined. (ii) The IRD maxima and minima will be diachronous. (iii) In the latitudes where the two curves intersect in Fig. 1a there must be a region where the IRD accumulation rate is essentially independent of the timing of glacial and interglacial periods. We call this the glacial debris conjugate region (GDCR). Here there will be minimal correlation between IRD accumulation rates and paleotemperatures, in contrast to expected strong positive and strong inverse correlation south and north of the GDCR, respectively.

As a simple test of this model, we have examined the IRD and paleotemperature record for the Brunhes to Matuyama epoch in five sedimentary cores collected by the R.V. *Eltanin* from the southeastern Indian Ocean (Fig. 2a). The cores were selected because of their coherent paleomagnetic signatures and known sedimentation rates. The coarse fractions (> 62 μm) are siliceous and calcareous planktonic microfossils, IRD, and minor amounts of manganese micronodules and volcanic glass shards. Siliceous components become subordinate in the lower latitudes. The IRD is subangular to subrounded quartz and feldspar grains, and rock fragments. Samples of volume 8 cm³ were taken from all cores at intervals of 5 to 10 cm. Conventional sedimentological methods were used to separate the fraction in the size range 62 to 250 μm, which we believe provides the optimum IRD signal. The percentage of IRD was obtained from 300-grain counts with transmitting and binocular microscopes. The criterion for diagnosing IRD is well established (12). An apparent accumulation rate, in milligrams per square centimeter per 1000 years, is then derived by using the volume, average density, and paleomagnetically and micropaleontologically determined sedimentation rates.

The most practical paleoclimatic index available is the percentage variation of the cold water radiolarian *Antarctissa strelkovi*, which Keany (13) has shown to be closely related to paleoclimatic variations identified by using planktonic foraminifera (14). Earlier, Petrushevskaya (15) showed that the surface sediment distribution of *Antarctissa strelkovi* in the sub-Antarctic corresponds closely to the observed sea surface distribution. Slides of the fraction > 62 μm were prepared by conventional microfaunal methods, and entire slide populations, ranging from 300 to 1000 specimens, were counted for all samples. The results for three of the five cores examined are illustrated in Fig. 2, b to d. Core E48-03 proved to be almost totally devoid of IRD, indicating the general absence of melting coastal icebergs in the area around latitude 40°S for the past 0.3 million years.

The paleoclimatic curves for the three cores do not correlate with those for equatorial and North Atlantic regions (*16, 17*). Ruddiman (*17*) mentioned that Pleistocene Antarctic sea temperatures seem to have been either out of phase or independent of those of the Northern Hemisphere and equatorial oceans. Differences in the paleoclimatic curves of E50-12 and E49-24 are due to a disconformity at the top of the latter, so that the upper 0.4 million years is absent.

The data for core E50-12 (Fig. 2b) exhibit a clear and consistent positive correlation between warmer waters (or interglacial periods) and increased IRD accumulation rates, as required by the segment of our model (Fig. 1) poleward of the GDCR. Exactly the same relationship has been observed throughout core E49-30 (Fig. 2a).

In core E49-24 (Fig. 2c) relatively large and frequent changes in IRD accumulation rates have been detected, but there is no correlation with the inferred temperature changes. We suspect that this core may be in the GDCR, and because of its close proximity to the present Antarctic Convergence (Fig. 2a) we believe that the positions of the Antarctic Convergence and the GDCR may be intimately related, particularly since iceberg occurrences could be expected to diminish north of the Convergence. Support for this proposal can be found in similar data for site 278 of the Deep-Sea Drilling Project (*18*), located on the present Antarctic Convergence in the Emerald Basin at latitude 56°33.4'S. The number of glacial quartz grains per sample shows virtually no relation to the percentage of *Antarctissa strelkovi* throughout 65 m of core, except in limited instances where an inverse correlation is present: This could be explained by occasional southward migration of the GDCR or Antarctic Convergence, with the site at other times being mostly in the GDCR. Core E49-24 (Fig. 2c) has much higher IRD accumulation rates than the cores to the south, conceivably because the core is in a region of rapid iceberg melting.

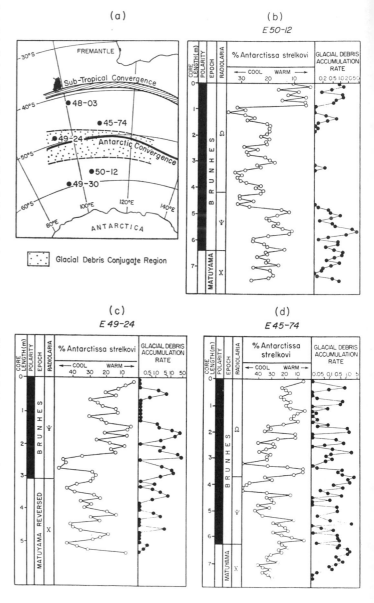

Fig. 2. (a) Map showing the locations of the five deep-sea sedimentary cores examined; the cores were obtained by the R.V. *Eltanin.* The exact core locations and water depths are: E49-30, 59°0.3'S, 95°13.8'E, 2339 fathoms (1 fathom ≈ 1.8 m); E50-12, 57°57.2'S, 105°01'E, 2407 fathoms; E49-24, 47°59.3'S, 95°02.2'E, 1757 fathoms; E45-74, 47°33.1'S, 114°26.4'E, 2080 fathoms; and E48-03, 41°01.2'S, 100°00.7'E, 2149 fathoms. The present positions of the Antarctic Convergence and Subtropical Convergence (*21*) are added. The approximate position of the possible glacial debris conjugate region (see Fig. 1) is added. (b to d) Glacial debris accumulation rates and paleoclimatic curves of the selected cores. The diagrams show core length, paleomagnetic stratigraphy, polarity epoch (*22*), radiolarian zones (*7, 23*), percentage of *Antarctissa strelkovi* (*13, 15*), and glacial debris accumulation rate for the fraction in the size range 62 to 250 μm (in milligrams per square centimeter per 1000 years).

which could be expected to result from the temperature rise at the Antarctic Convergence.

In core E45-74 (Fig. 2d) the IRD accumulation rate and the water temperature are positively correlated in most of the lower 500 cm. This segment of the core is therefore consistent with a location south of the GDCR, the approximate geographic limits of which are added to Fig. 2a. Similar strong positive correlations (between percentage of glacial quartz and inferred water temperature) have been documented by Margolis and Kennett (4) for two Pleistocene cores just north of the present Antarctic Convergence in the south-central Pacific. These occurrences, and their contrast with the data for E50-12 and E49-30, must be accepted as strong support for the major concepts involved in the model. The upper 300 cm of core E45-74 (Fig. 2d) exhibits a strong inverse relation between IRD and temperature, consistent with the core location being north of the GDCR. This requires the GDCR to have migrated southward at the time involved which, if constant sedimentation rates are assumed, is estimated paleomagnetically as about 0.33 million years ago.

In summary, we believe that our model (Fig. 1) may provide a means to more fully utilize the IRD signal recorded in all subpolar sediments, although we realize that it may be subject to modification (19). The limited tests we have applied and the earlier published data merely support the principles employed. Simple correlations between IRD maxima in deep-sea sediments (11) may not be valid unless the cores involved are all located on one side of the GDCR. Previously noted incoherence in correlations between cores (6) may be at least partially explained by the latitudinal range of the cores involved. It will be unrewarding to attempt to make paleoclimatic inferences from IRD abundance variations in cores spanning large latitude ranges. On the other hand, detailed and diverse analyses of closely spaced south-north traverses may yield definitive data on the timing, rate of change, extent, and intensity of the glacial and interglacial periods in the Southern Hemisphere (20).

References and Notes

1. J. Murray and A. F. Renard, *Proc. R. Soc. Edinb.* **12**, 474 (1884); E. Phillipi, *Geol. Geogr.* **2**, 415 (1910).
2. G. H. Denton, R. L. Armstrong, M. Stuiver, in *The Late Cenozoic Glacial Ages*, K. Turekian, Ed. (Yale Univ. Press, New Haven, Conn., 1971), p. 267.
3. K. R. Geitzenauer, S. V. Margolis, D. S. Edwards, *Earth Planet. Sci. Lett.* **4**, 173 (1968).
4. S. V. Margolis and J. P. Kennett, *Am. J. Sci.* **271**, 1 (1971).
5. J. R. Connolly and M. Ewing, *Science* **150**, 1822 (1965); *Nature (Lond.)* **208**, 135 (1965); *Geol. Soc. Am. Mem.* **126**, 219 (1970); N. D. Opdyke, B. Glass, J. D. Hays, J. Foster, *Science* **154**, 349 (1966); J. P. Kennett and C. A. Brunner, *Geol. Soc. Am. Bull.* **84**, 2043 (1973); Goodell *et al.* (6); Hays and Opdyke (7). It is now becoming clear that the terms "glacial" and "interglacial" may not be satisfactory, particularly when applied to Antarctica, since it is virtually certain that the continent has been almost entirely covered with ice for most of the upper Tertiary.
6. H. G. Goodell, N. D. Watkins, T. T. Mather, S. Koster, *Palaeogeogr. Palaeoclimatol. Palaeoecol.* **5**, 41 (1968).
7. J. D. Hays and N. D. Opdyke, *Science* **158**, 1001 (1967).
8. R. H. Fillon, *Nat. Phys. Sci.* **238**, 40 (1972).
9. J. B. Anderson, *ibid.* **240**, 189 (1972).
10. D. A. Warnke, *Am. J. Sci.* **269**, 276 (1970).
11. D. Kent, N, D. Opdyke, M. Ewing, *Geol. Soc. Am. Bull.* **82**, 2741 (1971); R. Von Huene, J. Crouch, E. Larson, *Geol. Soc. Am. Mem.*, in press. The latter authors specifically state that between-core correlation of IRD horizons is a critical means of evaluating data quality.
12. D. H. Krinsley and J. Donahue, *Geol. Soc. Am. Bull.* **79**, 743 (1968); D. H. Krinsley and S. V. Margolis, *Trans. N.Y. Acad. Sci.* **31**, 457 (1969); W. B. Whalley and D. H. Krinsley, *Sedimentology* **21**, 87 (1974).
13. J. Keany, *Nature (Lond.)* **246**, 139 (1970).
14. J. P. Kennett, *Deep-Sea Res.* **17**, 125 (1970); J. Keany and J. P. Kennett, *ibid.* **19**, 529 (1972).
15. M. G. Petrushevskaya, in *Biological Reports of the Soviet Antarctic Expedition, 1955–1958*, A. P. Andriyaskev and P. V. Ushakov, Eds. (Program for Scientific Translations, Jerusalem, Israel, 1968), vol. 3, p. 69.
16. A. McIntyre and R. Jantzen, *Resumes Commun. 7th Int. Quaternary Assoc. Congr.* **11**, 68 (1969).
17. W. F. Ruddiman, *Geol. Soc. Am. Bull.* **82**, 283 (1971).
18. S. V. Margolis, in *Initial Reports of the Deep-Sea Drilling Project, Leg 29* (Government Printing Office, Washington, D.C., in press).
19. We subscribe to the philosophy that there are three natural stages in the compilation of models: an initial simple stage, to provide the essential working model; a second stage involving the emergence of data which frequently clash with the initial model requirements, thus forcing changes in the model; and a final stage where data and model blend in a more complex fashion, or where the model collapses. In this report, we describe our first steps in the second stage.
20. See the models illustrated as figure 8 of Denton *et al.* (2, p. 293).
21. A. L. Gordon, in *Studies in Physical Oceanography—A Tribute to G. Wüst on His 80th Birthday*, A. L. Gordon, Ed. (Gordon & Breach, New York, in press).
22. A. Cox, *Science* **163**, 237 (1969). For the methods used, see N. D. Watkins and J. P. Kennett, *Antarct. Res. Ser.* **19**, 273 (1972).
23. J. D. Hays, in *Progress in Oceanography*, M. Sears, Ed. (Pergamon, Oxford, 1967), p. 117.
24. Supported by NSF grant GV25400.

17 July 1974; revised 26 August 1974   ■

Oxygen Isotope and Palaeomagnetic Stratigraphy of Equatorial Pacific Core V28-238: Oxygen Isotope Temperatures and Ice Volumes on a 10⁵ Year and 10⁶ Year Scale *

NICHOLAS JOHN SHACKLETON[1] AND NEIL D. OPDYKE[2]

Core Vema 28-238 preserves an excellent oxygen isotope and magnetic stratigraphy and is shown to contain undisturbed sediments deposited continuously through the past 870,000 yr. Detailed correlation with sequences described by Emiliani in the Caribbean and Atlantic Ocean is demonstrated. The boundaries of 22 stages representing alternating times of high and low Northern Hemisphere ice volume are recognized and dated. The record is interpreted in terms of Northern Hemisphere ice accumulation, and is used to estimate the range of temperature variation in the Caribbean.

INTRODUCTION

The application of magnetic stratigraphy (Hays *et al.*, 1969) has demonstrated that long undisturbed Pleistocene sequences may be obtained from the floor of the Equatorial Pacific. On the other hand oxygen isotope analysis, giving information about past ice volumes and sea surface temperatures, has only been applied in the Atlantic region. The present study places the established oxygen isotope stratigraphy within the palaeomagnetic framework.

The Pacific Ocean is more suitable than the Atlantic for setting up an oxygen isotope stratigraphy, because the complicating effect of temperature change is smaller, so that the glacially induced ocean isotopic changes dominate. Moreover, it is possible to utilize the data drawn from the Pacific to derive temperatures in the Atlantic re-gion, and this we also attempt in the present study. This is of particular interest in relation to the continuing work of CLIMAP following Imbrie and Kipp (1971).

PALEOMAGNETISM OF CORES V28-238 AND 239

During the twenty-eighth cruise of the R.V. Vema two piston cores V28-238 (01°01′ N, 160°29′ E) and V28-239 (3°15′ N, 159°11′ E) were raised from the Solomon Plateau from water depths of 3120 and 3490 m, respectively. Both cores are highly calcareous. On the basis of the preliminary paleontological study of the bottom samples, both cores were selected for paleomagnetic study.

Samples from V28-238 were taken at 10-cm intervals throughout the length of the core, and from V28-239 initially at 50-cm intervals. After reversals were discovered in them, closer sampling was carried out near reversal boundaries in order to better define their positions. The sampling procedure used is described in Foster and Opdyke (1970). Measurements of the

---

[1] Subdepartment of Quaternary Research, University of Cambridge in association with CLIMAP.

[2] CLIMAP (Lamont-Doherty Geological Observatories).

* Contribution No. 1970 of the Lamont-Doherty Geological Observatories.

direction and intensity of the remanent directions were carried out on a 5-cps spinner magnetometer described by Foster (1966).

Magnetic Stability

Alternating field partial demagnetization was done on three specimens from V28-238, and one from V28-239, using equipment similar to that described by McElhinny (1966). Figure 1 shows the a.f. demagnetization curves obtained from these specimens. It can be seen that the percentage of remanent intensity drops very slowly with increasing fields, indicating that the magnetization has very high coercivity. The median destruction field (the field required to reduce the NRM by 50%) ranges from 170 to 320 Oe, values which indicate very good stabilities. A small secondary component is, however, present in the specimen from V28-239 from below the Brunhes/Matuyama boundary which is shown by the initial rise in the value of remanent intensity in low fields. In all the samples studied the change in direction on a.f. demagnetization is insignificant. The high stability of magnetization of cores from the western equatorial Pacific has previously been noted by Kobayashi *et al.* (1971).

Magnetic Stratigraphy

Figure 2 shows the change in declination plotted against depth for V28-238 and 239. The internal orientation of the cores has been preserved and the declination change shown is relative only to the split face of the core. The only complete reversal in V28-238 occurs at 1200 cm and is interpreted to correlate to the reversal marking the Brunhes/Matuyama boundary 700,000 y.a. (Dalrymple, 1972). The core apparently terminates just above the Jaramillo event at 16 m.

Core V28-239 has a slower rate of sedimentation than V28-238 and consequently a larger segment of time is represented in this core. Five 180° reversals of polarity are observed at 735 cm (B/M), at 879 (J/M), 943 (M/J), 1550 (O/M), and 1779 (M/O) and have been determined to within ±5 cm. The top of the core down to 3 m is severely disturbed, probably during the extrusion process on the ship. The interpretation of the magnetic stratigraphy is shown in Fig. 2 and is supported by faunal and floral analysis (Saito and Geitzenauer, personal communication). This core probably contains one of the best and most complete records of the entire Pleistocene that is known. Unlike this core, other cores from the eastern equatorial Pacific often show severe solution effects in the interval between the Jaramillo and Olduvai events, severely restricting their usefulness over this segment of time (Hays *et al.*, 1969).

Principles of Oxygen Isotope Analysis

The idea of using oxygen isotopic abundance ratio changes as a means of estimat-

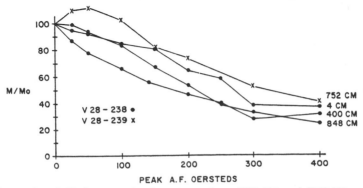

Fig. 1. Alternating field demagnetization curves from V28-238 and V28-239 expressed as a percentage of the original NRM.

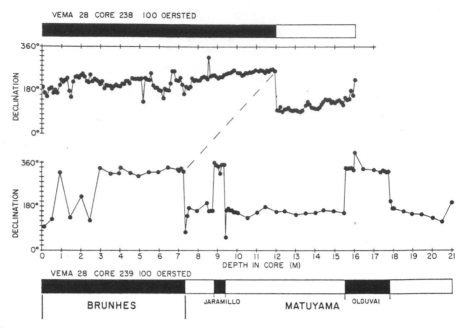

Fig. 2. Plot of the change in declination for cores V28-238 and V28-239 plotted against depth in core and its correlation with the standard magnetic time scale. The cores are not oriented with respect to true north.

ing past temperatures was first proposed by Urey (1947). He pointed out that since the small isotopic fractionation which takes place when a carbonate is deposited slowly from aqueous solution is temperature-dependent, the temperature of deposition may be estimated by measuring the extent of the isotopic fractionation. In the 25 yr that have elapsed since Urey's proposal was published, a number of workers in laboratories throughout the world have used his idea to study geological temperatures in a wealth of contexts. The study of palaeo-temperatures in the Pleistocene was pioneered by Emiliani; many of the ideas in this paper may be found in his classic "Pleistocene Temperatures" (1955). This, together with his later major paper (1966a) will be referred to frequently; together they constitute a sufficiently concise review of the field to obviate the need for any further introduction here.

Emiliani (1955) modified the Epstein equation from which palaeotemperatures are derived to

$$T = 16.5 - 4.3(\delta = A) + 0.14(\delta - A)^2, \quad (1)$$

where δ represents the difference between the ^{18}O content of the sample analysed and a standard carbonate PDB-1 (see Appendix 1) in parts per thousand (‰);

$$\delta = 1000 \left\{ \frac{(^{18}O/^{16}O)_{sample}}{(^{18}O/^{16}O)_{standard}} - 1 \right\}. \quad (2)$$

Equation (1) differs from that derived by Epstein et al. (1951, 1953) in the incorporation of the term A, representing the difference between the isotopic composition of the ancient sea water in which the fossil analysed lived, and average marine water today. A is defined in an analogous manner to δ in Eq. (2). The primary objective of Emiliani's work was to derive a palaeo-temperature sequence through the Pleistocene; he introduced the term A into Eq. (1)

so as to be able to correct for the changing isotopic composition during each glacial–interglacial cycle. For this purpose Emiliani (1955) estimated A in Eq. (1) to have changed by 0.4‰ between glacial and interglacial stages.

The magnitude of the correction factor was criticised by Olausson (1965) on theoretical grounds; he estimated the isotopic composition of the Pleistocene ice sheets and evaluated the effect on the oceans as about 1.1‰. Shackleton (1967) proposed that the isotopic composition of benthonic foraminifera from the ocean depths should provide the best experimental derivation of the changing isotopic composition of the oceans. He also noted that from the stratigraphic point of view a record of ocean isotopic composition change through the Pleistocene is of even more value than is a temperature record, in view of its direct relationship to ice volume and sea level. It is now widely accepted that the ocean was just over 1‰ isotopically more positive at the last glacial maximum than today (Shackleton, 1967, 1968; Dansgaard and Tauber, 1969; van Donk and Mathieu, 1969; Duplessy *et al.*, 1970) but no previous study has measured ocean isotopic composition changes over a longer interval.

Oxygen Isotope Analyses: the Past 130,000 yr

The upper 230 cm of core V28-238 have been sampled at 5-cm intervals. *Globigerinoides sacculifera* has been selected as a representative planktonic foraminiferal species for analysis, since it is known to yield good results (Emiliani 1955, 1966a). Eight specimens were used for each analysis; most horizons were analysed in triplicate so as to obtain a good precision as well as a reliable estimate of the overall analytical errors. The precision of a single analysis may be estimated at ±0.11‰ (1-σ) on the basis of the repeat analyses, so that the precision of the mean at each level should be ±0.07‰ (1-σ). An alternative method of obtaining an estimate of overall precision is to consider the average magnitude of the difference between analyses at adjacent core horizons on the assumption that away from the regions of very rapid isotopic change the main contribution to the apparent noisiness of the record is analytical error. This method also yields an estimated precision of ±0.07‰ (1-σ). Analyses are referred to Emiliani's standard B-1; the precision of this comparison is better than ±0.02‰. B-1 appears to be about (+0.2 ± 0.1)‰ with respect to the now used up PDB1 Chicago standard; interlaboratory calibrations are not at present sufficiently reliable to warrant a more precise calibration to PDB-1, for which reason we prefer not to quote our analyses with respect to this standard (see Appendix).

Analyses of *G. sacculifera* are listed in Table 1[3] and plotted against depth in core in Fig. 3. The record obtained shows de-

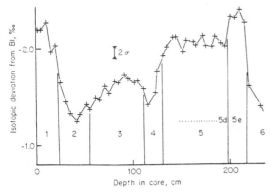

FIG. 3. Oxygen isotopic composition of *G. sacculifera* in core V28-238, top 230 cm, expressed as deviation ‰ from Emiliani B1 standard. Stages 1–6 are numbered on the basis of correlation with the stages defined by Emiliani (1955). Stage boundaries are defined here taking into account information contained in Fig. 4. Substage 5d/5e transition defined on the basis of Shackleton (1969).

[3] For Tables 1 and 2, order NAPS Document No. 02064 from ASIS/NAPS, c/o Microfiche Publications, 305 E. 46 St., New York, NY 10017, remitting $1.50 for microfiche or $5.00 for photocopies, with checks payable to Microfiche Publications.

tailed correlation with those obtained from the Atlantic and Caribbean by Emiliani; this correlation may be seen by comparing Fig. 3 with Fig. 5, taken from Emiliani's review of 1961. It is our contention that the reason for this detailed correlation is that the fine structure of the record, as well as the overall pattern, is controlled by changes in the isotopic composition of the oceans brought about by terrestrial ice volume changes. If this is the case, then the only limitations in interoceanic correlation are ocean mixing (less than 10^3 yr) and sediment mixing.

It was not possible to use a single benthonic species for analysis as no one species is sufficiently abundant. On the other hand it is clear from the work of Duplessy *et al.* (1970) that since some benthonic species (if not all) deposit carbonate slightly away from isotopic equilibrium with sea water at the temperature of test secretion, it is important to analyse monospecific samples. The record shown in Fig. 4 is based on three species: *Pyrgo murrhina*, *Uvigerina* spp., and *Planulina wuellestorfi*. We find the first two to yield indistinguishable results. *Planulina* was found by Duplessy *et al.* to deposit carbonate about 1‰ isotopically lighter than mixed *Pyrgo* spp. We find a difference of 0.64‰ between *Planulina* and

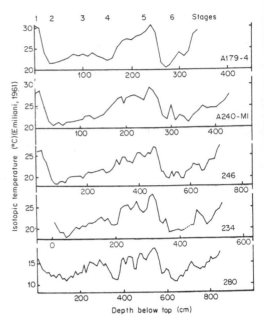

FIG. 5. Oxygen isotope stratigraphy of selected cores containing stages 1–6, after Emiliani (1961). Temperature values quoted uncorrected for ocean isotopic composition changes. Cores A179-4 and A240-M1 from Caribbean; cores 246 and 234 from Equatorial Atlantic; core 280 from North Atlantic.

the other two; Duplessy (personal communication) considers that the variability in the value he obtained at different core horizons was due to the fact that a mixture of different *Pyrgo* spp. was used; we also have evidence that not all *Pyrgo* spp. yield the same values. In obtaining mean values for each level we have added 0.64‰ to the *Planulina* values in order to correct for this "vital effect." The analyses are listed in Table 2[3].

It became apparent in the course of extracting the foraminifera for the present analyses that there is significant variation down the core in overall abundance of benthonic foraminifera, in proportional representation of different benthic species, and in the mean size of individuals of particular species. Benthic species in general, and the species mentioned above in particular, are scarce in stage 5 (from about 120 to 210 cm). Moreover, a consequence of this fact

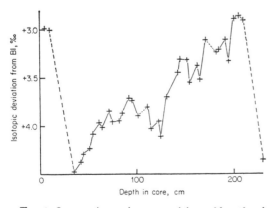

FIG. 4. Oxygen isotopic composition of benthonic foraminifera in core V28-238, top 230 cm, expressed as deviation ‰ from Emiliani B1 standard. Data from Table 2; precision varies with number of analyses utilized in deriving mean.

is that a significant proportion of those that are present prove from their isotopic composition to have been reworked from glacial levels.

Exactly what factor controls the abundance of benthonic foraminifera has yet to be determined, but it is possible that our inability to obtain consistent results at a few levels may be due to an overrepresentation of specimens which lived in higher abundance during relatively brief periods of glacial advance, such as is implied by the work of Steinen *et al.* (1973) to have occurred about 110,000 y.a. Such episodes would be largely or entirely obscured in the record of planktonic foraminifera by mixing in the sediment.

In view of this possibility, it is clear that the stage 5 section of the record cannot be discussed in detail until it has been examined in a core with higher sedimentation and greater abundance of benthic foraminifera during interglacial times.

Comparison of Benthonic and Planktonic Records

In Fig. 6 the planktonic and benthonic records from the top 220 cm of V28-238 are compared, by plotting them using different zero-points for the δ-scale axis. If the variations in both curves were due only to iso-topic composition changes in the Pacific Ocean, they would be coincident at all points. That this is almost true supports our contention that the greater part of the variation derives from this mechanism. We observe two types of deviation from this simple model.

First, the range of variation in the benthonic curve exceeds that in the planktonic curve by about 0.2‰; the surface-to-bottom difference is greater by this amount during stages 2–3–4 than it is today. This could arise from one of three causes: the surface water could have been warmer during the glacial, the bottom water could have been even cooler during the glacial, or the foraminifera analysed as planktonic (*G. sacculifera*) could have shifted to slightly warmer (less deep) water during the glacial.

The first explanation seems least likely on intuitive grounds although Ericson and Wollin (1970) have been able to contemplate the possibility.

The second explanation would imply that the bottom water, at present near 1.3°C, dropped in temperature to about 0.5°C. Production of bottom water at an increased rate could lead to such a drop in temperature; the reduced temperatures at high latitudes implied by isotope data from the

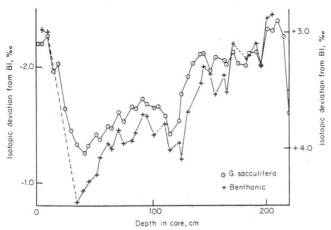

FIG. 6. Comparison of planktonic and benthonic oxygen isotopic record from core V28-238. The two sequences are plotted to the same scale of isotopic change, but with scale zero-points differing by 5.3‰, the present-day planktonic–benthonic difference.

Byrd Station Ice Core (Johnson *et al.*, 1972) might be expected to increase the production of bottom water, in addition to which the presence of an ice sheet on North America could have led to a second important source in the North Pacific. This possibility should be investigated by other means.

The possibility of variations in depth habitat has been suggested before (Emiliani, 1953; Lidz *et al.*, 1968; Shackleton, 1968) and cannot be excluded here since the isotopic composition of the *G. sacculifera* in the top of the core implies a temperature significantly below surface temperature, so that upward migration is feasible.

The second manner in which the curves differ is that around 35 and 125 cm, the planktonic–benthonic difference increases further. We believe that the cause of this phenomenon is mixing: the sample of *G. sacculifera* at 35 cm, for example, contains some younger specimens which may in fact be Holocene in age. On the other hand the benthonic specimens are not so contaminated, because they became much scarcer during the Holocene so that few were available to mix with the stage 2 specimens.

Comparison of Implied Ice Volume Curve with Evidence for Sea Level Variation

In Fig. 7 the new measurements are displayed as a sea-level curve, as it would be recorded without any isostatic effects whatever. This is based on the rough equivalence of 0.1‰ to 10 m sea level change (1.2‰ for a 120-m max sea-level drop about 17,000 y.a.). Figure 7 is based on the analyses of *G. sacculifera*, but is modified in the region of 35 and 230 cm to take account of the fact that the analyses of the benthonic species at these horizons probably give a truer picture (see discussion above).

We also plot on Fig. 7 the estimated position of sea-level maxima at about 120,000, 100,000, and 80,000 y.a. deduced by

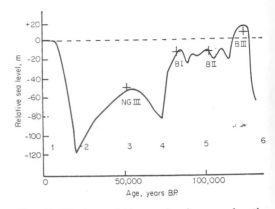

FIG. 7. Glacio–eustatic sea level curve for the past 130,000 yr derived from oxygen isotopic measurements in V28-238, compared with estimated sea levels derived from work on Barbados (B1, B11, B111) and New Guinea (NG 111).

Broecker *et al.* (1968) on the basis of work on Barbados, and about 50,000 deduced by Veeh and Chappell on the basis of similar work on New Guinea (Veeh and Chappell, 1970; Shackleton, 1971). The satisfactory agreement gives strong support for our contention that ice volume, or sea level, may be reliably estimated from the isotope data. Important continuing work in Barbados and New Guinea (Broecker, personal communication) will permit a more accurate conversion factor between isotopic composition and sea level to be determined, using other calibration points in addition to those mentioned relating to the past 120,000 yr. This in turn will enable us to use the isotopic variations in the earlier part of the Pleistocene as a standard sea level curve against which to calibrate long-term uplift rates in these same areas.

It is interesting to consider the important implications of the evidence compared in Fig. 7. First, the remarkable agreement between the relative positions of sea level 80,000, 100,000, and 120,000 y.a. derived by two entirely independent lines of reasoning must quell any remaining doubt as to the correctness of this correlation. In particular, the fact that we now have evidence from the analysis of benthonic foraminifera that there was less terrestrially stored ice at the beginning of stage 5 (substage 5e of Shackleton, 1969) than there is now, is in excellent accord with the widespread evidence of a sea-stand above present level some 120,000 y.a. This well-dated episode remains the most important calibration-point in the time interval between the Brunhes–Matuyama magnetic reversal and the ^{14}C dating range (reviewed by Shackleton as Items 367–394 in "The Phanerozoic Time-scale—a supplement"). The chronology set up by Evans (1971) on the basis of a miscorrelation of Barbados 11 with substage 5e is obviously incorrect.

The possibility that there may have been less stored ice during the last interglacial was discussed by Mercer (1968), who suggested that the West Antarctic ice may have melted at that time. The oxygen isotope sequence in the Byrd Station ice core suggests that perhaps the isotopic composition of that ice stored in Antarctica may have been a little less negative during the last interglacial than now, or alternatively that the ice was a little thinner (Johnsen et al., 1972). Thus there is corroborating evidence from Antarctica that an explanation for the peculiar situation 120,000 y.a. may lie in that continent.

If the comparison between the ice volume curve and the radiometric age determinations from the Barbados coral terraces sets the seal on later Pleistocene chronology, the oxygen isotope sequence between 115 and 35 cm in Fig. 7 sheds light on a no less thorny problem, that of sea level during the last glacial. Ever since the first appearance of the often-reproduced sea-level curve of Curray (1961) there has been a tendency for workers to assume that any ^{14}C age determination whatever appearing to relate to sea level prior to 15,000 y.a. constitutes further reason to remove the question mark which was quite properly put on the original thought-provoking curve. Figure 7 constitutes the first attempt to present a sea-level curve on the basis of a closely spaced series of analyses. It is clear that it offers no comfort for those who have based stratigraphies on the supposed evidence for a marine transgression in this time-range.

The only significant limitation on the reliability of Fig. 7 is that imposed by sediment mixing and the limited stratigraphic resolution of the core. It is perfectly possible that a short episode of high sea level could be obscured in a record with a resolution of a few thousand years—but any supposed evidence for such an event must be read in conjunction with the curve of averaged sea level given in Fig. 7. There was clearly substantial ice accumulated in the Northern Hemisphere from the stage 4 glacial maximum about 70,000 y.a. until the rapid melting at the end of stage 2, between 16,000 and 6,000 y.a.

Caribbean and Atlantic Temperatures, Stages 5–1

As Emiliani (1955) has shown, it is only possible to derive Pleistocene temperatures from the oxygen isotopic analysis of planktonic foraminifera after making due allowance for the change in ocean isotopic composition. Shackleton (1967) was only able to draw gross conclusions from his attempt to use the analysis of benthonic foraminifera to provide a time-varying correction for ice volume change; it is only thanks to the important work of Duplessy et al. (1971) that it is now possible to provide a more usable correction curve. In Fig. 8, an oxygen isotope sequence from Caribbean core P6304-8 (Emiliani, 1966a) is plotted together with our sequence for V28-238. Uniform sedimentation in both cores is assumed; the δ-value scales are adjusted to yield best agreement in stage 1. Three important deductions may be made even on the basis of this crude comparison.

First, we see that as far as temperature is concerned, substage 5e and stage 1 (Holocene) are identical. The difference noted by Shackleton (1969) between these episodes is now seen to be entirely due to isotopic change in the ocean, and not to a higher temperature during 5e as was previously suggested.

Second, we see that during the latter part of stage 5, Caribbean surface temperatures were measurably (approx 1°C) lower than today. This is an important observation, because it has previously been conceivable that the Eemian interglacial stage of North–West Europe, thought by Shackleton (1969) to be time-equivalent to substage 5e, might in reality have been more recent and associated with the high sea-level stand in 5c or 5a; the similarity between 5e and 1 (Holocene) might have been illusory. Thus the demonstration that the sea was cooler in the Caribbean and Atlantic region during the latter part of stage 5 constitutes perhaps the final assurance that correlation between the Eemian and 5e is valid.

Third, we may deduce a temperature difference of some 3°C between glacial stages 2, 3, and 4 in the Caribbean, and stage 1 (Holocene). This value is in good agreement with the value obtained by Imbrie and Kipp (1971). The continued use of both oxygen isotope techniques, and the methods developed by Imbrie and Kipp, to derive past temperatures should be stimulated by the knowledge that there is good agreement between the physical and the biological results now that they are both being used in a sophisticated manner.

We remain unable to resolve the question as to the exact time-relationship between temperature change and ice accumulation. The substage 5e–5d transition still appears to hold the key to this question, although the present work together with that of Steinen *et al.* (1973) shows that the solution proposed earlier (Shackleton, 1969) was incorrect in an important respect: there *was* significant ice accumulation over this important transition.

Oxygen Isotope Analyses: the Past 900,000 yr

Below from the top 250 cm, core V28-238 has been sampled at 10-cm intervals. *Globigerinoides sacculifera* is available in sufficient abundance throughout; the top

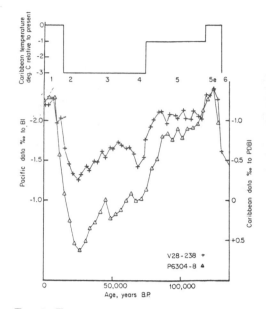

FIG. 8. Temperature record for Caribbean implied by comparison between Caribbean core P6304-8 (Emiliani, 1966) and Pacific core V28-238. No attempt has been made to allow for possible small variations in sedimentation rates, so that the temperature record is only schematic.

half of the core has been analysed in triplicate, and the remainder in duplicate. Analyses are listed in Table 1 and are displayed in Fig. 9. Stages 1–16 may be easily recognized by correlation with core P6304-9 or with Emiliani's generalized sequence (Emiliani, 1966a). The Brunhes–Matuyama boundary falls within stage 19, at 1200 cm. The following evidence may be adduced to support our belief that V28-238 is continuous to this point:

1. The coccolith *Pseudoemiliania lacunosa* becomes extinct within stage 13 (Geitzenauer, in press) and also at the same stage in Caribbean core P6304-8 (Gartner, 1972).

2. Hays *et al.* (1969) show consistent records of carbonate maxima and minima in Equatorial Pacific cores, which Arrhenius (1952) shows are indicators of glacial–interglacial climatic fluctuations. Carbonate minimum B3 corresponds to stage 5, and minimum B17 falls at the Brunhes–Matuyama boundary. Thus we

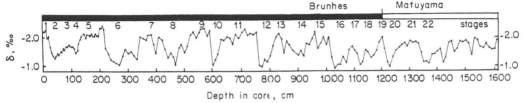

Fig. 9. Oxygen isotopic composition of *G. sacculifera* in core V28-238 complete record to 1600 cm, expressed as deviation ‰ from Emiliani B1 standard. Data from Table 1. Magnetic stratigraphy from Fig. 2. For depths of stage boundaries see Table 3.

record the same number of glacial cycles as these workers have established.

3. The sedimentary rate for the core appears to be very uniform: a rate of 17.1 cm/10^4 yr, derived from the position of the Brunhes–Matuyama boundary, yields an age of 123,000 yr for the peak of substage 5e, indistinguishable from the generally accepted age of this interval (Shackleton, 1969; Broecker and Van Donk, 1970).

Although correlations from Atlantic to Pacific can be, and have been, proposed on numerous bases, this is undoubtedly the most complete correlation that has so far been possible for sediment covering the past 10^6 yr. Moreover, it is soundly based inasmuch as it depends on a phenomenon, isotopic change in the oceans, that must occur essentially synchronously. Thus it is highly unlikely that any superior stratigraphic subdivision of the Pleistocene will ever emerge. Even more important, this subdivision is a convenient one to use because the underlying variable which we are using to correlate is the volume of terrestrially stored ice. We propose that the stages set up in this core be adopted as standard for the latter half of the Pleistocene.

Emiliani (1955) first divided the Pleistocene record into climatic stages on the basis of his work in the Caribbean. The numbering system which he adopted is well established, and there seems to be no difficulty in extending it to stage 22, the fully glacial cycle between the top of the Matuyama Magnetic epoch and the base of the core. This clearly corresponds with car-

bonate maximum M2 of Hays et al. (1969). We have not extended the stage numbering beyond this point; if the character of the record in the lowest 2 m were continued, the definition of stage boundaries might become difficult.

In many sequences it is difficult to distinguish stages 2, 3, and 4 from one another; it seems that the glacial advance which occurred during stage 4 was rather brief, and its record is not always seen. Moreover, there was not complete ice melting during stage 3. In one sense it is a fortunate coincidence that three stages were established over this interval. The availability of ^{14}C dates facilitates detailed correlation in this time range, and Richmond (1970) has recently suggested that in the type area in the European Alps, the Riss–Wurm interval has an age corresponding to a part of stage 3 rather than stage 5. Workers in Russia have also defined a local interglacial stage in this time-interval (Serebryanny et al., 1970).

Although the one-to-one relationship between the Pacific carbonate minima (Arrhenius, 1952; Hays et al., 1969) and the stages established by Emiliani on the basis of oxygen isotope analysis appears to be well established for the past 900,000 yr, the precise position of the carbonate minima within each stage remains to be established. This is but one of the very many variables that have previously been used to establish climatic cyclicity in the Pleistocene, and that may now be followed in detail through each stage within the stratigraphic framework provided by an $^{18}O/^{16}O$ sequence.

Chronology of Stages 1–22

Table 3 shows the depth in core V28-238 of each stage boundary, and the age which may be derived on the basis of a uniform rate of sedimentation. It will be noticed that the ages are on the whole slightly greater than those proposed by Broecker and van Donk (1970), which in turn are greater than those proposed by Emiliani (1966a). Broecker and van Donk suggested an age for their Termination V, equivalent to the stage 12/11 boundary, of 380,000 ± 20,000 yr. This was based partly on extrapolation on the assumption of a uniform sedimentation rate, assuming the age of the stage 6/5 boundary to be known, and partly on interpolation to the U/V zone boundary of Ericson and Wollin (1968), who estimated its age as 400,000 ± 20,000 yr. This age was in turn based on interpolation in five cores which penetrated the Brunhes–Matuyama magnetic reversal at depths between 2.7 and 4.9 m. In cores with such slow sedimentation rates, solution plays a dominant part in determining the rate of accumulation of sediment. The confidence limits of ±20,000 yr placed on the age of the U–V boundary certainly do not adequately cover our lack of confidence as to the constancy of bottom carbonate dissolution during the Pleistocene.

In the Pacific, Arrhenius (1952) showed that to some extent this difficulty may be alleviated by using the rate of noncarbonate accumulation, measured by the TiO_2 content, as a basis of age estimation, since this is not affected by solution. Using the correlations demonstrated by Emiliani (1966b, Fig. 7), the age of stage 12 may be estimated in cores 58, 59, and 62 by using Arrhenius' TiO_2 data to interpolate between this horizon and the Brunhes–Matuyama boundary which falls in Arrhenius' minimum 19 (see Hays *et al.*, 1969, Fig. 14). The results (Table 4) show excellent agreement with our age for this stage. Outside the ^{14}C dating range, the most

280

TABLE 3

DEPTHS AND ESTIMATED AGES OF STAGE BOUNDARIES[a]

Boundary	Depth (cm)	Age (yr)
1–2	22	13,000
2–3	55	32,000
3–4	110	64,000
4–5	128	75,000
5–6	220	128,000
6–7	335	195,000
7–8	430	251,000
8–9	510	297,000
9–10	595	347,000
10–11	630	367,000
11–12	755	440,000
12–13	810	472,000
13–14	860	502,000
14–15	930	542,000
15–16	1015	592,000
16–17	1075	627,000
17–18	1110	647,000
18–19	1180	688,000
19–20	1210	706,000
20–21	1250	729,000
21–22	1340	782,000

[a] Ages are estimated on the basis of a uniform sedimentation rate of 1.71×10^{-3} cm per year, calibrated by the presence of the Brunhes-Matuyama magnetic epoch boundary, age 700,000 yr, at 1200 cm.

reliably dated material of Middle Pleistocene age is probably the "Tuff with Black Pumices" of Italy (Evernden and Curtis, 1965). If this eruption could be confidently placed in a climatic sequence, it is possible that it could be used to render our chronology even more accurate. For the present it seems that neither this, nor any other presently available data, can improve upon the chronology proposed on the basis of uniform sedimentation in a single core. Clearly such a chronology cannot be regarded as definitive, but it may well be that a definitive chronology will emerge from the study of sediments in this region of the Pacific.

Implication for Earlier Marine Terrace Deposits

It appears that there may have been significantly more water in the oceans during substage 5e than during the Holocene. The

TABLE 4

APPROXIMATE AGE OF STAGE 12 IN SOME OTHER CORES FROM THE PACIFIC[a]

Core no.	Depth to stage 12 (cm)	TiO$_2$ to stage 12 (mg cm^{-2})	Depth to stage 19 (cm)	TiO$_2$ to stage 19 (mg cm^{-2})	Age (yr)
58	230	360	400	540	467,000
59	570	260	970	380	461,000
62	320	190	600	300	443,000

[a] Data from Arrhenius (1952) using correlations from Emiliani (1966b), assuming linear interpolation on the basis of uniform rate of accumulation of TiO$_2$ within each core, and assuming that the age of stage 19 is 700,000 yr as shown in the present paper.

mean isotopic difference is (0.13 ± 0.04) per mil on the basis of ten stage 1 measurements and nine from 5e. An examination of the earlier part of the record raises an interesting and important point. While the isotope record, interpreted as an ice-volume or ocean-volume record, shows stage 5 protruding above the stage 1 (Holocene), stage 7 does not even achieve the Holocene level; it stands 0.15 per mil below, and hence about 0.28 per mil below stage 5. This means that no marine terrace of this age is likely to survive above present sea level unless on an uplifted coast. Considering the effect of a uniform rate of uplift, a rate of say 10 m/10^5 yr might raise such a deposit from a hypothetical 10 m below present sea level to 15 m above (assuming an age of 2.5×10^5 yr); the terrace formed during 5e would then have a present-day altitude $6 + (1.2 \times 10)$, or 18 m, so that the earlier terrace would in fact still be below the more recent one. The reality of this comparison between the peaks of stages 7, 5, and 1 is of utmost importance and must be investigated in other cores, but the present indication is that it is unlikely that stage 7 can be correlated with a terrestrial interglacial stage associated with a sea level significantly higher than that associated with the last interglacial, unless local uplift of a nonuniform nature is invoked. Specifically, if there is any widespread reality in the concept of a "penultimate interglacial" associated with marine deposits some 30 m above the present level, it is not of the age of stage 7, but older.

The peak in stage 9 appears to be 0.05 per mil above the mean Holocene level, which is a barely measurable difference. To raise a marine terrace of this age 30 m above present sea level would require an uplift of the order of 10 m/10^5 yr (or an equivalent secular change in ocean floor levels) and would bring the terrace deposited 120,000 y.a. to some 15 m above sea level; possibly the deposits formed during substages 5a or 5c might also be found above present sea level.

Such a reconstruction is at present speculative, but it illustrates the point that if oxygen isotope measurements yield information about the amount of water abstracted from the oceans, and if the measurements can be refined sufficiently to distinguish between different interglacials in this respect, then the information is extremely valuable and is susceptible to refined interpretations.

Potential Terrestrial Correlations

Unlike either of the two interglacial stages preceding it, stage 7 is sharply broken, apparently by a substantial ice accumulation (marked by the measurement at 400 cm). This may perhaps be correlated with the "minimum 5" noted by McIntyre *et al.* (1972), an important point in view of the fact that these workers were using different techniques in a different ocean. The minimum is also visible in most of Emiliani's isotope records; it has the consequence that even in cores with poor stratigraphic resolution, stage 7 is distinguished by the fact that its peak does not closely follow upon a glacial minimum. This may be particularly significant from the point of view of terrestrial correlations. The Hoxnian interglacial stage of Britain is particularly well known from a lake deposit at Marks Tey, Essex (Turner, 1970), which can only be satisfactorily understood in terms of the filling of a channel cleared out by ice. This ice was at the extreme edge of the Fennoscandia Ice Sheet, which is unlikely to have extended so far in the brief episode within stage 7; thus the Hoxnian Stage is unlikely to be the time equivalent of the peak in stage 7. It is now a real possibility that the Hoxnian of Britain, and the Holsteinian of North–West continental Europe, must be correlated not with stage 7 but with an earlier stage.

It is now known from borings in the Netherlands that deposits of the so-called "Cromerian 111" interglacial and the Needian Interglacial both occur within the Urk Formation. The base of this formation is thought to have an age of about 400,000 yr (van der Hammen *et al.*, 1971). This date is based on the K/Ar dating of Frechen and Lippolt (1965) for volcanic rocks in the Eifel region; augite from these rocks occurs throughout the Urk formation. The precise relationship between the terraces of the Rhine with these volcanic eruptions on the one hand, and with the climatic sequence which is at least partly responsible for the terraces, has been much disputed, but if this limit to the age of the "Cromer 111" and Needian interglacials is even approximately correct, and if in addition we are correct in deducing that isotope stage 7 is not the marine correlative of either of these interglacial stages, then presumably they must be correlated with stages 11 and 9.

In the past it has been believed that the past 400,000 yr included some six interglacials as seen in the isotope record. If this number is to be reduced to four, and if these are as clearly differentiated one from another as appears from this discussion,

marine–terrestrial correlations are at last coming into focus.

It is hazardous to attempt detailed correlations or even discussion of the remaining part of the core in the absence of independent checks on the validity of its record. However, it is worth making the point that in this core, as in the records published over many years by Emiliani, stage 14 is the only one which descends to a significantly lesser extent than the others. It is the only glacial stage in which the ice sheets clearly did not reach approximately the same maximum size. This is strong support for our correlation to this point. Perhaps the Westerhoven interglacial ("Cromerian 11") is represented in isotope stage 15, inasmuch as this is the only well-developed interglacial stage in our record which is time-equivalent to any part of the period between the Brunhes–Matuyama Boundary and the base of the Urk Formation.

If we are correct in assuming that the record of V28-238 is essentially complete, 16 and 18 correspond to the high-carbonate peaks B14 and B16 just above the Brunhes–Matuyama magnetic reversal in the studies of Hays *et al.* (1969). In continental terms they are not separated by full deglaciation, and one of the longest glaciations of the Pleistocene is represented. If this interpretation is correct, the base of Brunhes is tantalizingly close to the base of Emiliani's sequences, many of which bottom in stage 16. The correctness or otherwise of our chronology will not be difficult to investigate by taking slightly longer cores in the Caribbean.

We have analysed 9 samples of benthonic foraminifera from stage 22 with a view to checking that this section really does represent a fully glacial stage. The results (Table 2) demonstrate conclusively that this is so. The interesting question as to whether this is in any sense a "first glaciation" marking the end of the "pre-glacial Pleistocene" can be investigated only by the study of longer cores. If this is in fact a first glaciation, perhaps it succeeded the so-called "Cromerian 1," an interglacial characterized by the presence of relics of the Tertiary vegetation of Europe of which the best-known is *Eucommia* (van der Hammen *et al.*, 1971); in this case some part of the bottom of the core may be time-correlative with "Cromerian 1." On the other hand, it is equally possible that there was extensive glaciation during the Menapian Stage preceding "Cromer 1," in which case the lowest part of the core may represent a part of the Waalian temperate stage of Northwestern Europe.

The reason for the use of the name "Cromerian" for three interglacials, the earliest of which is perhaps twice the age of the youngest, is that this name has been generally applied to the pre-Elsterian interglacial. The scheme proposed above would appear to restrict the Elsterian to isotope stage 10. Although this at first sight amounts to equating a very short glacial stage in the marine sequence with a very extensive European glaciation, two points must be remembered. First, stage 10 is demonstrably as extreme as its neighbors, so that sufficient ice may have accumulated to reach a wide extent in Europe. Second, it cannot be said to be certain that all deposits mapped as of Elsterian age do in fact postdate "Cromerian 111." It may be that the "Cromerian Complex" and the "Elsterian Complex," in the loose sense in which these names are used, *both* extend from 350,000 y.a. back into the Matuyama Epoch!

APPENDIX

Experimental methods: Comparison Standards

The team which set out to evaluate the potential of Urey's idea (Epstein *et al.*, 1951, 1953) evolved a technique for comparing the isotopic composition of their samples with an arbitrary calcite belemnite standard which they named PDB. However, it is now generally felt that PDB is to be regarded not as a carbonate standard, but as the carbon dioxide which is released from the belemnite carbonate under specified conditions. The chief reason for this is that it enables the standard to be defined in relation to the oxygen isotope standard for water samples, which is also a carbon dioxide gas, equilibrated with a standard water under specified conditions.

Although this is a good principle, it is a poor practice since it means that the isotopic composition of a sample is referred to PDB1 in publications regardless of experimental method. Since the isotopic composition of the gas evolved from a sample is well known to vary according to the experimental conditions (Emiliani, 1966a), we feel that there is little point in relating the result to PDB.

A more useful procedure in our view is to use a system which enables samples and a standard carbonate to be analysed under identical conditions using an arbitrary gas as a running standard; this means that whatever the experimental system, the comparison should remain the same. The results given in this paper were obtained using acid (H_3PO_4) decomposition at 50°C; with each nine samples a standard carbonate is included. The procedure will be detailed elsewhere. Over the past year reproducibility of comparisons between standard carbonates and the running standard gas has been about ±0.07‰ (1-sigma). Repeated analyses of a number of carbonate standards enables their isotopic composition to be interrelated with a precision of about ±0.02‰. It is not at present known whether intercomparison can be made more precise with the present analytical set-up: the observed precision is apparently very similar to that dictated by the mass spectrometer itself. Analyses quoted in this paper are related to Emiliani's standard B-1; this is a belemnite standard whose calibration to PDB is (+0.2 ± 0.1)‰. Since the uncertainty in its relationship to PDB is greater than our uncertainty in a single sample analysis, and far greater than any systematic error in our comparison between samples and B-1, there seems little point in introducing extra uncertainty by relating it to PDB.

Sampling Procedure

All core samples are dispersed in distilled water (about 50 ml for about 4 g sample). After a brief disturbance in an ultrasonic vibrator they are washed through a 180-μm mesh sieve with flowing distilled water. After rinsing with reagent grade methanol the sieve is roughly dried using a paper towel on the underside, and left to dry in an oven at about 50°C. Dried samples are stored in glass bottles with foil-lined plastic tops.

Foraminifera for analysis are extracted under $\times$25 magnification on a moving tape operated by a foot-switch. If a complete sample is searched for benthonic species much time and effort is saved by this procedure. Specimens are removed using a paint brush wetted with distilled water. Where possible about 0.3 mg sample is preferred, although samples as small as a tenth of this have been analysed successfully on occasions. Routinely we have extracted three aliquots of eight specimens each of *G. sacculifera* from every core sample; the mean weight of all samples was about 0.2 mg after cleaning (estimated from gas yields).

Samples are picked into small quartz thimbles. Cleaning comprised the following stages. First the thimble is filled with AR grade methanol and vibrated ultrasonically; the methanol is then decanted using absorbent paper to draw out the liquid. Methanol is used in preference to water mainly because it wets more easily. The specimens are checked under the microscope, lightly crushed under methanol using a glass rod, and recleaned ultrasonically. For benthonic species this cleaning is extremely important; for the planktonics it is perhaps less important as the average isotopic composition of the adhering fine-grained carbonate is probably not so distant from that of the foraminifera. Finally the crushed foraminifera are

roasted *in vacuo* at 450°C for 30 min. Whether or not this step is necessary to precise analysis is not at present clear; we have some evidence that it leads to greater reproducibility, but it is not conclusive. Emiliani (1966a) considered that it made little difference; at the same time the apparently higher stratigraphic noise he found in core P6304-9 as compared to P6304-8 could be due to the samples not having been roasted.

In core V28-238 we have made triplicate analyses from the top to 800 cm, and duplicate analyses from 800 to 1600 cm, so that we have a very good estimate available of the precision of a single analysis in its widest sense: given a sample of sediment we know with what confidence we can express the isotopic composition of the *G. sacculifera* in the sediment on the basis of a single analysis. This is ±0.11‰ (68% confidence limits). We consider that this is not really adequate to solve some of the problems of interest in the Pleistocene; by making triplicate analyses we improve the precision to ±0.06‰.

From the palaeoclimatological .and stratigraphic point of view, the maximum information is extracted from the core only if it is sampled sufficiently closely that the biggest contribution to the measured difference between successive analyses is analytical uncertainty. In the top 200 cm we have made intermediate analyses at 5 cm. In regions away from rapid change, the mean difference between successive analyses is about 0.07‰, indistinguishable from the contribution from analytical uncertainty. This is probably the ultimate criterion by which analytical uncertainty may be assessed; from the geological point of view, the next question must be, "How closely does the record from another core compare with this one?"

ACKNOWLEDGMENTS

Oxygen isotope analysis supported by N.E.R.C. Grants GR/3/768 and GR/3/1762 to N.J.S. Coring supported by Grants ONR (N00014-67-A-

0108-0004) and NSF-GA-29460 to Lamont-Doherty Geological Observatories. CLIMAP supported under the IDOE program by NSF to Lamont-Doherty, Brown University and Oregon State University.

N.J.S. is most grateful to M. A. Hall for operating the mass spectrometer with meticulous care over long periods. He also thanks Jim Hays and George Kukla in particular out of his CLIMAP colleagues, as well as Cesare Emiliani and Richard West from outside CLIMAP, for stimulating discussion and helpful comments. The final manuscript was much improved by detailed attention from Zelda Murray.

REFERENCES

ARRHENIUS, G. (1952). Sediment cores from the East Pacific. *Reports of the Swedish Deep-Sea Expedition* 5.

BROECKER, W. S., THURBER, D. L., GODDARD, J., KU, T-L., MATTHEWS, R. K., AND MESOLELLA, K. J. (1968). Milankovitch hypothesis supported by precise dating of coral reefs and deep-sea sediments. *Science* 159, 297–300.

BROECKER, W. S. AND DONK, J. VAN (1970). Insolation Changes, Ice Volumes, and the O^{18} Record in Deep-Sea Cores. *Reviews of Geophysics and Space Physics* 8, 169–198.

CURRAY, J. R. (1961). Late Quaternary sea level: a discussion. *Geological Society of America Bulletin* 72, 1707–1712.

DALRYMPLE, G. B. (1972). Potassium argon dating of geomagnetic reversals and North American glaciations in calibration of hominoid evolution. (Bishop, W. W. and Miller, J. A., Eds.). Scottish Academic Press.

DANSGAARD, W. AND TAUBER, H. (1969). Glacier Oxygen-I8 content and Pleistocene ocean temperatures. *Science* 166, 499–502.

DONK, J. VAN AND MATHIEU, G. (1969). Oxygen isotope compositions of foraminifera and water samples from the Arctic Ocean. *Journal of Geophysical Research* 74, 3396–3407.

DUPLESSY, J. C., LALOU, C., AND VINOT, A. C. (1970). Differential isotopic fractionation in benthic foraminifera and palaeotemperatures reassessed. *Science* 168, 250–251.

EMILIANI, C. (1954). Depth habitats of some species of pelagic foraminifera as indicated by oxygen isotope ratios. *American Journal of Science* 252, 149–158.

EMILIANI, C. (1955). Pleistocene temperatures. *Journal of Geology* 63, 538–578.

EMILIANI, C. (1961). Cenozoic climatic changes as indicated by the stratigraphy and chronology of deep-sea cores of Globigerina facies. *Annals of the New York Academy of Science* 95, 521–536.

EMILIANI, C. (1966a). Palaeotemperature analysis of Caribbean cores P 6304-8 and P 6304-9 and a generalised temperature curve for the last 425,000 years. *Journal of Geology* 74, 109–126.

EMILIANI, C. (1966b). Isotopic Palaeotemperatures. *Science* 154, 851–857.

EPSTEIN, S., BUCHSBAUM, R., LOWENSTAM, H. A., AND UREY, H. C. (1951). Carbonate–water isotopic temperature scale. *Geological Society of America Bulletin* 62, 417–426.

EPSTEIN, S., BUCHSBAUM, R., LOWENSTAM, H. A., AND UREY, H. C. (1953). Revised carbonate–water isotopic temperature scale. *Geological Society of America Bulletin* 64, 1315–1326.

ERICSON, D. B. AND WOLLIN, G. (1968). Pleistocene climates and chronology in deep-sea sediments. *Science* 162, 1227–1234.

ERICSON, D. B. AND WOLLIN, G. (1970). Pleistocene climates in the Atlantic and Pacific Oceans: a comparison based on deep-sea sediments. *Science* 167, 1483–1485.

EVANS, P. (1971). Towards a Pleistocene timescale. *In* "The Phanerozoic Time-scale—a supplement." Part 2, pp. 123–356. *Geological Society* (London) *Special Publication* 5.

EVERNDEN, J. F. AND CURTIS, G. H. (1965). The potassium–argon dating of late Cenozoic rocks in East Africa and Italy. *Current Anthropology* 6, 343–364.

FOSTER, J. H. (1966). A paleomagnetic spinner magnetometer using a fluxgate gradiometer. *Earth and Planetary Science Letters* 1, 463–466.

FOSTER, J. H. AND OPDYKE, N. D. (1970). Upper Miocene to recent magnetic stratigraphy in deep-sea sediments. *Journal of Geophysical Research* 75, 4465–4473.

FRECHEN, J. AND LIPPOLT, H. J. (1965). Kalium–Argon-Daten zum Alter des Laacher Vulkanismus, der Rheinterrassen und der Eiszeiten. *Eiszeitalter und Gegenwert* 16, 5–30.

GARTNER, S. (1972). Late Pleistocene calcareous nannofossils in the Caribbean and their interoceanic correlation. *Paleogeography, Palaeoclimatology, Palaeoecology* 12, 169–191.

HAMMEN, T. VAN DER, WIJMSTRA, T. A., AND ZAGWIJN, W. H. (1971). The floral record of the late Cenozoic of Europe. *In* "The Late Cenozoic Glacial Ages." (K. K. Turekian, Ed.) pp. 391–425. Yale University Press.

HAYS, J. D., SAITO, T., OPDYKE, N. D., AND BURCKLE, L. H. (1969). Pliocene–Pleistocene Sediments of the Equatorial Pacific: their Palaeomagnetic, Biostratigraphic, and Climatic Record. *Geological Society of America Bulletin* 80, 1481–1514.

IMBRIE, J. AND KIPP, N. G. (1971). A new micropalaeontological method for quantative palaeoclimatology: application to a late Pleistocene Caribbean core. *In* "The Late Cenozoic Glacial Ages." (K. K. Turekian, Ed.), pp. 71–183. Yale University Press.

JOHNSEN, S. J., DANSGAARD, W., CLAUSEN, H. B., AND LANGWAY, C. C. (1972). Oxygen Isotope profiles through the Antarctic and Greenland Ice Sheets. *Nature (London)* **235**, 429–434.

KOBAYASHI, K., KITAZAWA, KANAYA, T., AND SAKAI, T. (1971). Magnetic and micropaleontological study of deep-sea sediments from the west-central equatorial Pacific. *Deep Sea Research* **18**, 1045–1062.

LIDZ, B., KEHM, A. AND MILLER, H. (1968). Depth habitats of pelagic foraminifera during the Pleistocene. *Nature (London)* **217**, 245–247.

McELHINNY, M. W. (1966). An improved method for demagnetizing rocks in alternating magnetic fields. *Geophysical Journal Royal Astronomical Society* **10**, 369–374.

McINTYRE, A., RUDDIMAN, W. F., AND JANTZEN, R. (1972). Southward penetrations of the North Atlantic polar front: faunal and floral evidence of large-scale surface water mass movements over the last 225,000 years. *Deep-Sea Research* **19**, 61–77.

MERCER, J. H. (1968). Antarctic ice and Sangamon sea level. *International Association of Scientific Hydrology Publication* **79**.

OLAUSSON, E. (1965). Evidence of climatic changes in North Atlantic Deep-sea cores, with remarks on isotopic palaeotemperature analysis. *Progress in Oceanography* **3**, 221–252.

"The Phanerozoic Time-scale—a Supplement." Special Publication No. 5, Geological Society (London), 1971.

RICHMOND, G. M. (1970). Comparison of the Quaternary stratigraphy of the Alps and Rocky Mountains. *Quaternary Research* **1**, 3–29.

SEREBRYANNY, L., RAUKAS, A., AND PUNNING, J. (1970). Fragments of the natural history of the Russian plain during the late Pleistocene with special reference to radiocarbon datings of fossil organic matter from the Baltic region. *Baltica* **4**, 351–366.

SHACKLETON, N. J. (1967). Oxygen isotope analyses and Pleistocene temperatures re-assessed. *Nature (London)* **215**, 15–17.

SHACKLETON, N. J. (1968). Depth of pelagic foraminifera and isotopic changes in Pleistocene oceans. *Nature (London)* **218**, 79–80.

SHACKLETON, N. J. (1969). The last interglacial in the marine and terrestrial records. *Proceedings of the Royal Society London B* **174**, 135–154.

SHACKLETON, N. J. (1971). New Guinea Reef Complex 111. *In* "The Phanerozoic Time-scale—a supplement." Item 390, pp. 106–107. *Geological Society* (London) *Special Publication* **5**.

STEINER, R. P., HARRISON, R. S. AND MATTHEWS, R. K. (1973). Eustatic low stand of sea level between 105,000 and 125,000 B. P.: evidence from the subsurface of Barbados. *Geological Society of America Bulletin* **84**, in press.

TURNER, C. (1970). The middle Pleistocene deposits at Marks Tey, Essex. *Philosophical Transactions of the Royal Society of London B* **257**, 373–440.

UREY, H. C. (1947). The thermodynamic properties of isotopic substances. *Journal of the Chemical Society*, 562–581.

VEEH, H. H. AND CHAPPELL, J. (1970). Astronomical theory of climatic change: support from New Guinea. *Science* **167**, 862–865.

23

Reprinted from *Geol. Soc. America Mem. 145*, 1976, pp. 449–464

Oxygen-Isotope and Paleomagnetic Stratigraphy of Pacific Core V28-239 Late Pliocene to Latest Pleistocene

N. J. SHACKLETON
Sub-department of Quaternary Research
University of Cambridge.
5 Salisbury Villas, Station Road
Cambridge, England CB1 2JF

AND

N. D. OPDYKE
Lamont-Doherty Geological Observatory
Columbia University
Palisades, New York 10964
and
Department of Geological Sciences
Columbia University
New York, New York 10027

ABSTRACT

V28-239 core from cruise 28 of R/V *Vema* preserves a detailed oxygen-isotope and paleomagnetic record for all of the Pleistocene Epoch. The entire 21-m-long core has been analyzed at 5-cm intervals. Glacial stage 22, above the Jaramillo magnetic event, may represent the first major Northern Hemisphere continental glaciation of middle Pleistocene character. Prior to this, higher frequency glacial events extend to near the level of the Olduvai magnetic event. Glacial events of less regular frequency extend to the bottom of the core, which represents late Pliocene time. Fluctuations in carbonate dissolution intensity occur throughout the core with a similar frequency to the oxygen-isotope fluctuations.

287

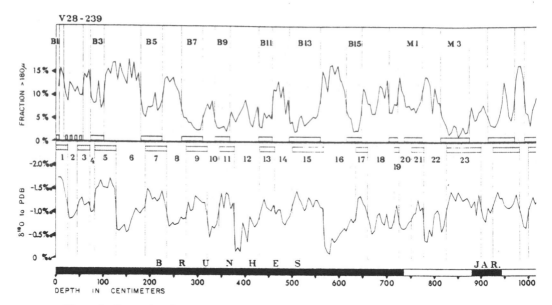

Figure 1. Coarse-fraction record (above), oxygen-isotope record (below), and paleomagnetic record in core V28-239. Dissolution zones in the coarse-fraction record are numbered after

INTRODUCTION

The combination of oxygen-isotope and paleomagnetic stratigraphy in core V28-238 (Shackleton and Opdyke, 1973) has provided an excellent framework within which to investigate the history of events in the western equatorial Pacific during the past 800,000 yr and to correlate this history with events elsewhere. We have now extended this study to about 2.1 m.y. by analyzing a stratigraphically longer core, V28-239, taken relatively close to core V28-238.

In the upper part of the record, comparison between the two cores provides valuable insight into the effects of postdepositional solution and mixing at the sea floor on the oxygen-isotope record and accumulation-rate variations. The lower part of the core provides, for the first time, detailed information on early Pleistocene climates.

ANALYTICAL RESULTS

Core V28-239 was raised from the Solomon Rise at lat 3°15′N, long 159°11′E from a depth of 3,490 m during cruise 28 of the R/V *Vema*. Preliminary magnetic stratigraphy of the core has already been reported (Shackleton and Opdyke, 1973). Analyses have been performed continuously across magnetic reversal boundaries in samples approximately 3 cm across, thus sharply constraining the magnetic boundaries. The Brunhes-Matuyama boundary is located at 726 cm, the Jaramillo event between 877 and 940 cm, and the Olduvai event between 1,553 and 1,781 cm.

Samples for oxygen-isotope analysis were taken at 5-cm intervals throughout the core. Magnetic stratigraphy indicates an average accumulation rate of 1 cm/10³ yr, giving one sample every 5,000 yr; by comparison, the sampling interval of 10 cm in core V28-238 is one sample every 6,000 yr.

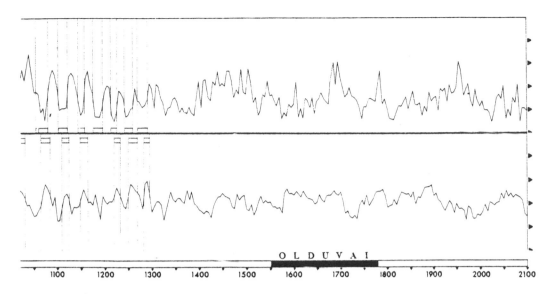

Hays and others (1969); stages in the oxygen-isotope record are numbered after Emiliani (1955, 1966) and Shackleton and Opdyke (1973).

Sediment samples were disaggregated in distilled water; foraminifers were selected for analysis from the >180-μm fraction after sieving and ultrasonic cleaning. Sample pretreatment and chemical processing were identical to those used for core V28-238 (Shackleton and Opdyke, 1973). Isotope analysis was performed in a new V.G. Micromass 602C mass spectrometer. Analyses are referred to the PDB standard (Epstein and others, 1951, 1953) using a value of +0.29‰ for the Emiliani B-1 standard (Shackleton, 1974). This calibration is accurate to better than ±0.05‰. Analytical results in Shackleton and Opdyke (1973) were referred to the B-1 standard and must be corrected by +0.29‰ before comparison with the data presented in this paper.

A single analysis has been made at each level in the core. For each analysis, 15 specimens of *Globigerinoides sacculifer* were selected (in the lower part of the core *G. fistulosus* was used in some samples, three samples contained insufficient specimens for analysis, and a few contained less than 15). Analytical precision is estimated to be ± 0.05‰, the standard deviation for 100 analyses of a standard carbonate performed during the first six months of instrument operation. However, the uncertainty in analysis of a single sample from the sediment is ±0.11‰ (Shackleton and Opdyke, 1973). Isotopic variability among the specimens and analytical precision are combined in this figure. Analytical results are given in Table 1. Figure 1 shows the percentage by weight retained on the 180-μm sieve for each sample, the oxygen-isotope record, and the paleomagnetic record.

Character of the Isotopic Record

Jaramillo Magnetic Event to Present. Figure 1 suggests that the oxygen-isotope record may be divided into three episodes of differing character. The upper part, all of which is represented in core V28-238 (Fig. 2) as well as core V28-239, contains glacial stages at approximately 100,000-yr intervals. Apparently, the isotopic composition of the ocean changed by almost the same extent in every glaciation

TABLE 1. OXYGEN–ISOTOPIC COMPOSITION OF *GLOBIGERINOIDES SACCULIFER* IN CORE V28-239

Depth	δ¹⁸O	Depth	δ¹⁸O	Depth	δ¹⁸O	Depth	δ¹⁸O	Depth	δ¹⁸O	Depth	δ¹⁸O	Depth	δ¹⁸O
5	-1.72	305	. .	605	-0.58	905	-0.88	1205	-0.99	1505	-0.86	1805	-0.95
10	-1.72	310	-1.20	611	-0.66	910	-0.79	1210	-1.05	1510	-0.99	1810	-0.97
15	-1.64	315	-1.20	615	-0.60	915	-1.10	1215	-1.04	1515	-0.76	1814	-1.08
20	-1.36	320	-0.92	620	-0.73	920	-0.99	1220	-1.13	1520	-0.87	1820	-0.97
25	-1.10	325	-0.45	625	-0.87	925	-1.12	1225	-1.35	1525	-1.04	1825	-1.13
29	-0.84	330	-0.70	630*	-0.78	930	. .	1230	-1.23	1530	-1.17	1830	-1.24
35	-0.86	335	-0.65	635	-0.87	935	-1.44	1235	-1.11	1535	-1.01	1835	-1.22
40	-0.92	340	-0.74	640	-1.31	940	-1.30	1240	-0.95	1540	-1.07	1840	-1.19
45	-1.03	345*	-1.06	645	-1.35	945	-1.31	1245	-0.90	1545	-1.02	1845	-1.21
50	-1.21	350*	-1.40	651	-1.18	950	-1.15	1250	-1.19	1550	-1.04	1850	-1.13
55	-1.30	355*	-1.07	655	-1.14	955	-1.10	1255	-1.42	1555	-0.93	1855	-1.25
60	-1.12	360*	-1.42	660	-0.94	960	-1.09	1260	-1.37	1560	-0.94	1860	-0.98
65	-1.15	365	-1.36	665	-0.78	965	-1.13	1265	-1.28	1564	-0.90	1865	-1.14
70	-1.20	370	-1.06	669	-0.53	970	-1.21	1270	-1.24	1570	-0.92	1870	-1.12
75	-0.98	375	-1.19	675	-0.61	975	-1.39	1274	-1.18	1575	-0.80	1875	-1.23
80*	-1.01	380	-0.15	680	-0.93	980	-1.13	1280	-0.90	1580	-1.23	1880	-1.34
85	-1.48	385	-0.24	685	-0.99	985	-0.96	1285	-1.42	1585	-1.32	1885	-1.35
90	-1.53	390	-0.14	690	-0.79	990	-0.70	1290	-1.48	1590	-1.21	1890	-1.36
95	-1.68	395	-0.77	695	-0.73	995	-0.66	1295	-1.21	1595	-1.22	1895	-1.39
100*	-1.53	400	-0.61	700	-0.98	1000	-1.02	1300	-0.82	1600	-1.18	1900	-1.15
105	-1.51	405	-0.35	705	-0.67	1005	. .	1306	-0.95	1605	-1.26	1905	-1.21
110	-1.50	410	-0.91	710	-0.67	1010	-1.33	1310	-1.05	1610	-1.20	1910	-1.07
115	-1.72	415	. .	715	-0.84	1015	-1.17	1315	-1.05	1615	-1.14	1915	-1.03
120	-1.57	420	-0.67	720	-1.16	1020	-1.29	1320	-0.98	1620	-1.27	1920	-0.93
125	-0.98	425	-0.78	726	-0.77	1025	-1.26	1325	-0.65	1625	-1.18	1925	-0.84
130	-0.61	430	-0.90	730	-0.66	1030	-1.14	1330	-0.84	1630	-1.11	1930	-0.95
135	-0.59	435	. .	735	-0.64	1035	-0.96	1335	-1.05	1635	-1.10	1935	-1.05
140	-0.65	440	-1.28	739	-0.66	1040	-0.98	1340	-1.26	1640	-1.08	1940	-0.97
146	-0.72	445	-1.11	746	-0.79	1045	-0.87	1345	-1.10	1645	-0.99	1945	-0.79
150	-0.56	450	-1.06	750	-0.82	1050	-0.73	1350	-1.04	1650	-1.03	1950	-0.99
155	-0.60	455	-1.02	756	-0.97	1055	-0.75	1355	-1.12	1655	-1.19	1955	-0.88
160	-0.80	460	-1.15	760	-0.98	1060	-0.86	1360	-1.12	1659	-1.09	1960	-0.89
165	-0.94	465	-0.95	765	-1.27	1065	-0.96	1365	-1.04	1665	-1.03	1966	-0.84
170	-1.07	470	-1.05	770	-1.04	1070	-1.25	1369	-1.01	1670	-1.17	1970	-0.64
175	-1.01	475	-1.05	775	-1.11	1074*	-1.41	1375	-1.29	1675	-1.18	1975	-0.94
180	-0.96	480	-0.78	780	-0.39	1080	-1.30	1380	-1.08	1679	-1.18	1980	-0.88
185	-1.05	485	-0.72	785	-0.37	1085	-1.17	1384	-1.26	1686	-1.28	1985	-0.67
190	-1.10	490	-0.83	790	-0.64	1090	-0.95	1390	-0.99	1690	-1.21	1990	-0.67
195	-1.50	495	-0.90	795	-0.49	1095	-1.14	1395	-0.96	1695	-1.17	1994	-0.77
200	-1.42	500	-1.11	800	-0.96	1100	-0.63	1400	-0.87	1700	-1.25	2000	-0.73
205	-1.26	505	. .	807	-1.05	1105	-0.64	1405	-0.84	1705	-1.06	2005	-0.82
210	-1.31	510*	-1.45	810	-0.97	1110	-0.93	1410	-0.79	1710	-0.79	2010	-0.97
215	-1.22	515	-1.29	815	-0.58	1115	-1.15	1415	-0.85	1715	-0.84	2016	-0.82
220	-1.28	519	-1.22	820	-0.71	1120	-1.21	1420	-0.82	1720	-0.72	2020	-1.02
225	-1.40	526	-1.03	825	-1.04	1125	-1.09	1425	-0.71	1725	-0.69	2025	-1.00
230	-1.24	529*	-1.07	830	. .	1130	-0.77	1430	-0.93	1730	-0.75	2030	-0.95
235	-0.99	535	-1.32	835	-1.45	1135	-0.93	1435	-1.06	1735	-0.67	2035	-1.01
240	-0.68	540	-1.24	840	-1.34	1140	-0.92	1440	-1.15	1740	-0.95	2040	-1.03
248	-0.77	545	-1.11	845	-1.25	1145	-0.94	1445	-1.16	1746	-1.07	2045	-1.21
251	-0.72	550	-1.26	850	-1.44	1150	-1.13	1450	-1.22	1750	-0.97	2050	-1.09
255	-0.73	554	-1.25	855	—	1155	-1.30	1455	-1.16	1755	-1.18	2055	-1.27
260	-0.76	561	-1.10	860	-1.25	1160	-1.16	1460	-1.14	1761	-0.95	2060	-1.14
265	-0.89	565	-1.09	865	. .	1165	-1.03	1465	-1.11	1765	-1.02	2065	-1.21
270	-0.84	570	-0.27	870	-1.31	1170	. .	1470	-0.93	1770	-1.11	2070	-1.11
275	-0.87	575	-0.15	874	-1.25	1175	. .	1475	-0.90	1775	-1.24	2075	-1.18
280	-1.21	580	-0.10	879	-1.10	1180	-1.20	1480	-1.21	1780	-1.26	2080	-1.15
285*	-1.02	585	-0.45	886	-1.24	1185*	-0.94	1485	-1.08	1785	-1.26	2085	-1.12
290	-1.05	590	-0.35	890	-0.84	1190*	-0.66	1490	-1.15	1790	-1.30	2090	-1.06
295*	-1.35	595	-0.41	895	-1.38	1195	-1.17	1495	-1.18	1795	-1.17	2095	-0.77
300	. .	602	-0.54	900	-1.13	1200	-1.06	1500	-0.88	1800	-0.91	2100	-0.71

Note: Composition is expressed as a deviation per mil from the PDB standard. Samples contained 15 individuals.

*Less than 15.

during this interval. The rather large variability among glacial extreme isotopic values in core V28-239 is an artifact of sedimentation processes. This is evident from the fact that the extreme isotopic values in successive glaciations are both less variable and more distant from the Holocene value in cores with higher accumulation rates. In core V28-239 (1.0 cm/10^3 yr) the extreme isotopic values in glacial stages 2 and 6 to 22 differ from the Holocene value by 1.22 ± 0.24‰. In core V28-238 (1.7 cm/10^3 yr) the same ten glacial extreme values differ from the Holocene value by 1.04 ± 0.14‰. In core V19-28 (4.0 cm/10^3 yr) the last five glacial extreme values differ from the Holocene value by 1.62 ± 0.11‰. (Ninkovich and Shackleton, 1975).

Figure 1 shows cyclic changes in the percentage of sediment that is greater than 180 μm as well as in oxygen-isotope composition. Thompson and Saito (1974) documented correlative cyclic variations in dissolution intensity in cores V28-238, V28-239, and RC11-210. The latter core is in the region where Hays and others (1969) defined dissolution zones on the basis of changing carbonate percentage. Thus, we may confidently ascribe the observed variations in coarse-fraction percentage to changes in dissolution intensity and assign them to zones according to the definition of Hays and others (1969). Figure 1 shows these dissolution zones. The delay between the climatic change recorded in the oxygen-isotope record and the dissolution change, noted by Luz and Shackleton (1975) and by Ninkovich and Shackleton (1975), is preserved throughout the sequence.

Characteristically, the transition from glacial to interglacial extreme occurred very rapidly (Broecker and van Donk, 1970); indeed, 12,000 yr ago deglaciation took place so fast that its record in sediment cores is almost invariably determined by the sediment-mixing depth rather than by the actual rate of change in the isotopic composition of the ocean (at least 0.3‰/10^3 yr). Glacial stages 2, 6, 10, 12, 20, and 22 terminated in this manner. Stages 4, 8, and 14 probably did

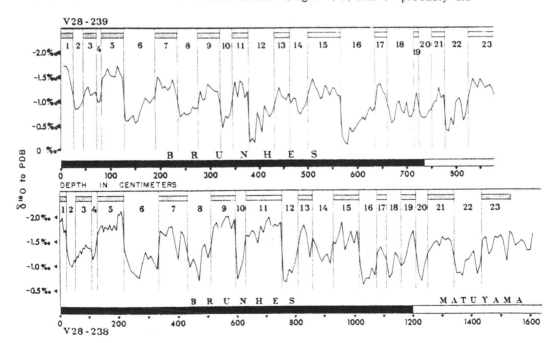

Figure 2. Oxygen-isotope and paleomagnetic record in upper 880 cm of cores V28-239 (above) and V28-238 (below).

291

not, whereas the evidence from stage 18 is inconclusive. The character of the record in other respects does not seem susceptible to generalization. There is a significant amount of fine structure that can be reliably appreciated only in cores with a higher sedimentation rate. Much potential information for marine-terrestial correlation is contained in this detail.

In core V28-239 glacial stage 22 is the first glaciation of duration and intensity similar to the glacial stages of the Brunhes epoch.

Mid-Matuyama Region. Between 0.8 to 1.4 m.y. ago (800 and 1,400 cm) isotopic fluctuations occurred at about 40,000-yr intervals. Because of the effects of mixing, it is not possible to say whether the true amplitude of glacial-interglacial change was smaller in mid-Matuyama time. The record of changing coarse fraction (Fig. 1) has a similar frequency, and again, changes in coarse fraction appear about 5 cm above oxygen-isotope changes, representing a lag of about 5,000 yr.

Among the stages so far defined, the isotopic difference between adjacent stages generally exceeds the isotopic variations within a single stage. In order to extend this principle to the section between 800 and 1,400 cm, one would need to define shorter stages; yet, with present coring techniques this degree of refinement could seldom be utilized. For the present we defer extending the scheme for numbering stages.

The pattern of coarse-fraction percentage may be zoned easily by reference to Hays and others (1969) as far back as their M3 zone (Fig. 1). Their M11 zone probably falls between 1,330 and 1,390 cm. Between M3 and M11, the records they described lack the resolution to display events of the frequency that we demonstrate.

Olduvai Region. The lower third of the core, from 1,400 to 2,100 cm, records lower frequency oxygen-isotopic changes with an amplitude no greater than in the middle section. The sediment recording the Olduvai magnetic event contains a well-marked "glaciation" lasting about 25,000 yr near the base of the event and another at the top of the event. A glaciation of similar magnitude is found at 2,000 cm (about 2 m.y. B.P., late Pliocene). Assuming that the Pliocene-Pleistocene boundary is in or near the Olduvai magnetic event, there is no associated climatic event that would enable late Pliocene to be distinguished from earliest Pleistocene time on the basis of climate. However, there is sufficient information in the isotopic record to facilitate detailed long-distance correlation in the region of this boundary. The relationship between the oxygen-isotope record and the record of changing coarse-fraction percentage is less obvious in this part of the core.

Comparison with Atlantic Core V16–205

The only other core extending to the base of the Pleistocene record that has been analyzed using the oxygen-isotope technique is North Atlantic core V16-205 (van Donk, this volume). This core has a sedimentation rate of only about 0.55 cm/10^3 yr, so that the distortion in the core V28-239 isotopic record, discussed below, is present to a considerably greater extent in core V16-205. In the Brunhes section of the core the major climatic cycles can still be distinguished with confidence, but in the section below the Jaramillo magnetic event, core V16-205 shows scarcely any change over an interval interpreted to represent 300,000 yr. The frequency of isotopic changes that we observe in core V28-239 is about one cycle per 40,000 yr, corresponding to a wavelength in core V16-205 of only 22 cm. It would be expected that mixing would largely obscure the record of these fluctuations, although high-precision analyses at closer sampling intervals might possibly detect a residual record. An additional impediment to correlation between the two cores is that

V16-205 seems to contain even greater changes in accumulation rate than V28-239. The rate between the Jaramillo and the top of the Brunhes in core V16-205 is ,reported to be 0.25 cm/10^3 yr, although the average rate through the entire core is 0.55 cm/10^3 yr. It is to be hoped that interoceanic correlations for the Matuyama epoch will become more reliable with the analysis of more cores in both the oceans.

ISOTOPE STRATIGRAPHY AND ITS LIMITATIONS

Oxygen-Isotope Stages: Terminology

Emiliani (1955, 1966) used numbers 1 to 16 to designate stages that he recognized in oxygen-isotope records he obtained in sediment cores from the Caribbean Sea and Atlantic Ocean. We (Shackleton and Opdyke, 1973) recognized 22 stages in core V28-238, the first 16 coinciding with those used by Emiliani. As a step toward formalizing this nomenclature, stage boundaries were defined by the depth at which they were located in core V28-238 (Fig. 2). For core V28-239, 23 stages are shown in Figure 1; the depths of stage boundaries, placed by correlation with core V28-238, are given in Table 2.

Before considering the extension of this terminology, it is important to consider the assumptions on which use of the oxygen-isotope record as a stratigraphic tool is based and the limitations of its usefulness. It is universally agreed that at the

TABLE 2. STAGE BOUNDARIES IN CORE V28-239 AS DETERMINED BY
CORRELATION WITH CORE V28-238

Boundary	Depth in core* (cm)	Age† (B.P.)	Termination§
1-2	25	13,000	I
2-3	45	32,000	
3-4	72	64,000	
4-5	82	75,000	
5-6	127	128,000	II
6-7	190	195,000	
7-8	235	251,000	III
8-9	275	297,000	
9-10	320	347,000	IV
10-11	345	367,000	
11-12	377	440,000	V
12-13	430	472,000	
13-14	462	502,000	
14-15	500	542,000	
15-16	567	592,000	VI
16-17	635	627,000	
17-18	660	647,000	
18-19	715	688,000	
19-20	725		
20-21	750		
21-22	777		
22-23	825		

* Determined by correlation with core V28-238.

† Ages are those estimated by Shackleton and Opdyke (1973) by linear interpolation in core V28-238 using a rate of 1.7 cm/10^3 yr.

§ Terminations from Broecker and van Donk (1969). They defined terminations on the basis of their interpretation of the saw-toothed character of the oxygen-isotope record. Owing to a possible hiatus in core V12-122, it appears that the event labeled termination VI by them is the stage 16-15 boundary.

maximum of the most recent glaciation, between 20,000 and 15,000 yr ago, sea level was lowered about 100 m as the result of ice storage on the continents. The same situation prevailed at each glacial maximum. This ice was certainly depleted in O^{18} relative to ocean water, and the result of its removal on the isotopic composition of the remaining ocean water has been discussed many times since its effect on the isotope record in fossil Foraminifera was first considered (Emiliani, 1955; Olausson, 1965; Shackleton, 1967; Dansgaard and Tauber, 1969; Shackleton and Opdyke, 1973). The resemblance between the actual (at present unknown) record of mean isotopic composition of oceans as a function of time and any particular measured record of isotopic composition in calcareous fossils from a sediment core depends on a number of factors which are discussed below.

Ocean Mixing and the Oxygen-Isotope Record

When isotopically light water is removed from or added to a particular part of the ocean, the resulting change will be noticed in another part of the ocean only after a delay resulting from the finite mixing time of the oceans. This has been variously estimated, but it is probably under 1,000 yr (Gordon, 1975). Thus, the effect of a sudden change in continental ice volume could be a transient spike in isotopic composition of the Atlantic, followed a few hundred years later by an isotopic change in the Pacific.

In reality this effect is limited by the finite response time of an ice sheet (Weertman, 1964). Nevertheless, transient effects can be detected in restricted areas such as the Mediterranean (Olausson, 1965) and the Gulf of Mexico (Kennett and Shackleton, 1975). This is the ultimate limit to the precision of oxygen-isotope stratigraphy.

Deep-Ocean Temperatures and the Ideal Oxygen-Isotope Record of Benthic Foraminifera

The isotopic composition of the calcite tests of foraminifera depends on both the isotopic composition and the temperature of the water they inhabited (Urey, 1947). In the ocean, surface-temperature changes probably were not synchronous over the entire globe. However, deep-ocean temperature changes during Pleistocene time were probably small, and the stability requirements of the ocean suggest that they should have occurred essentially synchronously. Thus, a record of the isotopic composition of the tests of suitable benthic foraminifers in a core from a depth of 3,000 m (and with a sufficiently high sedimentation rate to provide a stratigraphic resolution of about 1,000 yr) would approach the ideal with which the oxygen-isotope record in other cores might be correlated. Consideration of the available sections of record approaching this ideal suggests that in favorable circumstances such a record would contain enough information to permit a correlation accuracy of better than 2,000 yr throughout the past 900,000 yr.

Because the ideal isotopic record for benthic foraminifers has not yet been obtained, it is not possible to say in what manner the isotopic records of planktonic foraminifers in cores V28-238 and V28-239 differ from the ideal. We consider below the effects of mixing and burrowing, postdepositional solution, and surface temperature change on such isotopic records.

Sediment Mixing and Burrowing

Table 3 compares the peak-to-peak amplitude of isotopic changes in cores V28-238 and V28-239. The 5-cm sampling interval in core V28-239 approaches the mixing depth (Berger and Heath, 1968; Ruddiman and Glover, 1972), and the extremes analyzed must therefore approximate the extremes present in the core. The mean amplitude of isotopic fluctuations is less in core V28-239 than in V28-238, probably because the accumulation rate is lower in comparison with the mixing depth.

We have argued that the observed peak-to-peak amplitude of oxygen-isotope changes in core V28-238 was reduced by mixing (Shackleton and Opdyke, 1973). Thus, the full range of isotopic variation is attenuated in the sediment in core V28-238 and even more so in core V28-239. However, both cores preserve sufficient record that successive stages can be unambiguously recognized.

Carbonate Dissolution and the Oxygen-Isotope Record

Core V28-239 was taken at a depth of 3,490 m, compared to 3,120 m for core V28-238. This accounts for more intense dissolution occurring in core V28-239. Savin and Douglas (1973) pointed out that increasing dissolution not only progressively removes the more solution-susceptible species (often those that lived in shallower water), but it also selectively removes from the population of a single species those members that lived closer to the surface. Thus, the fossil population that has suffered more dissolution registers a lower isotopic temperature as a consequence of that dissolution.

TABLE 3. STAGE-BY-STAGE ISOTOPE EXTREMES FOR CORES V28-239 AND V28-238

Stage	Interval	V28-239 A (δ, ‰)	B (range)	V28-238 C (δ, ‰)	D (range)	Between-core Difference (A − C)*
1	1-2	−1.72	0.88	−1.98	1.01	0.26
2	2-3	−0.84	0.46	−0.97	0.47	0.13
3	3-4	−1.30	0.32	−1.44	0.59	0.14
4	4-5	−0.98	0.74	−0.85	1.26	−0.13
5	5-6	−1.72	1.16	−2.11	1.37	0.39
6	6-7	−0.56	0.94	−0.74	1.05	0.18
7	7-8	−1.50	0.82	−1.79	1.10	0.29
8	8-9	−0.68	0.67	−0.69	1.30	0.01
9	9-10	−1.35	0.90	−1.99	1.28	0.64
10	10-11	−0.45	0.97	−0.71	1.26	0.26
11	11-12	−1.42	1.28	−1.97	1.32	0.55
12	12-13	−0.14	1.14	−0.65	1.16	0.51
13	13-14	−1.28	0.56	−1.81	0.76	0.53
14	14-15	−0.72	0.73	−1.05	0.82	0.33
15	15-16	−1.45	1.35	−1.87	1.28	0.42
16	16-17	−0.10	1.25	−0.59	0.77	0.49
17	17-18	−1.35	0.82	−1.36	0.64	0.01
18	18-19	−0.53	0.63	−0.72	0.97	0.19
19	19-20	−1.16	0.52	−1.69	1.03	0.53
20	20-21	−0.64	0.63	−0.66	0.87	0.02
21	21-22	−1.27	0.90	−1.53	0.76	0.26
22		−0.37		−0.77		0.40

Note: Column A, extreme oxygen-isotopic composition in each stage in core V28-239, from Table 1. Column B, isotopic difference between adjacent stages in core V28-239. Mean 0.84 ± 0.28. Column C, extreme oxygen-isotopic composition in each stage in core V28-238, from Shackleton and Opdyke (1973, Table 1), corrected to PDB standard. Column D, isotopic difference between adjacent stages in core V28-238. Mean 1.00 ± 0.27

*Difference between the extreme reached in cores V28-239 and V28-238 for each stage. Mean 0.29 ± 0.21.

Samples from the two core tops differ isotopically by about 0.2‰. The isotopic extremes in each stage differ between the cores by 0.29 ± 0.21‰ (Table 3). This is probably due to greater solution in core V28-239, which is evident in the degree of preservation of the foraminifers. Part of the between-stage variation in core V28-239 may be due to the effect of fluctuating carbonate dissolution on the isotopic composition, and this clearly limits the veracity of a record of ocean oxygen-isotope composition derived from measurements in a deep-water core.

Sediment Accumulation-Rate Changes and Their Effect on the Oxygen-Isotope Record

Despite the distortions in the shape of the record of oxygen-isotopic changes in core V28-239 caused by dissolution and by sediment mixing discussed above, the position of the "terminations" (glacial-interglacial stage boundaries) of Broecker and van Donk (1970) may be placed confidently. Thus, the sediment thickness deposited during each climatic cycle may be measured and compared with the thickness deposited over the same time intervals in other cores.

Figure 3 compares the thickness of sedimentary deposits representing these major climatic cycles in cores V28-238 and V28-239 (data from Table 4). If the accumulation had remained constant in both cores, the points in Figure 3 would lie on a single straight line. The fact that they do not shows that in one or both cores the accumulation rate varied. In Figures 4 and 5, the thicknesses of sedimentary deposits representing the climatic cycles in cores V28-239 and V28-238 are compared with the average thicknesses of the same cycles in a suite of four cores analyzed by Emiliani (1966, 1972). Core V28-238 (Fig. 5) displays better agreement with the Caribbean suite. Therefore, the scatter in Figure 3 is probably due to stage-to-stage variations in accumulation rate in core V28-239.

In Figure 5 the greatest discrepancy between the Caribbean cores and core V28-238 is in stages 2 to 5. This suggests that in one of these areas there was a systematic change in accumulation rate about 130,000 yr ago (end of stage 6), as suggested by Emiliani and Shackleton (1974). The existence of a regional change in accumulation rate for any area at the end of stage 6 serves as a warning that linear extrapolation

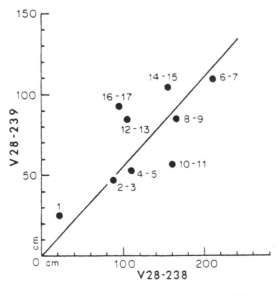

Figure 3. Comparison of thickness of climatic cycles (odd- and succeeding even-numbered stages) in cores V28-239 and V28-238. Data from Table 4.

TABLE 4. THICKNESS OF CLIMATIC CYCLES IN A SUITE OF CARIBBEAN CORES AND IN PACIFIC CORES V28-238 AND V28-239

Stages	Cores						
	P6304-4	P6304-7	P6304-8	P6304-9	Mean P6304 suite	V28-238	V28-239
1	25	30	30	30	29	22	25
2-3	130	145	150	165	148	88	47
4-5	175	185	170	145	169	110	53
6-7	240	275	280	240	259	210	110
8-9	160	195	210	200	191	165	85
10-11	145	180	190	180	174	160	57
12-13	155	150	. .	130	144	105	85
14-15	85	80?	. .	200	200*	155	105
16-17						95	93

Note: Data for Caribbean cores from Emiliani (1955, 1972); for Pacific cores from Shackleton and Opdyke (1973); for core V28-239 from this paper.
*Data for stages 14-15 from P6304 suite is inconsistent. Value for 6304-9 has been adopted, because it is more consistent with the Pacific cores (Figs. 3, 4).

performed on the basis of an assumed uniform accumulation rate may be in error even if the extrapolation is based on average accumulation rates from numerous cores.

Effect of Varying Surface Temperature

In the Caribbean, Emiliani (1966) has shown that the oxygen-isotope composition of *G. sacculifer* in recent sediment implies deposition at or near surface temperature. Vincent and Shackleton (1975) have shown that this is also true in the Indian Ocean. While this situation holds, changes in surface temperature during Pleistocene time, if present, should affect the isotopic composition of *G. sacculifer* populations in Pleistocene sediment. Hence, it is generally assumed that changes in surface temperature may be estimated by subtracting the component that is ascribed to glacially induced changes in ocean isotopic composition from the total record of oxygen-isotopic change (Imbrie and others, 1973).

In the Pacific a different situation prevails: The oxygen-isotope composition of *G. sacculifer* corresponds to a temperature several degrees below sea-surface temperature in many core-top samples (Savin and Douglas, 1973). To what extent this figure is an indication of a difference in the depth distribution of calcification in the species and to what extent it is a function of selective dissolution of the individuals from shallower depths in the water column (Savin and Douglas, 1973) remain to be evaluated. However, we (Shackleton and Opdyke, 1973) argued that the close similarity between the isotopic records of *G. sacculifer* and benthic species in core V28-238 implies that both records depict the history of the isotopic composition of the ocean, and that changes in surface temperature, temperature-depth structure, depth distribution of *G. sacculifer*, and selective dissolution all play minor roles. Discrepancies between the planktonic and benthic records in core V28-238 were ascribed by us (Shackleton and Opdyke, 1973) to the effects of postdepositional sediment mixing by burrowing organisms rather than to the factors mentioned above.

Long-Term Trends in the Oxygen-Isotope Record

There is no general agreement regarding any long-term trends in climate during Pleistocene time; trends that emerge from studies based on the present-day ecological requirements of any organisms may be due to the effect of evolution or of evolutionary adaptation. It might be hoped that oxygen-isotope analysis would provide a means of discerning real trends. Unfortunately, in the case of core V28-239 it seems certain that the trends observed are due to factors that alter the amount of smoothing the record has undergone, and not to trends in isotopic composition of the ocean. A careful comparison between the records of cores V28-238 and V28-239 shows that there is a relationship between trends in isotopic composition and changes in accumulation rate. For example, in core V28-239 the most extreme isotopic values are found in glacial stages 12 and 16. This might be taken to imply that these were the stages of greatest glacial advance. However, Table 2 shows that these two glacial stages are represented by almost the same sediment thickness in core V28-239 as in V28-238, whereas most units in V28-239 are about two-thirds of the thickness found in V28-238. Thus, these two glaciations appear more extreme only because sediment mixing has had a smaller blurring effect. It may be the case that the accumulation rate changed in response to climate, in which case the trend would have some second-order relationship to climate. In the lower part of the core, there may be long-term changes in dissolution intensity and perhaps evolutionary changes in calcite secretion depth. The *fistulosus* form of *G. sacculifer* is isotopically indistinguishable from the usual form, but it may prove to be related to a change in calcite secretion depth.

We conclude that it would be unwise to deduce from available oxygen-isotope measurements any trend either in surface temperature in the equatorial Pacific or in the amplitude of glacial advances in the Northern Hemisphere. In more favorable circumstances (higher accumulation rate, shallower water) oxygen-isotope analysis should be capable of yielding information on such trends.

CALCIUM CARBONATE DISSOLUTION STRATIGRAPHY

During Pleistocene time the extent of solution on the Pacific floor varied cyclically (Arrhenius, 1952; Hays and others, 1969) as climate changed. Luz and Shackleton (1975) have shown that during stage 5, dissolution systematically lagged behind

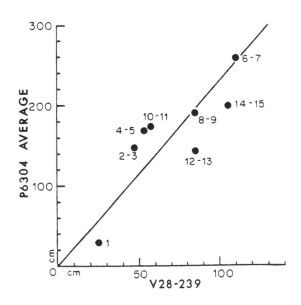

Figure 4. Comparison of thickness of climatic cycles (odd- and succeeding even-numbered stages) in core V28-239 and in suite of cores P6304-4, P6304-7, P6304-8, and P6304-9 from Emiliani (1966, 1972). Data from Table 4.

the climatic record. The intensity of dissolution rose not at the boundary between stages 6 and 5 (termination II of Broecker and van Donk, 1970) but a few thousand years later. We now show that this relationship has held through the past 1.5 m.y.

Figure 1 indicates the boundaries of the dissolution zones, numbered according to the scheme of Hays and others (1969), and their relation to the oxygen-isotope stages. The dissolution zones are not manifested as carbonate fluctuations, and the carbonate content is high throughout the core (Thompson, 1976). However, Thompson and Saito (1974) have shown that change in dissolution intensity is the dominant factor in determining the downcore changes in foraminiferal faunal composition in this area.

Figure 1 clearly indicates that Hays and others (1969) were correct in their assertion that changes in carbonate content in eastern equatorial Pacific sediments could be correlated with the Northern Hemisphere climatic record and with oxygen-isotope records from the Caribbean. However, use of carbonate cycles as a precise stratigraphic tool may be misleading. Figure 1 shows that the base of each dissolution zone in the sediment is not found at the same position as the glacial-to-interglacial isotopic transition, but rather some 5 to 20 cm above. This represents a delay of a few thousand years between the climatic change and its effect on bottom-water chemistry in the equatorial Pacific. This delay may not be constant from one latitude to another or from one climatic cycle to another.

CHRONOLOGY

The record of changes in the oxygen-isotope composition of the world oceans may be readily used as a stratigraphic tool in Pleistocene deep-sea sediments of all oceans (the Atlantic and Caribbean, Emiliani, 1955; the Arctic, van Donk and Mathieu, 1969; the Indian Ocean, Oba, 1969; the Pacific Ocean, Shackleton and Opdyke, 1973; the sub-Antarctic regions, Hays and others, this volume). Moreover, since the primary mechanism giving rise to these changes is the growth and retreat of continental ice sheets in the Northern Hemisphere, the record is of considerable

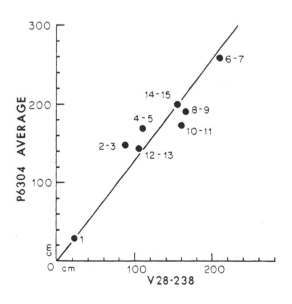

Figure 5. Comparison of thickness of climatic cycles (odd- and succeeding even-numbered stages) in core V28-238 and in suite of cores P6304-4, P6304-7, P6304-8, and P6304-9 from Emiliani (1966, 1972). Data from Table 4.

299

value as a basic Pleistocene stratigraphic tool to which the less continuous fragments of the Pleistocene record from the continents may be correlated (for example, the Netherlands, van der Hammen and others, 1967; Central Europe, Kukla, 1970).

Hence, the absolute dating of the oxygen-isotope record is of wide interest. Moreover, accurate dating is essential if oxygen-isotope records are to be used to test the hypothesis that variations in the Earth's orbital parameters caused climatic changes.

Isotopic stage boundaries in core V28-238 were first dated by assuming a constant rate of sediment accumulation above the Brunhes-Matuyama magnetic reversal boundary at 1,200 cm (Shackleton and Opdyke, 1973). Although more accurate ages will be obtained in the future by making use of several cores, it is important that this be done critically. We have argued that the accumulation rate within the Brunhes magnetic epoch was more uniform in core V28-238 than in V28-239. Thus, ages derived by assuming uniform accumulation rates may be more reliable if based on core V28-238 alone than they would be if based either on V28-239 alone or on a combination of the two cores.

For this reason, we do not offer new estimates of the ages of stage boundaries but consider those estimates made earlier (Shackleton and Opdyke, 1973) as the best available (Table 2). We emphasize again that these cannot be regarded as definitive.

CONCLUSIONS

Oxygen-isotope analysis provides an excellent stratigraphic tool for the past 900,000 yr. Oxygen-isotope stratigraphy has been used to locate the timing of carbonate dissolution zones relative to major climatic changes. Dissolution increased a few thousand years after each deglaciation.

Oxygen-isotope stratigraphy enables us to compare sediment accumulation rates. Cores V28-238 and V28-239 and a suite from the Caribbean analyzed by Emiliani have been compared in detail. Core V28-238 had a more constant accumulation rate from stage to stage than the deeper core V28-239.

In lower Pleistocene sediment, oxygen-isotope stratigraphy will not be useful in cores with an accumulation rate lower than $1 \text{ cm}/10^3$ yr, because the isotopic changes were more frequent than in the Brunhes epoch and therefore more readily obscured by mixing in the sediment.

ACKNOWLEDGMENTS

Coring was supported by Office of Naval Research Grant N00014-67-A-0108-0004 and National Science Foundation Grant GA-29460 to Lamont-Doherty Geological Observatory. The research was supported by National Science Foundation IDOE Grant IDO71-04204. Oxygen-isotope analysis was supported by Natural Environment Research Council Grant GR3/1762 to N. J. Shackleton at Cambridge. We gratefully thank M. A. Hall for his operation of the mass spectrometer with consistent care. We are grateful to W. A. Berggren, K. King, and G. J. Kukla for making instructive comments on an early version of this paper, and to S. M. Savin for a very constructive final review.

300

REFERENCES CITED

Arrhenius, G., 1952, Sediment cores from the East Pacific: Swedish Deep-Sea Exped. Repts., v. 5, p. 6–227.

Berger, W. H., and Heath, G. R., 1968, Vertical mixing in pelagic sediments: Jour. Marine Research, v. 26, p. 135–143.

Broecker, W. S., and van Donk, J., 1970, Insolation changes, ice volumes and the O^{18} record in deep-sea sediments: Rev. Geophysics and Space Physics, v. 8., p. 169–198.

Dansgaard, W., and Tauber, H., 1969, Glacier oxygen-18 content and Pleistocene ocean temperatures: Science, v. 166, p. 499–502.

Emiliani, C., 1955, Pleistocene temperatures: Jour. Geology, v. 63, p. 538–578.

——1966, Palaeotemperature analysis of Caribbean cores P6304-8 and P6304-9 and a generalized temperature curve for the past 425,000 years: Jour. Geology, v. 74, p. 109–126.

——1972, Quaternary paleotemperatures and the duration of the high-temperature intervals: Science, v. 178, p. 398–401.

Emiliani, C., and Shackleton, N. J., 1974, The Brunhes epoch: Palaeotemperatures and geochronology: Science, v. 183, p. 511–514.

Epstein, S., Buchsbaum, R., Lowenstam, H.A., and Urey, H. C., 1951, Carbonate-water isotopic temperature scale: Geol. Soc. America Bull., v. 62, p. 417–426.

——1953, Revised carbonate-water isotopic temperature scale: Geol. Soc. America Bull., v. 64, p. 1315–1326.

Gordon, A. L., 1975, General ocean circulation, in Numerical models of ocean circulation: Washington, D.C., Natl. Acad. Sci., p. 39–53.

Hays, J. D., Saito, T., Opdyke, N. D., and Burckle, L. H., 1969, Pliocene-Pleistocene sediments of the equatorial Pacific: Their paleomagnetic, biostratigraphic, and climatic record: Geol. Soc. America Bull., v. 80, p. 1481–1514.

Hays, J. D., Lozano, J., Shackleton, N., and Irving, G., 1976, Reconstruction of the Atlantic Ocean and western Indian Ocean sectors of the 18,000 B.P. Antarctic Ocean, in Cline, R. M., and Hays, J. D., eds., Investigation of late Quaternary paleoceanography and paleoclimatology: Geol. Soc. America Mem. 145 (this volume).

Imbrie, J., van Donk, J., and Kipp, N. G., 1973, Paleoclimatic investigation of a Late Pleistocene deep-sea core: Comparison of isotopic and faunal methods: Quaternary Research, v. 3, p. 10–38.

Kennett, J. P., and Shackleton, N. J., 1975, Latest Pleistocene melting of the Laurentide ice sheet recorded in deep-sea cores from the Gulf of Mexico: Science, v. 188, p. 147–150.

Kukla, J., 1970, Correlations between loesses and deep-sea sediments: Geol. Fören. Stockholm Förh., v. 92, p. 148–180.

Luz, B., and Shackleton, N. J., 1975, $CaCO_3$ solution in the tropical East Pacific during the past 130,000 years: Cushman Found. Foram. Research Spec. Pub. 13, p. 142–150.

Ninkovich, D., and Shackleton, N. J., 1975, Distribution, stratigraphic position and age of ash layer "L," in the Panama Basin region: Earth and Planetary Sci. Letters, v. 27, p. 20–34.

Oba, T., 1969, Biostratigraphy and isotopic paleotemperatures of some deep-sea cores from the Indian Ocean: Tohoku Univ. Sci. Repts., 2nd ser. (Geology), v. 41, p. 129–195.

Olausson, E., 1965, Evidence of climatic changes in North Atlantic deep-sea cores, with remarks on isotopic palaeotemperature analysis: Prog. Oceanography, v. 3, p. 221–252.

Ruddiman, W. F., and Glover, L. K., 1972, Vertical mixing of ice-rafted volcanic ash in North Atlantic sediments: Geol. Soc. America Bull., v. 83, p. 2817–2836.

Savin, S. M., and Douglas, R. G., 1973, Stable isotope and magnesium geochemistry of recent planktonic Foraminifera from the South Pacific: Geol. Soc. America Bull., v. 84, p. 2327–2342.

Shackleton, N. J., 1967, Oxygen isotope analyses and paleotemperatures re-assessed: Nature, v. 215, p. 15–17.

——1974, Attainment of isotopic equilibrium between ocean water and the benthonic foraminifera genus *Uvigerina*: Isotopic changes in the ocean during the last glacial: Paris, Centre National de la Recherche Scientifique, Colloquium 219.

Shackleton, N. J., and Opdyke, N. D., 1973, Oxygen isotope and palaeomagnetic stratigraphy of equatorial Pacific core V28-238: Oxygen isotope temperatures and ice volumes on a 10^5 and 10^6 year scale: Quaternary Research, v. 3, p. 39-55.

Thompson, P.R., 1976, Planktonic foraminiferal dissolution and the progress towards a Pleistocene equatorial Pacific transfer function: Jour. Foram. Research (in press).

Thompson, P. R., and Saito, T., 1974, Pacific Pleistocene sediments: Planktonic foraminifera dissolution cycles and geochronology: Geology, v. 2, p. 333-335.

Urey, H. C., 1947, The thermodynamic properties of isotopic substances: Chem. Soc. Jour., p. 562-581.

van der Hammen, T., Maarleveld, G. C., Vogel, J. C., and Zagwijn, W. H., 1967, Stratigraphy, climatic succession and radiocarbon dating of the last glacial in the Netherlands: Geologie en Mijnbouw, v. 46e, p. 79-95.

van Donk, J., 1976, An O^{18} record of the Atlantic Ocean for the entire Pleistocene, *in* Cline, R. M., and Hays, J. D., eds., Investigation of late Quaternary paleoceanography and paleoclimatology: Geol. Soc. America Mem. 145 (this volume).

van Donk, J., and Mathieu, G., 1969, Oxygen isotope compositions of foraminifera and water samples from the Arctic Ocean: Jour. Geophys. Research, v. 74, p. 3396-3407.

Vincent, E., and Shackleton, N. J., 1975, Oxygen and carbon isotope composition of recent planktonic foraminifera from the Southwest Indian Ocean: Geol. Soc. America Abs. with Programs, v. 7, no. 7, p. 1308.

Weertman, J., 1964, Rate of growth or shrinkage of nonequilibrium ice sheets: Jour. Glaciology, v. 5, p. 145-158.

MANUSCRIPT RECEIVED BY THE SOCIETY DECEMBER 18, 1974

REVISED MANUSCRIPT RECEIVED JUNE 11, 1975

MANUSCRIPT ACCEPTED JUNE 25, 1975

LAMONT-DOHERTY GEOLOGICAL OBSERVATORY CONTRIBUTION NO. 2277

24

Oxygen isotope and palaeomagnetic evidence for early Northern Hemisphere glaciation

N. J. Shackleton
Subdepartment of·Quaternary Research, University of Cambridge, Free School Lane, Cambridge, UK

N. D. Opdyke
Lamont-Doherty Geological Observatory, Columbia University, Palisades, New York 10964

THE stratigraphically longest detailed oxygen isotopic records from the oceans extend to sediments about 2.1 Myr old[1,2] and indicate Pleistocene-like glacial events well below the horizon stratigraphically equivalent to the base of the Quaternary as it is defined in Italy[3]. Here we extend the isotopic record and identify the onset of these quasi-cyclic glacial–interglacial fluctuations. For this we required a core of sediment that accumulated somewhat more slowly than in the area of our previous studies.

Cores from the Equatorial Pacific were described by Hays *et al.*[4], who showed that north of the belt of maximum biological productivity along the equator it is possible to take piston cores containing a record extending well into the Pliocene. They discussed the lithostratigraphy and magnetostratigraphy of several cores, and the biostratigraphic record for foraminifera, diatoms and radiolaria (expanded in refs 6 and 7) while Gartner[5] presented the nannofossil stratigraphy. Since then several cruises have recovered cores from the same general area. Of these we selected V28–179 (collected by N.D.O.), for isotope analysis because it contained the thickest accumulation of sediment for the Gauss magnetic epoch.

Core V28–179 was taken at 4° 37′N, 139° 36′W in 4,509 m water depth. The core comprises alternating layers of fora- miniferal chalk ooze, marl ooze and marl with some dominantly diatomaceous layers. Most of the core is intensely burrowed, and many of the plugs of sediment washed for analysis could be seen to contain components of slightly different colour mixed together by burrowing. This strongly suggested that the depth through which burrowing took place was frequently greater than the typical thickness of a single layer deposited under uniform environmental conditions. We assume, following evidence already available[9], that the chief source of variability in the sediment derives from climatically controlled variations in the proportion of carbonate remaining in the sediment. As these variations occurred on a depth scale that was smaller than the depth of plainly imperfect homogenisation by burrowing organisms, it shows that we are limited to an investigation of the general character of the ocean oxygen isotopic record within the good time-framework provided by the palaeomagnetic record of the core.

The record of changing ocean oxygen istopic composition in the Late Pleistocene has been investigated in some detail in cores with accumulation rates up to about 8 cm kyr⁻¹ (ref. 10),

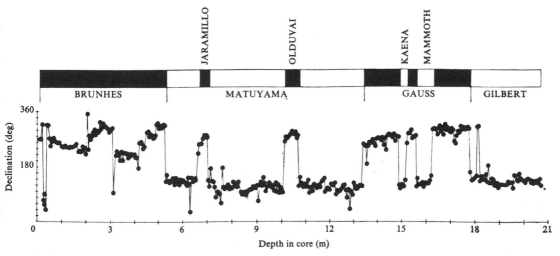

Fig. 1 Magnetic record for core V28–179, 150 Oe and interpretation in terms of the standard palaeomagnetic polarity scale.

303

and it has been shown that some cores with accumulation rates of over 3 cm kyr^{-1} preserve a sufficiently good record to allow the identification of the effect of the changing tilt of the Earth's orbit (period 40 kyr) and of precession of the equinoxes (period about 23 kyr)[11]. To investigate the Lower Pleistocene in a piston core, it is necessary to work in an area of rather lower sediment accumulation. We have already discussed[1] the loss of information which characteristically occurs under these conditions. Even if the mixing in the sediment of core V28–179 were homogeneous, climatic fluctuations with a 40 kyr period would have a wavelength in the sediment of only about 20 cm and would be scarcely detectable. Since the mixing is clearly not homogeneous, and since several samples proved too low in carbonate for an analysis to be made at all, it is likely that the detailed character of the climatic record is inaccessible. What we can do is to identify the date of inception of the isotopic fluctuations which have characterised at least the whole of the Pleistocene[1]. Although there are other types of evidence which indicate periods of climatic deterioration, there is as yet no information pertaining to the date of the earliest accumulation in the Northern Hemisphere of continental ice sheets that would have been of sufficient scale to produce a detectable effect on the oxygen isotopic composition of the oceans[12].

The core was sampled for magnetic measurement at 5 cm intervals along its length. The measurements were performed on a slow speed fluxgate magnetometer of the type described by Molyneaux[13]. In order to ascertain the proper alternating field to use for blanket demagnetisation and to determine the magnetic stability of the core 10 samples spaced throughout the core were progressively demagnetised in peak alternating fields of up to 250 Oe. The magnetic stability was found to vary throughout the length of the core with values of the median destructive field ranging from 25 to 225 Oe. Natural remanent magnetic intensities ranged from 0.2 to 20 $\times$ 10^{-6} e.m.u. cm^{-3}. A field of 150 Oe peak value was selected for blanket demagnetisation. The resulting plot of declination change against depth is shown in Fig. 1 plotted relative to an arbitrary fiducial mark scribed on the ship during the extrusion process[14]. The correlation of the resulting magnetic pattern to the standard time scale is easily done since the reversals are clearly delineated. This correlation has been confirmed by foraminiferal studies[15] using the foraminiferal zonation previously determined for equatorial Pacific sediments[4,6,7].

An important fact is that the pattern of coiling changes in *Pulleniatina* and in *Globorotalia tumida* shows that the lowest reversely magnetised section of sediment below about 18m can only represent the interval from the Gilbert–Gauss boundary to a horizon somewhat later than the Cochiti event; the bottom of the core probably has an age between 3.5 and 3.6 Myr (ref. 15).

The placement within the core of the major palaeomagnetic transitions are given in Table 1. The only departure of the magnetic stratigraphy from the standard magnetic time scale is the presence of a short normally magnetised interval at 1,800 cm, below the Gauss normal magnetic epoch. It is possible that this represents a short reversal of the magnetic field since the core does not show any apparent physical disturbance at this level.

Table 1 Magnetic boundaries V28–179

Brunhes–Matuyama	527 ± 2.5 cm
Matuyama–Jaramillo	657 ± 4 cm
Jaramillo–Matuyama	700 ± 2.5 cm
Matuyama–Olduvai	1,007 ± 2.5 cm
Olduvai–Matuyama	1,068 ± 3 cm
Matuyama–Gauss	1,358 ± 2.5 cm
Gauss–Kaena	1,488 ± 2 cm
Kaena–Gauss	1,517 ± 2 cm
Gauss–Mammoth	1,558 ± 2 cm
Mammoth–Gauss	1,622 ± 2.5 cm
Gauss–Gilbert	1,779 ± 2.5 cm
Unnamed short interval (1,808–1,817 ± 2.5 cm)	

Table 2 Oxygen and carbon isotope data for *Globocassidulina subglobosa* from core V28–179

depth δ(cm)	^{18}O	^{13}C
1,011	+4.30	−1.44
1,031	+3.79	−1.35
1,051	+4.03	−1.24
1,061	+3.52	−1.10
1,070	+3.67	−0.73
1,080	+4.21	−1.05
1,090	+3.89	−0.87
1,129	+3.86	−0.85
1,140	+4.14	−1.14
1,150	+3.72	−0.83
1,170	+3.48	−0.67
1,180	+4.31	−1.11
1,190	+3.77	−0.62
1,220	+3.79	−1.24
1,241	+3.80	−0.85
1,271	+4.15	−1.29
1,280	+3.46	−0.96
1,291	+4.02	−1.28
1,310	+4.30	−1.61
1,321	+3.65	−0.98
1,341	+3.98	−1.40
1,351	+3.99	−0.87
1,361	+3.41	−0.54
1,370	+4.05	−1.10
1,381	+3.94	−1.08
1,391	+3.29	−0.65
1,420	+3.92	−0.69
1,442	+3.55	−0.85
1,450	+3.77	−0.92
1,480	+4.02	−1.04
1,510	+3.42	−0.67
1,520	+3.87	−0.92
1,530	+3.36	−0.60
1,580	+3.93	−0.96
1,590	+3.80	−1.18
1,600	+3.92	−1.01
1,610	+3.35	−0.92
1,620	+3.76	−0.61
1,631	+3.69	−0.88
1,660	+3.66	−0.94
1,670	+3.60	−0.74
1,680	+3.52	−0.68
1,690	+3.13	−0.78
1,700	+3.50	−0.88
1,711	+3.16	−1.04
1,721	+3.19	−0.57
1,731	+3.19	−0.71
1,741	+3.37	−0.53
1,771	+3.13	−0.70
1,780	+3.43	−0.42
1,791	+3.53	−0.38
1,800	+3.14	−0.33
1,810	+3.56	−0.88
1,820	+3.19	−0.47
1,830	+3.28	−0.55
1,840	+3.30	−0.73
1,870	+3.25	−0.48
1,890	+3.20	−0.15
1,900	+3.41	−0.91
1,911	+3.29	−0.82
1,920	+3.20	−0.51
1,950	+3.40	−0.39
1,940	+3.24	−0.83
1,961	+3.36	−0.96
1,990	+3.40	−0.79
2,030	+3.30	−0.67
2,040	+3.51	−0.62
2,051	+3.45	−0.76
2,060	+3.14	−0.96

Oxygen isotope analyses were made at 10 cm intervals in the samples previously used for magnetic studies. The samples were dispersed in distilled water in a mechanical shaker, and after brief ultrasonic treatment, they were sieved on a 125 μm mesh and dried in a cool oven on the sieves. Foraminifera for isotopic analysis were taken from this fraction further cleaned ultrasonically and purified by roasting *in vacuo* at 400 °C for 30 min. Carbon dioxide was released from the carbonate by the action

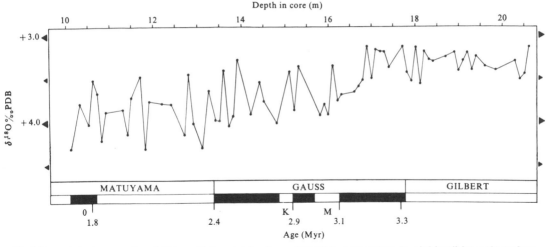

Fig. 2 Oxygen isotopic composition of *Globocassidulina subglobosa* in core V28–179 from 10m to 21m. *G. subglobosa* living on the sea floor at the coring site today would have an ¹⁸O content of about +3.50‰. Magnetic events are indicated by O (Olduvai), K (Kaena) and M (Mammoth). Ages indicated are based on current estimates of the chronology for the magnetic record[26].

of 100% orthophosphoric acid at 50 °C and analysed in a VGMicromass 602C mass spectrometer; analyses are reported to the PDB standard on the basis of multiple analyses of circulating standard carbonates.

The preservation of foraminifera is very poor at the water depth at which core V28–179 was taken (4,509 m). It was, therefore, generally not possible to obtain for analysis *Globigerinoides sacculifer*, the species used in our previous studies[1,8]. Instead, we chose to analyse benthonic foraminifera, which are less susceptible to dissolution. As many species of benthonic foraminifera do not deposit their carbonate in isotopic equilibrium with the sea water they inhabit[16], it would clearly be desirable to analyse a single species throughout. *Uvigerina*, known to deposit carbonate in isotopic equilibrium, is extremely rare, but in the majority of samples we were able to extract sufficient specimens of *Globocassidulina subglobosa* for analysis.

To calibrate *G. subglobosa* for possible departure from isotopic equilibrium, we analysed the top part of core V19–28 from which we obtained excellent isotopic data for *Uvigerina*[10,17]. Our measurements show that *G. subglobosa* may be even closer to isotopic equilibrium than *U. senticosa*, the species analysed in core V19–28. Comparisons of the oxygen isotopic composition of *G. subglobosa* and of *U. spinulosa* in Oligocene samples confirms the suitability of this species for oxygen isotope work[18,19]. The bottom temperature at the site of core V19–28 is about 1.8 °C; at the site of V28–179 it is about 1.1 °C. The mean ¹⁸O content of *Globocassidulina subglobosa* in the upper part of core V19–28 is +3.30‰ (five analyses); using the relationship $T = 16.9 - 4.4\delta + 0.10\delta^2$ (refs 20, 21) enables us to estimate that the temperature difference between the two sites should be equivalent to an isotopic difference of about 0.2 ‰, so that the expected value at the V28–179 site is about +3.50 ‰. Since the sediment accumulation rate at the coring site is only about 0.6 cm kyr⁻¹, the core top would be likely to contain a mixture of glacial and recent specimens and would not be useful for calibration purposes.

Analytical results for *G. subglobosa* are given in Table 2, and are plotted in Fig. 2. The record may best be described as a caricature of the true course of events, in that it is certainly not possible to derive an accurate record of events in such gradually accumulating sediment. It is, however, quite clear that the uppermost 17 m of the core are predominantly glacial in character; below 17 m, isotopic values are consistently close to the present-day value. Taking this lower section first, we may

conclude that between about 3.5 Myr and 3.2 Myr ago no big ice sheets accumulated in the Northern Hemisphere. Isotopically, the ocean was in an interglacial state, with more or less constant isotopic composition. The standard deviation among the measurements for this part of the core is 0.13 ‰. If this were entirely due to real variability, it would represent glacial events of a magnitude up to the equivalent of a 26 m range in eustatic sea-level fluctuation[8]. The variations are on so small a scale that if they derive from real glacial events, then they could be caused by changes in the Antarctic ice sheet, or by small Northern Hemisphere glaciations. It is conceivable, however, that this variability stems entirely from analytical error. More accurate measurements in a section with more rapid sediment accumulation will be needed to investigate possible small-scale glacial events earlier than 3.2 Myr ago.

Just below the Mammoth event at 1,680 m we see a positive excursion in isotopic composition that is well above possible analytical uncertainty and is of such a magnitude (0.4 ‰) as to represent about a 40 m sea-level equivalent in stored ice (assuming its isotopic composition was about −35 ‰). This date is in substantial agreement with the inception of ice-rafting observed in sediments from site 116 in Leg 12 of the Deep-Sea Drilling Project[22] and with evidence of glaciation in Iceland[23] and with ref. 12, but the present data give for the first time the scale of fluctuations which have continued ever since.

At the base of the Matuyama (sample at 1,310 cm) a glacial isotopic excursion of about 1.0 ‰ is observed. This is a characteristic extreme value over the lower Matuyama section. In deep-Pacific cores with a high accumulation rate the maximum difference between glacial and interglacial isotopic values in the late Pleistocene is about 1.60 per mil (ref. 10). Thus we have clear evidence that glaciations of a magnitude of at least two-thirds that of the late Pleistocene glacial maxima were occurring in the time interval from 2.5 to 1.8 Myr ago. This was probably the scale of glaciation throughout the Lower Pleistocene[1]. Evidently a major change in the character of glaciations occurred at about 2.5 Myr ago. The substantial carbon isotopic event at that point in the record may represent a large drop in the continental biomass[24] and may have been associated with significant floral extinctions under severe environmental pressure[25].

Few localities are likely to preserve a sea-level record for the past 3 Myr. But, it may be deduced from Fig. 2 that the last lengthy period of stable sea level and low ice volume terminated about 3.2 Myr ago, so that it is likely that a major constructional coral terrace feature is present in some areas dating from this time; marine highstands in the remaining Pliocene, like those of the Pleistocene, were probably brief. On the basis of this analysis one might advance the hypothesis that before 3.2 Myr ago, continental environments were relatively stable over a long period of time; since then the Earth has been subjected to the stress of continually fluctuating climate, with pleasant climates like our own occurring infrequently and lasting only about 10,000 years.

We thank NERC and NSF for financial support, M. A. Hall for his operation of the mass spectrometer, and W. A. Berggren and G. J. Kukla for helpful discussions.

Received 23 May; accepted 9 September 1977.

1. Shackleton, N. J. & Opdyke, N. D. *Geol. Soc. Am. Mem.* **145**, 449–464 (1976).
2. van Donk, J. *Geol. Soc. Am. Mem.* **145**, 147–164 (1976).
3. Haq, B. U., Berggren, W. A. & van Couvering, J. A. *Nature* (in the press).
4. Hays, J. D., Saito, T., Opdyke, N. D. & Burckle, L. H. *Geol. Soc. Am. Bull.* **80**, 1481–1514 (1969).
5. Gartner, S. *Geol. Soc. Am. Bull.* **84**, 2021–2034 (1973).
6. Saito, T., Burckle, L. H. & Hays, J. D. in *Late Neogene Epoch Boundaries* (eds Saito, T. & Burckle, L. H.) 226–244 (Am. Mus. nat. Hist., New York, 1975).
7. Saito, T. *Geology* 307–309 (1976).
8. Shackleton, N. J. & Opdyke, N. D. *Quat. Res.* **3**, 39–55 (1973).
9. Arrhenius, G. *Swed. Deep-Sea Exped. Rep.* **5**, 6–227 (1952).
10. Shackleton, N. J. *Phil. Trans. R. Soc.* **B280**, 169–179 (1977).
11. Hays, J. D., Imbrie, J. & Shackleton, N. J. *Science* **194**, 1121–1132 (1976).
12. Shackleton, N. J. & Kennett, J. P. in *Init. Rep. DSDP* **29**, 801–807 (1975).
13. Molyneux, L. *Geophys. J. R. Astr. Soc.* **24**, 1–5 (1971).
14. Opdyke, N. D. *Rev. Geophys. Space Phys.* **10**, 213–249 (1972).
15. Sciarrillo, J. R., thesis, Rutgers Univ. (1976).
16. Duplessy, J. C., Lalou, C. & Vinot, A. C. *Science* **168**, 250–251 (1970).
17. Ninkovich, D. & Shackleton, N. J. *Earth Planet. Sci. Lett.* **27**, 20–34 (1975).
18. Boersma, A. & Shackleton, N. J. *Init. Rep. DSDP* **39** (in the press).
19. Boersma, A. & Shackleton, N. J. *Init. Rep. DSDP* **40** (in the press).
20. O'Neil, J. R., Clayton, R. N. & Mayeda, T. K. *J. Chem. Phys.* **51**, 5547–5558 (1969).
21. Shackleton, N. J. *Colloques Int. Cent. natn. Rech. scient.* **219**, 203–210 (1974).
22. Berggren, W. A. *Init. Rep. DSDP* **12**, 953–963 (1972).
23. McDougall, I. & Wensink, H. *Earth Planet. Sci. Lett.* **1**, 232–236 (1966).
24. Shackleton, N. J. in *The Fate of Fossil Fuel CO_2 in the Oceans* (eds N. R. Andersen & A. Malahoff) Plenum, London and New York, 1977.
25. Zagwijn, W. H. *Boreas* **3**, 75–97 (1974).
26. Klitgord, J. L., Heustis, S. P., Mudie, J. D. & Parker, R. L. *Geophys. J. R. Astr. Soc.* **43**, 384–424 (1975).

Part VI

DATING OF OTHER PROCESSES
RECORDED IN OCEANIC SEDIMENTS

Editor's Comments
on Papers 25 Through 28

Magnetostratigraphy is not only applicable to the dating and correlation of historical oceanographic events derived from the study of microfossils; clearly, other processes or phenomena recorded in the sedimentary sequences can similarly be dated. These include changes in the general or specific characteristics of the sediments themselves, which are discussed in various chapters elsewhere in this volume. Additionally, however, other parameters have been investigated, a few of which are considered in the following four chapters including the dating of tektite distributions, volcanic ash layers, and deep-sea unconformities.

TEKTITE FIELDS

Shortly after it had been discovered that magnetostratigraphy could be applied to deep-sea sedimentary sequences, some tantalizing stratigraphic relations began to emerge between paleomagnetic reversals and other geological phenomena recorded in deep-sea cores. Among the earliest of these was the discovery by Billy Glass and Bruce Heezen in 1967 at the Lamont Geological Observatory (Glass and Heezen, 1967; Glass, 1967) that small (<1 mm) glassy objects known as *microtektites* are stratigraphically associated with the Brunhes-Matuyama paleomagnetic boundary (0.7 m.y. B.P.) in Indian Ocean sediments. Highest concentrations were discovered in the Indian Ocean area adjacent to Western Australia, where an extensive late Quaternary tektite field had been known for many years. The origin of such tektite fields preserved on land has been a problem for many years. Glass's Ph.D. research at the Lamont Observatory involved a search for tektites or microtektites in oceanic sediments adjacent to known occurrences on land and if discovered, to map their geographic and stratigraphic distributions in deep-sea sediments, to investigate their distinguishing characteristics, and to date their formation using the best available approaches. According to Glass (1968), the event that produced the Australasian tektite field spread a minimum of 10^{14} g of tektite material over 1/10 of the earth's surface. Although the origin of tektites is still unsolved, the three principal theories of tektite origin involve a meteorite's impact and resulting disintegration on the moon, on the earth's surface, or in the earth's atmosphere (Glass, 1968). Glass and Heezen (1967) speculated that paleomagnetic reversals may be at times related to collisions of cosmic bodies with the earth. This still remains as speculative as when first proposed. There is still insufficient data relating microtektites with paleomagnetic reversals and a lack of understanding of the source of the earth's magnetic field and the factors controlling reversals. The theory still rests only on the single stratigraphic observation of Glass and Heezen. A microtektite layer has also been documented in a few cores in the Indian Ocean adjacent to the Ivory Coast tektite fall, which has been paleomagnetically dated at between 0.85 and 0.9×10^6 years. This layer does not seem to be associated with any magnetic reversal, although too little stratigraphic work has yet been carried out to dismiss a possible relation with the Jaramillo Normal Event, which is dated at 0.9 m.y. B.P.

Paleomagnetic stratigraphy has thus been important for the dating and stratigraphic placement of microtektite horizons in deep-sea sediments. Such studies are summarized by Gentner et al. in Paper 25. The paleomagnetic dating of Indian Ocean microtektite layers, furthermore, has demonstrated that the Australasian and Ivory Coast tektite fields are of different age rather than a part of the same strew field as earlier proposed on geochemical evidence by Chapman and Schreiber (1969).

309

VOLCANIC ASH LAYERS

Magnetostratigraphy has also been valuable in providing a solid stratigraphic and chronological framework for volcanic ash (glass) layers in deep-sea sedimentary sequences. The study of the chronology of volcanic sequences is known as tephrachronology. The potential of using magnetostratigraphy to date and correlate ash layers in deep-sea sequences was recognized by Dragoslav Ninkovich of Lamont Geological Observatory very soon after paleomagnetic stratigraphy was first employed on deep-sea sediments. His first application of this approach, with Lamont colleagues, was carried out on a set of north Pacific cores containing layers of volcanic ash derived from the Aleutian Islands (Ninkovich et al., Paper 7). The approach is a simple but powerful one and has since been extended to other regions of the oceans such as the southwest Pacific to the east of New Zealand where Ninkovich (1968) dated a sequence of late Quaternary ash producing volcanic eruptions in New Zealand.

Likewise, ash horizons have been correlated and dated by magnetostratigraphy over vast areas of the Southern Ocean by Kennett and Watkins (Paper 33) and Huang et al. (Paper 26). The ability to date and correlate ash layers great distances from the volcanic source has also demonstrated that in some regions of high wind velocity such as in the Southern Ocean or during eruptions of vast magnitude coarse ($>61 \mu$), volcanic ash can be distributed up to 3000 km from the eruptive source. Such tephra is not necessarily deposited in the form of discrete layers but is represented as disseminated ash zones. The paleomagnetic and micropaleontological dating of cores containing volcanic ash either as layers or zones, in combination with sediment grain-size analysis, can provide an accurate measure of average accumulation rates and hence a means of determining the explosivity (Watkins et al., 1978) and duration (Ledbetter et al., 1978) of individual eruptions.

DEEP-SEA EROSION

Up until the late 1960s, when the study of deep-sea sediments increased dramatically, it was generally believed that the environment of sediment deposition in the deep sea was characterized by relative tranquility, therefore providing an ideal opportunity for continuous, uninterrupted deposition of deep-sea sediments. As the Deep Sea Drilling Project got underway, many paleontologists and stratigraphers expected to be furnished with continuous microfossil and sedimentary sequences from most areas, in turn providing an uncomplicated stratigraphic basis for the study of the evolution of life and environments in the oceans. When the extensive analysis of the deep-sea sedimentary sequences began as a result of the study of deep-drilled sequences, in addition to large piston core collections, the myth of the highly quiescent deep-sea sedimentary environment was soon to be demolished. The occurrence in the deep oceans of continuous sedimentary sequences spanning long intervals of time through the late Phanerozoic has been found to be the exception rather than the rule. Instead, almost all sections have been marked by one or more conspicuous interruptions in sedimentation during the time of sediment deposition. These hiatuses have been caused by active deep-sea erosion and/or calcareous biogenic dissolution, resulting from bottom currents of sufficient velocity flowing through the deep oceanic basins. The distribution of these currents is strongly controlled by deep-sea topography and the Coriolis effect. The present-day distribution of such active bottom currents is now well known from the study of bottom photographs, surface sediment ages and sediment characters, and from the measurement of bottom-water flow by current meters.

Magnetostratigraphy has been very important in the study of such breaks in sedimentation by providing a sound stratigraphic and chronological basis for both the identification of unconformities in single cores and the definition of the regional extent of such unconformities. The dating of these unconformities by magnetostratigraphy has been limited to sequences younger than 5 m.y. Older hiatuses have been dated using microfossils alone (for summary, see Moore, et al., 1978, and Kennett, 1977). Among the first studies concerned with the age and geographic definition of deep-sea hiatuses were those of Watkins and Kennett (Paper 27) and Kennett and Watkins (Paper 28). These studies established the magnetostratigraphy and biostratigraphy in large suites of late Cenozoic piston cores from various sectors in the Southern Ocean to provide a chronological basis for the definition of regional unconformities. The combination of magnetostratigraphy with biostratigraphy enhances the rapid dating of large numbers of cores because the biostratigrapher who is working from a magnetostratigraphic basis can work much more rapidly. This speed is facilitated because, except in the case of cores complicated by numerous unconformities, paleomagnetic age assignment can normally be provided from the biostratigraphic analysis of only a small number of sediment samples. Because the approach allows large suites of cores to be examined, extensive regional unconformites can be more readily identified and dated as demonstrated in Papers 34 and 35. The definition of the geographic and stratigraphic distribution of these unconformities has provided important present and past information on distributions of bottom currents and on the changing tempo of bottom-water activity in various oceanic regions. This in turn appears to have significant paleoclimatic implications. The paleomagnetic examination of large suites of cores from the ocean basins reflects the evolution of magnetostratigraphy from initial studies involving only a few selected cores to those that have utilized the approach on a more routine basis using large core collections, and providing definition of broad-scale regional oceanographic and paleoceanographic features.

REFERENCES

Chapman, D. R., and L. C. Schreiber. 1969. Chemical investigation of Australasian tektites. *Jour. Geophys. Research* **74**:6737–6776.

Glass, B. P. 1967. Microtektites in deep-sea sediments. *Nature* **214**:372–374.

Glass, B. P. 1968. Microtektites and the origin of the Australasian tektite strewn field. *Publication No. 6 by the Center for Meteorite Studies.* Tempe: Arizona State University, 22 pp.

Glass, B., and B. C. Heezen. 1967. Tektites and geomagnetic reversals. *Nature* **214**:372.

Kennett, J. P. 1977. Cenozoic evolution of Antarctic glaciation, the Circum-Antarctic Ocean, and their impact on global paleoceanography. *Jour. Geophys. Research* **82**:3843–3860.

Ledbetter, M. T., and R. S. J. Sparks. 1978. The duration of large-magnitude explosive eruptions deduced from graded bedding in deep-sea ash layers. *Geology* **7**:240–244.

Moore, T. C., T. H. van Andel, C. Sancetta, and N. Pisias. 1978. Cenozoic hiatuses and pelagic sediments. *Micropaleontology* **24**:113–138.

Ninkovich, D. 1968. Pleistocene volcanic eruptions in New Zealand recorded in deep-sea sediments. *Earth and Planetary Sci. Letters* **4**:89–102.

Watkins, N. D., R. S. J. Sparks, H. Sigurdsson, T.-C. Huang, A. Federman, S. Carey, and D. Ninkovich. 1978. Volume and extent of the Minoan Tephra from Santorini: New evidence from deep-sea sediment cores. *Nature* **271**:122–126.

ADDITIONAL READINGS

Cassidy, W. A., B. Glass, and B. C. Heezen. 1969. Physical and chemical properties of Australasian microtektites. *Jour. Geophys. Research* **74**:1008–1025.

Chapman, D. R. 1971. Australasian tektite geographic pattern, crater and ray of origin, and theory of tektite events. *Jour. Geophys. Research* **76**:6309–6338.

Davies, T. A., O. E. Weser, B. P. Luyendyk, and R. N. Kidd. 1975. Unconformities in the sediments of the Indian Ocean. *Nature* **253**:15–19.

Doake, C. S. M. 1977. A possible effect of ice ages on the earth's magnetic field. *Nature* **267**:415–417.

Durrani, S. A. 1972. Are microtektites the result of cometary impacts with the earth? *Nature* **235**:383.

Durrani, S. A., and H. A. Khan. 1971. Ivory Coast microtektites: Corrected values of uranium content. *Nature Phys. Sci.* **232**:175.

Fillon, R. H. 1972. Evidence from the Ross Sea for widespread submarine erosion. *Nature Phys. Sci.* **238**:40–42.

Frey, F. A., C. M. Spooner, and P. A. Baedecker. 1970. Microtektites and tektites: A chemical comparison. *Science* **170**:845–847.

Glass, B. P. 1969. Reworking of deep-sea sediments as indicated by the vertical dispersion of the Australasian and Ivory Coast microtektite horizons. *Earth and Planetary Sci. Letters* **6**:409–415.

Glass, B. P. 1972. Australasian microtektites in deep-sea sediments. In *Antarctic Oceanology II: The Australian–New Zealand Sector, Antarctic Research Series, Vol. 19,* edited by D. E. Hayes. Washington, D. C.: American Geophysical Union, pp. 335–348.

Glass, B. P., and B. C. Heezen. 1967. Tektites and geomagnetic reversals. *Sci. American* **217**:32–38.

Hays, J. D., and D. Ninkovich. 1970. North Pacific deep-sea ash chronology and age of present Aleutian underthrusting. *Geol. Soc. America Mem. 126,* pp. 263–290.

Huang, T.-C., and N. D. Watkins. 1977. Contrasts between the Brunhes and Matuyama sedimentary records of bottom water activity in the South Pacific. *Marine Geology* **23**:113–132.

Johnson, D. A., M. Ledbetter, and L. H. Burckle. 1977. Vema Channel paleo-oceanography: Pleistocene dissolution cycles and episodic bottom water flow. *Marine Geology* **23**:1–33.

Kennett, J. P., and C. A. Brunner. 1973. Antarctic Late Cenozoic glaciation: Evidence for initiation of ice-rafting and inferred increased bottom water activity. *Geol. Soc. America Bull.* **84**:2043–2052.

Kennett, J. P., and N. D. Watkins. 1975. Deep-sea erosion and manganese nodule development in the Southeast Indian Ocean. *Science* **188**:1011–1013.

King, E. A. 1977. The origin of tektites: A brief review. *Am. Scientist* **65**:212–218.

Ledbetter, M. T., D. F. Williams, and B. B. Ellwood. 1978. Late Pliocene climate and South-West Atlantic abyssal circulation. *Nature* **272**:237–239.

McColl, D. H., and G. E. Williams. 1970. Australite distribution pattern in Southern Central Australia. *Nature* **226**:154–155.

Ninkovich, D., and J. H. Robertson. 1975. Volcanogenic effects on the rates of deposition of sediments in the Northwest Pacific Ocean. *Earth and Planetary Sci. Letters* **27**:127–136.

O'Keefe, J. A. 1969. The microtektite data: Implications for the hypothesis of the lunar origin of tektites. *Jour. Geophys. Research* **74**:6795–6804.

Shaw, D. M., N. D. Watkins, and T.-C. Huang. 1974. Atmospherically transported volcanic glass in deep-sea sediments: Theoretical considerations. *Jour. Geophys. Research* **79**:3087–3094.

Taylor, S. R. 1969. Criteria for the source of australites. *Chem. Geology* **4**:451–459.

Taylor, S. R., and M. Kaye. 1969. Genetic significance of the chemical composition of tektites: A review. *Geochim. et Cosmochim. Acta* **33**:1083–1100.

Watkins, N. D., and J. P. Kennett. 1972. Regional sedimentary disconformities and Upper Cenozoic changes in bottom water velocities between Australia and Antarctica. *Antarctic Oceanology II, Antarctic Research Series 19.* Washington, D.C.: American Geophysical Union, pp. 273–293.

Watkins, N. D., and J. P. Kennett. 1977. Erosion of deep-sea sediments in the Southern Ocean between longitudes 70° E and 190° E and contrasts in manganese nodule development. *Marine Geology* **23**:103–111.

25

Reprinted from *Science* **168**:359–361 (1970)

FISSION TRACK AGES AND AGES OF DEPOSITION OF DEEP-SEA MICROTEKTITES

W. Gentner, B. P. Glass, D. Storzer, and G. A. Wagner

Microtektites have been found in deep-sea sediments adjacent to the Australasian and Ivory Coast tektite occurrences on land. Their deposition ages have been estimated by relating the microtektite-bearing layers to the paleomagnetic stratigraphy of the deep-sea cores. One of us (B.P.G.) reported deposition ages of 700,000 years for the Australasian microtektites and approximately 800,000 years for the Ivory Coast microtektites (*1*). Here we report the first direct age determinations of the microtektites, and we discuss the deposition ages based on paleomagnetic studies of the cores.

Since only small quantities of microtektites are available, the ages could not be measured by the K/Ar method. However, by applying the technique of fission track dating (*2*), we could measure ages with less than 5 mg of microtektites. Twenty microtektites (from cores V19-153, RC8-52, and RC8-53) from the Australasian tektite strewn

field and 40 microtektites (from core K9-56) from the Ivory Coast strewn field were selected for fission track dating.

The density of spontaneous fission tracks in microtektites is relatively low (of the order of 100 per square centimeter). Therefore, an area at least as large as ~ 0.5 cm² should be counted. By alternate grinding (in steps of 10 μm) and etching (for 10 seconds in 48 percent HF), each microtektite sphere was systematically checked for fission tracks. After about 50 to 100 spontaneous fission tracks had been counted, small remainders of the microtektite spheres were irradiated in the Karlsruhe reactor with a dose of 1.79×10^{15} thermal neutrons per square centimeter. The irradiated samples were handled in the same manner as the unirradiated ones. The densities of spontaneous and induced fission tracks were determined independently by three different persons.

The fission track ages were calcu-

lated with the decay constant for spontaneous fission of uranium-238: $\lambda_f = 8.42 \times 10^{-17}$ per year (*3*). The measured track densities and the calculated ages are given in Table 1; the Australasian microtektites have a fission track age of 0.71 ± 0.10 million years and the Ivory Coast microtektites have a fission track age of 1.09 ± 0.20 million years.

To check the possibility that these fission track ages were thermally lowered, the diameters of the etched spontaneous fission tracks were compared with the induced tracks. In both microtektite samples no difference could be seen in the size distribution of the etch pits between the spontaneous and the induced fission tracks. According to Storzer and Wagner (*4*), this result indicates thermally unaffected fission track ages. Consequently, these fission track ages are interpreted as formation ages.

To date, approximately 60 cores from many parts of the world's oceans (see Fig. 1) have been investigated by one of us (B.P.G.) for the presence or absence of microtektites associated with the Matuyama-Brunhes geomagnetic reversal boundary. Microtektites have now been found in 12 cores (see Table 2 and Fig. 1). Most of the cores are adjacent to the Australasian tektite strewn field as that field is defined by tektite occurrences on land. However, since three cores containing Australasian microtektites have been found southeast of Madagascar, the strewn field is now known to extend across the Indian Ocean (Fig. 1).

The paleomagnetic stratigraphies have been determined for nine of the twelve cores containing Australasian microtektites. In each core the microtektites are concentrated in a relatively thin layer (20 to 40 cm thick) that occurs at or just above the Matuyama-Brunhes (M-B) boundary (Fig. 2). In all but

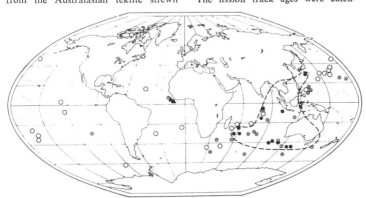

Fig. 1. Geographical distribution of deep-sea cores containing microtektites. Solid circles represent location of cores containing Australasian microtektites. Solid triangles represent location of cores containing Ivory Coast microtektites. The open circles represent location of cores that do not contain microtektites at the Matuyama-Brunhes boundary. Open circles with crosses represent cores that do not contain microtektites but that have a confusing magnetic stratigraphy. The boundary indicated for the Australasian strewn field should be regarded as provisional, since many of the cores used to define it have confusing magnetic stratigraphies; therefore, the absence of microtektites in the region where the core was taken cannot be regarded as firmly established.

Table 1. Fission track ages of microtektites. In the fourth column, the ratio of induced tracks of 10^{11} neutrons is proportional to the uranium content of the samples, and the range of values observed is given in parentheses; m.y., million years.

Strewn field	Tracks			Fission track age (m.y.)
	Spontaneous (No. per square centimeter)	Induced (No. per square centimeter)	Induced (No. per 10^{11} neutrons)	
Australasian	179	22.400	1.38 (0.79–2.52)	0.71 ± 0.10*
Ivory Coast	93.5	7.630	0.40 (0.21–0.85)	1.09 ± 0.20*

* Statistical counting error.

Table 2. Microtektite data from the Australasian and Ivory Coast tektite strewn fields.

Core	Latitude	Longitude	Microtektite zone (cm)	Depth to M-B boundary (cm)	Source of magnetic stratigraphy
Australasian tektite strewn field					
E35-9	45°02'S	128°00'E	230–240	233	*(15)*
LSDH 23G	31°14'S	62°58'E	40– 70		
MSN 48G	15°15'S	81°08'E	100–140		
RC8-52	41°06'S	101°25'E	340–400	390	*(14)*
RC8-53	39°23'S	104°22'S	150–190	185	*(14)*
RC9-137	45°01'S	123°45'E	390–420	445	
RC9-143	41°21'S	114°08'E	330–360	355	*(14)*
V16-70	29°27'S	56°06'E	140–170	155	*(14)*
V16-76	25°09'S	59°54'E	1020–1050	1045	*(14)*
V19-153	08°51'S	102°07'E	490–520	430	*(14)*
V19-171	07°04'S	80°46'E	330–360	350	*(14)*
V20-138	28°52'N	135°33'E	300–340	325	*(16)*
Ivory Coast tektite strewn field					
K9-56	03°23'N	15°37'W	Below 1425	1180	*(7)*
V19-297	02°37'N	12°00'W	510–550	490	*(1)*

one core (RC9-137) the peak abundance occurs within 10 cm of the M-B boundary. [In core V19-153 the maximum number of the microtektites were found about 20 cm above the M-B boundary, but in this core a thick (~ 10 cm) ash layer occurs between the M-B boundary and the microtektite layer. If this ash layer were removed, the microtektite layer would be within 10 cm of the M-B boundary.] The rela-

tively wide separation between the M-B boundary and the center of the microtektite layer in core RC9-137 may be due to a shifting downward of the magnetic stratigraphy. If the microtektite layer were appreciably younger than the M-B boundary, there would be a consistent relationship between the distances between the microtektite layer and the M-B boundary (that is, the higher the sedimentation rate, the

greater the separation). This consistent relationship was not observed; therefore, since the M-B boundary has a reported age of 0.69 million years (5), the Australasian microtektites must have fallen to the ocean floor approximately 0.7 million years ago.

The Ivory Coast microtektites have been found in two cores (V19-297 and K9-56). Paleomagnetic stratigraphy and paleontology indicate that core V19-297 contains an almost complete record of sedimentation for approximately 2 million years. The microtektites are concentrated in a thin layer (~ 15 cm thick) that occurs about 40 cm below the M-B boundary (Fig. 2). Unfortunately, the Jaramillo event was not found in this core. However, by extrapolation from the M-B boundary, it has been suggested that these microtektites were deposited about 800,000 years ago (1). Ericson and Wollin (6) have recently reported on the correlation of paleontology with paleomagnetic stratigraphy in core V19-297 and several other equatorial Atlantic cores. They show a peak in abundance of the *Globorotalia menardii* complex that correlates with the Jaramillo event in four cores. According to Cox (5) the Jaramillo event has an age of 0.89 to 0.95 million years. The microtektite-bearing layer coincides with this peak in abundance of the *Globorotalia menardii* complex in core V19-297; therefore, these microtektites probably fell to the ocean floor approximately 0.9 million years ago.

In core K9-56 several thousand microtektites have been recovered from the sediment sucked in by the piston when the core was pulled out of the bottom. The undisturbed portion of the

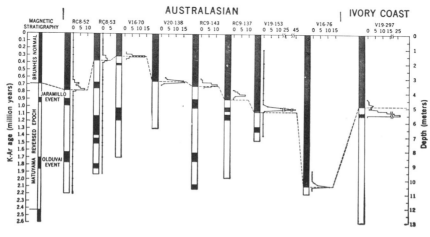

Fig. 2. Correlation of the microtektite horizon with the paleomagnetic stratigraphy. The scale at the top right of each core indicates the number of microtektites per 2 cm² of sample. The location of the Jaramillo event in the Ivory Coast core (V19-297) is based on paleontological data from Ericson and Wollin (6). The paleomagnetic stratigraphies of cores RC8-52, RC8-53, RC9-143, V16-70, V16-76, and V19-153 are discussed in greater detail by Opdyke and Glass (14).

core ends at 1425 cm. Three microtektites were recovered from the bottom of the undisturbed core and none were found above that depth. Thus the microtektite layer occurs below 1425 cm in the location at which this core was taken. The paleomagnetic stratigraphy of this core is not clear; however, paleontological correlations with other cores containing better paleomagnetic stratigraphies indicate that the M-B boundary probably occurs at 1180 cm in this core (7). Extrapolation from this boundary indicates that the Ivory Coast microtektites fell to the ocean floor more than 840,000 years ago. Thus evidence from both core V19-297 and core K9-56 indicates an age of deposition distinctly older than that of Australasian microtektites.

Potassium/argon ages of ash layers in core V19-153 (8) indicate that the microtektites in this core (Fig. 2) were deposited about 1.2 million years ago rather than 0.7 million years ago as suggested by the magnetic stratigraphy. We have just shown, however, that the microtektites from this core have a fission track age of 0.71 million years, thus confirming the age of deposition determined by paleomagnetics.

Australasian tektites have K/Ar (9) and fission track ages (10) of about 0.7 million years, and Ivory Coast tektites have K/Ar ages (9) and fission track ages (11) between 1.0 and 1.2 million years. The fission track ages of the microtektites agree very well with K/Ar and fission track ages of tektites from the tektite strewn fields to which they

were previously linked; hence, the fission track ages of these microscopic glassy objects strongly support the conclusion (1,. 12) that they are microtektites and that they belong to previously known strewn fields.

For each microtektite group the measured fission track ages and the stratigraphic ages agree within their limits of error. Thus, there appears to be no difference between the age of formation and the age of deposition.

Despite K/Ar, fission track, and stratigraphic ages indicating that the tektites from the Ivory Coast strewn field are distinctly older than the tektites from the Australasian strewn field, Chapman and Scheiber (13) have proposed (on the basis of chemical similarity between Ivory Coast tektites and a small group of Australasian tektites) that the Ivory Coast tektite strewn field is part of the Australasian strewn field. However, the measured fission track ages of 0.71 ± 0.10 million years for the Australasian microtektites and 1.09 ± 0.20 million years for the Ivory Coast microtektites are distinctly different beyond what may be attributed to experimental errors. It is thus clearly demonstrated that the Ivory Coast tektite strewn field is older than the Australasian strewn field.

References and Notes

1. B. P. Glass, *Nature* **214**, 372 (1967); *Science* **161**, 891 (1968).
2. For experimental details of glass dating by the fission track method see, for example, G. A. Wagner, *Z. Naturforsch. A* **21**, 733 (1966).
3. A. Spadavecchia and B. Hahn, *Helv. Phys. Acta* **40**, 1063 (1967).
4. D. Storzer and G. A. Wagner, *Earth Planet. Sci. Lett.* **5**, 463 (1969).
5. A. Cox, *Science* **163**, 237 (1969).
6. D. B. Ericson and G. Wollin, *ibid.* **162**, 1227 (1968).
7. The paleontological and paleomagnetic stratigraphies of core K9-56 were obtained from W. F. Ruddiman of GOFAR, U.S. Naval Oceanographic Office.
8. J. Dymond, *Earth Planet Sci. Lett.* **6**, 9 (1969).
9. J. Zähringer, in *Radioactive Dating* (International Atomic Energy Agency, Vienna, 1963), p. 289.
10. R. L. Fleischer and P. B. Price, *Geochim. Cosmochim. Acta* **28**, 755 (1964); ———, R. M. Walker, *ibid.* **29**, 161 (1965); G. A. Wagner, D. Storzer, G. A. Wagner, *Geochim. Cosmochim. Acta* **33**, 1075 (1969).
11. W. Gentner, B. Kleinmann, G. A. Wagner, *Earth Planet. Sci. Lett.* **2**, 83 (1969).
12. B. Glass, *Geochim. Cosmochim. Acta* **33**, 1135 (1969); W. A. Cassidy, B. Glass, B. C. Heezen, *J. Geophys. Res.* **74**, 1008 (1969); F. E. Senftle, A. N. Thorpe, S. Sullivan, *ibid.*, p. 6825.
13. D. R. Chapman and L. C. Schreiber, *J. Geophys. Res.* **74**, 6737 (1969).
14. N. D. Opdyke and B. Glass, *Deep-Sea Res.* **16**, 249 (1969).
15. The paleomagnetic stratigraphy of core E35-9 was determined by N. D. Watkins of Florida State University.
16. The paleomagnetic stratigraphy of core V20-138 was obtained from N. D. Opdyke of Lamont-Doherty Geological Observatory.
17. We thank M. Turner of the National Science Foundation and D. S. Cassidy and N. D. Watkins of Florida State University for samples from *Eltanin* core 35-9; W. R. Riedel and Phyllis B. Helms of Scripps Institution of Oceanography for samples from LSDH 23G and MSN 48G; and the Global Ocean Floor Analysis and Research Group of the U.S. Naval Oceanographic Office for taking core K9-56 for the purpose of obtaining additional information about the Ivory Coast microtektite horizon. E. D. Schneider and W. F. Ruddiman of GOFAR have been most helpful. The remainder of the cores are from the Lamont-Doherty Geological Observatory collection. The assistance of J. D. Hays and R. Capo of Lamont and A. Haidmann of Max Planck-Institut is acknowledged. For the irradiation of the microtektites with thermal neutrons and the calibration of the dose, we thank Mr. Rottmann of the Gesellschaft für Kernforschung, Karlsruhe.

22 December 1969; revised 17 February 1970

26

Reprinted from *Earth and Planetary Sci. Letters* **20**:119–124 (1973)

ATMOSPHERICALLY TRANSPORTED VOLCANIC DUST IN SOUTH PACIFIC DEEP SEA SEDIMENTARY CORES AT DISTANCES OVER 3000 KM FROM THE ERUPTIVE SOURCE

T.C. HUANG, N.D. WATKINS, D.M. SHAW and J.P. KENNETT

Graduate School of Oceanography, University of Rhode Island, Kingston, Rhode Island 02881, USA

Analyses of deep-sea sedimentary cores from the southwest and southcentral Pacific reveal very similar time patterns in the accumulation of atmospherically transported volcanic dust. A series of volcanic eruptions on the Balleny Islands between 1.8 and 1.6 my ago will be recorded in high latitude marine sediments of the entire South Pacific. The results of this study indicate that the history of the world's atmospheric volcanic activity is clearly amenable to definition by study of selected deep-sea sedimentary core traverses.

Volcanic dust and dust from desert regions constitute the major sources of atmospheric particulate matter. Many speculations have recently been made about the possible role of volcanic dust in changes of climate [1], atmospheric chemistry [2], and corresponding water chemistry and faunal activity [3]. Monitoring these effects in recent eruptions [4] presents great experimental difficulties, and any accumulative effects will require such a duration of observations that direct methods may be unfeasible. In this communication we describe the results of our indirect monitoring of atmospheric volcanic activity, as recorded in deep-sea sediments of the South Pacific.

Atmospherically transported volcanic sediments (or tephra) in deep-sea sediments can be examined in three size groups: as coarse, megascopically distinguishable layers relatively close to the eruptive sources [5]; as finely dispersed coarse fractions such as those observed during conventional separation of the $>61 \mu$ fraction for micropaleontological study [6]; and as the finest fractions (below 40μ). Separation and particle size analysis of this finest fraction has not hitherto been attempted. Development of a technique to accomplish this [7], when coupled with paleomagnetic and micropaleontological dating of the same sediment, can provide an accurate measurement of average accumulation rates, in appropriate units, which we designate

as milligrams per thousand years per square centimeter ($mg/1000 \ yr/cm^2$). In the absence of means to detect fluctuation in sedimentation rates, it becomes necessary to assume constant sedimentation rates between the paleomagnetically defined isochrons.

Fig. 1 shows the location of the two cores selected for study. They are composed of terrigenous components (ice rafted debris and clay minerals), silicious and carbonate biogenic components, manganese micronodules, and minor amounts of volcanic glass. Kennett and Watkins [6] have already shown that core E27-3 contains discrete layers of finely-dispersed coarse volcanic glass. Similar examination of other sedimentary cores in the southwest Pacific shows by simple means that the Balleny Islands are almost certainly the eruptive source region of the volcanic material observed. Little is known about the geology of the islands, except that they are volcanic [8]. E27-3 has been dated by paleomagnetic and micropaleontological means [9], the results of which are included in fig. 2. Core E11-9 was also selected because of previous dating and sedimentation rate data [10]. Our detailed examination of this core, including X-ray radiography observations (to evaluate the possible effects of bottom currents, turbidity currents, and bioturbation), and determination of the paleomagnetic stratigraphy, and biostratigraphy, has revealed the

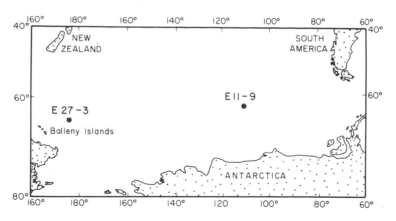

Fig. 1. Map showing locations of the two cores examined, and the source area for the volcanic glass (the Balleny Islands). Exact geographic coordinates and water depths are: lat. 66°04′ South, long. 176°30′ East, 3531 m (E27-3); and lat. 62°41.0′ South, long. 115°04′ West, 4932 m (E11-9). The distances of the cores from the source are 592 km (E27-3) and 3718 km (E11-9).

existence of a disconformity representing a period of about 0.7 my (fig. 2). The disconformity is marked by what appears to be a turbidite layer (fig. 2), occurring with a very well defined erosional base, with mean particle size decreasing upward from a lower segment of horizontal laminae. The volcanic glass in this possible turbidite is concentrated in its upper part. We do not favor an in-situ (or residual) origin for the layer. Association of the possible turbidite emplacement with the substantial erosion required to produce the disconformity is not surprising.

The volcanic glass has been separated, and counted [7] in a total of 227 specimens. It ranges from 0 to 12% by weight of the sample. The results are expressed in terms of the coarse (88 to 36 μ) and fine (36 to 11 μ) fractions, in mg/1000 yr/cm^2 (fig. 2). The volcanic glass maximum associated with the turbidite is the only segment of the observed volcanic glass variation in both cores which we believe to be representative of sea floor rather than atmospheric processes. Simple comparison of the abundance curves shows that the same series of volcanic eruptions are undoubtedly recorded in both cores for the period 2.4 to 1.4 my ago. It is impossible for surface or deep-sea currents to produce similar distributions of volcanic glass in deep-sea sediments over a distance of 3000 km [11]: only atmospheric transport can create this effect. Other observations, such as the complete lack of correlation

between volcanic glass and ice rafted debris concentrations, confirm this point. The important role of organisms in the deposition of fine particulate matter has been discussed by Smayda [12] and others: fecal pellets convert particles which would normally spend many years in the water column into much larger (rapidly-settling) particles which are in turn released as original fine particles during organic decay on the sea floor. In this way, atmospheric transport patterns may be faithfully recorded on deep-sea sediments, provided no subsequent dynamic processes occur. The volcanic glass distribution reflects up to seven separate periods of intensive volcanic activity (A to G), and within each period up to five possibly distinct eruptions or closely spaced series of eruptions (fig. 2). Periods of relative eruptive quiescence up to 0.37 my duration are indicated whereas periods of relatively sustained volcanic activity from 0.07 to 0.46 my duration can be recognized.

The age and concentrations of volcanic maxima, and the suspected between-core correlations are tabulated in table 1. Of 22 possibly separate eruptions inferred from the data in core E27-3, 13 are synchronous with eruptions recorded in E11-9. Considering the possibility that the sampling interval may have caused incomplete definition of minor peaks, and the possibility of vertical mixing by bioturbation [13] and other sea floor processes [14], the degree of

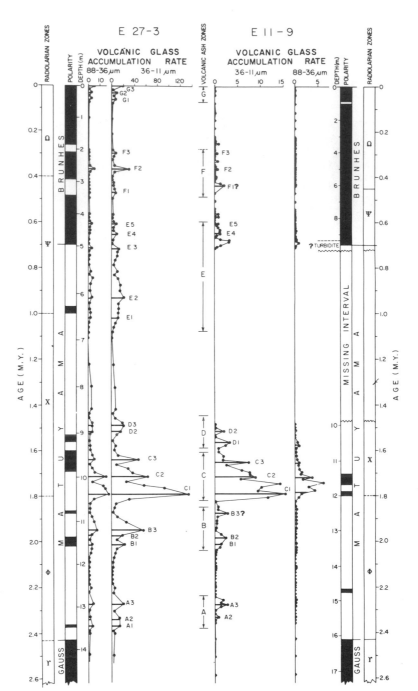

Fig. 2. Accumulation rate of atmospherically transported volcanic glass in cores E27-3 and E11-9 (fig. 1) as a function of age of sediment. Volcanic glass fraction is in two fractions (88 to 36 μ, and 36 to 11 μ) for each core. Accumulation rates are in mg/1000 yr/cm^2; (note the difference in scales between the two cores). Age of sediment in million years at left and right. Radiolarian zones after Hays [11]. Paleomagnetic polarities (black = normal; clear = reversed) after Kennett and Watkins [3]. Ages of Brunhes/Matuyama and Matuyama/Gauss boundaries from Cox [22]. Periods of maxima in volcanic glass accumulation rates are labelled A to G, with subscripts 1 to 5 indicating possible single eruptions. Note the disconformity in core E11-9, between 0.72 and 1.48 my. See text for discussion of the turbidite (?) layer in core E11-9.

T.C. Huang et al., Atmospherically transported volcanic dust

TABLE 1

Age and local volcanic glass accumulation rates of volcanic eruptions recorded in cores E27-3 and E11-9 (fig. 1)

Volcanic ash zone (fig. 2)	Period (my)	Core E27-3				Core E11-9				C_0	k
		Eruption (fig. 2)	Approx. age (my)	Volcanic glass accumulation rate (88–11μ)	Fine glass / Coarse glass	Eruption	Approx. age (my)	Volcanic glass accumulation rate (88–11μ)	Fine glass / Coarse glass		
A	2.38	A1	2.37	16.70	4.60	–	–	–	–	–	–
		A2	2.33	15.20	8.56	A2	2.33	0.89	146.67	26.40	8.87
	2.24	A3	2.27	23.67	6.51	A3	2.27	2.74	273.00	35.44	6.74
B	2.04	B1	2.01	25.71	9.85	B1	2.01	1.43	14.89	44.19	9.03
		B2	1.97	20.65	6.91	B2	1.98	2.24	111.00	31.71	6.94
	1.85	B3	1.95	61.65	8.75	B3	1.88	2.77	33.62	110.92	9.70
C	1.82	C1	1.79	150.03	7.57	C1	1.79	16.59	13.05	227.88	6.88
		–	–	–	–	C1	1.74	20.81	2.36	–	–
	1.61	C2	1.71	78.18	4.11	C2	1.72	12.52	2.76	106.41	5.72
		C3	1.63	51.19	9.49	C3	1.65	8.44	7.88	71.74	5.63
D	1.59	–	–	–	–	D1	1.56	3.62	5.96	–	–
		D2	1.52	19.43	4.12	D2	1.52	1.98	98.00	30.06	7.14
	1.49	D3	1.49	21.19	6.21	–	–	–	–	–	–
E	1.08	E1	1.02	13.96	6.46	–	–	–	–	–	–
		E2	0.93	23.74	7.51	–	–	–	–	–	–
		E3	0.72	16.32	4.63	–	–	–	–	–	–
		E4	0.65	10.65	12.18	E4	0.65	1.27	9.58	15.94	6.65
	0.60	E5	0.61	10.29	3.42	E5	0.61	0.93	153.33	16.24	7.51
F	0.49	F1	0.47	7.42	7.53	F1 [(?)]	0.44	2.17	53.25	–	–
		F2	0.37	37.32	5.39	F2	0.36	0.70	69.00	77.73	12.40
	0.28	F3	0.30	9.01	14.53	F3	0.30	0.61	11.20	14.96	8.41
G	0.07	G1	0.06	14.01	4.56	–	–	–	–	–	–
		G2	0.03	13.64	3.82	G2	0.04	0.04	20.00	45.19	18.20
	0.00	G3	0.01	24.97	3.62	–	–	–	–	–	·

Explanation of table. Volcanic glass accumulation rate in mg/1000 yr/cm^2. Fine glass = 36–11 μ; coarse glass = 88–36 μ: C_0 and k from $C_x = C_0 e^{-kx}$ where C_x = accumulation rate of fine volcanic glass at distance x from source, C_0 = accumulation rate at $x = 0$. Units of C_0 are in mg/1000 yr/cm^2. Units of k are km$^{-1} \times 10^4$.

duplication of eruptive character in the separate cores is remarkable.

The volcanic dust accumulation rates (fig. 2) varies from less than 7 to about 150 mg/1000 yr/cm^2 in core E27-3 (600 km from the source) and from less than 0.1 to about 20 mg/1000 yr/cm^2 in core E11-9 (3700 km from the source). The volcanic glass variation within cores could be, given normalized transportation dynamics, an indication of substantially varying eruptive intensities and corresponding heights of atmospheric injection, while the variation of concentrations of the

same eruption *between* cores would be a measure of the transportation dynamics, given normalized eruptive intensities. These inherent ambiguities are partially resolved by comparison of data between cores.

Several empirical models have been presented relating particle size, distance from source, and transporting wind velocity [15]. We propose that, to a first approximation, the accumulation rate (C) of fine volcanic glass at distances (x) greater than 500 km from the eruptive source would vary exponentially [16] as:

$$C_x = C_0 e^{-kx}$$

where C_0 = the hypothetical accumulation rate at the source, and k is the 'fallout coefficient' depending mainly on high altitude wind speed and local weather conditions. Given two values of C_x corresponding to the same eruption recorded in the two cores, then C_0 and k can be computed [17]. The results are given in table 1.

Analysis of the relationship between k and C_0 (table 1) yields a correlation coefficient of −0.012, which is not significantly different from zero. We interpret this to mean that the spreading of the volcanic dust clouds depended on factors other than eruptive intensity. If this proves to be true for other study areas we may have a technique for estimating functions of paleo-wind velocities.

The much higher ratios of the fine (36 to 11 μ) to coarse (88 to 36 μ) glass fractions in the core more distant from the source region (table 1) is a simple confirmation of the direction of the source. This ratio has narrow limits for the core close to the source, ranging from 3.4 to 14.5. The much lower accumulation rates of coarse fractions in the core 3700 km from the source provides a ratio which is higher and more variable.

A comprehensive analysis of the factors governing the distribution of atmospherically transported dust from volcanic eruptions would involve a large number of variables. We have defined the accumulation rates of two size fractions of deposits from the same eruptions in two separate cores. All the data are consistent with our preferred limited interpretation of a series of violent eruptions between t = 1.65 and 1.80 my. A figure of 10 mg/1000 yr/cm² would appear to be a conservative average accumulation rate for the fallout region in the South Pacific, for this period. The distance between the cores suggests that the fallout area may be at least 5×10^6 km² since an average north to south dimension of the fallout region is unlikely to be less than 1000 km. The measured average rate of accumulation therefore provides a figure of the order of 10^{11} million metric tons of volcanic dust (between 11 and 36 μ size) for the 150,000 yr period. It is not known if this increased volcanism together with the associated very fine fraction (less than 11 μ) could induce or trigger [18] the climatic cooling known to have occurred in the southern hemisphere between

2.1 and 1.6 my ago [19], since our experimental method cannot resolve the durations of sustained volcanic activity which would be essential to maintain the high atmospheric particulate concentrations necessary for climatic modification. Goldberg [20] has estimated that the upper limit on the mass of volcanic dust mobilized into the atmosphere is 1.6×10^{14} g/yr. This is 30 mg/1000 yr/cm², which is much higher than the rates observed even during the volcanic maxima in core E11-9 (fig. 2), so unless Goldberg's computation incorporates the much coarser fractions, we suggest that this estimate may be an order of magnitude too high.

Accumulation rates of fine volcanic glass from the Balleny Islands eruptions are sufficiently high in core E11-9 for the 1.65 to 1.80 million yr interval to enable confident prediction that the horizon is ubiquitous in undisturbed marine sediments throughout the high latitudes of the South Pacific, giving further emphasis to the unique stratigraphic value of tephra [21].

References and notes

[1] W.J. Humphreys, Physics of the air, 3rd ed. (McGraw-Hill, New York, 1940) 587–618; H. Wexler, Bull. Amer. Meteo Soc. 32 (1965) 10; H.H. Lamb, Phil. Trans. Royal Soc. Lo don A266 (1970) 425; J.M. Mitchell, in: Global effects of environmental pollution, Symp. Amer. Assoc. Advan. Sci., ed. S.F. Singer (D. Reidel Publ. Co., 1970) 135.

[2] R.E. Newell, Nature 277 (1970) 697; C.E. Jung, Tellus 18 (1966) 685; R.D. Cadle and J.W. Powers, Tellus 18 (1966) 176; J.P. Friend, Tellus 18 (1966) 465; R. Cadle and I.H. Blifford, Nature 230 (1971) 573.

[3] J.P. Kennett and N.D. Watkins, Nature 227 (1970) 930; T.C. Huang, R. Fillon and N.D. Watkins, in press (1973); U. Stefansson, J. Marine Res. 24 (1966) 241; H. Elderfield Marine Geology 13 (1972) 1.

[4] A.J. Dyer and B.B. Hicks, Nature 208 (1965) 131; A.J. Dyer, J. Geophys. Res. 75 (1970) 3007; G.P. Eaton, J. Geophys. Res. 68 (1963) 521; C.J. Fries, Trans. Amer. Geophys. Un. 34 (1953) 603; A.P. Lisitzin, Soc. Econ. Paleontologists Mineralogists, Spec. Publ. 17 (1972) 179; A.B. Meinel and M.S. Meinel, Nature 201 (1964) 657.

[5] O. Mellis, Deep Sea Res. 2 (1954) 89; D. Ninkovitch, B.C. Heezen, J.R. Connolly and L. Burckle, Deep Sea Res. 11 (1964) 605; Y.R. Nayudu, Marine Geol. 1 (1964) 194; D. Ninkovitch and B.C. Heezen, in: Submarine geology and geophysics, eds. W.F. Wittard and R. Bradshaw (Butterworths, London, 1965) 413; D. Ninkovitch, N.D. Opdy B.C. Heezen and J.H. Foster, Earth Planet. Sci. Letters 1 (1966) 476; D.R. Horn, M.N. Delach and B.M. Horn, Geol. Soc. Amer. Bull. 80 (1969) 1715.

[6] For example, J.P. Kennett and N.D. Watkins (ref. 3).

[7] T.C. Huang and N.D. Watkins, Separation and counting of fine volcanic glass from deep sea sediments (in preparation). Basically, our method is to use heavy liquids to separate the volcanic glass from the host sediment, and a Coulter counter and Cahn settling tube to determine the particle size fractions. Each determination incorporates a transmitted light optical calibration of the separation efficiency.

[8] T. Hatherton, E.W. Dawson and F.C. Kinsky, New Zealand J. Geol. Geophys. 8 (1965) 164. The coarse volcanic fraction increases drastically in those sedimentary cores within 200 km of the islands, to include ferromagnesian and opaque minerals, as well as glass.

[9] For methods and results see: N.D. Watkins and J.P. Kennett, Amer. Geophys. Un. (Antarctic Res. Ser.) 19 (1972) 273.

[10] J.D. Hays, Amer. Geophys. Un. (Antarctic Res. Ser.) 5 (1968) 125; N.D. Watkins and H.G. Goodell, Science 156 (1967) 1003.

[11] Surface currents of this area could not transport particles larger than 11 μ such a great distance: T.C. Huang, N.D. Watkins, D.M. Shaw and J.P. Kennett, Volcanism in the southeastern Pacific Region (in preparation).

[12] T.J. Smayda, Marine Geology 11 (1971) 105.

[13] J.N. Bramlette and W.H. Bradley, U.S. Geol. Surv. Prof. Papers 196A (1940) 34 pp.; W.F. Ruddiman and L.K. Glover, Geol. Soc. Amer. Bull. 83 (1972) 2817.

[14] T.C. Huang, N.D. Watkins and R. Fillon, Bull Amer. Assoc. Petrol. Geol. 57 (1973) 784.

[15] H. Williams, Carnegie Inst. Washington, Pub. No. 540 (1942) 162 pp.; G.P. Eaton, J. Geophys. Res. 68 (1963) 521 and J. Geology 72 (1964) 1; R.V. Fisher, J. Geophys. Res. 69 (1964) 341; M. Slaughter and M. Hamis, Geol.

Soc. Amer. Bull. 81 (1970) 961; J.B. Knox and N.M. Short, Bull. Volcanology 27 (1964) 5.

[16] This general relationship can be derived from the data of Eaton [15] and others.

[17] While data from cores at only two separate distances from the source are insufficient to verify the exponential decrease of volcanic glass accumulation rate with distance from source, we have nevertheless presented results (T.C. Huang, N.D. Watkins, D.M. Shaw and J.P. Kennett, EOS (Trans. Amer. Geophys. Un.) 53 (1972) 423) from single specimens from the same volcanic ash layer in at least four cores at different distances but similar latitudes across the South Pacific, to confirm the generally exponential decrease of volcanic glass accumulation rate with distance from the Balleney Islands. Detailed analyses of these additional cores at specimen separations similar to that shown in fig. 1 are in progress, and will be completed by late 1973.

[18] W.T. Humphreys (ref. 1); H.H. Lamb (ref. 1); J.M. Mitchell (ref. 1).

[19] J.P. Kennett, N.D. Watkins and P. Vella, Science 171 (1971) 276.

[20] E.D. Goldberg, Comments on earth sciences, Geophysics 1 (1971) 117.

[21] We have in fact detected this middle Matuyama eruption in three cores: E27-4 (68°03'S, 174°35'E), E17-10 (65°01'S, 134°52'W) and E33-16 (63°14'S, 120°01'W) which are 500 km, 2689 km and 3267 km from the source, respectively.

[22] A. Cox, Science 163 (1969) 237.

[23] This work is supported by NSF grant numbers GA28853 (Marine Geology Program) and GA25400 (Office of Polar Programs).

27

Reprinted from *Science* **173**:813–818 (1971)

ANTARCTIC BOTTOM WATER: MAJOR CHANGE IN VELOCITY DURING THE LATE CENOZOIC BETWEEN AUSTRALIA AND ANTARCTICA

N. D. Watkins and J. P. Kennett

Abstract. *Paleomagnetic and micropaleontological studies of deep-sea sedimentary cores between Australia and Antarctica define an extensive area centered in the south Tasman Basin, where sediment as old as Early Pliocene has been systematically eroded by bottom currents. This major sedimentary disconformity has been produced by a substantial increase in velocity of Antarctic bottom water, possibly associated with late Cenozoic climatic cooling and corresponding increased glaciation of Antarctica.*

Studies of past oceanic circulation patterns have been restricted almost entirely to those changes in surface-water distributions indicated by fossil planktonic organisms (*1*). Changes in distribution and activity of deep-water masses have received little attention except where they are related to broad-scale, seismically defined changes in sediment patterns (*2*). Antarctic bottom water, which is produced under glaciated Antarctic conditions, plays an important role in oceanic circulation, since its formation creates transportation of new, oxygen-rich waters to the deep ocean basins in the world, thus inhibiting stagnation (3).

As water velocity is related to grain size, structures, and erosional and depositional history of sediments (4), estimates can be made of variations among these parameters. Until recently, only surface sediments were amenable to regional comparisons of current velocities because of the unique, apparently isochronous plane of observation provided (4). Paleomagnetic dating with micropaleontological control now enables such studies to be extended to subsurface isochronous planes, or to fine definition of possible diachronous relations of such planes. In this report we examine a series of deep-sea sedimentary cores from the Southern Ocean between Australia and Antarctica (Fig. 1) for evidence of past changes in the velocity of Antarctic bottom water. This represents an ideal area for such an examination because of the strong regional topographic control of bottom waters.

The cores used were collected during cruises 27 to 39 of U.S.N.S. *Eltanin* (Table 1 and Figs. 1 and 2). These form less than 20 percent of the total number of cores from the southwest Pacific, which are included in a more comprehensive analysis to be published elsewhere (5). The relevant shipboard operations and the paleomagnetic and micropaleontological methods have been described previously (6).

Sediment type in the cores varies greatly. It includes lutite, siliceous and carbonate ooze, and glacial marine material. Each core was sampled at 10-cm intervals. All specimens were demagnetized in alternating magnetic fields prior to measurement of remanent magnetism. The resulting paleomagnetic data for each core are shown in Fig. 2 as a function of age, which has been micropaleontologically verified. Actual core lengths, from which sedimentation rates can be calculated, are given in Table 1. Clearly, the known polarity history for the past few million years (7) is not ideally recorded in any cores, a fact that emphasizes the critical role of micropaleontological control of the ages defined. Material as old as Gilbert epoch is identified. Both planktonic Foraminifera and Radiolaria were examined to provide the micropaleontological control. The correlation lines for the major polarity boundaries have been placed by applying the radiolarian zona-

tion established by Opdyke *et al.* (8) and Hays and Opdyke (9), and the foraminiferal zonation of Kennett (10).

The radiolarian species that were particularly valuable in our correlations were *Triceraspyris* sp. and *Lychnocanium grande* in the early and middle

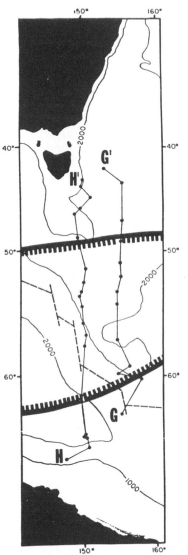

Fig. 1. Map showing location of core traverses *G–G'* and *H–H'*. For cruise and core numbers, see data in Fig. 2 and Table 1. Latitude is in degrees south; longitude is in degrees east. The 1000- and 2000-fathom bathymetric contours are shown. Fine dashed lines indicate the approximate positions of the ridge axis. Bold, hachured lines show the limits of the central zone of definite sediment removal by bottom currents.

Gilbert; *Desmospyris spongiosa* (the extinction of which coincides approximately with the Gauss-Matuyama boundary); *Eucyrtidium calvertense* and *Clathrocyclas bicornis* (the extinction of which occurs in the middle Matuyama); and *Saturnulus planetes*, which in sub-Antarctic waters disappeared near the Brunhes-Matuyama boundary ($t = 0.69$ million years). The foraminiferal species that were particularly valuable in our correlations were *Globorotalia inflata* and *G. crassaformis*, whose appearance in sub-Antarctic waters is closely associated with the Brunhes-Matuyama boundary. Paleontological control for the shorter duration magnetic polarity events (Fig. 2) is often not available, especially for the events within the Gilbert epoch, where detailed paleontological control has yet to be established. Correlations of events are therefore made subjectively with use of the magnetic data alone, by comparison with the established geomagnetic polarity scale (7). In many of the cores, micropaleontological samples from near the top, middle, and bottom were sufficient to confirm paleomagnetic dating. In others, samples were examined at closer intervals for adequate dating, or for defining the position of disconformities.

It is clear from Fig. 1 that disconformities defined both paleomagnetically and micropaleontologically have been detected in many cores. Moreover, about 35 percent of the cores from the two traverses contain no Recent or Brunhes sediment, whereas at least four cores contain two disconformities. Furthermore, the major break in sedimentation in the Brunhes and Matuyama (Fig. 2) is not a random or local effect but is centered in the southern part of the Tasman Basin and northern flank of the Indian-Pacific Rise (Fig. 1). Uninterrupted sedimentary sequences are present to the north of the traverses on the south Tasman Ridge and within the central Tasman Basin, and to the south of the traverses nearer Antarctica. Hiatuses in sedimentation have previously been detected paleomagnetically in isolated cores (11) but have not been previously defined regionally in deep-sea areas.

Disconformities in deep-sea regions can result from three main processes: slumping, lack of sediment deposition, or submarine erosion. The disconformity is far too extensive (12) to have resulted from slumping and, furthermore, is centered in a basinal area rather than being confined to ridges. The formation of the disconformity entirely through lack of sedimentation is also untenable, because the area is in latitudes characterized by high biogenic productivity throughout the late Cenozoic.

There is strong evidence to indicate that deep-sea erosion has produced the disconformity. For instance, the upper few centimeters of numerous cores lacking Recent-Brunhes sediments are coarse residual sands or coarse accretionary sands (4) commonly made up of a mixture of micromanganese nodules, ice-rafted debris, fish teeth, robust Radiolaria and Foraminifera, and clay aggregates. Several cores in these surface sands contain reworked Radiolaria older than the underlying horizons. Clearly, these surface sands have been formed either by winnowing or reworking of coarse debris, or by both. According to the data of Gordon (13), topographic control has made the south Tasman Basin a major junction area of bottom currents from the South Australian Basin, the northern and central Tasman Basin, and the South Indian Basin (Fig. 3). It is not surprising, therefore, that high-velocity bottom currents (greater than 15 cm sec^{-1}) have been noted by Gordon in this general area. Such velocities are sufficient to remove material finer than fine sand size (< 0.2 mm) (4), and possibly even coarser sand-sized sedimentary fractions (14). The only important flow of bottom

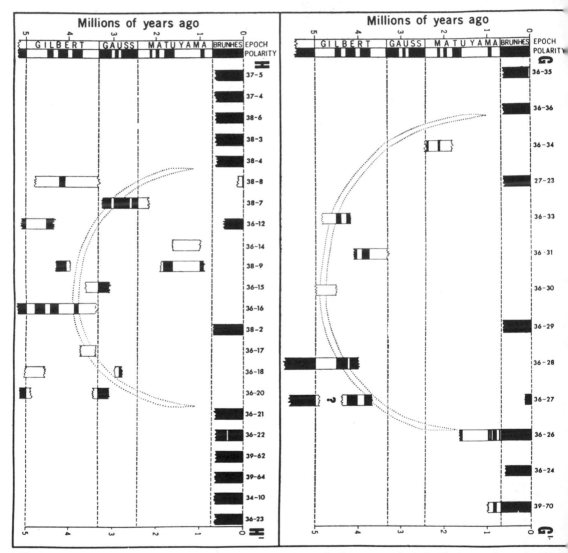

Fig. 2. Logs showing paleomagnetic data and time ranges of each core in the traverses shown in Fig. 1. At left are the known polarity history and polarity epochs (7, 9) (black, normal polarity; clear, reversed polarity). At the head of each log are the cruise and core numbers. Time ranges were determined by combining paleomagnetic and micropaleontological data [for limitations, see text and (23)]. The fine-dotted concave bowls represent the approximate ages of surface sediments based on cores from the two traverses (5).

water from the South Indian Basin to the Tasman Basin is across the Australian-Antarctic Rise immediately south of the Tasman Basin.

Sea-bottom features observable in bottom photographs (*15*) also provide substantial evidence of current activity in the region (Fig. 3). Sea-bottom photographs from the south Tasman Basin consistently show common to abundant manganese nodules and conspicuous ripple marks. In the central basinal region some scouring is evident. The photographs yield conspicuous evidence for bottom currents east and south of Macquarie Ridge (Fig. 3), where high-velocity currents have been measured directly (*16*). Other areas between Australia and Antarctica show evidence of much lower velocity bottom currents (Fig. 3). If 5 m of sediment is taken to represent the average amount of sediment removed over the area of the disconformity (approximately 3×10^6 km²), then 1.5×10^{19} cm³ of sediment has been removed. We speculate that the bulk of this vast volume of sediment has been transported by eastward-flowing bottom currents into the Emerald Basin of the southwest Pacific (Fig. 3). This speculation is clearly testable. Evidence of the scouring effect of the circumpolar current exists in the Bellingshausen Basin far to the east: Goodell and Watkins (*17*) have defined a west-to-east zone of low deposition rates by paleomagnetic means.

Despite the widespread nature of the sedimentary hiatus so defined, no sediments older than approximately 6 million years were found, and widespread sediments of Pliocene age, which accumulated at reasonable rates, are evident (Fig. 2). It is important to note that despite its vast areal extent, the disconformity represents the removal of only a few meters of sediment, most of which is of Brunhes and Matuyama epoch ($t = 0$ to 2.43 million years). We propose that only one explanation is possible for this situation: at some time in the latest Cenozoic, Antarctic bottom water has increased substantially in velocity. Studies of current velocities for erosion, transportation, and deposition (*4*) show that only particles larger than approximately 0.2 mm would remain deposited if current velocities are even only spasmodically greater than 20 cm sec^{-1}. A current system almost invariably less than 10 cm sec^{-1} is required for relatively continuous deposition of fine sediments. We therefore believe that there was an increase in bottom current velocities in this area from dominantly less than 10 cm sec^{-1} to dominantly more than 10 cm sec^{-1} during the latest Cenozoic, although it is highly probable that velocities have been spasmodically much greater during the period, to account for transportation of the coarsest sediments from the area.

A major factor contributing to preservation of sediment of Matuyama and older age but loss of much sediment of Brunhes age is, in our opinion, the change of cohesiveness of sediment with depth of burial. The older, more cohesive material would resist erosion by higher current velocities, which would nevertheless be effective in removing the younger, less cohesive material; a model of rapid sediment removal followed by marked slowing of the process is envisaged. Another factor contributing to the preservation of the older material is the creation of a manganese nodule pavement in the areas of nondeposition, which in turn would eventually inhibit erosion substantially. This factor is effective only in part of the area involved (Fig. 3) and must, therefore, be secondary. Core penetration is substantially less in the older sediment areas than in areas of Brunhes sedimentation (Fig. 1 and Table 1), and thus our concept of a critical increase in cohesiveness (and therefore resistance to erosion with depth) is supported. It appears that this critical increase must occur within 8 m of the surface in the cores under examination.

At this time insufficient evidence is available to determine with accuracy the time of initiation of the high bottom-current velocities that created the Recent disconformity. Variation in age of cores and disconformities throughout the region indicates that the

Table 1. Number, location, length, and water depth of cores used in this study.

Cruise No.	Core No.	Latitude (S)	Longitude (E)	Water depth (m)	Core length (m)
		Traverse G–G'			
36	35	62°44′	154°58′	3493	584
36	36	60°23′	157°32′	2817	572
36	34	59°59′	155°02′	2795	452
27	23	59°37.1′	155°14.3′	3182	955
36	33	57°45′	154°53′	3433	503
36	31	55°00′	155°00′	4271	461
36	30	54°04′	155°00′	4088	551
36	29	53°00′	155°10′	3905	500
36	28	51°35′	155°09′	4426	584
36	27	49°40′	155°	4536	1153
36	26	47°51′	155°08′	4691	1224
36	24	44°00′	155°00′	4549	569
39	70	42°23.8′	153°10′	4664	1673
		Traverse H–H'			
37	5	65°31′	147°26.3′	2990	684
37	4	64°49′	150°30′	3274	584
38	4	64°13.9′	150° 3.8′	3658	731
38	3	64°14.5′	150°01.3′	3493	943
38	6	64°17.5′	150°11′	3457	559
38	8	61°48.6′	149°54.2′	3292	544
38	7	61°49.3′	149°53′	3658	599
36	12	61°45′	149°33′	4057	603
36	14	58°06′	150°10′	3054	601
38	9	57°27.7′	150°06.3′	3173	610
36	15	56°34′	150°15′	3517	207
36	16	55°07′	150°00′	3823	590
38	2	54°14.1′	149°57.7′	4060	423
36	17	54°02′	150°05′	3951	612
36	18	53°00′	150°00′	3877	582
36	20	51°47′	150°27′	3863	541
36	21	49°27′	149°08′	3846	493
36	22	47°32′	148°01′	1103	422
39	62	46°56.8′	149°32.6′	3219	284
39	64	45°33.6′	150°21′	4653	453
34	10	44°31.6′	149°31.6′	2853	1186
36	23	43°53′	150°02′	533	542

history of bottom-current activity is not simple and is difficult to determine in detail because the minimum ages of most cores are a function of the degree of sediment erosion rather than a reflection of the initiation of current activities. Presence of disconformities with associated winnowed sediment horizons within the Gilbert epoch indicates that increased current activity did not commence suddenly but was active at times during the Gilbert, and that these periods of erosion were separated by periods of apparently normal biogenic sedimentation.

Based on the number of cores in the traverses containing normal biogenic and fine-grained sediments of Gilbert age (Fig. 2), the major late Cenozoic change in Antarctic bottom water probably took place during the post-Gilbert (since $t = 3.32$ million years) and possibly post-Gauss time (since $t = 2.43$ million years). Normal fine-grained biogenic sedimentation occurred during the Matuyama epoch on parts of the Australian-Antarctic Rise (cores E36-14, E38-9, and E36-34 in Fig. 2). This area, being more remote from the basinal areas, has been less affected by the bottom-current activities. Pockets of reworked Brunhes sediments occur within the basin but in some cases (for example, core E38-2, Fig. 2) consist of clays and robust reworked Radiolaria and Foraminifera of Brunhes age. Such cores indicate that high-velocity bottom currents have occurred during the Brunhes and that sediments containing Brunhes biogenic components have accumulated in isolated pockets. It appears, therefore, that Antarctic bottom water currents were active at times during the Gilbert epoch, became highly active at some. time during post-Gilbert times, and have lasted until the Recent. Sedimentological analyses will undoubtedly refine this suggested history.

The major question posed by the inferred increase in velocities of Antarctic bottom water in the late Cenozoic is the cause of the change. Despite the obvious relation between topography and current systems, insufficient time has elapsed since deposition for changes in currents to result from substantial changes in submarine topography. Instead we propose that this increase in average current velocity is associated with a large change in the production of Antarctic bottom water from Antarctic ice. There is strong evidence for extensive glaciation of Ant-

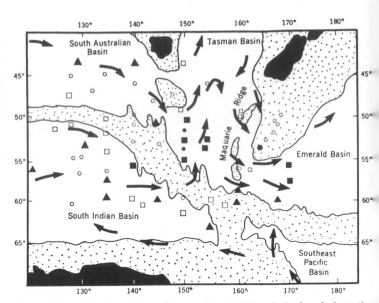

Fig. 3. Map of Southern Ocean between Australia–New Zealand and Antarctica showing circulation patterns of bottom waters and bottom-photograph (*15*) characteristics. Arrows represent bottom-water directions [from Gordon (*13*)]. Submarine ridge system is stippled. Symbols represent bottom-photograph characteristics as follows: □, abundant bioturbation; ○, bioturbation uncommon or absent; ▲, manganese nodules common; ■, manganese nodules abundant; ●, ripples and scour marks.

arctica in the late Cenozoic since at least the beginning of the Gilbert epoch ($t = 0$ to 5.0 million years) and for worldwide pre-Pleistocene cool climatic cycles assumed to be associated with increased glaciation of Antarctica (*18*). Periods of nonglaciation or partial glaciation of Antarctica result in decreased circulation of Antarctic waters, including decrease in magnitude of bottom circulation (*3*). It is, therefore, possible that ice development on the Antarctic continent reached a critical level during post-Gilbert or Gauss time. It is conceivable, for instance, that extensive, floating ice shelves developed at this time, and such shelves are likely to be important producers of Antarctic bottom water (*3*). It also follows that deep circulation may have experienced periods of even more highly intensified flow in response to more extreme climatic conditions associated with glacial periods and that deep circulation is now relatively quiescent (*19*). We suspect that the number of factors involved in our postulated change in current velocity is likely to be large: changes in latitudinal temperature gradients (*20*) and corresponding wind systems (*21*) would be substantial contributors to difference in current velocity with time.

We consider that, in addition to the observed late Cenozoic disconformities, other disconformities certainly must exist in older parts of the sedimentary basins between Australia and Antarctica. The sediments of that part of the Southern Ocean between Antarctica and Australia are highly susceptible to changes in bottom-current velocity because of the general circulation and bottom topography. Australia and Antarctica separated perhaps as recently as 45 to 50 million years ago (*22*). Therefore, for much of the lower and middle Cenozoic, circumpolar currents including bottom waters were restricted to a relatively narrow channel and, hence, must have had very high current velocities. As Antarctica was glaciated during this same interval (*17*), circulation was probably vigorous, and the prospects of high current velocities in the area were therefore even more enhanced. Sediments recovered in any future deep-sea drilling operations will, in our opinion, include much coarse material, including concentrations of manganese nodules, in addition to numerous and widespread disconformities.

References and Notes

1. For example, J. C. Ingle, Jr., *Bull. Amer. Paleontol.* **52**, 236 (1967); J. P. Kennett, *Deep-Sea Res.* **17**, 125 (1970); O. L. Bandy, R E. Casey, R. C. Wright, in *Antarctic Oceanology I*, J. L. Reid, Ed. (American Geophysical Union, Washington, D.C., 1971), p. 1; D. B. Ericson and G. Wollin, *Science* **162**, 1227 (1968).
2. There changes have recently been summarized by E. J. W. Jones, M. Ewing, J. I. Ewing, and S. L. Eittreim [*J. Geophys. Res.* **75**, 1655 (1970)].
3. A. L. Gordon, in *Antarctic Oceanology I*, J. L. Reid, Ed. (American Geophysical Union, Washington, D.C., 1971), p. 169.
4. B. C. Heezen and C. D. Hollister, *Marine Geol.* **1**, 141 (1964).
5. N. D. Watkins and J. P. Kennett, "Paleomagnetic and micropaleontological definition of a sedimentary disconformity of regional extent in the Southwest Pacific," in *Antarctic Research Series*, D. E. Hayes, Ed. (American Geophysical Union, Washington, D.C., in preparation).
6. N. D. Watkins and H. G. Goodell. *Earth Planet. Sci. Lett.* **2**, 123 (1967); N. D. Watkins, *ibid.* **4**, 341 (1968).
7. A. Cox, *Science* **163**, 237 (1969).
8. N. D. Opdyke, B. Glass, J. D. Hays, J. Foster, *ibid.* **154**, 349 (1966).
9. J. D. Hays and N. D. Opdyke, *ibid.* **158**, 1001 (1967).
10. J. P. Kennett, *Deep-Sea Res.* **17**, 125 (1970).
11. For example: core A185 [B. Glass, D. B. Ericson, B. C. Heezen, N. D. Opdyke, J. A. Glass, *Nature* **216**, 437 (1967)]; core E32-51 [J. P. Kennett and N. D. Watkins, *ibid.* **227**, 930 (1970)]; core E13-4 [K. Geitzenauer, S. V. Margolis, D. Edwards, *Earth Planet. Sci. Lett.* **4**, 173 (1958)]; core CH61(171) [J. D. Phillips, W. A. Berggren, A. Bertels, D. Wall, *ibid.*, p. 118]. Many other examples can be found in the literature.
12. The disconformity is defined by traverse *A′–A* to *M′–M* (5). These range from latitude 120°E to 160°W.
13. A. L. Gordon, "Spreading of Antarctic bottom waters, II," in *Studies in Physical Oceanography—A Tribute to Georg Wüst on His Eightieth Birthday*, A. L. Gordon, Ed. (Gordon & Breach, New York, in press).
14. J. B. Southard, R. A. Young, C. D. Hollister, "Experimental erosion of calcareous ooze," *J. Geophys. Res.*, in press.
15. S. S. Jacobs, P. M. Bruchhausen, E. B. Bauer, Eds., *Eltanin Reports—Hydrographic Status, Bottom Photographs, Current Measurements: Cruises 32–36, 1968* (Lamont-Doherty Geological Observatory, Palisades, N.Y., 1970) for the U.S. Antarctic Research Program (National Science Foundation).
16. S. Eittreim, A. L. Gordon, M. Ewing, E. M. Thorndike, P. Bruchhausen, "The nepheloid layer and observed bottom currents in the Indian Pacific-Antarctic Sea" in *Studies in Physical Oceanography—A Tribute to Georg Wüst on His Eightieth Birthday*, A. L. Gordon, Ed. (Gordon & Breach, New York, in press).
17. H. G. Goodell and N. D. Watkins, *Deep-Sea Res.* **15**, 89 (1968).
18. J. P. Kennett, *N.Z. J. Geol. Geophys.* **10**, 1051 (1967); S. V. Margolis and J. P. Kennett, *Amer. J. Sci.*, in press; O. L. Bandy, in *Tertiary Correlations and Climatic Changes in the Pacific*, K. Hatai, Ed. (Sasake, Sendai, Japan, 1967), p. 9; J. C. Ingle, Jr., *Bull. Amer. Paleontol.* **52**, 236 (1967).
19. D. A. Johnson, *Trans. Amer. Geophys. Union* **52**, 244 (1971).
20. C. H. Hendy and A. T. Wilson, *ibid.*, p. 229.
21. W. H. Munk, *J. Meteorol.* **7**, 79 (1950).
22. R. S. Dietz and J. C. Holden, *J. Geophys. Res.* **75**, 4939 (1970); *Sci. Amer.* **223** (4), 30 (1970); X. Le Pichon and J. R. Heirtzler, *J. Geophys. Res.* **73**, 2101 (1968).
23. In assigning time ranges for each core, nonlinear sedimentation rates (with which a perfect match of polarity scale and paleomagnetic data can obviously be forced) have not been assumed. This means that in the Gilbert epoch sediments in particular, a linear sediment rate based on two time-reference points has often resulted in imperfect matching of the polarity data and polarity scale. Therefore, although we believe that the time ranges are correct to ±0.4 million years, we are not certain of the correlation of many observed polarity events to the polarity scale, or the durations of the observed events.
24. Supported by NSF grants GA25400 and GV28305 (Office of Polar Programs).

30 April 1971; revised 28 June 1971

28

Reprinted from Geol. Soc. America Bull. **87**:321–329 (1976)

Regional deep-sea dynamic processes recorded by late Cenozoic sediments of the southeastern Indian Ocean

J. P. KENNETT
N. D. WATKINS } Graduate School of Oceanography, University of Rhode Island, Kingston, Rhode Island 02881

ABSTRACT

A large collection of USNS *Eltanin* deep-sea sedimentary cores and bottom photographs from the southeast Indian Ocean between long 70°E and 120°E, and between Antarctica and lat 30°S, were analyzed. Cores from the crest and flanks of the mid-ocean ridge are mostly late Quaternary in age, with only rare breaks in sedimentation. In great contrast, flanking this zone in deep basins immediately to the south of the ridge in the South Indian Basin and in a broad zone in the western sector of South Australian Basin, there are areas where bottom currents have systematically eroded or inhibited deposition of sediments ranging in age from Quaternary to Pliocene, and occasionally middle Tertiary. This regional deep-basin erosion extends northward between Broken Ridge and the Naturaliste Plateau to the Wharton Basin where sediments as old as Late Cretaceous are exposed. As indicated by disconformities, ocean-floor characteristics, and seismic-profile data, much of the shallower, north-trending Kerguelen Plateau has also undergone widespread erosion by bottom currents.

The erosional disconformities in the deep basins have been created by general increase in velocities of Antarctic Bottom Water during the last 2.5 m.y., apparently with major separate pulses during the Brunhes epoch (t = 0.69 m.y. to present) and part of the Matuyama epoch (t = 2.43 to 0.69 m.y.). Extensive areas of manganese nodules have developed in conjunction with this bottom-current activity, most spectacularly as a vast pavement in the northwestern sector of the South Australian Basin. This feature, which we name the "Southeast Indian Ocean Manganese Pavement," is approximately 10^6 km² in area.

The available evidence indicates long-term major erosion by the eastward-flowing Circumpolar Current across the central and southern parts of the Kerguelen Plateau. In the deep basins, high-velocity Antarctic Bottom Water flows eastward through the northern sector of the South Indian Basin, with important northward flow crossing the mid-ocean ridge at 110°E and 120°E into the South Australian Basin. This northward branch traverses the western sector of the South Australian Basin, the Southeast Indian Ocean Manganese Pavement, and then flows between Broken Ridge and Naturaliste Plateau into the Wharton Basin. Major Cenozoic to Late Cretaceous hiatuses in the Wharton Basin revealed by deep-sea drilling suggest that northward-flowing bottom water through this conduit has been a very long term feature. *Key words: marine geology, southeast Indian Ocean, Circumpolar Current, manganese pavement, hiatuses.*

INTRODUCTION

The Southern Ocean region is the site of the world's most active current system. This system includes high-velocity surface current flow associated with the Antarctic Circumpolar Current. Deeper parts of the Southern Ocean are also marked by widespread high-velocity bottom currents largely related to the flow of Antarctic Bottom Water and strongly controlled by deep-ocean topography (Gordon, 1971a). Effects of high-velocity deep-sea currents on abyssal geological features have now been well documented. Bottom photographs readily reveal features such as sediment-free manganese nodules, ripple marks, and sedimentary lineations (Heezen and Hollister, 1964, 1971; Hollister and Heezen, 1967). Similarly, paleontological and paleomagnetic dating of sedimentary cores obtained by piston coring and deep-sea drilling now yield unambiguous evidence of common and sometimes extensive regional disconformities resulting from erosion by deep-sea currents (Watkins and Kennett, 1971, 1972; Kennett and others, 1972; Johnson, 1972; Kennett and others, 1974). The widespread nature of deep-sea erosion can also be inferred from seismic profiles (Houtz, 1975; Johnson, 1972) and the character of selected redeposited deep-sea sediments (Johnson, 1972; Kennett and Brunner, 1973). It appears that the Circumpolar Current is by far the major system in the world's oceans, in terms of potential dynamic effects on sea-floor sediments.

In this contribution, we present two new major sets of observations from the southeastern Indian Ocean segment of the circum-Antarctic regime. Paleomagnetic and micropaleontological ages have been derived for 187 deep-sea sedimentary cores collected during 6 *Eltanin* cruises throughout the southeastern Indian Ocean (Fig. 1) between lat 70°E and 120°E and between Antarctica and long 30°S. Our second series of observations have been obtained from approximately 2,800 deep-sea bottom photographs from 143 *Eltanin* camera stations throughout the southeastern Indian Ocean (Fig. 2). Our motives are fivefold: (1) to provide first-order paleomagnetic and micropaleontological ages of the entire set of *Eltanin* cores from the southeastern Indian Ocean; (2) to determine late Cenozoic sedimentary patterns throughout the region; (3) to determine ocean-floor dynamic characteristics throughout the region from analysis of bottom photographs; (4) to determine the long-term bottom-water circulation in relation to bottom topography; and (5) to establish the late Cenozoic history of major pulses in the higher velocity bottom water.

There is now a substantial body of evidence, mainly as a result of the Deep Sea Drilling Project, for extensive past sediment erosion in the deep sea as a result of vigorous activity of deep-sea currents. Certain Cenozoic hiatuses are now known to be very widespread throughout the oceans of the world, reflecting episodic pulses in bottom-water activity of considerable magnitude (Rona, 1973; Kennett and others, 1972; Johnson, 1972; van Andel and others, 1975; Moore and others, 1975; Davies and others, 1975). The distribution of bottom currents is strongly controlled by deep-sea topography. Deep-sea sediment hiatuses representing erosion by such currents are now known to occur within certain deep basins, mid-ocean ridges, and even on shallow-water continental plateaus such as the Campbell Plateau south of New Zealand (Kennett, Houtz, and others, 1975).

Little is definitely known of the velocities required to create regional deep-sea erosion. Experimental work summarized by

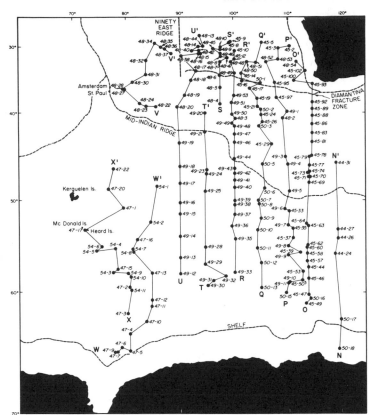

Figure 1.   Map showing locations of cores and selected traverses N-N' to X-X'. Letter symbols for the set of traverses follow on from the sequence of traverses previously described by Watkins and Kennett (1972) south of New Zealand and Australia. Number next to each core site is cruise number followed by core number. Dashed lines show locations of mid-oceanic ridge and continental shelves. The Diamentina fracture zone is also shown.

made by Frakes (1971, 1973). In addition, 13 sites have been drilled in the region during the Deep Sea Drilling Project. Four of these were drilled during Leg 28 in the southeastern part of the region (Hayes, Frakes, and others, 1973). The other sites, drilled during Legs 26, 27, and 28, occur in the northern part of the region and provide evidence of widespread disconformities of considerable duration both on shallow and deeper topographic features (Davies and others, 1975). Almost all previous micropaleontological studies of cores from the region concern late Quaternary planktonic microfaunal changes in response to changes in surface water distributions (Hays and others, 1976; Williams and Johnson, 1975; Vella and Watkins, 1975; and Vella and others, 1975). Weaver and McCollum (1974) discovered a disconformity spanning approximately $t = 2.1$ to 1.3 m.y., using detailed analyses of diatoms in four piston cores from the western sector of the South Indian Basin. Burckle and others (1973) have also used diatoms to investigate the distribution of bottom water in the Southeast Indian Ocean, employing transportation directions of Antarctic diatoms by the Antarctic Bottom Water. Quilty (1973) has described the planktonic foraminiferal biostratigraphy of a Late Cretaceous foraminifera-rich core (E54-7) from the northeast side of the Kerguelen Plateau.

METHODS

Paleomagnetism

Shipboard methods have been described previously (Watkins and Goodell, 1967). The cores are collected in plastic liners and are 6 cm in diameter. The liners are split and the sediment separated by using a nonmagnetic fiber. Specimens were taken for the paleomagnetic study by inserting oriented 10.7-cm^3 plastic boxes into the sediment face. Those cores showing signs of sediment disturbance and very limited recovery were not examined. All specimens were demagnetized in alternating magnetic fields of 100 or 150 Oe to minimize unstable components. The direction and intensity of remanent magnetism were measured by using a 5-cycle/sec spinner magnetometer, and the polarity of any specimen is interpreted as normal for upward inclination, reversed for downward inclination. Age assignments for each core are based on recognition of the major features of the known geomagnetic polarity history for the last 5 m.y. (Cox, 1969) and by biostratigraphic dating, using previous planktonic foraminiferal and radiolarian zonations (Hays and Opdyke, 1967; Kennett, 1970; Keany and Kennett, 1972). Since the purpose of this study was to identify and use

---

Heezen and Hollister (1964) was used by Watkins and Kennett (1971, 1972) in their suggestion that an increase in current velocities from very dominantly less than 10 cm/sec^{-1} to dominantly greater than 10 cm/sec^{-1} is required to explain the hiatuses. Maximum velocity values of bottom water are likely to be much greater over short intervals of time. Watkins and Kennett (1972) have shown that a substantial increase in deep-water velocities must have occurred at some time since $t = 3.5$ m.y. and that the erosion is strongly associated with major positive bathymetric elements.

Our study is essentially an extension of previous similar analyses conducted in the Southern Ocean south of Australia and New Zealand (Watkins and Kennett, 1971, 1972) and in the South Pacific region (Goodell and Watkins, 1968), during which

paleomagnetically defined sedimentation rates enabled the determination of regional sedimentary patterns from which long-term oceanographic controls could be inferred.

PREVIOUS WORK

Little has been published on the oceanography of the southeast Indian Ocean. Although the general bathymetry has been presented by Heezen and others (1972), no detailed charts have yet been published. The physical oceanography and seismic profiling data of the region have similarly not yet been presented. Nevertheless, cruises of USNS *Eltanin* between 1969 and 1973 yielded a large amount of diverse marine data which are in the process of being analyzed. Cursory descriptions of the sediment of our dated cores (Fig. 1) were the positions of the major polarity changes in the core collection, the fine detail of the magnetic data reflecting noise or possible undiscovered short polarity events was suppressed by subjecting most results to a five-point running average.

Micropaleontology

Because of the unusually large-scale nature of this study, detailed micropaleontologic investigations could not be made of all cores. Therefore, in the majority of cores, examination of three samples (top, middle, and bottom) was sufficient for unambiguous assignment of core segments to the known geomagnetic polarity history. Discovery of obvious disconformities in many cores, however, made examination at closer intervals necessary. In such cases, as many as ten samples per core were required for satisfactory interpretation of the paleomagnetic stratigraphy. This provided intervals rarely more than 200 cm or less than 100 cm, and so the position of disconformities not marked by a polarity change could be as much as 100 cm from their exact position. We should emphasize that sedimentary hiatuses of relatively short duration (up to about 0.5 m.y.) cannot be necessarily identified by our methods. More detailed micropaleontological analyses such as those carried out by Weaver and McCollum (1974) can be used to detect shorter hiatuses which might also be expected to exist in regions where our techniques have revealed major hiatuses.

The correlation lines for the major polarity boundaries have been placed by applying the radiolarian zonation of Opdyke and others (1966) and Hays and Opdyke (1967), and the foraminiferal zonation of Kennett (1970), and Keany and Kennett (1972). The radiolarian species that were particularly valuable in our correlations were *Triceraspyris* sp. and *Lychnocanium grande* in the early and middle Gilbert; *Desmospyris spongiosa*, the extinction of which coincided approximately with the Gauss-Matuyama boundary; *Eucyrtidium calvertense* and *Clathrocyclas bicornis*, the extinction of which occurred in the middle Matuyama; and *Saturnalus planetes*, which in subantarctic waters disappeared near the Brunhes-Matuyama boundary (t = 0.69 m.y.). The foraminiferal species that were particularly valuable in our correlations were *Globorotalia puncticulata*, the disappearance of which is close to the Brunhes-Matuyama boundary; *Globorotalia inflata* and *Globorotalia crassaformis*, the appearance of which in subantarctic waters is closely associated with the Brunhes-Matuyama boundary. Other available Southern Ocean biostratigraphic schemes based on diatoms (McCollum, 1975) and calcareous nannofossils (Geit-

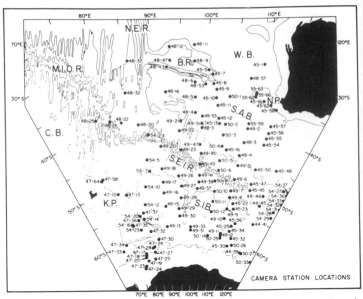

Figure 2. Map showing locations of bottom camera stations. Bathymetric contours, shown for 2,000, 3,000, and 4,000 m, are from Heezen and others (1972). Symbols indicating major physiographic features of the Southeast Indian Ocean are as follows: K.P. = Kerguelen Plateau; S.I.B. = South Indian Basin; S.A.B. = South Australian Basin; S.E.I.R. = Southeast Indian Ridge; M.I.O.R. = Mid–Indian Ocean Ridge; C.B. = Crozet Basin; W.B. = Wharton Basin; B.R. = Broken Ridge; N.P. = Naturaliste Plateau. Number next to each camera site is cruise number followed by station number (not camera-station number).

zenauer, 1972; Miyajima, 1974) have not been utilized in this regional study. Paleontologic confirmation of the shorter duration magnetic polarity events is often not available, especially for those events within the Gilbert epoch (t = 5.1 to 3.34 m.y.) where detailed paleontologic control has yet to be established. Correlations of events are therefore made subjectively by simply comparing these data with the established geomagnetic polarity scale (Cox, 1969) and using assumptions of constant sedimentation rates between any available known polarity boundaries. It is important to note that if it is assumed that sediment rates have varied throughout a core, then any magnetic data can be forced into equivalence with virtually any part of the polarity scale. This degree of freedom clearly cannot be legitimately adopted. Linear extrapolation of sediment rates between assigned polarity boundaries must be used consistently, even if the assigned ages result in a conflict between measured polarity and the details in the known polarity time scale. In our opinion, a precision of ± 0.4 m.y. can still be maintained for the Pliocene and Quaternary. Despite the dual methods used for determination of ages of the cores, several imperfections still exist. In almost all cases,

cores restricted to the Brunhes epoch (t = 0 to 0.69 m.y.) have not been assigned to a more specific time interval. Such cores are therefore arbitrarily assigned to reach an age of t = 0.6 m.y.

Cores taken from depths greater than about 5,000 m or deeper typically contain few or no microfossils, and ages are thus based entirely on the paleomagnetic polarities. The single method utilized in dating cores made up of red clay throughout is not entirely satisfactory, because it is known that red clay can be magnetically unstable (Kent and Lowrie, 1974). Similarly, few high-latitude cores closely adjacent to Antarctica commonly lack planktonic foraminifera and contain few biostratigraphically significant radiolaria and thus have been dated using only the paleomagnetic stratigraphy.

Bottom Photograph Analysis

The entire set of *Eltanin* bottom photographs from the southeast Indian Ocean taken during seven cruises (44, 45, 47, 48, 49, 50, 54) were supplied on microfilm by the Smithsonian Oceanographic Sorting Center, Washington, D.C. The bottom photographs were taken using a camera

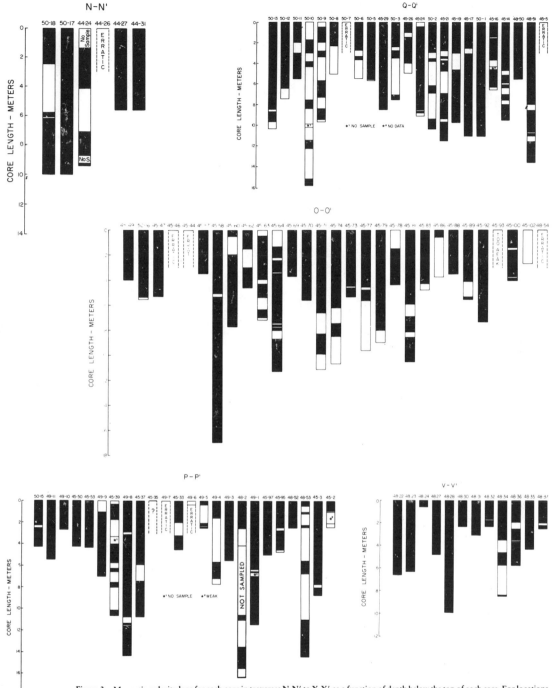

Figure 3. Magnetic polarity logs for each core in traverses N-N′ to X-X′ as a function of depth below the top of each core. For locations of traverses and cores, see Figure 1 and Table 1. In each log: black = normal polarity, clear = reversed polarity. See text for specimen intervals in each core and methods of interpretation of data.

Figure 3. (Continued).

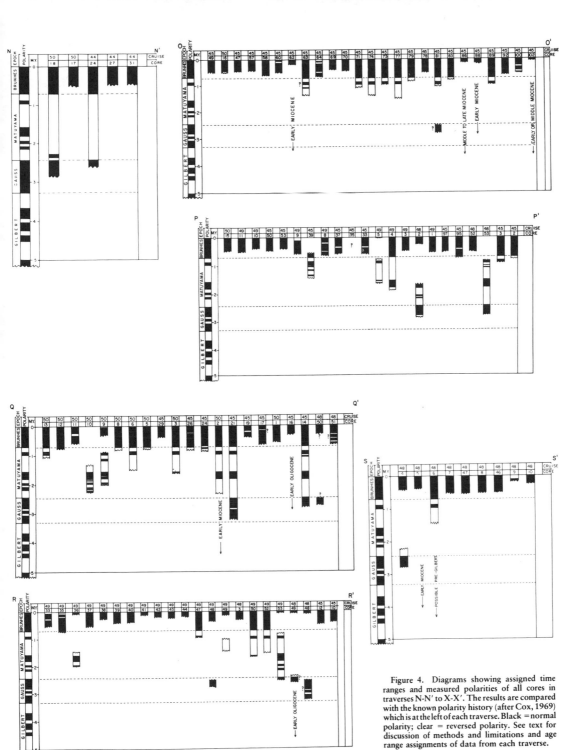

Figure 4. Diagrams showing assigned time ranges and measured polarities of all cores in traverses N-N' to X-X'. The results are compared with the known polarity history (after Cox, 1969) which is at the left of each traverse. Black = normal polarity; clear = reversed polarity. See text for discussion of methods and limitations and age range assignments of data from each traverse.

335

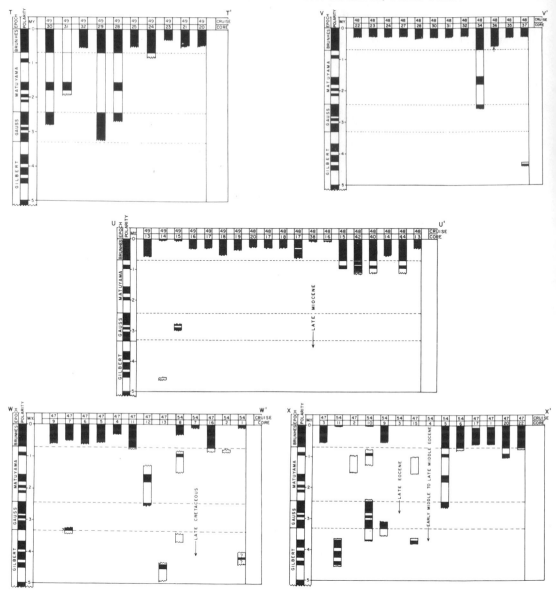

Figure 4. *(Continued).*

system involving bottom-contact multiple exposure and a strobe light (Jacobs and others, 1970, 1972). For repeated exposures, the system is raised a few meters and lowered again as the ship drifts. Thus the camera is bounced across the bottom in a series of short arcs, taking a photograph at each hit (Simmons and Landrum, 1973).

The area covered during the exposure of a usual set of 20 to 30 frames will vary with the drift speed of the ship but can be expected to cover a distance of about 1 km. An area of approximately 5 m² is recorded by each photograph. Each camera station typically has 20 good individual frames per station.

The bottom photographs were examined using a microfilm reader, and bottom features for each station were placed within six progressively less dynamic categories as follows: (1) manganese nodule pavement to abundant manganese nodules, (2) common to scattered manganese nodules, (3) ripple marks and extensive scour, (4) distinct current-formed lineations on an otherwise smooth sediment surface, (5) smooth bottom to rare bioturbation, and (6) abundant to common bioturbation. A seventh category is rocks that include submarine lavas and large ice-rafted boulders. In a limited number of cases where different bottom features were found in different exposures at the same camera station, the dominant characteristic bottom feature from the total number of frames was chosen as representative. In the small percentage of the photographs which featured distinct sedimentary lineations, it was possible to determine current directions, using the magnetic orientation system. In most cases, where current-formed sedimentary lineations were apparent, however, the compass was obscured by mud in suspension. At sites where readings were possible, regional magnetic declinations of up to 40° (Vestine and others, 1948) were used to correct the observations into true geographic bearings.

RESULTS

The paleomagnetic properties of a total of more than 12,000 specimens from 185 cores have been measured. The results are given in Figure 3. Polarities for all data points are included. Intensities of magnetization range from 10^{-5} to 10^{-7} emg/g^{-1}.

The age ranges of all cores in each traverse are shown in Figure 4. The measured magnetic polarities of each core are located in their assigned relationships to the known geomagnetic polarity history for the last 5 m.y. Only 13 of the cores consist entirely or in part of sediment older than 5 m.y.; ages from the Late Cretaceous to the late Miocene have been detected. The possibility that cores consisting of red clay show spurious magnetic data is supported by the fact that despite expected very low sedimentation rates (~5 mm/1,000 yr), these cores usually display normal polarity throughout even in those with lengths of up to 12 m. Cores of this length with such low sedimentation rates should extend well into the Matuyama reversed epoch. Therefore, it seems likely that such cores have not recorded the polarity history at time of deposition, and their ages are too young. Additional notes on the cores are given in Appendix 1, where they are shown in the order of general north-south traverses from east to west (Fig. 1).

DISCUSSION

Regional Sedimentary Trends: Distribution of Disconformities

Cores containing disconformities of various ages occur in two well-defined zones in the southeast Indian Ocean (Fig. 5), separated by the Southeast Indian Ridge. The zone south of the ridge encompasses the

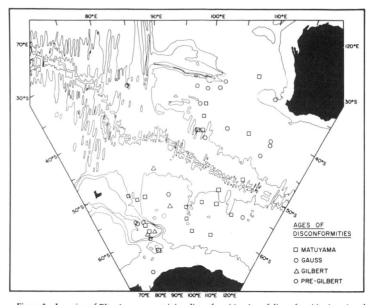

Figure 5. Location of *Eltanin* cores containing disconformities. Age of disconformities is assigned in terms of the polarity epoch of the sediment immediately below the disconformity. If more than one disconformity is present, the age of the younger is shown.

southern sector of the Kerguelen Plateau and the northern sector of the South Indian Basin. The zone north of the ridge includes the western sector of the South Australian Basin and the southern part of Wharton Basin between the Naturaliste Plateau and Broken Ridge. Most of the few other cores containing disconformities are scattered throughout other parts of the South Australian and South Indian Basins. Therefore, virtually all cores containing disconformities occur in deep basinal areas and on the southern sector of the shallower Kerguelen Plateau and Broken Ridge. The mid-oceanic ridge system and Broken Ridge, which separate the deep basins of the region, are almost completely devoid of any cores containing disconformities.

It has now been well established that disconformities of wide extent occurring in deep-sea sedimentary sequences have resulted from erosion by increases in velocity of bottom water (Watkins and Kennett, 1971, 1972; Kennett and others, 1972; Johnson, 1972, 1974). A discussion of the relation between inferred bottom-water flow and the regional disconformities will be made below. We present here mathematical expressions of the regional sedimentary breaks and related sedimentation rates and sediment thickness.

In order to detect the regional configuration of erosional features, a trend surface has been fitted to the age (in mil-

lions of years) of the sediment at the top of each core. Excluded, when present, is the few centimetres of veneer of Brunhes epoch sediments, so that the major erosional surface provides the age data used. The configuration of the surface is controlled by minimizing the root mean square residual, and the order of the surface has been selected by conventional comparisons of the rate of improved surface fit to the data, with increasing order. The results are shown, as a sixth-order surface, for the entire area of study (Fig. 1) in Figure 6. The surface clearly shows the presence of areas on either side of the ridge, with old sediment exposed at or very near the sediment-ocean interface. The region displaying the oldest surfaces is clearly centered over the southern part of the Kerguelen Plateau and extends throughout much of the Kerguelen Plateau and the northwestern sector of the South Indian Basin. The other area is centered in the western sector of the South Australian Basin. The mid-ocean ridge system is devoid of the same erosional feature.

Regional Sedimentary Trends: Variation in Sedimentation Rates

Combining core lengths with assigned age ranges (Fig. 4) yields average sedimentation rates for each core. In the cores that do not reach the Brunhes-Matuyama boun-

TABLE 1. LOCATION, LENGTH, AND WATER DEPTH FOR EACH CORE

Cruise	Core	Length (cm)	Latitude (S)	Longitude (E)	Water depth (m)
			Traverse N-N'		
50	18	1021	64°25.47'	119°58.64'	3074
50	17	1011	63°00.06'	120°03.04'	4019
44	24	1020	56°02.35'	119°53.47'	4320
44	26	1110	54°01.12'	119°46.41'	4185
44	27	583	53°02.22'	119°44.03'	3450
44	31	581	46°00.00'	119°53.00'	4203
			Traverse O-O'		
45	49	423	61°21.04'	113°44.79'	4167
45	16	581	61°02.63'	114°48.81'	4239
45	47	542	60°45.20'	114°15.30'	4222
45	46	119	59°44.00'	114°57.00'	4347
45	44	403	58°29.90'	114°05.70'	4370
45	57	382	57°06.00'	114°01.00'	4384
45	58	1713	56°35.88'	114°06.21'	4348
45	60	803	55°03.57'	114°09.16'	3997
45	62	483	55°04.75'	114°07.11'	4131
45	63	1052	52°26.20'	114°15.40'	3762
45	64	1143	52°28.92'	114°05.96'	3747
45	69	411	48°50.46'	114°36.18'	3357
45	70	583	48°30.00'	114°28.60'	3639
45	71	1133	48°01.50'	114°29.20'	3619
45	74	1083	47°33.10'	114°26.40'	3744
45	73	543	47°33.24'	114°26.44'	3690
45	77	983	46°26.90'	114°25.00'	3942
45	79	913	45°03.40'	114°22.00'	4050
45	78	453	45°03.98'	114°21.20'	4158
45	81	1063	43°58.62'	114°21.68'	4248
45	83	493	42°35.30'	114°26.40'	4257
45	86	403	41°29.70'	114°26.00'	4356
45	88	383	40°31.07'	114°29.00'	4527
45	89	563	39°30.90'	114°28.40'	4284
45	92	753	38°02.74'	114°30.47'	5117
45	93	283	35°36.20'	114°34.00'	4392
45	100	420	35°11.60'	111°46.30'	3049
45	102	267	33°36.80'	113°35.10'	4784
48	54	522	32°03.22'	111°50.38'	
			Traverse P-P'		
50	15	451	60°02.	109°58.97'	4203
49	11	561	59°38.89'	110°09.41'	4233
49	10	542	59°01.81'	110°07.97'	4303
45	50	443	59°25.10'	113°53.93'	4334
45	53	448	58°27.52'	113°54.44'	4368
45	9	721	56°58.25'	110°05.34'	4260
45	39	1083	56°00.00'	112°43.00'	4302
45	8	1441	55°04.24'	110°01.10'	3636
49	37	1101	54°49.18'	111°57.00'	4086
45	35	1112	53°29.50'	111°20.63'	3780
49	7	1133	53°02.19'	110°02.82'	3537
49	33	532	51°57.00'	110°27.00'	3369
49	6		51°00.44'	109°59.21'	2276

TABLE 1. (Continued)

Cruise	Core	Length (cm)	Latitude (S)	Longitude (E)	Water depth (m)
			Traverse S-S'		
48	4	1122	38°53.30'	97°57.92'	4023
48	5	902	36°26.67'	97°28.22'	4221
48	6	1070	34°00.25'	97°32.55'	4212
48	7	1212	32°32.09'	97°38.59'	5112
48	47	562	32°02.73'	98°30.97'	1668
48	8	322	31°36.86'	97°42.68'	2253
48	46	542	31°09.10'	97°26.82'	2484
48	45	562	30°33.90'	96°44.97'	2943
48	9	62	30°27.76'	97°34.75'	3038
48	10	92	30°28.06'	97°36.20'	3060
45	8	46	29°58.90'	99°53.10'	3474
45	9	76	29°58.90'	99°59.10'	3474
			Traverse T-T'		
49	30	1222	59°00.31'	95°13.81'	4212
49	31	671	58°48.58'	96°20.77'	4374
49	32	1061	58°22.07'	98°28.06'	4077
49	29	1111	57°05.72'	94°57.25'	4172
49	28	1111	55°10.87'	94°51.20'	4467
49	25	531	49°22.71'	94°49.90'	3285
49	24	542	47°59.25'	95°02.17'	3164
49	23	581	47°07.69'	95°04.78'	3207
49	21	921	42°11.06'	94°53.13'	3240
49	20	471	40°05.08'	94°53.39'	3492
			Traverse U-U'		
49	12	872	58°22.09'	89°59.24'	4419
49	13	1202	56°50.22'	89°44.37'	4050
49	14	108	54°50.03'	90°08.36'	4590
49	15	581	52°46.97'	90°00.51'	3996
49	16	1191	50°25.98'	90°10.60'	3987
49	17	1301	48°16.84'	90°14.82'	3448
49	18	1551	46°02.96'	90°09.28'	3204
49	19	551	43°53.19'	90°06.03'	2988
48	20	352	39°10.70'	89°22.74'	3393
48	19	452	37°39.07'	90°41.59'	3691
48	18	322	35°25.52'	91°52.55'	3605
48	17	1142	34°02.07'	92°40.32'	4302
48	38	342	31°58.22'	90°09.96'	3708
48	16	42	31°20.23'	93°37.15'	4489
48	15	1582	31°18.68'	93°33.06'	4489
48	42	2462	31°18.34'	93°28.87'	4491
48	40	402	31°25.60'	88°58.24'	3047
48	14	302	30°30.64'	93°30.21'	2115
48	44	262	30°28.19'	94°55.35'	1764
48	13	492	28°30.95'	93°30.30'	3326
			Traverse V-V'		
48	22	682	39°53.65'	85°24.62'	3324
48	23	132	39°30.96'	83°43.40'	3459
49			39°26.16'	83°10.19'	3396

48	2	1652	41°01.47'	109°54.56'	4167
49	1	1158	39°59.24'	110°03.82'	4491
45	97	520	37°26.70'	108°57.00'	4842
45	95	493	35°21.00'	108°13.10'	4734
48	52	262	32°03.86'	106°42.22'	4780
48	53	1452	31°58.52'	108°24.60'	5103
45	3	1000	29°35.00'	108°18.00'	4908
45	2	240	29°21.70'	110°15.00'	4908

Traverse Q-Q'

50	13	1054	59°59.06'	105°00.02'	4145
50	12	751	57°57.18'	105°00.99'	4334
50	11	571	55°56.73'	104°56.73'	3862
50	10	1604	53°58.67'	104°56.21'	3618
50	9	994	52°01.28'	105°00.61'	3186
50	8	971	50°56.01'	104°54.52'	3177
50	7	1641	50°01.53'	104°55.03'	3132
50	6	571	48°01.58'	105°14.61'	3009
50	5	591	46°06.86'	105°01.76'	3405
45	29	1072	44°53.00'	104°27.00'	3758
54	3	771	42°00.81'	104°53.77'	4014
45	26	512	41°46.10'	105°00.41'	3753
45	24	932	40°09.00'	104°14.10'	3983
50	2	1061	39°57.47'	104°55.69'	4242
45	21	1182	39°00.00'	103°33.00'	4237
45	19	992	37°38.66'	103°06.17'	4507
45	17	1111	36°30.00'	102°40.30'	4500
50	1	1121	35°51.73'	105°09.54'	6042
45	16	680	35°07.00'	101°58.00'	4312
45	14	963	32°12.62'	101°39.10'	4816
48	50	572	32°11.96'	102°49.64'	5054
48	51	1362	32°48.66'	104°38.48'	5175
45	5	588	29°44.00'	105°58.00'	5009

Traverse R-R'

49	33	1702	57°45.83'	100°02.48'	3978
49	35	801	54°23.18'	100°01.56'	3715
49	36	542	52°21.56'	99°49.15'	3366
49	37	1182	51°41.67'	100°03.37'	3486
49	38	1141	50°50.91'	100°05.26'	3564
49	39	1191	50°03.64'	100°13.14'	3344
49	40	582	49°04.33'	100°04.71'	3166
49	41	142	48°14.20'	100°03.36'	2997
49	42	572	47°14.46'	100°07.76'	2853
49	43	348	46°29.31'	100°03.10'	3024
49	44	411	45°39.57'	100°07.22'	3411
49	46	672	44°04.11'	100°01.01'	3562
49	47	882	43°20.25'	100°01.10'	3618
49	48	201	42°26.76'	100°01.91'	3600
49	49	201	41°31.33'	99°58.46'	3906
48	3	622	41°01.15'	100°00.67'	3868
49	50	778	40°36.66'	99°54.82'	3996
49	52	722	39°00.28'	99°56.90'	4131
49	53	901	37°51.57'	100°01.73'	4248
49	49	572	34°29.97'	100°01.89'	4347
48	48	582	33°03.33'	99°15.80'	4203
48	12	1043	32°41.00'	100°58.00'	4816
45	10	722	31°38.00'	100°33.00'	2899

48	31	312	34°48.60'	84°07.56'	3582
48	32	452	32°47.55'	83°52.21'	2926
48	34	882	29°57.55'	85°34.19'	4068
48	35	452	30°21.15'	86°29.06'	3240
48	36	602	30°55.81'	87°46.35'	1344
48	37	272	31°25.60'	88°58.24'	3047

Traverse W-W'

47	9	1108	66°22.76'	78°01.18'	2394
47	7	220	66°39.34'	77°53.97'	1404
47	6	1224	66°06.74'	78°26.60'	2889
47	5	570	65°32.59	80°25.53'	2880
47	4	60	64°07.10	80°23.89'	3540
47	10	518	63°57.54	83°59.50'	3549
47	11	298	62°58.84	84°11.64'	2552
47	12	537	61°56.68'	84°02.46'	2721
47	13	571	58°47.04'	84°14.40'	2844
54	8	582	56°52.49'	81°11.20'	4158
47	7	483	55°52.78'	81°07.09'	4021
54	16	749	54°51.04'	82°50.54'	4419
54	2	32	52°04.96'	84°32.18'	4226
54	1	1102	48°07.13'	86°11.50'	3801

Traverse X-X'

47	3	380	62°23.09'	80°47.33'	2736
54	11	322	59°46.89'	81°00.93'	1800
47	2	559	59°41.82'	80°49.04'	1769
54	10	513	57°45.41'	80°39.69'	1706
54	9	452	57°44.27'	81°16.44'	1634
54	3	22	57°25.74'	77°49.81'	1890
47	15	498	57°17.27'	78°48.47'	1603
54	4	82	57°26.47'	77°52.75'	1828
54	5	542	56°52.59'	74°32.39'	2890
54	6	542	55°28.11'	76°01.01'	2120
47	17	578	53°21.14'	72°10.92'	946
47	1	620	51°34.83	78°58.10'	3528
47	20	408	49°11.48'	72°08.70'	734
47	22	150	47°28.83'	73°55.63'	2628

dary, only minimum sedimentation rates can be calculated, using an arbitrarily assigned maximum age of 0.6 m.y. Where such cores are less than 6 m, however, such minimum sedimentation rate estimates may be substantially in error, and such estimates are thus not included in any regional analysis. This problem will be even further exaggerated in areas beneath the Antarctic Convergence because of particularly high sedimentation rates associated with high biogenic productivity. The error will have the effect of distorting any regional analysis, particularly if specific cruises can be characterized by minimal attempts to obtain longer cores. The number of cores recovered (Fig. 1) is so large, however, that such bias may be efficiently averaged out.

In Figure 7, we present a 5th-order trend surface map (derived using the same rationale employed to construct Fig. 6) of the calculated average sedimentation rates for the entire area for the period t = 0 to 2.5 m.y. This excludes those cores with hiatuses during that period. According to this analysis, sedimentation rates in the region vary on average from less than 0.25 cm per thousand years to about 1.5 cm per thousand years: this is obviously less than real in the area of the Antarctic Convergence. The regional trends in sedimentation rates indicated by the trend surface analysis are clearly related to bottom topography. Except near the Antarctic continent, highest sedimentation rates generally occur over the major part of the mid-ocean ridge system, with lower sedimentation rates evident in the South Australian Basin, the westernmost part of the South Indian Basin and the Kerguelen Plateau. Thus the sedimentation rate trend surface (Fig. 7) shows a rough inverse relationship to the sediment age-trend surface map (Fig. 6), suggesting a similar causative agent.

Nature of Ocean Floor from Sea-Bottom Photographs

The *Eltanin* bottom photographs (Fig. 2) have been examined to delineate regional variations in microtopography and infer long-term bottom-current activity. The simple interpretation techniques employed have been summarized by Heezen and Hollister (1964, 1971) and Hollister and Heezen (1967). In Figures 8 through 10, typical examples of the Southeast Indian Ocean bottom photographs are shown. At lowest current velocities, muddy sea bottom characteristically has an abundance of distinct faunal tracks (Fig. 8, D, F) which become erased with increased bottom-current velocities until the bottom develops a smooth appearance, although evidence of life in the form of partly erased tracks may still be abundant (Fig. 8, B, C, E). As velocities are further increased, surfaces develop with elongate sedimentary deposits

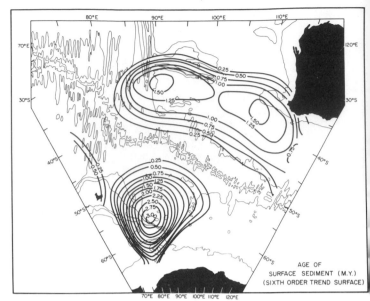

Figure 6. Map of sixth-order trend surface on the age (in millions of years) of the sediment at the top of each core (Fig. 1). See text for discussion of data limitations. Note that any surface veneer of Brunhes age sediment is excluded from the trend surface analysis.

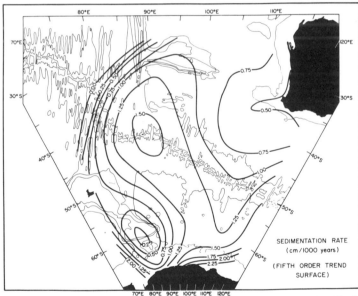

Figure 7. Map of fifth-order trend surfaces on the sedimentation rate (in cm/1,000 yr) for the period t = 0 to 2.5 m.y. The sedimentation rates used are those calculated for all cores (Figs. 1 and 4) except when hiatuses were encountered for the period involved. In cores that do not reach the Brunhes-Matuyama boundary (t = 0.69 m.y.), only minimum sedimentation rates can be calculated, using an arbitrarily assigned maximum age of 0.6 m.y. See text for limitations of the method.

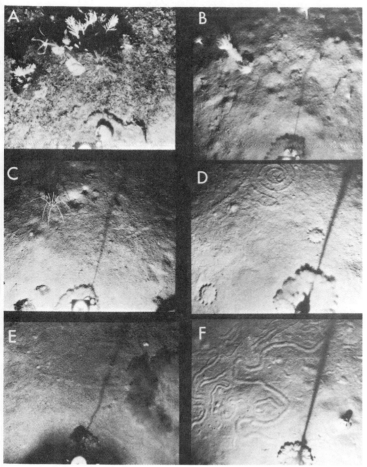

Figure 8. Bottom photography of the Southeast Indian Ocean indicative of weak to moderate bottom currents. A. Abundant organisms on firm, sandy, and rocky bottom (*Eltanin* 47-64 camera station 23, frame 7; 47°43'S, 73°26'E; 1,422 m). B. Organisms bowed over by bottom currents over moderately smoothed, sandy bottom with abundant organisms but little bioturbation (E47-24-9, f 4; 66°48'S; 77°54'E; 3,528 m). C. Smoothed, sandy bottom, manganese-coated rocks in lower segment, pycnogonid at upper left (E47-25-10, f 13; 66°22'S; 78°01'E; 2,412 m). D. Bioturbation marks on otherwise smooth featureless muddy bottom (E48-20-9, f 21; 39°55'S; 85°22'E; 3,231 m). E. Smooth, featureless bottom (E45-13-10, f 12; 40°09'S; 104°13'E; 3,947 m). F. Abundant bioturbation on muddy bottom (E48-43-17, f 21; 31°22'S; 92°36'E; 4,217 m).

(streamers) behind positive features such as lumps or feces (Fig. 9, D). Further velocity increase may lead to the development of ripple marks (Fig. 9, A, B, C), although this feature is relatively rare in the Southeast Indian Ocean bottom photographs. Some ripples exhibit winnowed sand, gravel, and manganese micronodules in their troughs (Fig. 9, A, B). Increasing current velocities are also associated with increased numbers of manganese nodules (Fig. 9, E, F; Fig. 10)

which lie on a sediment bottom. Invariably where sediment separates individual nodules, strong evidence exists of associated moderate to high current velocities with the smoothing of sediment (Fig. 10, C) or the development of crag and tail (Fig. 9, E; Fig. 10, E, F) or scoured moats (Fig. 10, E). Manganese nodules on the ocean floor may be common (Fig. 9, F), abundant (Fig. 10, C, E), or developed to a maximum to form a pavement surface (Fig. 10, A, B, D).

In this study, we infer high bottom currents associated with the occurrence of nodules; the highest velocities are inferred in areas with manganese pavements. Previous studies have shown clear relationships between manganese nodule development in the Southern Ocean and high bottom-current velocities. Hollister -and Heezen (1967) describe a linear belt of manganese nodules in the Drake Passage related to high-velocity, latitudinally restricted easterly flowing bottom currents. In the South Pacific, a belt of manganese nodules in the vicinity of the Antarctic Convergence is related to flow of the Circum-Antarctic Current (Goodell and Watkins, 1968). In the South Tasman Sea south of Tasmania, high-velocity currents have been directly responsible for the development of a vast manganese pavement (Watkins and Kennett, 1971, 1972; Conolly and Payne, 1972).

The qualitative use of bottom photograph features to enhance the mapping of current velocities has been successfully applied by several authors, including Heezen and Hollister (1964), Heezen and others (1966), and Watkins and Kennett (1972). We have plotted six progressively more dynamic categories of inferred bottom-current velocities in Figure 11. Areas with abundant bioturbation or with smooth bottoms occur throughout almost all of the mid-ocean ridge system and on Broken Ridge. These bottom features also occur in less extensive areas in the southeastern sector of the South Indian Basin and in areas close to the Antarctic continent (Fig. 11). In contrast, areas of sea floor that typically display distinct current lineations, scour, rippling, or manganese-nodule development occur extensively throughout the northwestern sector of the South Indian Basin and the South Australian Basin. These features also occur on the flanks and crest of the Kerguelen Plateau.

Direction of bottom-current flow based on compass bearings on oriented sediment features at 22 locations show quite random directions even in some closely adjacent sites. Thus the current direction measurements have been of little assistance in determining the general bottom current patterns in the area. The relative randomness of the current directions may merely reflect temporary directions on a constantly changing, small-scale pattern of bottom-current flow. Hollister and Heezen (1967), for instance, have shown that orientation of deep-sea ripples may differ radically in the same area and even in the area of one photograph. Bottom-current directions in some areas may be strongly influenced by very local topographic features which we infer must be superimposed on broad regional patterns.

Of particular interest in this regional analysis is the discovery of vast concentrations of manganese nodules in the northwestern sector of the South Australian Basin immediately south of Broken Ridge and Naturaliste Plateau (Fig. 11). A large part of this region is characteristically covered by dense packings of manganese nodules to form a manganese pavement. We name this feature the "Southeast Indian Ocean Manganese Pavement." The manganese nodules that make up much of this surface appear to be quite spherical, although in some areas, botryoidal nodules occur. If continuous distribution is assumed between the bottom photographs displaying these features (Fig. 11), the concentrated nodules may occur over an area as large as 10^6 km^2. The relative proximity of this pavement to the Perth-Fremantle industrial area of Western Australia must make it of some prospective economic interest. We believe that this pavement, like that of the Tasman Manganese Pavement (Kennett and Watkins, 1971, 1972; Conolly and Payne, 1972), has accumulated as a result of prolonged, high-velocity bottom currents. Other occurrences of abundant manganese nodules in the Southeast Indian Ocean are more scattered and are found mostly in the northwestern sector of the South Indian Basin. These do not appear to form large pavements as is the case in the northwestern sector of the South Australian Basin.

Bottom Currents
in the
Southeast Indian Ocean

Although bottom photographs and cores were not usually taken at the same locations, cores with sedimentary hiatuses, with few exceptions, occur in regions where there is photographic evidence of strong to very strong bottom current activity (Fig. 12). The availability of both parameters facilitates long-term principal paths of bottom water to be determined through the region. A major path of bottom water flow (Fig. 12) is inferred to cross the relatively shallow Kerguelen Plateau. In deeper areas, we infer major eastward flow of Antarctic Bottom Water traversing the northern part of the South Indian Basin and flowing northward over the Southeast Indian Ridge at lat 110°E and 120°E into the South Australian Basin. In this basin, the flow splits into eastward and northward branches, the latter of which provides Antarctic Bottom Water to the Wharton Basin by flowing between Broken Ridge and the Naturaliste Plateau. Directions of flow (Fig. 12) are largely inferred from known water mass-distribution patterns in basins farther to the east in the Southern Ocean (Gordon,

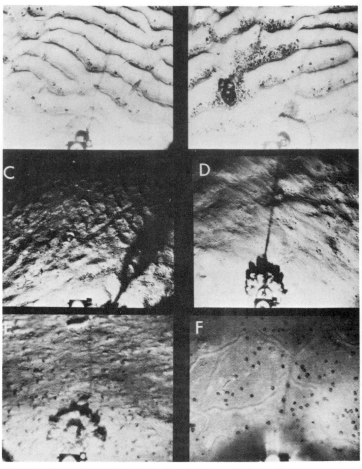

Figure 9. Bottom photography of the Southeast Indian Ocean indicative of strong to very strong bottom currents. A. Short crested asymmetrical ripple marks with a few manganese micronodules in troughs. Holothurian at right middle (E47-32, camera station 16, frame 22; 58°47'S; 84°13'E; 2,750 m). B. Short crested asymmetrical ripple marks with concentrations of manganese micronodules and sand in troughs (E47-32-16, f 8; 58°47'S; 84°13'E; 2,750 m). C. Ripple marks (E54-12-5, f 16; 53°45'S; 83°52'E; 4,574 m). D. Current-formed sediment lineations (E54-36-17, f 12; 48°25'S; 123°45'E; 4,241 m). E. Sediment tails formed behind partially buried manganese nodules. Holothurian at upper center (E47-36-20, f 19; 56°26'S; 80°13'E; 2,974 m). F. Common manganese nodules on sandy bottom with distinct bioturbation (E45-12-9, f 14; 38°58'S; 103°42'E; 4,302 m).

1971a) and from deep-sea topographic considerations.

It is clear (Fig. 12) that major bottom-current activity occurs over much of the crest and flanks of the Kerguelen Plateau to depths of 1,700 m, at least. The presence of several cores containing disconformities, some of which are of substantial age (Eocene to Cretaceous), is strong evidence for major bottom-water erosion over the

Kerguelen Plateau. While sediment of this age at or near the sediment-water interface could result from tectonic exposure, seismic profiles taken over the Kerguelen Plateau confirm that major sediment erosion has occurred (R. Houtz, 1975, personal commun.). Furthermore, bottom photographs from the crest and flanks of the Kerguelen Plateau display features created by strong to very strong currents (Fig. 11), even at

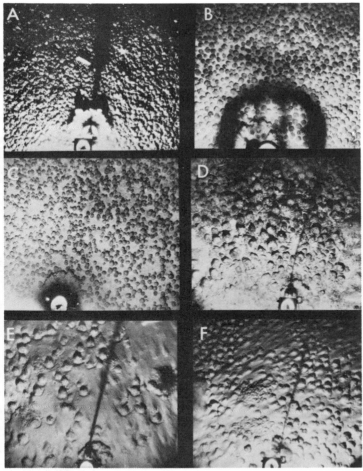

Figure 10. Bottom photography of manganese nodule fields in the Southeast Indian Ocean. All frames considered to be indicative of very strong bottom currents. A,D. Manganese pavement with extremely tightly packed nodules. A. Photo from the Southeast Indian Ocean Manganese Pavement. Holothurian near center (A. E54-20 camera station 8, f 10; 55°50′S; 81°02′E; 4,178 m. D. E49-29-24, f 15; 55°11′S; 94°50′E; 4,541 m). B,C. Very abundant spherical manganese nodules within sandy ooze sediment B. Ophiuroid at upper right (E45-57-30, f 5; 38°05′S; 114°31′E, 4,428 m. C. E44-48-33, f 48; 46°06′S; 119°53′E; 4,207 m). E,F. Abundant manganese nodules partially buried by sediment and forming distinct tails. Moating has formed around individual nodules in E (E. E49-16-14, f 19, 52°45′S; 90°0′S; 4,003 m. F. E49-28-23, f 20; 53°37′S; 95°09′E; 3,798 m).

ble northwest-to-southeast flow over the Kerguelen Plateau, based on salinity measurements. They further suggest that despite the relatively shallow water depths, some Antarctic Bottom Water component derived originally from the Weddell Sea area may be involved in the erosion of the Kerguelen Plateau. A major west-to-east fracture zone at a depth of about 5,000 m occurs south of the Kerguelen Plateau. Current meter evidence shows no major present-day bottom-water flow through this gap (M. R. Rodman and A. L. Gordon, 1975, personal commun.), possibly because it lies within the zone of the East Wind Drift (Fig. 12). This observation is critical because this fracture zone would seem to be an obvious passage for westward flow of Antarctic Bottom Water derived originally from the Weddell Sea region and feeding the South Indian Basin. Although no such flow is indicated by the current meter measurements, it does not rule out the possibility of such flow at times in the past. Luyendyk and Davies (1974) indicated the presence of bottom waters derived from the Weddell Sea in the Crozet Basin to the west of the Kerguelen Plateau.

In the South Indian Basin, major eastward current flow occurs in the northern sector (Fig. 12). Because of the considerable depth of this basin, this flow is almost certainly related to movement of Antarctic Bottom Water. Antarctic Bottom Water in the South Indian Basin originates almost entirely from the Ross Sea (Gordon, 1971c; M. R. Rodman and A. L. Gordon, 1975, personal commun.). This initially flows westward close to the Antarctic continent as part of the East Wind Drift (Gordon, 1971a; M. R. Rodman and A. L. Gordon, 1975, personal commun.). We infer a relatively weak westerly flow along the upper continental rise of Antarctica as in areas farther to the east (Hollister and Heezen, 1967; Watkins and Kennett, 1972; Gordon, 1971a). On contact with the north-trending Kerguelen Plateau, the Antarctic Bottom Water is initially diverted to the north and then to the east (Heezen and others, 1972) where the geologic evidence indicates strong erosive capacity in northern parts of the South Indian Basin. Strong eastward flow of Antarctic Bottom Water may be enhanced by the addition of high-velocity Antarctic Circumpolar Water. Such an important eastward-flowing bottom current is similar to that described by Hollister and Heezen (1967) in the northern part of the Bellingshausen Basin.

The South Indian Basin is separated from the South Australian Basin to the north by the Southeast Indian Ridge (Fig. 12). A major problem in mapping deep bottom-water circulation in the Southeast Indian

these relatively shallow depths. For instance, camera stations E47-32 and E47-36 from the Kerguelen Plateau reveal the presence of strong asymmetric ripple marks and sedimentary lineations that clearly indicate very strong, long-term, bottom-current action. The Kerguelen Plateau forms a major north-south–trending barrier to the meridional flow of deep Antarctic Bottom Water.

Thus the relatively shallow water erosion observed on the Kerguelen Plateau has probably resulted from high current velocities associated with eastward flow of the Antarctic Circumpolar Current. Although no bottom current data are available from the crest of the Kerguelen Plateau (Gordon, 1971b), M. R. Rodman and A. L. Gordon (1975, personal commun.) suggest a possi-

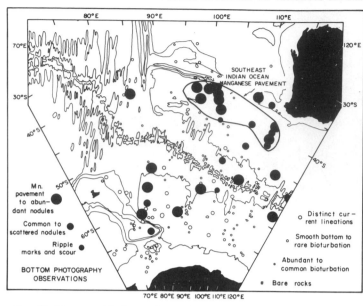

Figure 11. Map summarizing bottom photograph observations at camera-station sites. Symbols 1 to 6 indicate six inferred progressively less "dynamic" categories. Category 7 indicates bare rocks. Limits of the Southeast Indian Ocean Manganese Pavement discovered as the result of these observations are shown.

Ocean is the location of the supply route of bottom waters northward to the western sector of the South Australian and Wharton Basins, where the geologic evidence shows major erosive activity by deep bottom waters. The two major basins in the northern part of the Indian Ocean, the Bengal and Wharton Basins, receive separate flows of Antarctic Bottom Water from the south because of separation by the continuous north-trending Ninetyeast Ridge. As the Bengal Basin water is nearly a degree warmer, it must be separately derived (Heezen and others, 1972). There appear to be two possible supply routes to the South Australian–Wharton Basins. The first possibility is from the Crozet Basin via a broad sector of the mid-ocean ridge northeast of the Kerguelen Plateau where water depths range from about 3,000 to 5,000 m (Fig. 12). The alternative for bottom-water transport is northward from the South Indian Basin through some sector of the east-trending Southeast Indian Ridge (Fig. 12). No evidence suggests any major bottom-water transport northeast of the Kerguelen Plateau. On the contrary, this area appears to have experienced relatively quiescent bottom conditions which extend as far eastward as 110°E. Evidence for strong bottom current activity does occur, however, in narrow zones on the mid-ocean ridge at 110°E and at 120°E. This suggests that the

Figure 12. Diagram showing proposed long-term circulation patterns of bottom waters in the Southeast Indian Ocean. Inferred relative bottom current activities (Fig. 11) indicated by symbols. Core locations with dated disconformities (Fig. 5) are added. The location of Deep Sea Drilling Project sites in areas (Legs 26, 27, 28) are indicated, with distinction made between most cores with or without major disconformities (Luyendyk, Davies, and others, 1974; Veevers, Heirtzler and others, 1974; Hayes, Frakes, and others, 1973). Camera stations considered to represent very strong bottom-current activity include categories 1, 2, and 3 of Figure 11; strong activity includes category 4. See text for discussion of evidence for the circulation patterns of bottom water, as indicated by arrows.

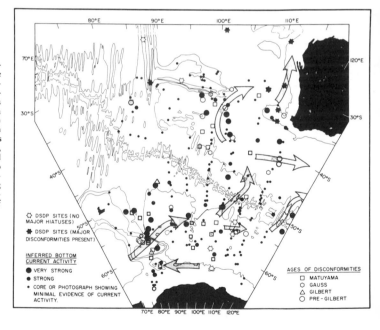

supply route of bottom water to the South Australian Basin may occur in this area (Fig. 12). This differs from the interpretation of Burckle and others (1973). On the basis of distribution of northward dispersion of Antarctic diatoms by Antarctic Bottom Water, they postulated a major northward flow between 95°E and 100°E; but all of the bottom photographs and piston cores we have analyzed from this area show no evidence of major bottom-current activity. On the other hand, if the proposed conduit is very narrow, its effect may not be observable in our data set.

In the western part of the South Australian Basin, Antarctic Bottom Water flows in two directions. A major eastward flow through the basin eventually supplies highly erosive bottom waters to the South Tasman Basin south and east of Tasmania (Watkins and Kennett, 1972). Another major current flows northward through the western sector, creating substantial erosion and the development of the Southeast Indian Manganese Pavement. This northward flow continues through a gap between the east-west-trending Broken Ridge and Naturaliste Plateau, which form major barriers to any northward deep flow. Broken Ridge does not appear to experience any major bottom-water activity, which probably all funnels through the Diamantina Fracture Zone at its southern edge. The Naturaliste Plateau is in contrast, however, in that it has experienced strong currents recently (according to our ocean-bottom photographic analysis) and in the past (according to DSDP evidence of disconformities; Davies and others, 1975; Veevers, Heirtzler and others, 1974). Northward flow of Antarctic Bottom Water through the gap between Broken Ridge and Naturaliste Plateau supplies the Wharton Basin with bottom water (Fig. 12). Within this basin and on the continental margin of Western Australia, the presence of highly condensed Cenozoic sequences or major Cenozoic disconformities (Davies and others, 1975; Veevers, Heirtzler and others, 1974) indicates that northward-flowing, erosive bottom waters have been a major long-term feature of the Southeast Indian Ocean.

History of Bottom-Water Activity in the Southeast Indian Ocean

In the Southeast Indian Ocean, sediments of the Gilbert epoch (t = 5.1 to 3.34 m.y.) are poorly represented because either erosion has not been adequate to allow sediments of this age to be sampled, or in some cases, because sediments of greater age were the youngest present. Thus we have been unable to estimate bottom-water history for the late Cenozoic prior to the Gauss epoch.

Ages of disconformities indicate increases in bottom-water activity during the last 2.5 m.y. Fillon (1972) showed an increase in bottom-current activity throughout the Ross Sea since near the end of the Gauss epoch (t = 2.4 m.y. B.P.). Disconformities of more than one age in several cores indicate at least two major pulses of bottom-water activity during this interval. One of these includes part of the Matuyama and late Gauss (t = 2.5 to 1.5 m.y.) while another pulse occurred during the Brunhes (t = 0.69 to 0 m.y.). In the present study of the 44 cores containing disformities, 24 have a veneer of Brunhes sediment. This sediment probably represents a relatively recent temporary waning of bottom-water activity, and renewed activity in the future may cause further erosion. In contrast, nine cores which lack even a 4-cm veneer of Brunhes sediment are evidence that erosion in these areas has continued to the present day. Six of these cores occur in the vicinity of the Kerguelen Plateau, suggesting that very strong currents may have persisted over the Kerguelen Plateau, in contrast to pulses of strong current, for which evidence exists in the adjacent deeper basins. This may be because the erosion observed on the Kerguelen Plateau is probably more closely related to Antarctic Circumpolar Current flow than to activity of Antarctic Bottom Water. During the late Cenozoic, the net long-term activity of the Circumpolar Current is likely to be relatively constant, since no radical continental configurations would have occurred to cause such changes. Erosion in the deeper basins, on the other hand, almost certainly reflects pulses in the production and velocity of Antarctic Bottom Water on a relatively short term basis because of changes in the glacial activity of Antarctica.

The presence of middle and early Cenozoic sediments exposed by the deep-sea erosion implies long-term bottom-current activity, because it seems unlikely that sediments of early Miocene or older age would be exposed by bottom-current pulses restricted to the latest Cenozoic. Moderate compaction of these sediments indicates that some unknown amount of overburden has been removed by the erosion. In the Wharton Basin, the presence of major disconformities which have removed much or all of the Cenozoic or that have limited Cenozoic sedimentation rates (Veevers, Heirtzler, and others, 1974) indicates that bottom-current activity in the region has prevailed for a considerable time during the Cenozoic. The late Cenozoic disconformities which we have defined using piston cores thus almost certainly represent the latest of a series of bottom-water events that have occurred for much of the Cenozoic. It is possible that deep-sea erosion in the Southeast Indian Ocean due to

activity of Antarctic Bottom Water has occurred spasmodically since at least the beginning of the Oligocene (38 m.y. B.P.); oxygen isotopic evidence suggests that formation of cold Antarctic Bottom Water, much like that in the present-day Antarctic region, commenced then (Kennett and others, 1972; Shackleton and Kennett, 1975; Luyendyk and Davies, 1974). If so, the hiatuses representing the early Cenozoic–Cretaceous in the Wharton Basin must have been formed by bottom waters from sources other than the Ross Sea, because the Australian-Antarctic passage did not open until about the late Oligocene (Kennett and others, 1974). The Weddell Sea area was probably the source of such bottom water, and thus the Ross Sea bottom waters have not always been dominant in the deep basins of the eastern Indian Ocean as they appear to be in the present day. The mechanisms for formation of Antarctic Bottom Water and for periodic pulses are not well understood. The extent of sea ice around Antarctica, the permanency of sea ice, and the development and extent of floating ice shelves may all play some role in governing the amount of Antarctic Bottom Water that eventually forms and flows through the deep basins adjacent to the Antarctic continent.

CONCLUSIONS

1. Paleomagnetic and micropaleontological dating of 187 deep-sea sedimentary cores in the Southeast Indian Ocean reveals a total of 44 cores containing disconformities of various ages and durations caused by erosion. These cores occur in three regions: the Kerguelen Plateau, the northern sector of the South Indian Basin, and the western sector of the South Australian Basin. Undisturbed sequences occur throughout most of the mid-ocean ridge system separating these two major basins.

2. Trend surface analysis on ages of surface sediments and sedimentation rates for sediments deposited during the past 2.5 m.y. reveal complementary patterns, with evidence of erosion and lower sedimentation rates occurring in the three regions marked by sedimentary hiatuses. Zones of erosion in deep seas clearly grade laterally into zones representing simple decreased sedimentation rates and grade upward and downward in time into sequences with corresponding sedimentation-rate changes.

3. Analysis of bottom photographs at 143 camera stations throughout the region also reveals that bottom features characteristic of very strong to strong bottom currents occur in the three regions marked by disconformities and that the mid-ocean ridge shows little evidence of dynamic bottom conditions. There can no longer be any doubt that sediments are far superior in

practice to current meters in yielding evidence of long-term current activities. The duration of direct observations required for integrating current meter observations to a meaningful long-term signal may well be orders of magnitude greater than feasible.

4. A major manganese nodule pavement with an area of about 10^6 km^2 has been found in the northwestern sector of the South Australian Basin. This is named the "Southeast Indian Ocean Manganese Nodule Pavement." Its development is associated with prolonged and intensive bottom-current activity in the area.

5. A model of long-term bottom-water flow is inferred from the distribution of erosional features determined from dating of cores and from bottom-photograph analysis. Important erosion occurs over much of the relatively shallow Kerguelen Plateau and is considered to be largely a result of scour by the Antarctic Circumpolar Current. In the deep South Indian Basin to the east, major eastward flow of Antarctic Bottom Water occurs in the northern sector, creating erosional features. Antarctic Bottom Water breaches the mid-ocean ridge at about 110°E and 120°E and flows into the South Australian Basin. Within this basin, a branch of Antarctic Bottom Water flows northward in the western sector to enhance development of the Southeast Indian Ocean Manganese Nodule Pavement and continues to flow north between Broken Ridge and Naturaliste Plateau into the Wharton Basin.

6. The presence of sediments of Late Cretaceous through middle Cenozoic age exposed by this deep-sea erosion suggests that highly erosive deep-sea currents have been active in the Southeast Indian Ocean for considerable parts of the Cenozoic. Sequences in the majority of the cores, which are Pleistocene and Pliocene in age, indicate an increase in velocities of deep bottom water during the past 2.5 m.y. This appears to have occurred in two main pulses: one during the late Gauss to early Matuyama, and the other during the late Brunhes.

7. Sea-floor dynamic processes and resulting geological features in the Southeast Indian Ocean are thus fairly typical of the Southern Ocean, resulting from the unique character of the Antarctic Circumpolar Current and the flow of Antarctic Bottom Water, although the latter effects will be found in the basins of the deep oceans of the world.

ACKNOWLEDGMENTS

The performances of the large number of personnel involved with the *Eltanin* program is greatly appreciated. The careful supervision and curating of the sediment cores by Dennis Cassidy of Florida State University have been critical for this and other core studies. We thank Betty Landrum and other personnel of the Smithsonian Oceanographic Sorting Center for kindly and efficiently providing large numbers of bottom photographs. We also thank Nancy Healy, Linda Steere, Nancy Meader, and Rosemarie Raymond for technical assistance.

These studies have been supported by the National Science Foundation: Grant OPP 71-04027 A04 (Office of Polar Programs) and DES74-19370 (Geological Oceanography) to James P. Kennett and GV-25400 and OPP 74-18529 (Office of Polar Programs) to Norman D. Watkins.

APPENDIX 1.
ADDITIONAL DATA ON INDIVIDUAL CORES

Traverse N–N'. No detectable disconformities are present. Because of a lack of biogenic material in E50-18 at all levels below the top sample, no paleontological age could be assigned. The paleomagnetism, however, indicates a Brunhes to Gauss age. Cores E50-17 and E44-24 from high latitudes contain rich radiolarian faunas, and the more northern cores E44-27 and E44-31 contain abundant planktonic foraminiferal faunas, facilitating good biostratigraphic control.

Traverse O–O'. Five cores have disconformities, all overlain by a thin layer of Brunhes age sediment. All other cores appear to show relatively constant sedimentation rates. All cores south of E45-69 have rich radiolarian faunas throughout, while most cores to the north of this are rich in planktonic foraminifera. The lower part of E45-89 contains little biogenic material, and the age is based on the measured polarities alone. In E45-78, the lowest part of the core contains *Globorotalia puncticulata*, suggesting a Matuyama age, and it therefore conflicts with the paleomagnetic data which are normal throughout, suggesting restriction within the Brunhes epoch. The intensity of magnetization is too low for measurement in E45-93, but the presence of *Globorotalia truncatulinoides* indicates a Quaternary age. The general lack of biogenic material in E45-92 prevented a dual age determination on this core. Normal magnetic polarity throughout suggests that the core is restricted to the Brunhes epoch, but suspected low sedimentation rates in such red clay suggest that the magnetic data for this 7.5-m-long core may reflect normal polarity overprint (Watkins and others, 1974). Four cores contain sediment older than 5 m.y. In E45-62, a disconformity that occurs between the top of the core and 240 cm probably lies close to 150 cm, where a polarity change occurs. This disconformity separates radiolarian-rich Brunhes epoch sediments that are found above from an early Miocene sequence below. The latter, despite much dissolution, contains an abundant low-diversity planktonic foraminiferal fauna, including *Catapsydrax unicavis, Globorotaloides sp., Globigerina bulloides, Globigerina quinqueloba, Neogloboquadrina pachyderma*, and *Globorotalia sp.* In E45-88, a disconformity occurs between 20 and 100 cm so that Brunhes epoch sediment overlies early Miocene sediment with a fauna dominated by *C. unicavis* and *Globigerina cf. woodi*. In E45-102, a thin veneer of Brunhes sediments is underlain disconformably by sponge-spicule-rich sediments of early or middle Miocene age that contain *Catapsydrax dissimilis*. No ice-rafted debris was observed in any of the pre-Pliocene core samples examined.

Traverse P–P'. Three of the cores (E49-1, E45-97, and E45-95) are from very deep water and contain little biogenic material. Ages are therefore based on the polarity stratigraphy. In E45-35, the magnetic data were erratic, but a rich radiolarian fauna of Brunhes age occurs throughout. Four cores contain hiatuses. A disconformity occurs in E49-5 between the top and 80 cm, at which level the presence of *G. puncticulata* indicates a Matuyama age. The base of the core contains manganese micronodules and other coarse debris with little fine material, indicating that the core terminated at the surface of another disconformity. Abundant sponge spicules occur throughout E48-53. Much dissolution of calcium carbonate occurs throughout this core, although the base has an abundant planktonic foraminiferal fauna of Pliocene age, including *Globigerinoides obliquus, Globoquadrina altispira, G. puncticulata, Globorotalia crassaformis*, and *Globorotalia cf. inflata.*

Traverse Q–Q'. Six cores contain hiatuses. A disconformity occurs in E50-2 within the 900- to 1,050-cm depth interval. Beneath this, an early Miocene sequence contains *Globoquadrina dehiscens, C. dissimilis*, and *G. cf. woodi.* In E45-16, the position of the disconformity is difficult to place because of rare biogenic material at critical levels, but we believe that it most probably occurs above a depth of 260 cm. The lower part of this core is early Oligocene in age, as indicated by the presence of *Globigerina angiporoides* and *Chiloguembelina cubensis* and the absence of characteristic Eocene forms (Jenkins, 1971). A middle Matuyama disconformity occurs in E50-13 (Weaver and McCollum, 1974). No magnetic data were obtained for the lower part of this section. Six cores in the northern part of this traverse are from deep water and contain very little biogenic material, and ages are based on the magnetic polarities alone. Although the lower part of E50-9 contains a late Matuyama radiolaria assemblage, this section is assigned to the early Matuyama–middle Matuyama based on the magnetic stratigraphy. The intensity of magnetization of the sediments in E45-5 is too low for satisfactory measurement, but the presence of *G. truncatulinoides* indicates a Quaternary age.

Traverse R–R'. Six of the cores in this traverse contain disconformities, five of which occur in the deep basin to the north. Although E49-48 has no polarity change, the radiolaria show the existence of a disconformity between 210 and 330 cm, separating Brunhes and Gauss age sediments. In E48-49, the biostratigraphic assignment is difficult because of scarce biogenic materials. The middle part of this core is assigned to the Gauss primarily because of the paleomagnetic data. Below this, sediments of early Oligocene age are identified by the rare presence of *G. angiporoides* and *C. dissimilis* and on the absence of characteristic Eocene forms. E48-48 is poorly dated: the lower segment is assigned to the Gauss epoch on the basis of a combination of normal polarity and an absence of *G. trun-*

catulinoides. Rich radiolarian faunas occur in all cores south of E49-4, but to the north, foraminiferal faunas dominate.

Traverse S–S'. The three northernmost cores in this traverse contain hiatuses. In E48-5, a disconformity between 350 and 450 cm separates Brunhes age from early Miocene age sediments which contain the planktonic foraminifera *G. dehiscens, Globoquadrina venezuelana, G. cf. woodi,* and rare *C. dissimilis.* In E48-6, the presence of a major lithologic change at approximately 700 cm suggests a possible disconformity. Above the boundary, planktonic foraminifera are present; below this, very little biogenic material is present except fish teeth and rare benthonic foraminifera. The age of the sediments below this sediment break could not be determined, although tentative assignment to the pre-Gilbert is made because, as described above, similar lithologic changes in this vicinity have been shown to represent major hiatuses. The assignment of a Brunhes age to E48-7 is based on its normal magnetic polarity in view of the fact that biogenic material is virtually absent. All other cores contain sufficient numbers of planktonic foraminifera for reliable age assignments.

Traverse T–T'. The cores in this traverse appear from the paleomagnetic polarities (Fig. 3) to exhibit fairly linear sedimentation rates with no apparent disconformities. This is, however, not the case, as Weaver and McCollum (1974) have demonstrated from detailed analyses of the diatom biostratigraphy. Three of the cores (E49-28, E49-29, and E49-30) contain a distinct disconformity of relatively short duration encompassing the middle Matuyama (about t = 1.3 to 2.0 m.y. B.P.). As stressed earlier, our methods will not allow detection of this. Three diatom zones are recognizable within the Matuyama epoch (Weaver and McCollum, 1974). This compares with two using radiolaria (Hays and Opdyke, 1967) and a single zone for the foraminifera (Keany and Kennett, 1972). Thus, diatoms provide a superior method for delineation of middle Matuyama disconformities. Abundant biogenic material occurs throughout each core, with rich radiolarian faunas in cores south of E49-28 and rich planktonic foraminiferal faunas in cores to the north.

Traverse U–U'. Rich radiolarian assemblages occur in cores south of E49-17; north of E49-16, planktonic foraminifera tend to dominate. Three cores in this traverse exhibit hiatuses. In E48-38, a disconformity exists above the depth of 165 cm, with Brunhes sediments above and late Miocene sediments below. The latter contain a planktonic foraminiferal fauna, including *Globigerina nepenthes, G. dehiscens,* and *Globorotalia conomiozea* but lacking Pliocene *Globorotalia.* The lower part of core E48-44 contains a reworked planktonic foraminiferal fauna of possible late Miocene age, a mixture of *G. dehiscens* and *Neogloboquadrina acostaensis* with a typical temperate Quaternary assemblage.

Traverse V–V'. This traverse includes cores rich in planktonic foraminifera throughout, although substantial calcium-carbonate dissolution exists in E48-34. Most cores are restricted to the Brunhes epoch. Presence of only tiny planktonic foraminifera in samples examined from E48-30 prevented a paleontological age determination; a Brunhes epoch age assignment is based on the measured polarities. A lack of *G. trun-*

catulinoides in the lower part of E48-36 indicates a possible pre-Quaternary age. A Brunhes age is preferred, however, because of the normal magnetic polarity and the presence of an otherwise Brunhes planktonic foraminiferal fauna. Only one core in this traverse (E48-37) displays a detectable disconformity. This occurs between 135- and 190-cm depth and separates Brunhes sediments from a sequence containing *G. nepenthes, G. conomiozea,* and *Sphaeroidinella subdehiscens.* The presence of *G. conomiozea* indicates an early Gilbert age (Kennett and Watkins, 1974).

Traverse W–W'. Seven cores in this traverse, the southern and central parts of which occur over the Kerguelen Plateau, contain hiatuses. The oldest sediments found are of Late Cretaceous age, in E54-7. A major disconformity in this core separates a veneer of Brunhes sediments from calcareous ooze of late Cenomanian and Turonian age (Quilty, 1973). An abundant and diverse planktonic foraminiferal fauna is dominated by *Hedbergella* and *Globigerinelloides.* As shown by Quilty, a high diversity of the planktonic foraminifera and high abundance of coccoliths indicate sediment accumulation below warm surface waters. No ice-rafted debris was observed in the samples examined. Two of the cores (E47-7 and E47-12) with disconformities are dated primarily on the basis of the magnetic data. In E47-12, although rich assemblages of *N. pachyderma* are present at various intervals, the radiolarian species diagnostic of the Matuyama epoch were not observed. The presence of abundant *D. spongiosa* at the base of the core requires, however, sediment of Gauss age. The middle and lower segments of core E47-7 contain a radiolarian fauna of relatively low diversity that makes faunal age assignments difficult. The presence of *Prunopyle cf. titan* and *Clathrocyclas cf. bicornis* suggests an age range within the Gilbert to Gauss epochs, which is used to interpret the paleomagnetic data as representing earliest Gauss to latest Gilbert polarities. Core E54-1 is from relatively shallow depths and thus contains abundant planktonic foraminifera and radiolaria. Much of this core contains an association of the planktonic foraminifer *G. puncticulata* with the radiolarians *Lychnocanium grande, Desmospyris spongiosa, Eucyrtidium calvertense,* and *Clathrocyclas bicornis.* Previously, *G. puncticulata* has been recorded only within the Matuyama epoch; it defines the *Globorotalia puncticulata* Zone (Keany and Kennett, 1972). The associations in this core therefore now suggest that this zone extends through the Gauss and into at least the latest part of the Gilbert epoch.

Traverse X–X'. This traverse is entirely over the Kerguelen Plateau. Half of the cores contain disconformities which are micropaleontologically distinct because of abundant, well-preserved radiolaria and foraminifera. Two of the cores (E54-10 and E45-15) display more than one disconformity separated by late Matuyama sediments. Two additional cores (E54-3 and E54-4) are of Eocene age and have no young sediment even in the upper 2 cm. E54-3 contains a moderately diverse, abundant planktonic foraminiferal assemblage of late Eocene age as indicated by an association of *Globigerapsis index, Globigerina linaperta,* and *Chiloguembelina cubensis* (Jenkins, 1971). A few

well-preserved radiolaria are also present. No ice-rafted debris was observed. In E54-4, the lower and middle segments contain a diverse and abundant fauna, including *Acarinina primitiva, Globigerina angiporoides, Truncorotaloides collactea, C. cubensis,* and *G. linaperta,* of early middle Eocene age (Jenkins, 1971). The upper part of this core contains a similar fauna which, with the addition of *G. index,* indicates a late middle Eocene age (Jenkins, 1971).

Two of the cores in this traverse have ages based on magnetic polarities alone: E47-3 contains no foraminifera or radiolaria which can be used for age assignment, and E47-22 contains only radiolaria of Brunhes age throughout, so that the late Matuyama age at the base of the core is deduced from the measured reversed polarity.

REFERENCES CITED

Burckle, L. H., Abbot, W. H., and Maloney, J., 1973, Sediment transport by Antarctic bottom water in the Southeast Indian Ocean: EOS (Am. Geophys. Union Trans.), v. 54, p. 336.

Conolly, J. R., and Payne, R. R., 1972, Sedimentary patterns within a continent–mid continent, ridge-continent profile: Indian Ocean south of Australia: Antarctic Research Ser., v. 19, p. 295–315.

Cox, A., 1969, Geomagnetic reversals: Science, v. 163, p. 237–245.

Davies, T. A., Weser, O. E., Luyendyk, B. P., and Kidd, R. B., 1975, Unconformities in the sediments of the Indian Ocean: Nature, v. 253, p. 15–19.

Fillon, R. H., 1972, Evidence from the Ross Sea for widespread submarine erosion: Nature Phys. Sci., v. 238, p. 40–42.

Frakes, L. A., 1971, USNS *Eltanin* sediment descriptions, cruises 32–45: Sedimentology Research Lab., Dept. Geology, Florida State Univ.

——1973, USNS *Eltanin* sediment descriptions, cruises 47 to 54: Sedimentology Research Lab., Dept. Geology, Florida State Univ.

Geitzenauer, K. R., 1972, The Pleistocene calcareous nannoplankton of the subantarctic Pacific Ocean: Deep-Sea Research, v. 19, p. 45–60.

Goodell, H. G., and Watkins, N. D., 1968, The paleomagnetic stratigraphy of the Southern Ocean: 20° West to 160° East longitude: Deep-Sea Research, v. 15, p. 89–112.

Gordon, A. L., 1971a, Oceanography of Antarctic waters, *in* Reid, J. L., ed., Antarctic oceanography I: Antarctic Research Ser., v. 15, p. 169–203.

——1971b, *Eltanin* physical oceanography, Cruises 44, 45, and 47: Antarctic Jour. U.S., v. 6, p. 166–167.

——1971c, Spreading of Antarctic bottom waters, 2, *in* Gordon, A. L., ed., Studies in physical oceanography — A tribute to Georg Wust on his 80th birthday: New York, Gordon and Breach, p. 1–18.

Hayes, D. E., Frakes, L. A., and others, 1973, Leg 28 deep-sea drilling in the southern ocean: Geotimes, v. 18, p. 19–24.

Hays, J. D., and Opdyke, N. D., 1967, Antarctic radiolaria, magnetic reversals and climatic changes: Science, v. 158, p. 1001–1011.

Hays, J. D., Lozano, J. A., Shackleton, N., and Irving, G., 1976, Reconstruction of the Atlantic and western Indian Ocean sectors of the 18,000 B.P.-Antarctic Ocean, in Cline, R. M., and Hayes, J. D., eds., Investigation of the late Quaternary paleoceanography and paleoclimatology: Geol. Soc. America Mem. 145 (in press).

Heezen, B. C., and Hollister, C. D., 1964, Deep-sea current evidence from abyssal sediments: Marine Geology, v. 1, p. 141–174.

——1971, The face of the deep: New York, Oxford Univ. Press, 659 p.

Heezen, B. C., Hollister, C. D., and Ruddiman, W., 1966, Shaping of the continental rise by deep geostrophic contour currents: Science, v. 152, p. 502–508.

Heezen, B. C., Tharp, M., and Bentley, C. R., 1972, Morphology of the Earth in the Antarctic and Subantarctic: Am. Geog. Soc., Antarctic Map Folio Ser., Folio 16.

Hollister, C. D., and Heezen, B. C., 1967, The floor of the Bellingshausen Sea, in Hersey, J. B., ed., Deep-sea photography: Johns Hopkins Oceanog. Studies 3, p. 177–189.

Houtz, R. E., 1975, South Tasman Basin and borderland: A summary of seismic data, in Kennett, J. P., Houtz, R. E., and others, Initial reports of the Deep Sea Drilling Project, Vol. 29: Washington D.C., U.S. Govt. Printing Office.

Jacobs, S. S., Bruchhausen, P. M., and Bauer, E. B., 1970, Eltanin reports, cruises 32-36, 1968: Hydrographic station, bottom photographs, current measurements: Lamont-Doherty Geol. Observatory, Columbia Univ., 460 p.

Jacobs, S. S., Bruchhausen, P. M., Rosselot, F. L., Gordon, A. L., Amos, A. F., and Belliard, M., 1972, Eltanin reports, cruises 37-39, 1969, 42-46, 1970: Hydrographic stations, bottom photographs, current measurements, nephelometer profiles: Lamont-Doherty Geol. Observatory, Columbia Univ., 490 p.

Jenkins, D. G., 1971, New Zealand Cenozoic planktonic foraminifera: New Zealand Geol. Survey Paleont. Bull. 42, 278 p.

Johnson, D. A., 1972, Ocean floor erosion in the Equatorial Pacific: Geol. Soc. America Bull., v. 83, p. 3121–3144.

——1974, Deep Pacific circulation: Intensification during the early Cenozoic: Marine Geology, v. 17, p. 71–78.

Keany, J., and Kennett, J. P., 1972, Pliocene–early Pleistocene paleoclimatic history recorded in Antarctic-Subantarctic deep-sea cores: Deep-Sea Research, v. 19, p. 529–548.

Kennett, J. P., 1970, Pleistocene paleoclimates and foraminiferal biostratigraphy in subantarctic deep-sea cores: Deep-Sea Research, v. 17, p. 125–140.

Kennett, J. P., and Brunner, C. A., 1973, Antarctic late Cenozoic glaciation: Evidence for initiation of ice-rafting and inferred increased bottom water activity: Geol. Soc. America Bull., v. 84, p. 2043–2052.

Kennett, J. P., and Watkins, N. C., 1974, Late Miocene–early Pliocene paleomagnetic stratigraphy, paleoclimatology, and biostratigraphy in New Zealand: Geol. Soc.

America Bull., v. 85, p. 1385–1398.

Kennett, J. P., Burns, R. E., Andrews, J. E., Churkin, M., Davies, T. A., Dumitrica, P., Edwards, A. R., Galehouse, J. C., Packham, G. H., and Van der Lingen, G. J., 1972, Australian-Antarctic continental drift, paleocirculation changes and Oligocene deep-sea erosion: Nature Phys. Sci., v. 239, p. 51–55.

Kennett, J. P., and others, 1974, Development of the Circum-Antarctic Current: Science, v. 186, p. 144–147.

Kennett, J. P., Houtz, R. E., and others, 1975, Initial reports of the Deep Sea Drilling Project, Vol. 29: Washington, D.C., U.S. Govt. Printing Office, 1197 p.

Kennett, J. P., Houtz, R. E., Andrews, P. B., Edwards, A. R., Gostin, V. A., Hajos, M., Hampton, M. A., Jenkins, D. G., Margolis, S. V., Ovenshine, A. T., and Perch-Nielsen, K., 1975, Cenozoic paleoceanography in the southwest Pacific Ocean, Antarctic glaciation, and the development of the Circum-Antarctic Current, in Kennett, J. P., Houtz, R. E., and others, Initial reports of the Deep Sea Drilling Project, Vol. 29: Washington, D.C., U.S. Govt. Printing Office, p. 1155–1169.

Kent, D. V., and Lowrie, W., 1974, Origin of magnetic instability in sediment cores from the central north Pacific: Jour. Geophys. Research, v. 79, p. 2987–3000.

Luyendyk, B. P., and Davies, T. A., 1974, Results of DSDP Leg 26 and the geologic history of the southern Indian Ocean, in Davies, T. A., Luyendyk, B. P., and others, Initial reports of the Deep Sea Drilling Project, Vol. 26: Washington, D.C., U.S. Govt. Printing Office, p. 909–943.

McCollum, D. W., 1975, Antarctic Cenozoic diatoms, Leg 28, Deep Sea Drilling Project, in Hayes, D. E., Frakes, L. A., and others, eds., Initial reports of the Deep Sea Drilling Project, Vol. 28: Washington, D.C., U.S. Govt. Printing Office, p. 515–571.

Miyajima, M. H., 1974, Absolute chronology of upper Pleistocene calcareous nannofossil zones of the southeast Indian Ocean: Antarctic Jour. U.S., v. 9, p. 261–263.

Moore, T. C., van Andel, Tj. H., Sancetta, C., and Pisias, N., 1975, Cenozoic hiatuses in pelagic sediments: Internat. Planktonic Cont., 3rd (in press).

Opdyke, N. D., Glass, B., Hays, J. D., and Foster, J., 1966, Paleomagnetic study of Antarctic deep-sea cores: Science, v. 154, p. 349–357.

Quilty, P. G., 1973, Cenomanian-Turonian and Neogene sediments from northeast of Kerguelen Ridge, Indian Ocean: Geol. Soc. Australia Jour., v. 20, p. 361–370.

Rona, P. A., 1973, Worldwide unconformities in marine sediments related to eustatic changes of sea level: Nature Phys. Sci., v. 244, p. 25–26.

Shackleton, N. J., and Kennett, J. P., 1975, Paleotemperature history of the Cenozoic and the initiation of Antarctic glaciation: Oxygen and carbon isotope analyses in DSDP Sites 279, 277, and 281, in Kennett, J. P., Houtz, R. E., and others, eds., 1975, Initial reports of the Deep Sea Drilling Project, Vol. 29: Washington, D.C., U.S. Govt.

Printing Office.

Simmons, K. L., and Landrum, B. J., 197 , Sea-floor photographs: Antarctic Jou U.S., v. 8, p. 128–133.

van Andel, Tj. H., Heath, G. R., and Moor T. C., 1975, Cenozoic history an paleoceanography of the central equatoria Pacific Ocean: Geol. Soc. America Mem 143 (in press).

Veevers, J. J., Heirtzler, J. R., and others, 197 Initial reports of the Deep Sea Drilling Project, Vol. 27: Washington, D.C., U.S. Gov Printing Office.

Vella, P., and Watkins, N. D., 1975, Middle an late Quaternary paleomagnetism and bio stratigraphy of four subantarctic deep-se cores from southwest of Australia (O. I Bandy Memorial volume): Los Angeles Univ. Southern California Press.

Vella, P., Ellwood, B. B., and Watkins, N. D 1975, Surface temperature changes in th Southern Ocean southwest of Australi during the last one million years: Roya Soc. New Zealand Spec. Bull. (A selectio of papers presented at the IX INQUA Cong., Christchurch, New Zealand, 1973)

Vestine, E. H., LaPorte, L., Cooper, C., Lange, I. and Hendrix, W. C., 1948, Description o the Earth's main magnetic field and its secu lar change, 1905-1945: Carnegie Inst Washington Pub., 578 p.

Watkins, N. D., and Goodell, H. G., 1967 Confirmation of the reality of the Gils geomagnetic polarity event: Earth an Planetary Sci. Letters, v. 2, p. 123–129.

Watkins, N. D., and Kennett, J. P., 1971, Antarc tic bottom water: Major change in velocit during the late Cenozoic between Australi and Antarctica: Science, v. 173, p 813–818.

——1972, Regional sedimentary disconformitie and upper Cenozoic changes in bottom water velocities between Australia and An tarctica: Antarctic Research Ser., v. 19, p 273–293.

Watkins, N. D., Kester, D. R., and Kennett, J. P 1974, Paleomagnetism of the typ Pliocene-Pleistocene boundary section a Santa Maria de Catanzaro, Italy, and the problem of post-depositional precipitation of magnetic minerals: Earth and Planetary Sci. Letters, v. 24, p. 113–119.

Weaver, F. M., and McCollum, D. W., 1974 Sedimentary hiatus in the south Indian Basin: Antarctic Jour. U.S., v. 9, p 250–251.

Williams, D. F., and Johnson, W. C., 1974, Late Pleistocene paleotemperature model for th southern Indian Ocean: Antarctic Jour U.S., v. 9, p. 260–261.

——1975, Diversity of recent planktonic foraminifera in the southern Indian Ocean and late Pleistocene paleotemperatures Quaternary Research, v. 5, p. 237–250.

MANUSCRIPT RECEIVED BY THE SOCIETY JANUARY 20, 1975
REVISED MANUSCRIPT RECEIVED JUNE 2, 1975
MANUSCRIPT ACCEPTED JUNE 23, 1975

Part VII

SPECULATIONS: ARE POLARITY CHANGES LINKED TO OTHER PHENOMENA?

Editor's Comments
on Papers 29 Through 34

Part VII discusses several contributions that concern possible relationships between polarity changes and other phenomena. At this time these contributions can be considered to be of a more speculative nature, and the proposed interrelationships are not necessarily widely accepted. A healthy skepticism reigns over most of these ideas, a skepticism that in large measure results from a rather limited data basis in support of the suggested correlations. Several phenomena have been suggested to be related directly or indirectly to paleomagnetic reversals. These include faunal and floral extinctions at times of paleomagnetic reversals resulting either from increased cosmic radiation at the earth's surface or indirectly by paleoclimatic change triggered by the paleomagnetic reversals. Other suggested correlations include increased volcanism resulting from intensification of upper mantle activity, which in turn may lead to paleoclimatic changes by increasing volcanic ash in the atmosphere; precession or obliquity of the earth; and suggested correlations between variations in the intensity and inclination of the magnetic field, atmospheric radiocarbon activity, and climatic changes at the earth's surface. These speculations require some discussion here because to a large extent the problems posed form a basis for important future research in this field. Most of the problems require studies of stratigraphic sequences. Much new data still needs to be accumulated in the study of all these theories. Quantitative data are especially required, which thus severely reduce the speed of the research. Several parameters require study in the same stratigraphic sections to reduce intercalibration problems. If any of the associations are found to be valid, they are of great significance in the understanding of the development of the earth and its biota.

The first three papers (Papers 29, 30, and 31) concern the intriguing possibility that faunal and floral extinction may in part be related to changes in the earth's magnetic field. The more general question of the causes of animal and plant extinctions through geologic time has long been studied by many workers. The formulation of adequate explanations still ranks among the most important unanswered geological problems. During the Phanerozoic, times of dramatic extinctions of organisms occurred at the end of the Cambrian, Ordovician, Devonian, Permian, Triassic, and Cretaceous periods. Other intervals exhibit less dramatic but nevertheless severe faunal extinction. Indeed, the intervals marked by faunal change have provided the definitions of the boundaries of the stratigraphic subdivisions used by geologists during the last two centuries. Many theories have been proposed to explain these extinctions. A rather recent theory has suggested that these extinctions may have been linked in some way with the changing tempo of magnetic polarity changes of the earth. For instance, the faunal extinctions near the end of the Cretaceous and Permo-carboniferous periods also coincide with the close of long intervals of dominantly one magnetic polarity. This problem has in part been addressed in Paper 29 by Kent, who has examined magnetostratigraphy in relation to planktonic foraminiferal changes over an apparently continuous Cretaceous-Tertiary boundary section in the pelagic limestones of the Scaglia

Rossa at Gubbio, northern Italy. In this section, the Cretaceous-Tertiary boundary, as clearly marked by changes in planktonic foraminifera, occurs within polarity zone Gubbio G-, which is correlated to the reversed interval between anomalies 29 and 30 in the oceanic magnetic anomaly reversal time scale (Alvarez et al., Paper 17). The duration of this polarity zone is estimated at 4.7×10^5 years based on assumptions concerning the degree of linearity of the magnetic reversal sequence and on ages chosen for its calibration. The dramatic planktonic foraminiferal change at the Cretaceous-Tertiary boundary in this section was computed by Kent to be only 10,000 years in duration, an extremely brief event in light of its magnitude. Of particular importance, however, is that this planktonic foraminiferal change and hence the Cretaceous-Tertiary boundary, is not coincident with any particular polarity change in the Gubbio section, thus eliminating any simple explanation based on a relationship between faunal change and possible effects on organisms of such a reversal.

In the early 1960s, during the early phases of development of the polarity time scale, it was suggested by Uffen (1963) that evolutionary radiations should occur at every polarity reversal of the earth's magnetic field. He suggested that during a reversal, the field intensity is much reduced for a relatively short interval of time and, as a result, cosmic radiation would increase at the earth's surface, which in turn would produce high mutation rates and rapid evolution. This theory was formulated by Uffen in the absence of any stratigraphic data between paleomagnetic reversals and faunal extinctions. Uffen's work stimulated others to examine microfossil extinctions in polar to tropical deep-sea cores in relation to magnetic polarity history. This research resulted in a number of contributions involving the collaboration of paleontologists and geophysicists.

The first such contribution with data to address this question was that of Opdyke et al. (Paper 6), who showed clear relationships between the extinction of several radiolarian species and times of paleomagnetic reversals. Since this study, several others have appeared that also support the idea of at least some faunal and floral extinctions being related to paleomagnetic reversals. Much of this evidence has been summarized in Paper 30 and in Paper 31. Data and discussions on this subject are presented by Harrison and Funnell (Paper 5); Opdyke et al. (Paper 6); Hays and Opdyke (Paper 8); Hays and Donahue (1972); and Hays et al. (Paper 10). The contribution of Hays and others is of particular importance because it examined extinctions within several microfossil groups in relation to paleomagnetic reversals in equatorial Pacific cores of Quaternary and Pliocene age. They discovered that a high proportion of extinctions are stratigraphically closely related to paleomagnetic reversals, specifically 7 of 8 foraminiferal extinctions, 2 of 3 radiolarian extinctions, and 2 of 5 diatom extinctions.

The most critical and detailed examination of possible relationships between faunal extinctions and paleomagnetic reversals was conducted by Hays in 1971 (Paper 31). This contribution is the only one that has quantitatively examined microfossil changes in relation to the paleomagnetic reversals. Faunal counts of a minimum of 1000 individuals of radiolarians were made in selected high-latitude cores from the Northern and Southern Hemispheres and from the equatorial region. The statistical basis used in this study allowed a number of new and useful observations to be made. Species disappear abruptly, they do not decline gradually. The upper limit or final extinction of a species is nearly isochronous over broad areas and does not seem to depend on latitude or longitude. Six of the 8 radiolarian species that became extinct within the time interval examined do so close to stratigraphic levels also marked by reversals in the earth's magnetic field.

It needs to be emphasized that it is not only faunal extinction that is known to occur in close association with paleomagnetic reversals but also the appearance of certain species. Kennett (1970) and Keany and Kennett (Paper 20) have reported that in Subantarctic and northern Antarctic areas, the most important changes exhibited in planktonic foraminiferal assemblages during at least the last 2 m.y. occurred close to the Brunhes-Matuyama boundary. This boundary is marked by the extinction of a single planktonic foraminifer *Globorotalia puncticulata* and the appearance of two planktonic formainifera *Globorotalia inflata* and *Globorotalia crassaformis*. All these changes are rather abrupt, and none exhibit evolutionary gradations. The Brunhes-Matuyama boundary is also marked by the disappearance in the Subantarctic-Antarctic areas of two radiolarian species that persist in present-day warmer oceanic regions: *Saturnalis circularis* and *Pterocanium trilobum*. Keany and Kennett (Paper 20) have suggested that the disappearance of these two radiolaria in the Southern Ocean and their persistence in warmer areas is due to a loss of the previous environmental tolerance for Southern Ocean conditions. Why this should occur close to a paleomagnetic reversal is unknown. The faunal changes documented by Hays and Opdyke (1967), Kennett (1970), and Keany and Kennett (Paper 20) close to the Brunhes-

Matuyama boundary do not appear to be related to any paleo-climatic change that would be of sufficient magnitude to cause them.

How paleomagnetic reversals might have affected organisms in the past remains a matter of speculation at this time. Hays (Paper 31) critically discusses various theories that have been proposed, although the one that has received greatest attention is attributed to Uffen (1963). The theoretical basis of Uffen's proposal has been critically examined by Black (Paper 32), Waddington (1967), and Harrison (1968). All these workers have theoretically considered the increase in radiation expected if the geomagnetic field were removed and concluded that radiation increases at the earth's surface during a paleomagnetic reversal should be negligible. Black (Paper 32) correctly pointed out that an increased mutation rate is unlikely to create increased evolutionary changes in the absence of other environmental pressures.

Thus a theory was originally proposed in the absence of data predicting extinctions of organisms at times of paleomagnetic reversals. This stimulated the generation of a set of tantalizing stratigraphic data gained from the study of deep-sea sedimentary sequences supporting the theory. This, however, was followed by theoretical studies that have proposed that increases in cosmic radiation during polarity reversals are not high enough to cause significant faunal change. No compelling evidence has yet been presented to demonstrate that paleoclimatic change of sufficient magnitude at time of paleomagnetic reversals could have caused the extinctions as suggested by Harrison (1968). Harrison and Prospero (1974) suggested that paleoclimatic change may result from reversals of the earth's magnetic field. Previously Roberts and Olson (1973) had found that increased ionization in the upper atmosphere resulting from geomagnetic disturbances could cause ion-induced nucleation and hence enhanced formation of cirrus clouds. This observation led Harrison and Prospero (1974) to suggest that during the 1,000 to 10,000 year period of reduced field intensity during a paleomagnetic reversal, profound climatic change would result due to increased cloud cover at low latitudes. Surprisingly little detailed information exists on stratigraphic relations between paleoclimatic change and paleomagnetic reversals. These data need to be generated. At this time no workers have strongly questioned the data that indicate relations between extinctions and paleomagnetic reversals, and thus these observations still lack an acceptable theory to explain them.

Several other theories have been proposed that link paleomagnetic reversals with other terrestrial phenomena including paleoclimatic change, increased mantle activity, and increased volcanic activity. For instance, Heirtzler (1970) speculated that increased earthquake activity and related increased upper mantle activity may somehow be linked with paleomagnetic changes. Heirtzler suggested that a sufficiently large earthquake may cause a wobble of the earth's spin axis and create a magnetic reversal. Kennett and Watkins (Paper 33) extended the thesis of Heirtzler by suggesting that an increase in upper mantle activity should be manifested by increased volcanic activity, and thus evidence of such activity should be found at times of paleomagnetic reversals. Kennett and Watkins presented evidence from Southern Ocean late Cenozoic deep-sea sediment sequences indicating increased amounts of volcanic ash close to paleomagnetic reversals. They also presented other tantalizing, although inconclusive, evidence from other regions, including oceanic islands, that increased volcanic activity occurred close to paleomagnetic reversals. The theory was expanded by suggesting that if increased episodes of explosive volcanicity have occurred at specific times in the past, these possibly effected global climates and in turn faunal change. Since first suggested by Benjamin Franklin, it has long been considered that increased volcanic activity may have played some role in climatic change (Lamb, 1972; Budyko, 1974). However, at this time little evidence exists in support of significant paleoclimatic changes associated with paleomagnetic reversals, although as Hays (Paper 31) stated, such possible relationships cannot yet be dismissed.

Empirical data have been presented in support of yet other relationships between magnetic reversals and other phenomena. Wollin and his colleagues at Lamont-Doherty Geological Observatory have suggested that changes in magnetic intensity and inclination have affected paleoclimatic changes in response to changes and flux of cosmic rays into the upper atmosphere as the magnetic field strength decreased (Wollin et al., 1971a, b; 1977). Shortly after these ideas were proposed, Amerigian (1974) disputed them on the grounds that the paleomagnetic changes identified by Wollin and colleagues did not directly record changes in the earth's magnetic field at the time of sediment deposition. Instead, he suggested that intensity changes in the cores reflect paleoclimatically induced sediment compositional changes, while inclination changes reflect paleoclimatically induced hydrodynamic changes during sediment deposition. Another theory has linked the precession of the earth with changes in the dynamic elipticity of the mantle and core of the earth, and hence in turn suggests that the precession of the earth may have played some role in creating paleomagnetic reversals (Bucha, 1973). A useful critical examination of the components that make up some of these theories has been made by Chappel (Paper 34), who on theoretical grounds has largely dismissed most of the proposed interrelationships. Nevertheless, Chappel pleaded for the production of greater amounts of detailed data from deep-sea sedimentary sequences; for instance, integrating paleomagnetic, paleoclimatic, sedimentological, and volcanic data, and by further studies to eliminate chronological discrepancies between different sections. A healthy skepticism concerning most of these theories prevails in the scientific community. Nevertheless, some tantalizing correlations remain.

REFERENCES

Amerigian, C. 1974. Sea-floor dynamic processes as the possible cause of correlations between paleoclimatic and paleomagnetic indices in deep-sea sedimentary cores. *Earth and Planetary Sci. Letters* **21**: 321–326.

Bucha, V. 1973. Correlation between variations of the geomagnetic field and precession of the earth in the Quaternary. *Nature Phys. Sci.* **244**: 108–109.

Budyko, M. I. 1974. *Climate and Life*, D. H. Miller, trans. New York: Academic Press, 309 pp.

Harrison, C. G. A. 1968. Evolutionary processes and reversals of the earth's magnetic field. *Nature* **217**:46–47.

Harrison, C. G. A., and J. M. Prospero. 1974. Reversals of the earth's magnetic field and climatic change. *Nature* **250**:563–565.

Hays, J. D., and J. G. Donahue. 1972. Antarctic Quaternary climatic record and radiolarian and diatom extinctions. In *Antarctic Geology and Geophysics*, International Union of Geological Sciences, Ser. B, No. 1, edited by R. J. Adie. Oslo: Universitetsforlaget, pp. 733–738.

Heirtzler, J. R. 1970. The paleomagnetic field as inferred from marine magnetic studies. *Jour. Geomagnetism and Geoelectricity* **22**:197–211.

Kennett, J. P. 1970. Pleistocene paleoclimates and foraminiferal biostratigraphy in Subantarctic deep-sea cores. *Deep-Sea Research* **17**:125–140.

Lamb, H. H. 1972. *Fundamentals and Climate Now. Climate: Present, Past and Future*, Vol. I. London: Methuen and Co., 613 pp.

Roberts, W. O., and R. H. Olson. 1973. Geomagnetic storms and wintertime 300-mb trough development in the North Pacific-North America area. *Jour. Atmos. Sci.* **30**:135–140.

Uffen, R. J. 1963. Influence of the earth's core on the origin and evolution of life. *Nature* **198**:143.

Waddington, C. J. 1967. Paleomagnetic field reversals and cosmic radiation. *Science* **158**:913–915.

Wollin, G., D. B. Ericson, and W. B. F. Ryan. 1971a. Variations in magnetic intensity and climatic changes. *Nature* **232**:549–550.

Wollin, G., D. B. Ericson, W. B. F. Ryan, and J. H. Foster. 1971b. Magnetism of the earth and climatic changes. *Earth and Planetary Sci. Letters* **12**:175–183.

Wollin, G., W. B. F. Ryan, D. B. Ericson, and J. H. Foster. 1977. Paleoclimate, paleomagnetism and the eccentricity of the earth's orbit. *Geophys. Research Letters* **4**:267–270.

ADDITIONAL READINGS

Barbetti, M. 1972. Geomagnetic field behaviour between 25,000 and 35,000 yr B.P. and its effect on atmospheric radiocarbon concentration: A current report. *Proceedings of the 8th International Conference on Radio Carbon Dating*, Wellington, New Zealand, October, 1972, pp. A106–A113.

Hays, J. D. 1972. Faunal extinctions and reversals of the earth's magnetic field: Reply. *Geol. Soc. America Bull.* **83**:2215–2216.

Hays, J. D., and N. J. Shackleton. 1976. Globally synchronous extinction of the radiolarian *Stylatractus universus*. *Geology* **4**:649–652.

Kellogg, D. E., and J. D. Hays. 1975. Microevolutionary patterns in Late Cenozoic radiolaria. *Paleobiology* **1**:150–160.

King, J. W. 1974. Weather and the earth's magnetic field. *Nature* **247**:131–134.

Kulkarni, R. N. 1963. Relation between atmospheric ozone and geomagnetic disturbances. *Nature* **198**:1189–1191.

McCormac, B. M., and J. E. Evans. 1969. Consequences of very small planetary magnetic moments. *Nature* **223**:1255.

McElhinny, M. W. 1971. Geomagnetic reversals during the Phanerozoic. *Science* **172**:157–159.

Newell, N. D. 1963. Crises in the history of life. *Sci. American* **208**:77.

Olausson, E., and B. Svenonius. 1973. The relation between glacial ages and terrestrial magnetism. *Boreas* **2**:109–115.

Simpson, J. F. 1966. Evolutionary pulsations and geomagnetic polarity. *Geol. Soc. America Bull.* **77**:197–203.

29

Reprinted from Geology 5:769–771 (1977)

AN ESTIMATE OF THE DURATION OF THE FAUNAL CHANGE AT THE CRETACEOUS-TERTIARY BOUNDARY

D. V. Kent

An episode of drastic faunal change is recorded at the close of the Cretaceous Period and is characterized by the seemingly abrupt, possibly synchronous worldwide disappearance of a major group of marine and terrestrial organisms, particularly the dinosaurs, flying reptiles, great marine reptiles, ammonites, belemnites, a variety of marine phytoplankton and zooplankton, and many others. In many if not most stratigraphic sections, the apparent abruptness in the change of fossil biota from Cretaceous to Tertiary could be attributed to an imperfect stratigraphic record. That is, in most such sections there is a break in the continuity of sedimentation near the boundary, so that the biostratigraphic boundary is within an unconformity representing an imprecisely known length of time. Although the sedimentary record across the Cretaceous-Tertiary boundary may appear to be continuous in some sections, it is nevertheless difficult to prove that it is so, depending on the degree of time resolution one attempts to attain. Certain geologic events (for example, the accumulation of a volcanic ash layer) are known to have occurred very rapidly, but others are usually considered so only in a more relative sense (that is, in comparison to the amount of time since the event transpired), largely because of the lack of adequate present-day analogs and of the decreasing time resolution with increasing age inherent in most of the current dating techniques. Perhaps the marked faunal turnover at the Cretaceous-Tertiary boundary falls in this latter category, yet a fairly precise knowledge of its duration could provide some insight into its cause. What limits can therefore be placed on the interval of time during which this event occurred?

On the basis of available isotopic age data, the Cretaceous-Tertiary boundary lies near to 65 m.y. B.P., with an estimated uncertainty of 1 m.y. or perhaps 2 m.y. (Van Hinte, 1976). If the various isotopic age determinations compiled by Van Hinte are assumed to represent estimates of the *boundary* itself, which is of zero duration, then the uncertainty in the age merely reflects errors in the dating technique(s). In reality, the boundary is not dated directly; rather, samples for dating studies are obtained stratigraphically near the level at which the boundary is thought to be. Consequently, the amount of time represented by an uncomformity that may be associated with the boundary will also contribute to the uncertainty in its age. For this reason, the estimated uncertainty of 1 or 2 m.y. for the date of the Cretaceous-Tertiary boundary can be considered a maximum limit to the length of time represented by the apparent break in the fossil succession. In the light of the magnitude of experimental errors, typically a few percent of the calculated age, it is unlikely that isotopic age dating alone can resolve the duration of this event better than this.

The results of a recently reported lithostratigraphic, biostratigraphic, and magnetostratigraphic study of an essentially complete section of Upper Cretaceous to Paleocene pelagic calcareous sediments exposed at Gubbio, Italy (Alvarez and others, 1977) suggest an indirect method of estimating the length of time represented by the faunal event at the Cretaceous-Tertiary boundary that is relatively insensitive to the accuracy and precision of absolute age determination. The Gubbio section is thought to be complete across the boundary, because the highest recognized planktonic foraminifera zone of the Maestrichtian (*Abathomphalus mayaroensis* zone) and the lowest recognized planktonic foraminifera zone of the Paleocene (*Globigerina eugubina* zone) are present here in stratigraphic succession (Premoli Silva, 1977). This does not imply, however, that these zones are present in their entirety here. Magnetic studies have delineated zones of normal and reversed magnetization polarities within the section, and these zones have been correlated to the calibrated sequence of marine magnetic anomalies (Lowrie and Alvarez, 1977; Roggenthen and Napoleone, 1977). The Cretaceous-Tertiary boundary, as recognized by the planktonic foraminifera present, was found in polarity zone Gubbio G-, which is correlated to the reversed interval between anomalies 29 and 30 of the magnetic-reversal time scale (Alvarez and others, 1977).

According to the most recent calibration of marine magnetic anomalies (La Brecque and others, 1977), this reversed interval represents 470,000 yr. This estimate of its duration is dependent on the degree of linearity of the magnetic-reversal sequence and on the ages chosen for its calibration. The former is based on the assumption of a constant rate of sea-floor spreading in a given region of the ocean. The relative widths of the polarity intervals corresponding to anomalies 29 to 34, as determined from North Pacific profiles that were selected for the magnetic-reversal standard sequence, were shown to give an internally consistent pattern of sea-floor spreading in the world's ocean during this time (Cande and Kristoffersen, 1977). This suggests that the calibrated sequence of marine magnetic anomalies closely approximates a linear time sequence, even though error limits are difficult to define. Assuming a linear chronology uncertainty in the ages used for calibration proportionally affects

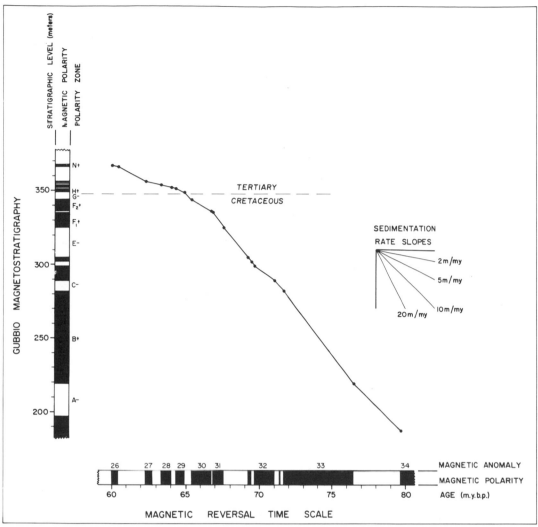

Figure 1. Sedimentation rates in Upper Cretaceous-Paleocene stratigraphic section of Scaglia Rossa at Gubbio, based on correlation of magnetic polarity zones in section (Alvarez and others, 1977; Lowrie and Alvarez, 1977; Roggenthen and Napoleone, 1977) to revised geomagnetic reversal time scale, based on calibrated sequence of marine magnetic anomalies (La Brecque and others, 1977). Shaded parts of polarity columns refer to normal magnetic polarity and unshaded part to reversed magnetic polarity. Stratigraphic level of Cretaceous-Tertiary boundary, base of Maestrichtian (at 270.2 m), and base of Campanian (at 196 m) determined by Premoli Silva (1977).

the age estimates for duration of magnetic polarity intervals. For example, if the age of 64.9 m.y. assigned to the base of anomaly 29 (La Brecque and others, 1977) is uncertain even by ±5% (more than the estimated uncertainty in the age of the Cretaceous-Tertiary boundary), then the duration of the reversed interval between anomalies 29 and 30 would be in error by only 23,500 yr. In other words, this indirect dating method of correlation to the magnetic-reversal time scale can be used advantageously to determine the duration of relatively short time intervals with a high degree of precision for sedimentation-rate analysis.

The stratigraphic positions of magnetic polarity zones in the Gubbio section are plotted against the correlative portion of the magnetic time scale (Heirtzler and others, 1968, as modified by La Brecque and others, 1977) in Figure 1. The calculated sedimentation rate (for the consolidated rock) decreased from an average of about 10 m/m.y. through the Campanian and Maestrichtian sequence to about 4 m/m.y. in the Paleocene sequence. This decrease in apparent sedimentation rate most likely reflects an overall change in lithology in an otherwise continuously deposited sequence (Arthur and Fischer, 1977), although it is quite possible that variations in sedimentation rate calculated for shorter time intervals can be due to intermittent diastems, resulting from hiatuses in deposition or diagenetic dissolution of carbonate. However, the presence in the section of the Gubbio G- magnetic zone, within which lies the Cretaceous-Tertiary boundary, obviously shows that any time gap near or at the boundary must be less than the duration of the correlative geomagnetic reversed interval, or about 470,000 yr. In fact, it is apparent that there was not even a marked change in sedimentation rate in the Gubbio G- magnetic zone compared to the adjoining parts of the section, implying that sediment accumulation was, to a good approximation, uniform. On the scale considered here, the highest sedimentation rate (14.9 m/m.y.) occurs in the Gubbio F_1+ magnetic zone, whereas the calculated sedimentation rate for the Gubbio G- magnetic zone is 12.8 m/m.y. If it is hypothesized that the limestone now present in Gubbio G- was actually deposited at the highest average rate observed in the section, then about 1 m of section, or 70,000 yr of record, could be missing. On the other hand, because the average sedimentation rate in Gubbio G- is near to and, in fact, greater than in the magnetic zones stratigraphically immediately below as well as above suggests that sediment accumulation was continuous and, hence, that little time is not represented in this part of the section.

On a finer scale, Arthur and Fischer (1977) described this part of the section as an alternation of more massive limestones and shaly interbeds of lesser thickness; geochemical studies suggest that the time represented by a bedding couplet is divided about equally between the thin shaly phase and the thicker calcareous phase. Thus, the calculated sedimentation rates shown in Figure 1 are an integrated average of the relatively rapidly deposited limestone beds and the more slowly deposited shaly interbeds. The limestone beds in Gubbio G- are reported to be on the average greater than 25 cm thick (Arthur and Fischer, 1977), so that each couplet represents 20,000 yr and more of record. Therefore, about one-half this value, or 10,000 yr, is the order of maximum resolution possible in estimating the time separation across the Cretaceous-Tertiary boundary as it occurs in the Gubbio section.

From the assumptions above, it can be concluded that the duration of a major faunal event correlated to the Cretaceous-Tertiary boundary in pelagic limestones of the Scaglia Rossa at Gubbio, Italy, is likely to be quite short, possibly on the order of 10,000 yr. Thus, it appears that the replacement of almost all Cretaceous planktonic foraminifera by typical earliest Paleocene forms was indeed abrupt, reflecting a small amount of elapsed time rather than a significant hiatus in the stratigraphic record.

Numerous mechanisms have been proposed to explain mass extinctions in general and at the end of the Cretaceous Period in particular (see Newell, 1967). The apparent rapidity in the overturn of planktonic foraminifera at the Cretaceous-Tertiary boundary might seem to favor a catastrophic cause, although some slower change reaching a critical threshold could produce a similar effect. Moreover, because the stratigraphic correlation of the foraminiferal zones at Gubbio with Upper Cretaceous and Paleocene stages defined on the basis of different fossil groups elsewhere is not yet precisely known (Premoli Silva, 1977), it is possible that a fine structure or detailed sequence, yet to be fully explored, may in fact characterize the faunal crises; this would require a more complex explanation of the underlying causal mechanism. For example, Butler and others (1977) presented evidence from the San Juan Basin of New Mexico suggesting that the Cretaceous-Tertiary boundary, based upon the last occurrence of dinosaurs, is significantly younger than in the marine deposits

at Gubbio. The last occurrence of dinosaurs in this terrestrial section is within the upper part of a normal polarity zone that they correlate with anomaly 29 of the magnetic-reversal time scale; after a barren interval, this is followed by the first occurrence of Paleocene mammal fossils within the succeeding reversed polarity zone (between anomalies 28 and 29). The Cretaceous-Tertiary boundary identified by Butler and others (1977), therefore, lies somewhere within the barren interval and is at least about 500,000 yr younger than at Gubbio.

If this temporal sequence of events is corroborated, the explanation of the faunal change at the end of the Cretaceous Period should lie in a process that initially and rapidly affected the marine faunal realm and only later, perhaps indirectly, caused the crisis in the terrestrial fauna. The Cretaceous-Tertiary boundary is not coincident with a geomagnetic polarity transition in the Gubbio section, thus eliminating an explanation based on any simple relationship between the faunal change and possible effects on organisms of a reversal of the Earth's magnetic field (Uffen, 1963). Although further speculation on the nature of the critical process may not be warranted on the basis of the available data, the expansion of this data base to place the Cretaceous-Tertiary boundary as identified elsewhere in the precise relative time framework that can be provided by magnetostratigraphic correlations should eventually give additional important constraints on the causal mechanism.

REFERENCES CITED

Alvarez, W., Arthur, M. A., Fischer, A. G., Lowrie, W., Napoleone, G., Premoli Silva, I., and Roggenthen, W. M., 1977, Upper Cretaceous-Paleocene magnetic stratigraphy at Gubbio, Italy. V. Type section for the Late Cretaceous-Paleocene geomagnetic reversal time scale: Geol. Soc. America Bull., v. 88, p. 383-389.

Arthur, M. A., and Fischer, A. G., 1977, Upper Cretaceous-Paleocene magnetic stratigraphy at Gubbio, Italy. I. Lithostratigraphy and sedimentology: Geol. Soc. America Bull., v. 88, p. 367-371.

Butler, R. F., Lindsay, E. H., Jacobs, L. L., and Johnson, N. M., 1977, Magnetostratigraphy of the Cretaceous/Tertiary boundary in the San Juan Basin, New Mexico: Nature, v. 267, p. 318-323.

Cande, S. C., and Kristoffersen, Y., 1977, Late Cretaceous magnetic anomalies in the North Atlantic: Earth and Planetary Sci. Letters (in press).

Heirtzler, J. R., Dickson, G. O., Herron, E. M., Pitman, W. C. III, and Le Pichon, X., 1968, Marine magnetic anomalies, geomagnetic field reversals, and motions of the ocean floor and continents: Jour. Geophys. Research, v. 73, p. 3119-3136.

La Brecque, J. L., Kent, D. V., and Cande, S. C., 1977, Revised magnetic polarity time scale for the Late Cretaceous and Cenozoic: Geology, v. 5, p. 330-335.

Lowrie, W., and Alvarez, W., 1977, Upper Cretaceous-Paleocene magnetic stratigraphy at Gubbio, Italy. III. Upper Cretaceous magnetic stratigraphy: Geol. Soc. America Bull., v. 88, p. 374-377.

Newell, N. D., 1967, Revolutions in the history of life: Geol. Soc. America Spec. Paper 89, p. 63-91.

Premoli Silva, I., 1977, Upper Cretaceous-Paleocene magnetic stratigraphy at Gubbio, Italy. II. Biostratigraphy: Geol. Soc. America Bull., v. 88, p. 371-374.

Roggenthen, W. M., and Napoleone, G., 1977, Upper Cretaceous-Paleocene magnetic stratigraphy at Gubbio, Italy. IV. Upper Maastrichtian-Paleocene magnetic stratigraphy: Geol. Soc. America Bull., v. 88, p. 378-382.

Uffen, R. J., 1963, Influence of the earth's core on the origin and evolution of life: Nature, v. 198, p. 143-144.

Van Hinte, J. E., 1976, A Cretaceous time scale: Am. Assoc. Petroleum Geologists Bull., v. 60, p. 498-516.

ACKNOWLEDGMENTS

Reviewed by Walter Alvarez. Supported by National Science Foundation Grant EAR75-18955. L. H. Burckle, J. C. Liddicoat, N. D. Opdyke, and T. Saito critically read the manuscript.

Contribution No. 2569, Lamont-Doherty Geological Observatory.

MANUSCRIPT RECEIVED JUNE 9, 1977
MANUSCRIPT ACCEPTED SEPTEMBER 26, 1977

30

Reprinted from *Science* **156**:1083–1087 (1967)

GEOMAGNETIC POLARITY CHANGE AND FAUNAL EXTINCTION IN THE SOUTHERN OCEAN

N. D. Watkins and H. G. Goodell

Abstract. Paleomagnetic polarity changes have been detected in nine deep-sea sedimentary cores (from the Pacific-Antarctic Basin) in which an extinction horizon of a radiolarian assemblage was previously independently determined. The depths of the polarity change 0.7 million years ago and the faunal boundary are closely correlated, confirming that the faunal extinction was locally virtually synchronous. Although the reason for the faunal extinction is unknown, the possibility of causal relationships between faunal extinction and factors directly involved with sedimentation rate, sedimentation rate variation, and sediment type appears to be excluded.

Uffen (1) suggested that during a geomagnetic polarity reversal the loss of magnetic shielding during the possible zero dipole field condition (2) and consequent excessive cosmic radiation at the earth's surface could lead to increased rates of genetic mutation. Although Simpson (3) believes that macropaleontological data, when combined with paleomagnetic polarity observations, support this possibility for the past 600 million years, the evidence is not convincing in view of lack of coverage and depth of appropriate data. At present the most promising sources for investigation of the hypothesis are microfossils and paleomagnetic data in deep-sea sedimentary cores.

Harrison and Funnell (4) noted that extinction of the radiolarian species *Pterocanium prismatium* and evidence for a geomagnetic polarity reversal occurred together in an equatorial Pacific core. Hays (5, 6) studied the distribution of radiolaria in the submarine sediments of the Southern Ocean and found four widespread well-defined successive assemblage zones, which he named, in order of increasing age, Ω, Ψ, χ, and Φ. On the basis of isotope data for the upper parts of a single core and on extrapolation, he inferred that boundaries between these four zones occurred at 0.4, 0.9, and 1.6 million years, respectively. Opdyke *et al.* (7) observed that the upper part of Hays's χ radiolarian zone, which is marked by the local extinction of several species (especially *Pterocanium tribolum*, *Saturnalus planetes*, and *Sethocorys* sp.) and the appearance of *Stylatractus* sp., *Lacopyle* sp., and *Prunopyle buspinigernum*, occurred in close association with a change in the magnetic polarity in four cores. Because of relatively low sedimentation rates in the cores concerned, it is not clear whether the observed faunal extinction horizon is associated with the polarity change at the Brunhes-Matuyama epoch boundary (8) 0.7 million years ago, or with the change marking the end of the Jaramillo event (9) about 0.85 million years ago (Figs. 1 and 3). The faunal horizon is apparently geologically synchronous. Opdyke *et al.* (7) did not dismiss the possibility that, in harmony with Uffen's hypothesis (1), there may exist a causal relationship between these extinctions and magnetic reversals. Much additional information is needed concerning this possibility (7, 10).

We have measured paleomagnetic polarity changes in the "Eltanin" core collection, primarily for use in marine stratigraphic studies (11). In this work, an astatic magnetometer with a maximum sensitivity of 10^{-7} emu/mm was used (12). Four measurements of the vertical component of NRM (natural remanent magnetism) yield reliable polarity determinations, due to the fact that the present ambient magnetic field direction in the area surveyed is as little as 20° from the vertical. In the particular cores concerned, minimizing of unstable magnetic components by application of an alternating magnetic field of 200 oersteds (13) to all specimens has not led to modification of the polarity data from untreated specimens. Such high magnetic stability is apparently characteristic of siliceous oozes, whereas our limited examination of brown silts ("red clays") has revealed magnetic instability.

Hays (6) gave the radiolarian assemblage data for a traverse of 13 "Eltanin" cores along the 115°W longitude, between 55° and 68° south latitudes, in the Pacific-Antarctic Basin (Fig. 1, inset). To extend the previous limited observations of the magnetic and faunal variations and to maximize the value of the available data, it is essential to compare our observations of the depth of the Brunhes-Matuyama and Matuyama-Jaramillo polarity changes in these cores with Hays' Ψ-χ faunal boundary. Although the geomagnetic polarity changes preceding the polarity change 0.7 million years ago may be far more complex than was at first realized (14), the general consensus is that no geomagnetic polarity change since that time would have gone undetected (15). Therefore, we feel that the first major polarity break in these very high magnetic latitude sedimentary cores can be identified as the base of the Brunhes epoch. Our correlation line between the cores, at the base of the Brunhes epoch, is included in Fig. 1, in which we present the NRM polarity logs of the cores in the traverse. Data points for each specimen, taken at 10-cm intervals, are not shown individually, but presentation is completely nonsubjective in that all polarity changes are indicated, even if only one specimen is involved. Intensities of magnetization for these sediments, which are dominantly siliceous oozes or clayey silts, are of the order of 10^{-5} emu gm^{-1}.

Figure 2 gives the result of the comparison of our determinations of the depth of the Brunhes-Matuyama boundary and Hays' (6) Ψ-χ faunal boundary depths in the nine cores where both boundaries occur. Four of the 13 cores do not provide sufficient data: in cores E11-2 and E11-15 the faunal and magnetic boundaries are both presumably below the maximum depth reached by the recovered cores; in core E11-12 no definite Ψ-χ faunal boundary is detected by Hays; in core E11-14 the faunal boundary is again absent, and in addition the magnetic data in this core are inconclusive because the long normal polarity sequence at the base severely weakens the possibility that the core penetrates material from the Jaramillo event. For various reasons, spurious data may result from limited parts of cores; such data are more likely to exist at the bottom (16) and top of a core or core section because of the nature of the coring operation. The strong correlation existing between the faunal extinction and the Brunhes-Matuyama polarity boundary, which ranges from 4 to 14 m in depth, is obvious. Comparison of appropriate data from the cores of higher sedimentation rate (in which the Brunhes epoch and Jaramillo event are separated by up to 2 m of sediment) demonstrates without question that the polarity changes due to the Jaramillo event are not correlated with the Ψ-χ faunal horizon. We conclude that the previous suspicion of a correlation between the geomagnetic polarity change 0.7 million years ago and the Ψ-χ radiolarian boundary (7) is confirmed. Therefore, Hays' (6) assumption of a geologically isochronous local faunal extinction is valid.

Because it is evident that the radiolarian assemblage was affected catastrophically 0.7 million years ago, we must consider the possibility of a direct connection between the geomagnetic polarity and faunal changes. The direct approach would demand a search for proof that the specific faunal species involved are anomalously susceptible to the hypothetical excess radiation at the earth's surface during dipole collapse, but the number of variables involved is probably great (17). We ca

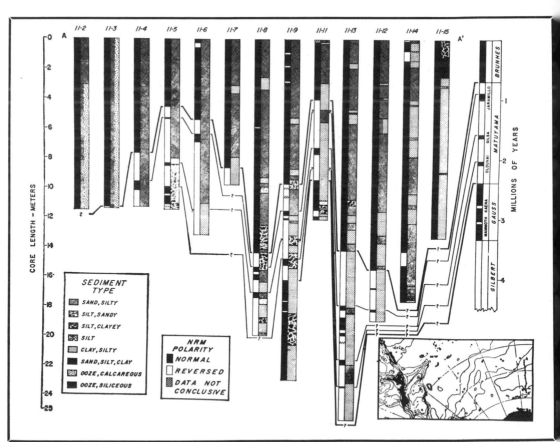

Fig. 1. Magnetic polarity and sedimentary variation in traverse E11. (Inset) Core locations. The line of traverse (A-A') extends from 58°S to 68°S along the 150°W longitude. The ordinate is the most recently defined geomagnetic polarity time scale (14).

only suggest an indirect approach to this problem: if the geomagnetic polarity change is not in any way connected with the faunal extinction, then it is reasonable to expect that a third variable might show a strong time correlation with the resulting faunal extinction and, only by coincidence, with the time of the magnetic polarity change. Of course, it is also possible that a dipole collapse may cause faunal extinction through another dependent effect rather than through the suggested excessive radiation. Initial evidence for either of these two possibilities would be a correlation between the faunal extinction and a third variable. Sedimentation rate and sediment type provide the only pertinent evidence which is available at this time.

The sedimentation rates in the cores examined have been determined by the use of paleomagnetic data (Fig. 3). Watkins and Goodell (*12*) presented data from some marine sedimentary cores strongly consistent with the reality of the Gilsa event, but the Kaena event has not been reliably detected in material other than the single Hawaiian lava which provided the first evidence of the event (*14*). The Brunhes-Matuyama depths, occurring in nine cores, are shown at the terrestrially defined age. We consider that the second major polarity boundary (the Matuyama-Gauss transition), which is found only in cores E11-11 and E11-9, is also reliably identified. These epoch boundaries establish average epoch sedimentation rates which are then used to date polarity events within the epoch. Considerable doubt obviously exists about the specific identification of an apparent event (which could be spurious for a number of reasons) if a significant change in sedimentation rate within an epoch must be proposed in order to force an apparent event to equiv-

alence with an established (terrestrially dated) event. Such an apparent event may be real, but the specific identification would involve circular reasoning, particularly because it is possible that the relevant part of the geomagnetic polarity scale may be incompletely known. We therefore rely heavily on the acceptability of the sedimentation rates and changes of sedimentation rate inferred by the data, because specific identification is compelling if a relatively constant sedimentation rate between the reliably chosen epoch boundaries incorporates an apparent event at the ages of known events. Of the known events during the Matuyama, the Jaramillo will be the least susceptible to misidentification, by virtue of its close relationship with the base of the Brunhes. A spacing of 10 cm between specimens may result in only partial delineation (or even omission) of events.

In seven of the eight cores which we interpret as having reached material deposited during the Jaramillo event, the sedimentation rate has not varied since at least the end of the Jaramillo event, although the actual sedimentation rate across the traverse ranges from 0.57 cm per 1000 years in core E11-11 (*18*) to 2.18 cm per 1000 years in core E11-12 (Fig. 3). Reliable determination of earlier average sedimentation rates is only possible in cores E11-9 and E11-11, which reach Gauss epoch sediments. The data strongly suggest a drop in average sedimentation rate to 0.3 cm per 1000 years for the period of 0.8 to 2.4 million years. Estimates of sedimentation rate could be made for at least part of this period in the other cores if the polarity events could be identified, but without the presence of the Matuyama-Gauss boundary, such identifications are not rigorously possible. The same problem exists for the possible Gauss events in E11-9 and E11-11 because no Gauss-Gilbert boundary is detected. We interpret the data in Fig. 3 to show identification of the Gilsa event in E11-11. In core E11-5, despite the absence of the base of the Matuyama, the apparent Gilsa and Olduvai events are convincingly distributed, as is the apparent Mammoth event in core E11-11. In all other cores, for several reasons, the apparent events are, at the very best, only tentatively identified (*19*). Despite the problems involved, however, we believe that a general characteristic of the area is a marked increase in sedi-

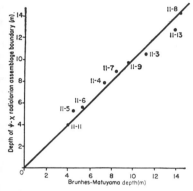

Fig. 2. Comparison of the depth of the Brunhes-Matuyama geomagnetic polarity change and the Ψ-χ radiolarian boundary in cores from traverse E11. Both axes are in meters below the top of the core. The faunal data is due to Hays (*6*). Core numbers correspond to the numbers shown in Figs. 1 and 3. The line corresponds to a 1 : 1 relationship (or the line of perfect correlation).

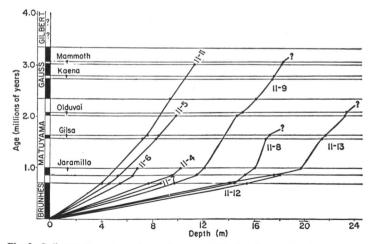

Fig. 3. Sedimentation rate analyses of cores which include the Brunhes-Matuyama boundary, in traverse E11 (A-A' in Fig. 1). Ordinate: geomagnetic polarity time scale of McDougall and Chamalaun (*14*); polarity epochs are in vertical print; polarity events are in horizontal print; normal polarity is shown in black; reversed polarity is shown in clear. Abscissa: depth from the top of the core in meters. Question marks are included where assigned identifications older than the Jaramillo event are tentative. The bottom 2 m of cores 11-8 and 11-9 are possibly "sucked" (*16*).

mentation rate either during or shortly after the Jaramillo event (7): we also propose that there existed far less variation of sedimentation rate across the traverse prior to the Jaramillo event compared to the last 0.8 million years. Neither the time of the change in sedimentation rate nor the drop in the variation of sedimentation rate across the traverse occurred within 0.1 million years of the χ faunal extinction. There is apparently no connection between sedimentation rate or regional sedimentation rate variation and the observed faunal extinction.

Our observation of a marked change in sedimentation rate emphasizes that, in the absence of a known defined polarity scale, the extrapolation into the subsurface of sedimentation rates based on near-surface data could yield very misleading interpretations of the earth's history.

A pronounced change of sediment type, from clay to diatomaceous ooze, occurs close to Hays' χ-Φ boundary (20). Hays thinks that this sediment change accompanies a relatively sudden increased vertical movement of water resulting from the initiation of iceberg production from the Antarctic ice cap. No sediment change occurs close to the Ψ-χ faunal boundary. There is apparently no relation between sediment type and the χ faunal extinction horizon.

Our data show some evidence, however meager, that the magnetic polarity change occurs slightly later than the faunal change. If the sedimentation rates for the interval between 0 and 0.8 million years are used to calculate any time differential between the faunal and magnetic change in each core, the average differential is close to 13,000 years, with the faunal change preceding

the magnetic change. This difference is statistical only and is on the fringes of the resolution of the methods, which depends dominantly on the sample spacing and cannot therefore be considered significant. Apparently, however, the precise time relationship of the faunal and magnetic boundaries, as well as the geographic limits of the correlation, must be accurately determined, because, if the hypothetical dipole collapse occurs after the faunal change or if the extinction becomes diachronous elsewhere, then the geomagnetic polarity change cannot be the cause of the faunal extinction. One must realize, however, that the detailed mechanism and duration of the effect of a polarity reversal is not known.

An independently determined faunal boundary has been confirmed as a geologically synchronous phenomenon, correlating strongly with the Brunhes-Matuyama geomagnetic polarity change. The sedimentation rate, although variable across the core traverse, has not varied significantly since at least 0.8 million years ago, shortly before which the sedimentation rate was lower by a minimum factor of 2 to 3 and probably much less variable between cores. No correlation exists between the sedimentation rate changes (or changes of sediment type) and the faunal extinction. If the faunal extinction is not directly due to the increased radiation from dipole collapse during polarity change, then the cause must lie in factors other than those that we have examined.

References and Notes

1. R. J. Uffen, *Nature* **198**, 143 (1963).
2. T. Rikitake, *Electromagnetism and the Earth's Interior* (Elsevier, Amsterdam, 1966), p. 81.
3. J. F. Simpson, *Geol. Soc. Amer. Bull.* **77**, 197 (1966).
4. C. G. A. Harrison and B. M. Funnell, *Nature* **204**, 556 (1964).
5. J. D. Hays, *Ant. Res. Ser.* (Am. Geophys. Un.) **5**, 125 (1965).
6. ———, *Progress in Oceanography* (Pergamon, London, in press).
7. N. D. Opdyke, B. Glass, J. D. Hays, J. Foster, *Science* **154**, 349 (1966).
8. A. Cox, R. R. Doell, G. B. Dalrymple, *ibid.* **144**, 1537 (1964).
9. R. R. Doell and G. B. Dalrymple, *ibid.* **152**, 1060 (1966).
10. A. Cox, G. B. Dalrymple, R. R. Doell, *Sci. Amer.* **216**(2), 54 (1967).
11. N. D. Watkins and H. G. Goodell, *Trans. Am. Geophys. Un.* **47**, 478 (1966); H. G. Goodell and N. D. Watkins, in *Abstracts*, Annual Meeting Geol. Soc. Am., 1966, p. 78; manuscript in preparation.
12. N. D. Watkins and H. G. Goodell, *Earth Planet Sci. Lett.* **2**, 123 (1967).
13. M. W. McElhinny, *Geophys. J.* **10**, 375 (1966).
14. I. McDougall and F. H. Chamalaun, *Nature* **212**, 1415 (1966).
15. ———, *ibid.*, p. 1416.
16. G. O. Dickson and J. H. Foster, *Earth Planet. Sci. Lett.* **1**, 460 (1966). This article describes a very large fluctuation of the NRM declination in the bottom 100 cm or so of a marine sedimentary core (V21-65) from 40° magnetic latitude. Dickson and Foster see this as an indication of true flow-in (or "sucking") of the sediments during coring. Presumably no sedimentary features indicated this flow-in, emphasizing the possible magnitude of the problem in paleomagnetic studies. The NRM declination in our 70° (or higher) magnetic latitude cores is a virtually meaningless parameter and cannot be used in the same fashion.
17. Hays (6) has pointed out that the radiolaria comprising the χ assemblage are usually thin-walled compared to those species in the Φ zone: it is conceivable that this is relevant.
18. Hays (6) refers to a personal communication from Dr. T. Ku who obtained an average sedimentation rate of 0.5 cm per 1000 years for the top 3 meters of core E11-11, using excess Th²³⁰ measurements. This agrees reasonably well with the figure of 0.57 cm per 1000 years from our paleomagnetic data.
19. We stress the tentative nature of our assignment of the normal polarity event at 1465 cm in core E11-9 to the Olduvai by pointing out that the apparent event could equally as well be assigned to the Gilsa event in Fig. 3.
20. Opdyke *et al.* (7) suspect that the χ-Φ faunal boundary may also be correlative with a magnetic polarity change. Their study indicates that the Olduvai event occurs close to the χ-Φ horizon. A comparison of Hays' χ-Φ boundary and our magnetic events cannot be satisfactorily made because of our difficulty in identifying specific events. We can state, however, that compared with the correlation in Fig. 2, the χ-Φ boundary is only poorly correlated with a polarity change due to any events, whether the event is identified or only apparent.
21. Supported by grants GA 523 and GA 602 from NSF. We thank Dr. J. D. Hays of Lamont Geological Observatory of Columbia University for a preprint of one of his papers (6).

31

Reprinted from *Geol. Soc. America Bull.* **82**:2433–2447 (1971)

Faunal Extinctions and Reversals of the Earth's Magnetic Field

JAMES D. HAYS *Lamont-Doherty Geological Observatory of Columbia University, Palisades, New York 10964*

INTRODUCTION

During the last several years the body of evidence has grown supporting the correlation between the time of extinction of certain marine protozoans and reversals of the Earth's magnetic field (Harrison and Funnel, 1964; Opdyke and others, 1966; Hays and Opdyke, 1967; Hays and others, 1969; Hays, 1970).

It has been shown that during the last 4.5 m.y. nine radiolarian species have become extinct and, of these, six disappeared from the fossil record close to reversals of the Earth's magnetic field. In the Equatorial Pacific eight species of Foraminifera became extinct, seven of which disappeared near magnetic reversals. Because of the potential significance of these data to our understanding of extinctions and the processes of evolution, it is important to know precisely the connection between the final decline of extinguished species and magnetic reversals. The purpose of this paper is to study in detail the relations between radiolarian extinctions and magnetic reversals during the last 2.5 m.y. and to examine the possible causes for these relations. Two possible causes have already been suggested, radiation and climatic change, and a third, biomagnetic effects, should be considered at this time.

RADIOLARIAN EXTINCTIONS DURING LAST 2.5 M.Y.

Although the magnetic stratigraphy of deep-sea sediments has been studied beyond 2.5 m.y.(Foster and Opdyke, 1970), I chose to closely examine the record of only this interval because the Lamont-Doherty core collection contains a number of piston cores from a variety of latitudes that include sediments of this age. Those few piston cores that contain records much longer than this have slow rates of accumulation and their highly condensed sections are less suitable for this kind of study than cores with higher accumulation rates.

The eight species of Radiolaria that became extinct during the last 2.5 m. y. represent the end of lineages and do not include species that evolved into other species (Table 1). A possible exception to this is *Eucyrtidium calvertense*. This species disappeared from Antarctic sediments near the base of the Olduvai event, about 2 m.y. ago (Opdyke and others, 1966). A similar form lived on in the North Pacific past the Ol-

TABLE 1. EIGHT SPECIES OF RADIOLARIA THAT BECAME EXTINCT DURING LAST 2.5 M. Y., THEIR
GEOGRAPHIC DISTRIBUTION, AGE AT EXTINCTION, AND RELATION TO MAGNETIC REVERSALS

Species	Age of upper limit in yrs B. P.	Nearest reversal	Geographical distribution
Druppatractus acquilonius	300,000	Well removed from any well-documented reversal	N. Pacific Ocean
Stylatractus universus	400,000	Well removed from any well-documented reversal	Cosmopolitan
Eucyrtidium matuyamai	1,000,000	Base of Jaramillo event	N. Pacific Ocean
Pterocanium prismatium	1,800,000	Top of Olduvai event	Equatorial Indo-Pacific
Clathrocyclas bicornis	1,800,000	Top of Olduvai event	Antarctic Ocean
Eucyrtidium calvertense	2,000,000	Base of Olduvai event	Antarctic Ocean (N. Pacific?)
Helotholus vema	2,400,000	Top of Gauss series	Antarctic Ocean
Desmospyris spongiosa	2,400,000	Top of Gauss series	Antarctic Ocean

duvai event and is found in Holocene sediments of this region (Hays, 1970). In gross morphology the two forms are similar; however, it is not possible to know if they are the same species. The disappearance of the Antarctic form was abrupt and marked an important event in the history of Antarctic Radiolaria, so for the purpose of this paper, I will consider the disappearance of *E. calvertense* in the Antarctic as an extinction, bearing in mind the possibility that the species continued to live after this in the North Pacific.

Four of the eight extinct species lived in the Antarctic Ocean, two in the North Pacific, one in the Equatorial Indo-Pacific, and one was cos-

mopolitan, having lived in all three regions (Table 1). Of these eight, six disappeared near geomagnetic reversals as previously reported (Fig. 1; Opdyke and others, 1966; Hays and Opdyke, 1967; Hays and others, 1969; Hays, 1970). These six species all belong to the Nassellaria, elongate forms with an elongate central capsule pored at one end.

The two species that do not become extinct near reversals (*Stylatractus universus* and *Druppatractus acquilonius*) became extinct in the middle of the Brunhes epoch (Fig. 1). They belong to the Spumellaria, basically spherical forms with a spherical uniformly pored central capsule. *Stylatractus universus* is cosmopolitan but *Drup-*

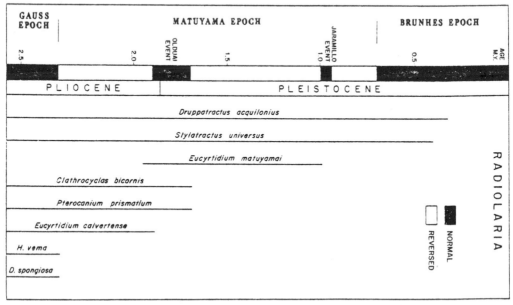

Figure 1. Ranges of eight radiolarian species that became extinct in the last 2.5 m.y. plotted against mag-   netic stratigraphy.

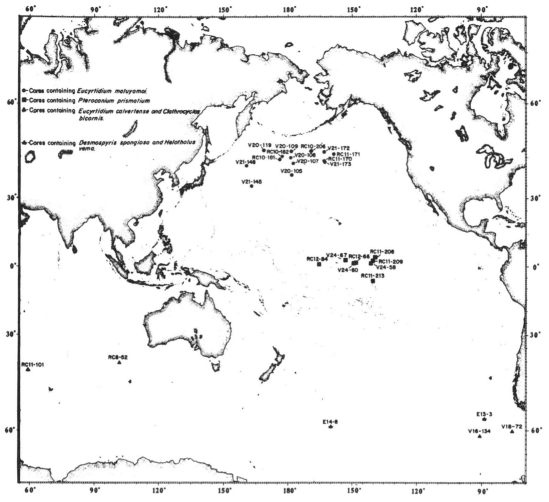

Figure 2.   Location of 28 long piston cores used in this study.

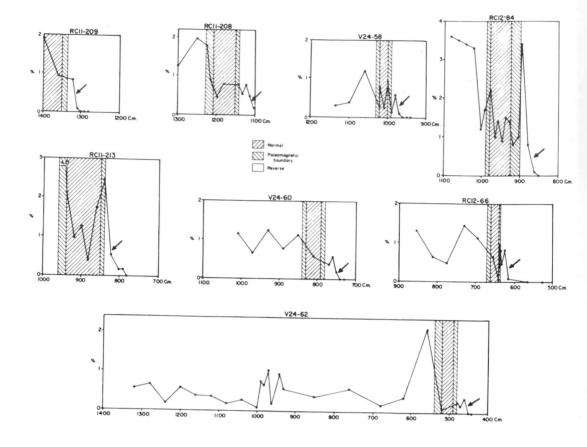

Figure 3. Abundance of *Pterocanium prismatium* through its termination in eight cores. Upper right-lower left diagonal hachuring represents normal polarity; upper left-lower right hachuring represents that interval between two oppositely polarized samples; no hachuring represents reversed polarity. Arrow indicates midpoint of estimated extinction interval.

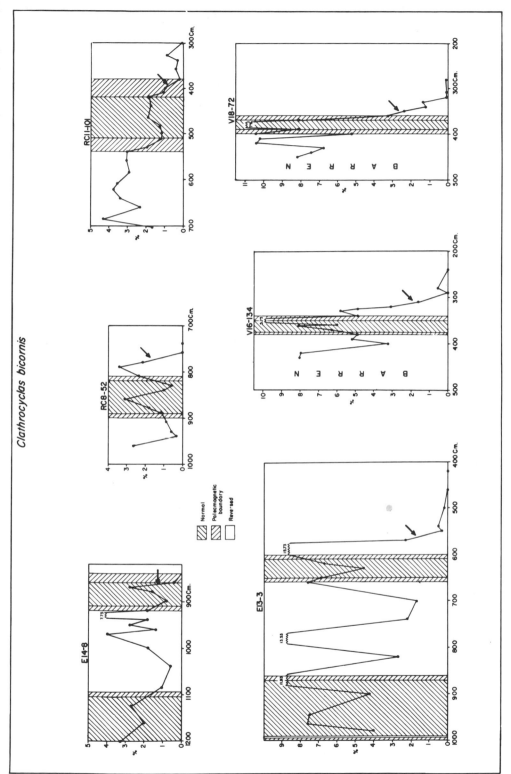

Clathrocyclas bicornis

Figure 4. Abundance of *Clathrocyclas bicornis* through its termination in six cores. Upper right–lower left diagonal hachuring represents normal polarity; upper left–lower right hachuring represents that interval between two oppositely polarized samples; no hachuring represents reversed polarity. Arrow indicates midpoint of estimated extinction interval.

369

Eucyrtidium calvertense

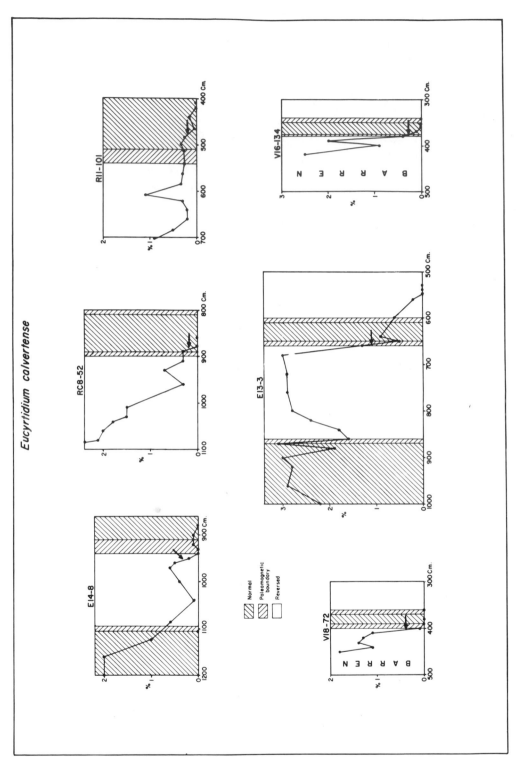

Figure 5.   Relative abundance of *Eucyrtidium calvertense* through its termination in six cores. Upper right–lower left diagonal hachuring represents normal polarity; upper left–lower right hachuring represents that interval between two oppositely polarized samples; no hachuring represents reversed polarity. Arrow indicates midpoint of estimated extinction interval.

patractus acquilonius is probably restricted to the North Pacific. There are no confirmed reversals in the neighborhood of the extinction points of these species, although some possible short events have been reported (Ryan, 1970). Until these are confirmed it can only be assumed that the time of extinction of these species is far removed from any geomagnetic reversals. No other radiolarian species than those considered here are known to have become extinct during the last 2.5 m.y.

To precisely determine the relationship between the extinction levels of the six species that disappear near reversals and the position of the neighboring reversals, 28 long piston cores were selected for study; 14 from the North Pacific containing *E. matuyamai*, eight from the Equatorial Pacific containing *P. prismatium*, six from the Antarctic with *C. bicornis* and *E. calvertense*, two of which also contain *D. spongiosa* and *H. vema* (Table 2; Fig. 2). The paleomagnetic stratigraphy of these cores has been determined by Opdyke (Opdyke and others, 1966; Hays and Opdyke, 1967; Hays and others, 1969; Opdyke and Foster, 1970).

Counts of each of the six extinct species versus all other radiolarian species were made at 10 to 20 cm intervals through at least the last meter of the extinct species range in all cores where this is possible. A minimum of 1,000

TABLE 2. LOCATION AND LENGTH OF PISTON CORES USED IN THIS STUDY

Core no.	Latitude	Longitude	Length (cm)
North Pacific			
V20-105	39°00'N	178°17'W	1237
V20-107	43°24'N	178°52'W	1282
V20-108	45°27'N	179°14.5'W	1671
V20-109	47°19'N	179°39'W	1452
V20-119	47°57'N	168°47'E	1170
V21-145	34°03'N	164°50'E	1225
V21-148	42°05'N	160°36'E	1448
V21-172	47°40'N	164°21'W	1081
V21-173	44°22'N	163°33'W	1218
RC10-181	44°05'N	176°50'E	1161
RC10-182	45°37'N	177°52'E	1130
RC10-206	47°13'N	170°25'W	1152
RC11-170	44°29.4'N	163°21.1'W	1002
RC11-171	46°36.2'N	159°39.7'W	1161
Equatorial Pacific			
V24-58	02°16'N	141°40'N	1140
V24-60	02°48'N	149°00'W	1256
V24-67	10°13'N	155°01'W	468
RC11-208	05°21'N	139°58'W	1636
RC11-209	03°39'N	140°04'W	1463
RC11-213	06°08'S	140°51'W	1068
RC12-34	02°20'N	165°12'W	2230
Antarctic			
E13-3	57°00'S	39°29'W	1603
E14-8	59°40'S	160°17'W	1830
V16-134	61°54'S	91°15'W	328
V18-72	60°29'S	75°57'W	562
RC8-32	41°06'S	101°25'E	1103
RC11-101	44°04'S	59°50'E	305

individuals were counted in each sample and the counts were found to be reproducible to about 0.5 percent. Counts of this number of species should reveal the presence of the extinct species if it had a true abundance in the fossil population of more than 0.05 percent (Shaw, 1964). Additional counts of up to 4,000 individuals were made on several samples above the last occurrence of an extinct species, as determined by the 1,000 specimen counts, and only two specimens were detected. In all cores slides of at least four samples taken at 20 cm intervals above the last occurrence as detected by the counts were scanned for the extinct species. These scans of entire slides containing in excess of 6,000 individuals only rarely detected one or two individuals of the extinct species, so the counts accurately detect the final occurrence of the species in the sediment section. A total of over 400,000 Radiolaria were counted in the 28 cores studied. These data are shown graphically in Figures 3 through 7.

In all cores studied the final decline of the one Equatorial species (*Pterocanium prismatium*) to become extinct near a reversal during the last 2.5 m.y. comes shortly above the reversal that marks the upper limit of the Olduvai event (Fig. 3). In the cores that have the highest accumulation rates (RC11-209, RC11-208, V24-58, RC12-84, and RC11-213) there is also a decline in abundance above the reversal that marks the base of the Olduvai event; however, the abundance does not drop to zero and increases again before the final decline.

The four Antarctic species that become extinct during the last 2.5 m.y. do so just above the top of the Gauss series (*H. vema, D. spongiosa*) at the base of the Olduvai event (*E. calvertense*) and the top of the Olduvai event (*C. bicornis*).

The final decline of *Clathrocyclas bicornis* (Fig. 4) occurs either at (V18-72, RC11-101), slightly above (RC8-52, E13-3, V16-134), or just below (E14-8), the reversal at the top of the Olduvai event. This species, like *P. prismatium*, also shows a decline after the penultimate reversal (base of Olduvai event) in two cores (RC11-101 and E14-8) and there is a suggestion of a decline across the antepenultimate reversal (top of Gauss) in E14-8.

The final decline of *Eucyrtidium calvertense* (Fig. 5) either coincides with the reversal at the base of the Olduvai event olduvai V16-134), falls slightly above it (RC8-52, RC11-101), or just below it (E14-8, V18-72). A decline just after the penultimate reversal can be seen in two of the abundance curves for this species (E13-3, E14-8). In these a decline followed by an increase in abundance occurs just above the top of the Gauss series (Fig. 5).

Of the six Antarctic cores in which the termination of the species *C. bicornis* and *E. calvertense* is recorded, RC11-101 has the highest rate of accumulation through the Olduvai event. Careful examination of the fauna of this core indicates that it has a large hiatus in its upper portion. This hiatus allowed the corer to reach the level of the Olduvai event in a region of rapid sedimentation. The Olduvai event in this core is over 1 m long and, using the estimates of Opdyke and Foster (1970) for the age of the top and bottom of this event, the rate of accumulation across this event is about 1 cm/1,000 yrs. This core therefore probably depicts most accurately the relationship between the abundance of these species and the reversals recorded in the sediment.

The two remaining Antarctic species to become extinct near reversals are *H. vema* and

D. spongiosa, and they both show a final decline near the top of the Gauss normal polarity series (Fig. 6). In the Antarctic, south of the Antarctic Polar Front, most cores become barren of Radiolaria or have reduced assemblages shortly below the upper limit of *E. calvertense* (Hays, 1965). Therefore, only two cores were available for this study that contained abundant Radiolaria within the range of *D. spongiosa* and *H. vema*. This occurrence of barren sediments is also the reason why counts of *E. calvertense* and *C. bicornis* could not be extended lower in cores V16-134 and V18-72 (Figs. 4 and 5).

Eucyrtidium matuyamai, the one North Pacific species that disappeared near a reversal, has been studied in some detail (Hays, 1970). Figure 7 shows its abundance curve in eleven North Pacific cores. The final decline of this species occurs at or slightly above the reversal at the base of the Jaramillo event.

The following generalizations can be made about these curves. (1) The species disappear abruptly and, within the interval sampled, do

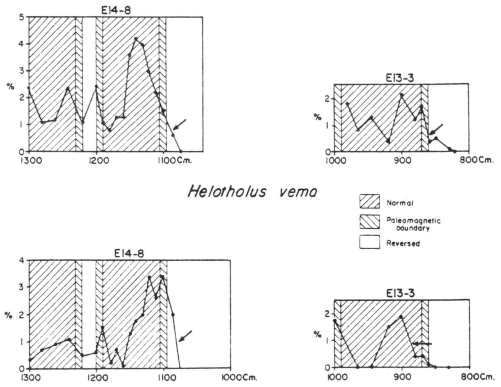

Figure 6. Relative abundance of *Helotholus vema* and *Desmospyris spongiosa* through their termination in two cores. Upper right–lower left diagonal hachuring represents normal polarity; upper left–lower right hachuring represents that interval between two oppositely polarized samples; no hachuring represents reversed polarity. Arrow indicates midpoint of extinction interval.

Eucyrtidium matuyamai

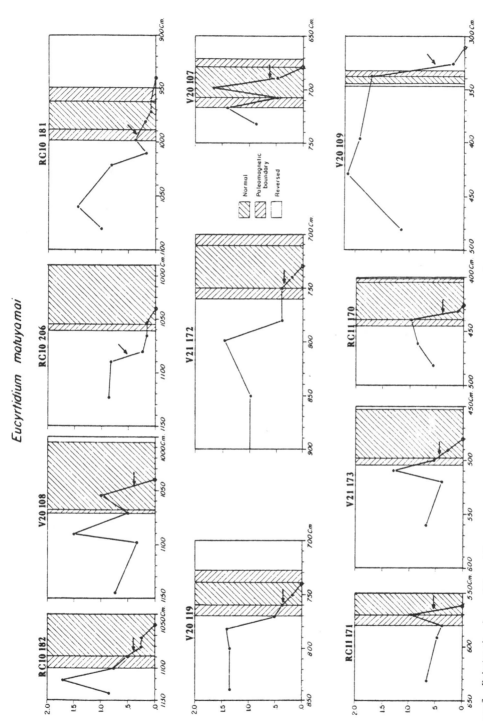

Figure 7. Relative abundance of _Eucyrtidium matuyamai_ through its termination in eleven cores. Upper right–lower left diagonal hachuring represents normal polarity; upper left–lower right hachuring represents reversed polarity. Upper left–lower right hachuring represents that interval between two oppositely polarized samples; no hachuring represents reversed polarity. Arrow indicates midpoint of extinction interval.

373

not appear to be declining gradually toward final disappearance. In fact, species in some cores are near a maximum just prior to their final decline (for example, *E. calvertense,* E13-3, Fig. 5; *C. bicornis* E13-3, V16-134, V18-72, Fig. 4; *P. prismatium,* RC11-213, Fig. 3; *D. spongiosa* and *H. vema* in E14-8, Fig. 6). (2) The upper limit or final decline of the species is nearly isochronous over a broad area and does not seem to be dependent on latitude or longitude. (3) For all species the final decline is in close proximity to a reversal of the Earth's magnetic field.

ACCURACY OF THE GEOLOGIC RECORD

There is apparently a high degree of correlation between the extinctions of these six species and certain reversals of the Earth's magnetic field. However, the degree to which the sedimentary record accurately represents the temporal relationship between the extinctions and reversals is dependent upon the amount of postdepositional alteration of both the reversal boundary and the upper limit of the species. Hiatuses are not uncommon in deep-sea sediments and could cause the apparent coincidence of two well-separated events. However, the number of cores in this study and their wide geographical distribution make the probability small that any of the species terminations and their corresponding reversals were ever separated by thick sedimentary sections. There may, however, be hiatuses that have altered the upper limits of some species. For example, the final occurrence of *C. bicornis* in all cores except E14-8 occurs above the Olduvai event (Fig. 4). In E14-8 it disappears abruptly at the reversal, suggesting there may be a hiatus in this core which has exactly superimposed the two events. This hiatus may be of short duration.

A second process that may alter a species final disappearance and the position of a magnetic reversal in the sediments is postdepositional activity of burrowing organisms.

If we assume the extinction of a species is instantaneous from a geological point of view (< 1,000 yrs and this extinction was not preceded by an interval of population decline), then in the absence of postdepositional mixing the species abundance should drop from normal abundance to zero in a centimeter or less. The activity of burrowing organisms will alter this pattern by reducing the abundance of individuals below the extinction point and carry them to higher levels in the sediments, thereby

producing a more gradual abundance decline in the sediment than occurred in the living population.

Berger and Heath (1968) have shown on theoretical grounds that the time of extinction in the sediment will be represented by the level at which the abundance of individuals falls to 37 percent of normal.

There is sufficient variation of the species considered in this study to make any firm estimate of normal abundance unwise. The final decline of all of the six species is sufficiently steep so that uncertainty in the abundances does not alter the level of the extinction point appreciably. Table 3 shows the interval in which the extinction is estimated to have occurred, based on 37 percent of abundance prior to final decline and its distance from the interval in which the adjacent reversal is recorded.

Postdepositional mixing may also alter the position of the imprint of a reversal in the sediments. In the highly fluid upper layers of ocean sediments it is conceivable that immediately following a geomagnetic reversal disturbance of some thickness of this layer by burrowing organisms, the previous direction of remnant magnetism may be erased and cause the magnetic particles to realign themselves in the direction of the new field (Watkins, 1967). This would cause a depression of the reversal boundary in the sediments. Presumably, the amount of depression would be a function of the thickness of the mixed layer. The consistent relationship between the terminations of the six species studied here and their adjacent reversals argues against the depression of reversals by distances much in excess of 35 cm. Differential depression of reversal boundaries a few tens of centimeters may in part explain the varying distances between reversal interval and extinction interval of a given species (Table 3).

STATISTICAL ANALYSIS

Since the correlation between the extinction point of the six species and the nearest reversal varies somewhat from core to core, and this variation may be accounted for, at least in part, by variations in mixing and other local events, a better approximation of the true correlation may be achieved by generating regression lines for the points produced by the relationship between each extinction and its corresponding reversal (Figs. 8 through 11). The size and shape of the data points in these figures are determined by the degree of accuracy with which both the position of the extinction point (horizontal dimension) and reversal (vertical dimension) could be measured. The slopes are in all cases near unity, the standard error of estimate small (7 to 22 cm) and, of some importance, the intercept is never zero but is always displaced along the Y axis from 11 cm to 26 cm, indicating that the extinction points fall above the reversals by about this much (Table 4). Since the standard deviation of the points around the regression line is small this line probably closely approximates the relationship between the reversal intervals and extinction intervals. It is interesting to note that the species with the largest number of data points (*Eucyrtidium matuyamai*) has the largest slope most nearly one, the next to the lowest standard deviation and the smallest intercept. Reduced major axes were computed for each extinction and its cor-

responding reversal and all had nearly the same slope and intercept as the regression lines.

The value of the intercept of the regression line on the Y axis is about the amount one would expect if the reversal and extinction were in fact coincident and the reversal was subsequently depressed by the activity of burrowing organisms. If this is the case then the correlation between the reversal intervals and extinction intervals is indeed striking. The data for the high sedimentation rate cores do not, on the average, lie farther from the regression line than those for low sedimentation rate cores. Therefore, the deviations of points from the regression line is probably not due to any real time difference.

The upper limit of the two Antarctic species *Desmospyris spongiosa* and *Helotholus vema* could

TABLE 3. RELATIONSHIP BETWEEN EXTINCTION POINT OF SIX RADIOLARIAN SPECIES AND NEIGHBORING REVERSAL

Core no.	Probable extinction interval (cm)	Reversal interval in cm	Average sedimentation rate cm/1,000 yrs	Distance between extinction interval and reversal interval (cm)	
		Pterocanium prismatium			
RC11-207	1,310-1,320	1,338-1,350	0.8	+18	22,500
RC11-208	1,100-1,110	1,140-1,150	0.7	+30	42,000
V24-58	970-980	990-1,000	0.6	+10	22,000
RC12-34	880-890	900-925	0.5	+10	2,000
RC11-213	800-820	840-850	0.5	+20	4,000
V24-60	750-760	780-790	0.5	+20	4,000
RC12-66	615-625	635-640	0.4	+10	2,500
V24-62	450-460	480-490	0.3	+20	7,000
		Eucyrtidium calvertense			
RC11-101	470-485	510-540	1.0	+25	25,000
E14-8	940-960	910-940	0.5	0	
RC8-52	880-890	880-890	0.5	0	
E13-3	650-660	650-680	0.4	0	
V18-72	400-410	390-400	0.2	0	
V16-134	375-380	370-390	0.2	0	
		Clathrocyclas bicornis			
RC-11-101	380-420	380-420	11.0	0	
E14-8	860-870	840-860	0.5	0	
RC8-52	780-790	810-820	0.5	+20	40,000
E13-3	570-580	600-610	0.4	-20	50,000
V18-72	350-360	360-370	0.2	0	
V16-134	330-340	340-350	0.2	0	
		Eucyrtidium matuyamai			
RC10-182	1,082-1,092	1,090-1,100	1.2	+10	< 10,000
V20-108	1,040-1,050	1,068-1,072	1.2	+10	< 10,000
RC10-206	1,080-1,090	1,055-1,062	1.1	-20	< 20,000
RC10-181	990-1,015	990-1,000	1.1	0	
V20-119	750-770	760-770	0.9	0	
V21-172	745-785	750-760	0.8	0	
V20-107	685-695	710-720	0.8	+15	20,000
RC11-171	560-570	570-580	0.6	0	
V21-173	490-500	495-510	6.0	0	
RC11-170	430-440	437-445	5.0	0	
V20-109	330-335	345-347	4.0	+10	2,500
		Helotholus vema			
E14-8	1,090-1,080	1,095-1,105	0.5	+5	10,000
E13-3	900-870	860-870	0.4	0	
		Desmospyris spongiosa			
E14-8	1,100-1,080	1,095-1,105	0.5	0	
E13-3	840-870	860-870	0.4	0	

(+) extinction above reversal

(-) extinction below reversal

be determined in only two cores (E13-3 and E14-8). Although regression lines based on only two cores have little meaning the relationship between the abundance curves of these species and the reversal at the top of the Gauss is similar to the other four species that disappear near reversals (Fig. 6).

Since the degree of correlation is so high between the extinction of each of these species and its corresponding reversal, what then is the probability of six out of eight Radiolaria becoming extinct this close to reversals of the Earth's magnetic field in the last 2.5 m.y.? Two standard errors of estimate above and below the regression lines will include 95 percent of the data. *Eucyrtidium calvertense* has the largest standard error of estimate (22.3 cm), so for this species two standard errors of estimate are $\pm$ 45 cm from the regression line. The core with the highest sedimentation rate that contains this species is RC11-101 (accumulation rate across Olduvai event about 1 cm/1,000 yrs).

The time interval included by $\pm$ 2 standard errors of estimate in this core is $\pm$ 45,000 yrs. This interval of $\pm$ 45 cm from the regression lines of the other three species easily includes $\pm$ 2 standard errors of estimate. Since the data for these regression lines also include cores with rates of accumulation similar to that of RC11-101, $\pm$ 45,000 yrs includes 95 percent of the cases for all four species.

The probability of any one species becoming extinct within 45,000 yrs of a reversal is

$$\frac{90,000 \times 6}{2,500,000} \text{ or approximately one-fifth, assum-}$$

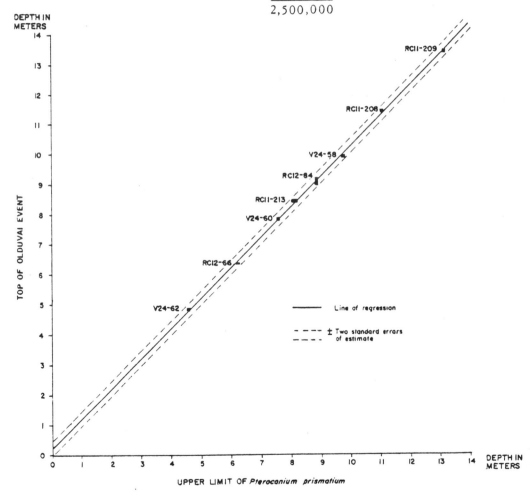

Figure 8.   Least-squares regression line of extinction point of *Pterocanium prismatium* against reversal at top of Olduvai event based on midpoints of extinction and reversal intervals.

376

ing that extinctions are equally likely to occur at any time and extinctions are independent of reversals.

Let $X_i = 1$ if a species becomes extinct within 45,000 yrs of a reversal, and X_o if not.

$$\text{Let } S = \sum_1^n X_i$$

What is the probability if $S \geqslant 6$, $N = 8$, and $p = 1/5$?

$$Ph_o(S \geqslant 6) = \sum_{\geqslant 6}^{8} (n)\, p^s\, (1-p)^{n-s} = 1.4 \times 10^{-3}$$

The data are strongly suggestive that reversals directly or indirectly exert a selective influence on Radiolaria. The foraminiferal data of

Saito (*in* Hays and others, 1969) provide additional evidence for this hypothesis. It is important to note that diatom extinctions during the last 2.5 m.y. in high and low latitudes (Hays and others, 1969; Donahue, 1970) show a low degree of correlation with magnetic reversals.

Whatever the selective force that magnetic field reversals may exert on animal life, the effect is such that it results in the extinction of only one or two species of those groups studied at any reversal and these same species may have survived many reversals before becoming extinct near one. Assuming that all reversals are similar, then species, if affected by reversals, must develop some vulnerability to them. The drop in abundance of some species at the penultimate reversal may be an indication of this.

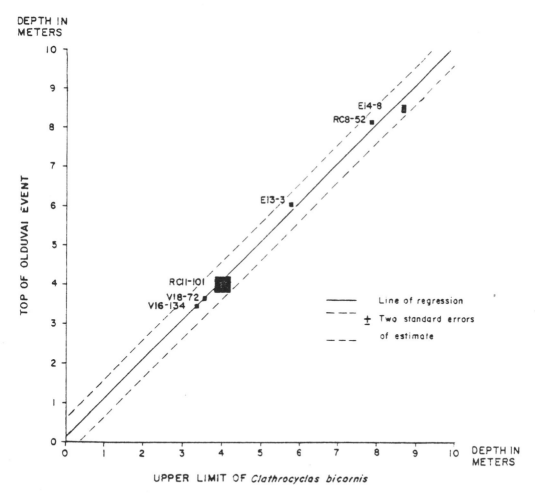

Figure 9.   Least-squares regression line of extinction point of *Clathrocyclas bicornis* against reversal at top of Olduvai event based on midpoints of extinction and reversal intervals.

The possible importance of species evolution to the development of vulnerability to a reversal is demonstrated by a study of the species *E. matuyamai* (Hays, 1970). It was shown that the rapidly evolving *E. matuyamai* became extinct near the reversal at the base of the Jaramillo event while its more conservative immediate forebearer survived to the Holocene.

It is also important to note that in the Equatorial Pacific during the last 2.5 m.y. only two species became extinct, while in the Antarctic five became extinct although the Equatorial fauna is far more diverse than the Antarctic fauna. Since there were, of course, the same number of reversals in both regions the response of the Antarctic fauna suggests that stresses other than magnetic reversals were important.

MASS EXTINCTIONS DURING THE PHANEROZOIC AND THE REVERSAL RECORD

If reversals of the Earth's magnetic field exert a selective force on organisms, then the frequency with which reversals occur may be important in the role reversals play in evolution (Simpson, 1966).

The Earth's magnetic field has reserved its polarity at least 171 times during the Cenozoic (Heirtzler and others, 1968). The average duration of each period of constant polarity is about 0.22 m.y. for the past 10 m.y. (Heirtzler and others, 1968) and somewhat longer (about 1.0 m.y.) for the early Tertiary (Heirtzler and others, 1968; Cox, 1968). All the data for the Tertiary supports the assumption that the field has spent about equal amounts of time in the

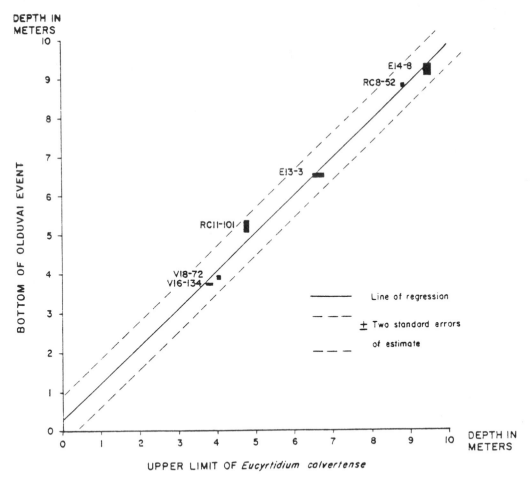

Figure 10.   Least-squares regression line of extinction point of *Eucyrtidium calvertense* against reversal at base of Olduvai event based on midpoints of extinction and reversal intervals.

reversed and normal states (Heirtzler and others, 1968). There is substantial evidence that the reversal pattern for the Tertiary is not representative of all geologic time, and there were several intervals during the Phanerozoic when the field maintained a constant polarity for long intervals interrupted only occasionally by short intervals of the opposite polarity.

One of the most striking of these is the late Paleozoic Kiamen magnetic interval (Irving, 1966; Irving and Perry, 1963), when a reversed field persisted during some 50 m.y. of late Carboniferous and Permian time. The uppermost Permian (Tartarian), however, like the Lower Triassic, has numerous reversals (Khramov and others, 1966).

In the Mesozoic, Helsley and Steiner (1969) have documented long intervals of constantly normal polarity in the Upper Cretaceous. The

sampling to date indicates the field was normal from the upper Albian to the Maestrichtian, with a short interval of reversed polarity in the Santonian. The Maestrichtian, like the lower Tertiary, contains intervals of both polarities.

DeBoer (1968) has suggested there may be a lengthy period of constant polarity within the Triassic; however, this has not yet been confirmed.

Extinctions of animal taxa during the Phanerozoic have been the subject of much discussion. Newell (1963) has summarized much of the information on extinctions and concludes that times of increased extinction rates occur at the close of the Cambrian, Ordovician, Devonian, Permian, Triassic, and Cretaceous. The extinctions at the close of the Permian and Cretaceous are the most profound. An important part of these times of extinction is the

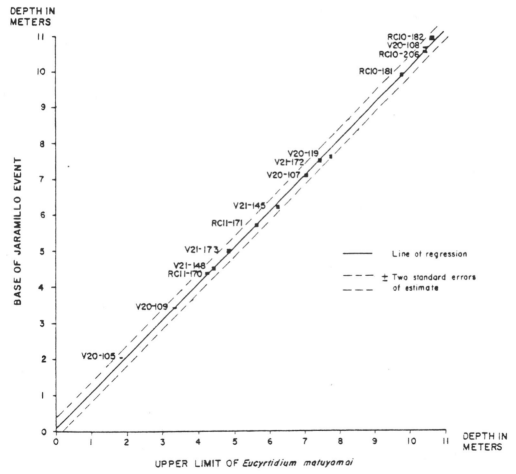

DEPTH IN
METERS

Figure 11.   Least-squares regression line of extinction point of *Eucyrtidium matuyamai* against reversal at base of Jaramillo event based on midpoints of extinction and reversal intervals.

TABLE 4.   STATISTICAL PARAMETERS OF REGRESSION LINES
GENERATED BY RELATIONSHIP BETWEEN THE EXTINCTION
POINTS OF FOUR RADIOLARIA SPECIES
AND THEIR NEIGHBORING REVERSALS

Species	Slope	Intercept	Std. error of estimate	Correlation coefficient
P. prismatium	1.01	22.7	7.5	0.9995
C. bicornis	0.99	13.3	17.1	0.9982
S. salvertense	0.96	26.2	22.3	0.9970
S. matuyamai	1.00	11.1	10.1	0.9995

diversity of groups involved. Toward the close of the Permian, for example, perhaps one of the largest mass extinctions occurred when nearly half of the known families of animals disappeared (Newell, 1963). The groups affected included the protozoans (Fusulinids), Ammonites (only three families survived from the Permian into the Triassic), and land dwelling tetrapods. Colbert (1965) has noted that 204 tetrapod genera occur in the Upper Permian, while only 48 have been found in the Lower Triassic. Although certainly preservation and the quantity of sediment of these ages exposed may influence these numbers, the figures give some measure of the magnitude of extinctions that occurred during the transition from Permian to Triassic times.

The close of the Cretaceous also witnessed many extinctions that included a number of planktonic protista, ammonites, rudistids, dinosaurs, marine reptiles, and flying reptiles. Some of these were on the decline before the end of the Cretaceous but others, such as the horned dinosaurs, were expanding just prior to their extinction (Colbert, 1965).

Not only did these times of extinction affect a wide variety of animals but they were worldwide in extent, affecting both the marine and terrestrial environment. These times of extinction were not geologically instantaneous but certainly covered many millions of years. It is possible that the causes of these diverse extinctions were also diverse yet, if so, it is surprising that they should have occurred at the same time. On the other hand, any single cause advanced to explain these extinctions must have affected all environments and acted on such a fundamental level that a number of diverse groups were vulnerable to its influence.

The depletion of groups at the close of the Permian and Cretaceous is striking. It may not be fortuitous that these times of extinction coincide with the re-establishment of frequent reversals following long intervals of nearly constant polarity. This relationship raises an important question. Is it possible, if magnetic reversals exert some selective force (as the radiolarian data suggest), that the removal of this selective force during long intervals of nearly constant polarity, such as the Permo-Carboniferous Kiamen interval or the Cretaceous, will allow organisms that are potentially susceptible to reversal to proliferate? These then would become extinct when a high frequency of reversals again occurred. If the number of reversal-susceptible species were large enough it could seriously affect other organisms dependent upon them which are not themselves vulnerable to field reversals.

The low degree of correlation between diatom extinctions and reversals and the absence of major depletions of plant taxa at times of mass animal extinctions suggests that plants are not vulnerable to the same factors that eliminated some animals.

POSSIBLE EFFECTS OF GEOMAGNETIC REVERSALS

How geomagnetic reversals might affect organisms must be a matter of speculation at this time. Two possibilities have already been proposed: (1) increased cosmic radiation at the time of a reversal (Uffen, 1963); and (2) that reversals may cause climatic changes that in turn influence organisms (Harrison, 1968). A third possibility, that organisms interact with the Earth's magnetic field, should also be considered.

Uffen (1963) suggested that if the Earth's dipole field is reduced to zero (Bullard, 1955) at the time of a magnetic reversal then the shielding effect of the field would be largely removed and the Earth's atmosphere and surface would be subjected to a higher incidence of cosmic radiation than when the field was at full strength, possibly inducing mutations and extinctions. Simpson (1966), following the thinking of Uffen, attempted to correlate intervals of high frequency reversals during the Phanerozoic with periods of accelerated evolution. Several investigators (Waddington, 1967; Black, 1967: Harrison, 1968) examined the theoretical increase in radiation expected if the Earth's dipole field were removed and concluded that the radiation increase at the Earth's surface during a reversal would be negligible. Their conclusions are supported by the fact that, if anything, there are more radiolarian extinctions occurring near reversals in high latitudes where the shielding effect of the field is minimal than in low latitudes.

Harrison (1968) suggested that reversals may be accompanied by climatic changes. He reasoned that the removal of the Earth's magnetic field would cause substantial changes in the upper atmosphere which might in return influence the climate at the Earth's surface. The mechanism by which fluctuations in the upper atmosphere would affect conditions at the Earth's surface, however, is not clear.

There is some evidence of climatic change near some reversals. At the Brunhes/Matuyama boundary, for example, in Antarctic (Hays and Opdyke, 1967) and Equatorial Pacific cores (Hays and others, 1969), there is evidence of cooling. In the Antarctic, again at the reversal that marks the boundary between the Gauss and the Matuyama epochs, there is also evidence of cooling (Hays and Opdyke, 1967). Not all reversals are accompanied by evidence of climatic change and there is no reason to believe there is a causal connection between the climatic change and the reversals. Nevertheless, the posssibility that there is some connection between reversals, climatic changes, and extinctions cannot be ruled out at this time.

A third possibility is that geomagnetic reversals directly affect organisms. A considerable literature now exists on the biological effects of magnetic fields (Barnothy, 1969). That organisms respond to magnetic fields of various intensities, frequencies, and directions has now been well established (Barnothy, 1964, 1969).

Magnetotropism of plants has been demonstrated (Audus, 1960; Audus and Whish, 1964; Pittman, 1964). Levengood (1967) concluded that *Drosophila melanogaster* pupae subjected to strong magnetic fields exhibited morphogenic anomalies which were transmitted for 30 generations. For our purposes the observations made of animals in null or weak magnetic fields is probably most important (Conley, 1969).

A variety of organisms have now been shown to be sensitive to very weak magnetic fields similar in intensity to the geomagnetic field. The mud-snail, *Nassarius,* is sensitive to a magnetic field as weak as 1.5 gauss and is able to distinguish between fields parallel and at 90° to its path of locomotion (Brown and others, 1960). A similar capacity has been reported for *Volvox* subjected to a 5 gauss field (Palmer,

1963). The planarian *Dugesia* also displays sensitivity to very weak magnetic fields (Brown, 1962), as does the fruit fly *Drosophila* (Barnwell and Brown, 1961, 1964).

Of considerable importance is the fact that the mud snail *Nassarius* is most responsive to the horinzontal components of a magnetic field of about 0.17 gauss similar to the horizontal component of the geomagnetic field (Brown and others, 1964), suggesting adaptation to the Earth's magnetic field.

This response of a number of unrelated organisms to magnetic fields similar in strength to the geomagnetic field suggests a possible biological importance of this geophysical parameter and its adaptive significance. The mechanism by which magnetic fields affect organisms is still a matter of speculation. Although more experimental work is needed if other organisms than those now studied also show a responsiveness to weak magnetic fields, then geomagnetic reversals could have had considerable evolutionary significance. It is quite conceivable that the collapse of the dipole field at the time of a geomagnetic reversal, and possible concomitant fluctuations in its intensity and frequency dissimilar to those now characteristic of the Earth's field, might have disturbing effects on some organisms that had grown accustomed and perhaps adapted to the natural strengths and rhythms of the field over hundreds of thousands of years. By comparison with climatic changes geomagnetic field reversals are swift. Ninkovich and others (1966) have estimated that the polarity of the field reverses in less than 2,000 yrs and the total duration of reduced field intensity is not more than 10,000 yrs before or after the polarity reversal.

An appeal to geomagnetic reversals as a possible cause of extinctions either of individual species, as the Radiolaria discussed above, or exerting an environmental stress important in mass extinctions is completely in accord with the principal of uniformitarianism. Polarity reversals have been a characteristic of the geomagnetic field for as far back as the remnant magnetism of rocks can be reliably measured. It is also compatible with the reasoning of Bretsky and Lorenz (1970), who argue that late Paleozoic extinctions were caused as much by mid-Paleozoic environmental stability as they were by late Paleozoic environmental instability.

Another possible explanation of the mass extinctions at the close of long intervals of constant polarity may be found in the suggestion by

Irving (1966) that periods of rapid polar wandering are accompanied by intervals of frequent reversals, while the geomagnetic polarity is more constant during intervals of slow polar wandering. If polar wandering and continental drift are the same phenomena, or at least closely related, then we might expect environmental change to accompany rapid continental drift.

This argument, however, could not be applied to the close correlation between individual reversals and species extinctions.

It is too early to place great emphasis on any of these possible explanations for the correlation between extinctions and geomagnetic reversals, even though the direct effect of the field on the organism is preferred. It may be best in closing to look at the problem in the broadest of terms. Through the processes of evolution organisms change physiologically to adapt to their environment and to variations within their environment. The Earth's magnetic field is a part of the environment and some organisms may adapt to it and thereby become progressively more vulnerable to the abrupt change of the field intensity and direction that occurs at a reversal.

The data gathered to date indicates that for most Radiolaria and Foraminifera during the last 2.5 m.y. the effects of reversals were not sufficient to cause their extinction, and for those that did become extinct near reversals previous reversals did not severely affect them. However, the data now in hand is strongly suggestive that for some species of Radiolaria reversals did influence their extinction. Mass extinctions at the close of long intervals of dominantly one polarity, such as the Cretaceous and Permo-Carboniferous, is again suggestive that the reversing field may have a selective influence. Experimental evidence has shown biological effects of null magnetic fields and fields approximating the strength of the Earth's magnetic field, raising the possibility that the action of the reversing magnetic field may cause extinctions.

ACKNOWLEDGMENTS

The author is greatly indebted to a number of staff members and students at the Lamont-Doherty Geological Observatory for very fruitful and stimulating discussions; in particular, N.D. Opdyke, W. S. Broecker, J. Nafe, K. King, and D. Kellogg, as well as B. Singer and J. Roth of the Mathematics Department of Columbia University. I would also like to acknowledge helpful discussions with J. Imbrie of Brown University and N. D. Watkins of the University of Rhode Island. The cores were gathered and the research supported with funds from grants GA-1193, GA-19690, GA-4499, and GA-21174 of the National Science Foundation, and Grant N00014-67-A-0108-0004 of the Office of Naval Research.

REFERENCES CITED

Audus, L. J., 1960, Magnetotropism, a new plant growth response: Nature, v. 185, p. 132.

Audus, L. J., and Whish, J. L., 1964, in Barnothy, M.F., ed., Biological effects of magnetic fields: Plenum Press, New York.

Barnothy, M. F., ed., 1964, Biological effects of magnetic fields: Plenum Press, New York.

—— ed., 1969, Biological effects of magnetic fields: Plenum Press, New York.

Barnwell, F. H., and Brown, F. A., Jr., 1961, Magnetic and photic responses in snails: Experientia, v. 4, p. 513.

—— 1964, Responses of planarians and snails, in Barnothy, M.F., ed., Biological effects of magnetic fields: Plenum Press, New York.

Berger, W. H., and Heath, G. R., 1968, Vertical mixing in pelagic sediments: Jour. Marine Research, v. 26, p. 134-143.

Black, D. I., 1967, Cosmic ray effects and faunal extinctions at geomagnetic field reversals: Earth and Planetary Sci. Letters, v. 3, p. 225-236.

Bretsky, P. W., and Lorenz, D. M., 1970, An essay on genetic-adaptive strategies and mass extinctions: Geol. Soc. America Bull., v. 81, p. 2449-2456.

Brown, F. A., 1962, Responses of the planarian Dugesia, and the Protozoan, Paramecium, to very weak horizontal magnetic fields: Biol. Bull., v. 123, p. 264-281.

Brown, F. A., Brett, W. J., Bennett, M. F., and Barnwell, F. H., 1960, Magnetic response of an organism and its solar relationship: Biol. Bull., v. 118, p. 367-381.

Brown, F. A., Barnwell, F. H., and Webb, H. M., 1964, Adaptation of the magnetoreceptive mechanism of mudsnails to geomagnetic strength: Biol. Bull., v. 127, p. 221-231.

Bullard, E. C., 1955, The stability of a homopolar dynamo: Cambridge Philos. Soc. Proc., v. 51, p. 744-760.

Colbert, E. H., 1965, in The age of reptiles: W. W. Norton & Co., New York.

Conley, C. C., 1969, Effects of near-zero magnetic fields upon biological systems, in Barnothy, M.F., ed., Biological effects of magnetic fields: Plenum Press, New York.

Cox, A., 1968, Lengths of geomagnetic polarity intervals: Jour. Geophys. Research, v. 73, no. 10, p. 3247-3260.

de Boer, J., 1968, Paleomagnetic differentiation and correlation of the Late Triassic volcanic rocks in the central Applachians (with special reference to the Connecticut Valley): Geol. Soc. America Bull., v. 79, p. 609-626.

Donahue, J. G., 1970, Diatoms as indicators of Pleistocene climatic fluctuations in high latitudes of the Pacific Ocean [Ph.D. dissert.]: Columbia Univ., New York.

Foster, J., and Opdyke, N. D., 1970, Upper Miocene to Recent magnetic stratigraphy in deep-sea sediments: Jour. Geophys. Research, v. 75, p. 4465-4473.

Harrison, C. G. A., 1968, Evolutionary processes and reversals of the Earth's magnetic field: Nature, v. 217, p. 46-47.

Harrison, C.G.A., and Funnel, B. M., 1964, Relationship of paleomagnetic reversals and micropaleontology in two late Cenozoic cores from the Pacific Ocean: Nature, v. 204, p. 566.

Hays, J. D., 1965, Radiolaria and late Tertiary and Quaternary history of Antarctic Seas: Biology of the Antarctic Sea, II: Antarctic Research, Ser. 5, p. 125-184.

—— ed., 1970, The stratigraphy and evolutionary trends of Radiolaria in North Pacific deep-sea sediments, in Geological investigations of the North Pacific: Geol. Soc. America Mem. 126, p. 185-218.

Hays, J. D., and Opdyke, N. D., 1967, Antarctic Radiolaria, magnetic reversals and climatic change: Science, v. 158, p. 1001-1011.

Hays, J. D., Saito, T., Opdyke, N. D., and Burckle, L., 1969, Pliocene/Pleistocene sediments of the Equatorial Pacific: their paleomagnetic, biostratigraphic, and climatic record: Geol. Soc. America Bull., v. 80, p. 1481-1514.

Heirtzler, J. R., Dickson, G. O., Herron, E. M., Pitman, W. C. III, and LePichon, X., 1968, Marine magnetic anomalies, geomagnetic field reversals, and motions of the ocean floor and continents: Jour. Geophys. Research, v. 73, p. 2119-2136.

Helsley, C. E., and Steiner, M. B., 1969, Evidence for long intervals of normal polarity during the Cretaceous period: Earth and Planetary Sci. Letters, v. 5, p. 325-332.

Irving, E., 1966, Paleomagnetism of some carboniferous rocks from New South Wales and its relation to geological events: Jour. Geophys. Research, v. 71, p. 6025-6051.

Irving, E., and Perry, L. G., 1963, Magnetism of some Permian rocks from New South Wales: Jour. Geophys. Research, v. 7, p. 395-411.

Khramov, A. N., Radinov, V. P., and Kosissarova, R. A., 1966, in The present and the past of the geomagnetic field: Nauka Press, Moscow, p. 206-213.

Levengood, W. C., 1967, Morphogenesis as influenced by locally administered magnetic fields: Biophysics Jour., v. 7, p. 297-307.

Newell, N. D., 1963, Crises in the history of life: Sci. American, v. 208, p. 76-92.

Ninkovich, D., Opdyke, N. D., Heezen, B. C., and Foster, J., 1966, Paleomagnetic stratigraphy, rates of deposition and tephra-chronology in North Pacific deep-sea sediments: Earth and Planetary Sci. Letters, v. 1, p. 476-492.

Opdyke, N. D., Glass, B., Hays, J. D., and Foster, J., 1966, A paleomagnetic study of Antarctic deep-sea cores: Science, v. 154, p. 349-357.

Opdyke, N. D., and Foster, J., 1970, The paleomagnetism of cores from the North Pacific, in Hays, J. D., ed., Geological investigations of the North Pacific: Geol. Soc. America Mem. 126, p. 83-119.

Palmer, J. D., 1963, Organismic spatial orientation in very weak magnetic fields: Nature, v. 198, p. 1061-1062.

Pittman, U. J., 1964, Magnetism and plant growth, II. Effect on root growth of cereals: Canadian Jour. Plant Sci., v. 44, p. 283-294.

Ryan, W., 1970, The floor of the Mediterranean [Ph.D. dissert.]: Columbia Univ., New York.

Shaw, A. B., 1964, Time in stratigraphy: McGraw-Hill Book Co., New York.

Simpson, J. F., 1966, Evolutionary pulsations and geomagnetic polarity: Geol. Soc. America Bull., v. 77, p. 197-204.

Uffen, R. J., 1963, Influence of the earth's core on the origin and evolution of life: Nature, v. 198, p. 143.

Waddington, C. J., 1967, Paleomagnetic field reversals and cosmic radiation: Science, v. 158, p. 913-915.

Watkins, N. D., 1967, Short period geomagnetic polarity events in deep-sea sedimentary cores: Earth and Planetary Sci. Letters, v. 4, p. 341-349.

MANUSCRIPT RECEIVED BY THE SOCIETY JANUARY 20, 1971

LAMONT-DOHERTY GEOLOGICAL OBSERVATORY CONTRIBUTION NO. 1677

32

Reprinted from *Earth and Planetary Sci. Letters* **3**:225 (1967)

COSMIC RAY EFFECTS AND FAUNAL EXTINCTIONS AT GEOMAGNETIC FIELD REVERSALS

D. I. BLACK

*Department of Geodesy and Geophysics, University of Cambridge,
Cambridge, England*

Studies of deep-sea cores show that the last reversal of the earth's magnetic field and the extinction of a group of Radiolarian species occurred simultaneously: the evidence suggests a possible causal relation. Some effects on the earth and atmosphere at a field reversal are quantitatively considered, but it is shown that no direct or indirect radiation effect from cosmic rays or solar flares, nor any solar wind effect, can be a significant factor in the extinction of the species. The grounds for believing in a causal relation between magnetic and faunal changes are examined.

[*Editor's Note:* The article itself has not been reproduced here.]

33

Reprinted from *Nature* **227**:930–934 (1970)

Geomagnetic Polarity Change, Volcanic Maxima and Faunal Extinction in the South Pacific

by
J. P. KENNETT
N. D. WATKINS
Graduate School of Oceanography,
University of Rhode Island,
Kingston, Rhode Island 02881

Studies of deep-sea sedimentary cores from Antarctic Pacific waters show that some volcanic maxima occurred when the geomagnetic polarity was changing. Upper mantle activity and geomagnetic polarity change may therefore be related. Coincidences of faunal extinction and geomagnetic polarity change may be explained by corresponding volcanically induced climatic changes.

GEOMAGNETIC field reversals occurred more than twenty times during the past 4 m.y. (ref. 1) and probably more than one hundred times during the Tertiary[2]. The most significant aspect of this discovery is the recognition of important features of the history of geomagnetic polarity[1] in the linear magnetic anomalies of the oceanic crust[3,4], and their convincing interpretation in terms of sea-floor spreading[5-8].

The geomagnetic polarity history has also been recognized in deep-sea sedimentary cores[9,10]. In such studies, several microfaunal extinctions and appearances have been recognized as synchronous, and virtually simultaneous with the period of transition between opposite geomagnetic polarities[9-13] which lasts about 5,000 years[14]. The reasons for these extinctions and appearances are unknown, but they may be the result of climatic change and polarity change being related[15,16], or of increased mutation rates during polarity changes[17], although several authors[15,18,19] oppose this possibility.

Heirtzler[20] has recently speculated on a relationship between earthquake activity (and by implication upper mantle activity) and geomagnetic polarity change. He reasons that because there is evidence to show that earthquakes of magnitude 7·5 or greater may cause wobble of the spin axis[21], it is therefore conceivable that an Earth wobble may be of a magnitude sufficient to cause reversal of the geomagnetic field. Here we present data pertinent to Heirtzler's speculations[20], and we expand such speculations to include our preferred explanation of the synchronous geomagnetic polarity changes and faunal extinctions.

We know of no method to detect the time of occurrence and frequency of large earthquakes in the geological past. Stress release is often manifested in volcanic activity, however, so a search may be profitably made for evidence of relationships between volcanic maxima and geomagnetic polarity change. Volcanism certainly may have more than one cause, and on a global scale must be virtually continuous, so we therefore emphasize our use of the term "volcanic maxima". If relationships exist between polarity change and upper mantle activity they are more likely to be found in oceanic areas, for there the mantle is most mobile and accessible, and (in all probability) more commonly at "threshold", in terms of potential volcanism. Our search therefore examines the character of volcanic activity during polarity change in some deep-sea sedimentary cores.

The cores used were collected during cruises of USNS Eltanin and their locations are included in Fig. 1 and Table 1. Core selection was based only on length, minimum age, and geographic position. Collection, transportation, and sampling procedures have been described elsewhere[22]. In this study, specimens for palaeomagnetic study were taken at 10 cm intervals, demagnetized at 150 oersteds using a four axis tumbler[23] and measured using a 5 Hz spinner magnetometer[24].

At 20 to 30 cm intervals the palaeomagnetic specimens were then weighed dry, disaggregated and washed on a 63 µm mesh Tyler screen. The sand-size fraction was examined to determine the relative abundance of volcanic glass shards and the nature of the micro-fossils. Restriction to the sand-sized fraction probably eliminates much

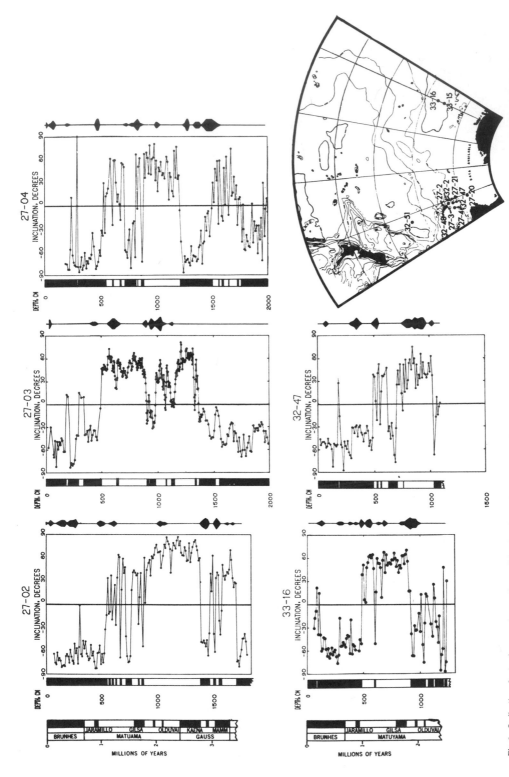

Fig. 1. Inclination of natural remanent magnetism, and fraction of sand-sized ash in five selected cores. Location of these and other cores (examined in Fig. 2) is given in the inset, and in Table 1 (core number at head of each plot). In each core, the inclination of natural remanent magnetism is shown for each specimen. Negative inclination (above horizontal) is normal polarity. Inclination without sign is positive (below horizontal), or reversed polarity. Depth of specimen in core shown next to each plot, where a polarity log is also shown: this is a non-subjective plot of the inclination data, with black = normal polarity; clear = reversed polarity. The polarity timescale at the left is that due to Cox; it consists, from left to right, of the age in m.y.; the names of the polarity epochs; the names of the polarity events, and the polarity history. The histogram to the right of each log expresses the fraction of sand-sized volcanic glass shards and other volcanic material in the given specimen. It results from examination using a binocular microscope and is defined using micropalaeontological conventions. The width definition is five-fold as follows: width 4 (widest) = very abundant volcanic material; 3 = abundant; 2 = common; 1 = uncommon; 0 (fine line) = very rare or absent. Note that the inclination plot of core E27–03 includes a sequence of data from specimens taken at 2 cm intervals (for a special study of the Matuyama epoch polarity history) and subjected to a five-point running average. The corresponding polarity log is a slightly subjective interpretation of these data.

386

TABLE 1. DETAILS OF CORE LOCATION AND STATISTICAL ANALYSIS OF THE RELATIONSHIP BETWEEN MAGNETIC POLARITY BOUNDARIES AND VOLCANIC ASH HORIZONS

Core No.	Latitude (degrees south)	Longitude (degrees east)	Water depth (m)	Core length (m)	Percentage ash (Pa)	No. of reversals	No. of coincidences	Pa,g
E27–02	63° 00′	177° 42′	3,358	16·97	42·6	8	4	0·469
E27–03	66° 04′	176° 30′	3,520	21·66	34·1	8	6	0·023
E27–04	68° 03′	174° 35′	3,433	19·89	37·3	8	5	0·137
E27–20	71° 57′	178° 36′	2,116	5·20	—	—	—	—
E27–21	69° 02′	179° 50′	3,440	16·07	34·1	8	3	0·572
E32–02	67° 22′	178° 44′	3,656	10·21	67·1	3	3	0·303
E32–47	68° 06′	176° 10′	3,413	10·93	45·4	5	5	0·019
E32–51	58° 50′	176° 52′	4,845	7·53	—	—	—	—
E33–15	65° 07′	199° 45′	4,823	11·52	21·6	8	5	0·158
E33–16	63° 14′	199° 59′	4,890	15·03	67·3	8	6	0·489

Percentage ash is based on Fig. 1: it is the fraction of the core in which any sand-sized ash is found, and is not a relative volume count. It follows that this is also the probability (Pa) of finding sand-sized ash in any part of each core. Number of reversals is the number of known polarity changes during the last 2·5 m.y., or less if the given core does not reach that age. Number of coincidences is the number of times that an ash occurrence is found at a polarity change in Fig. 2. A ±30 cm range is allowed for detection of such coincidences, because reworking and polarity recording imperfections are likely to be present. Pa,g is the probability that the observed number of coincidences of ash and geomagnetic polarity change are simply the result of coincidence of two randomly occurring phenomena. No analysis is given for E32–51: this is illustrated in Fig. 2 (right) and is not amenable to analysis.

very fine volcanic material, which simplifies the search for evidence of regional volcanic maxima.

Sediment type in the cores varies greatly. It includes lutite, silicious and carbonate ooze, and glacial-marine material. Most cores lack planktonic foraminifera, because of their solution at the depth of the cores, but when present they are restricted to one species, *Globigerina pachyderma*. Radiolaria are much more diverse, and are often abundant throughout the cores, and frequently offer the only micropalaeontological control possible.

The palaeomagnetic results are shown in part in Fig. 1 and completely in Fig. 2. Material as old as the Gilbert Epoch is identified. The correlation lines for the major polarity (epoch) boundaries have been readily confirmed by applying the radiolarian zonation established by Opdyke *et al.*[9] and Hays and Opdyke[13]. The species which were particularly valuable in our correlations were *Prunopyle titan* and *Desmospyris spongiosa* (the extinction of which coincides approximately with the Gauss–Matuyama boundary); and (in northern cores) *Eucyrti-*

dium calvertense whose extinction is, we believe, associated with the Gilsa event[1] and *Saturnulus planetes*, which disappeared near the Brunhes–Matuyama boundary (0·7 m.y.). Correlation of the shorter duration magnetic polarity events is shown in Fig. 2 only for the events in the Matuyama epoch. Palaeontological controls for these are simply not available locally, and the correlations are therefore made subjectively using the magnetic data alone and, where these are not clear, by extrapolation of sedimentation rates. Question marks are added, where appropriate, to Fig. 2.

Volcanic shards, which usually do not form megascopically distinct layers, were found in all the sedimentary cores examined for this study. Within each core large variations in the relative abundance of the shards occur, ranging from complete dominance to virtual absence (Figs. 1 and 2). The glass shards occur mostly as flaky fragments, very similar to those illustrated by Ninkovitch[25,26]. The data clearly show periods of volcanic maxima. A general increase in the size of the glass shards, and in the amount of other volcanic debris, such as small volcanic pellets, occurs in cores nearer the Balleny Islands, suggesting that the source is at least in part in this vicinity. Similar but finer volcanic glass is found at the same palaeomagnetic horizon in the central Pacific (core E33–16, Fig. 1) strongly suggesting either contemporaneous activity in the area, or (more likely) a very large fallout area, possibly larger than previous considerations allow[27].

Figs. 1 and 2 clearly show that when geomagnetic polarity changes were taking place, volcanic maxima were also occurring, at least locally. Volcanic fragments are most abundant and consistent in their occurrence close to the Brunhes–Matuyama boundary (0·7 m.y.), near the Matuyama–Gauss boundary (2·4 m.y.), and during at least two periods within the Brunhes epoch (0 to 0·7 m.y.). Other less well defined horizons are related in part to the sediments of the shorter duration events, and the Gauss–Gilbert boundary. Volcanism was clearly taking place locally during other periods. It is intriguing that the Brunhes–Matuyama and Matuyama–Gauss polarity changes are remote from evidence of

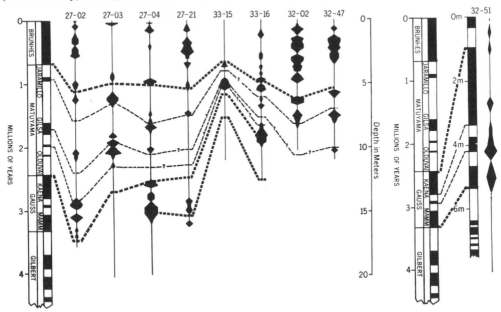

Fig. 2. Histograms of volcanic material fraction in each core and the geomagnetic correlation lines. For description of the volcanic fraction histogram construction see legend to Fig. 1. Geomagnetic polarity time scale after Cox[1] as in Fig. 1. The correlations at the three magnetic epoch boundaries are given by heavy dashed lines. These are confirmed micropalaeontologically. The correlations of the polarity events within the Matuyama epoch are based on magnetic and sedimentation rate data only. Core 32–51 features a major disconformity, which results in most of the upper 2·4 m.y. of the core being absent.

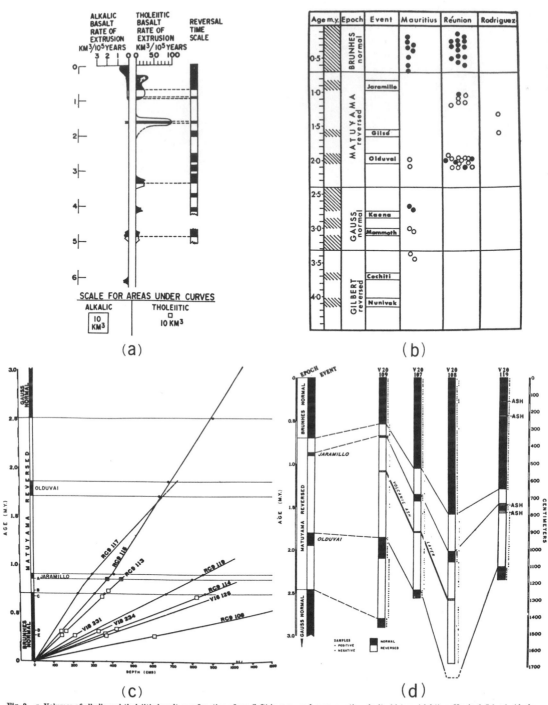

Fig. 3. *a*, Volumes of alkalic and tholeiitic basalt as a function of age (left) in m.y., and geomagnetic polarity history (right) on Nunivak Island, Alaska. Geomagnetic polarity scale is clear for reversed polarity; black for normal polarity. From Hoare *et al.*[31]. *b*, Plot of potassium-argon age and polarity for 53 igneous units on the islands of Mauritius, Réunion, and Rodriguez, from McDougall and Chamalaun[32]. Normal polarity ages are black; reversed polarity ages are open. The polarity time scale at the left is modified after McDougall and Chamalaun[32]; striped is normal polarity; clear is reversed polarity. It is emphasized that the grouping of data may reflect sampling biases. *c*, Plot of age versus depth in the core for seven deep-sea cores from east of New Zealand, from Ninkovitch[33]. The age determinations are the palaeomagnetic time lines, as shown by the polarity time scale. This polarity time scale is the earlier less-complete version of 1966 as discussed by Cox[1]. The depth of five ash layers (*A* to *E*) is shown in the cores. Core numbers are given on each plot. For further details see Ninkovitch[33]. *d*, Distribution of volcanic ash layers in four cores from the North Pacific from Ninkovitch *et al.*[34]. Polarity scale as in *c*. Core numbers at head of each palaeomagnetic polarity log (black=normal; clear=reversed polarity). Note that the ash layers in *c* and *d* are megascopically distinct, in contrast to those considered in Figs. 1 and 2, which are detected only under binocular microscope examination of the sand-sized sediment fraction.

volcanic activity in only one of the fourteen such examples contained in the eight cores shown in Fig. 2 (left). Nevertheless, we obviously cannot dismiss the possibility that the results are simple coincidence. Despite obvious experimental and analytical difficulties, we therefore offer a statistical analysis (Table 1). Our approach is to restrict analysis to sediments aged 2·5 m.y. or less if the cores do not reach this age, as is the case in E32–02 and E32–47. We assume that eight polarity changes have taken place: one each for the Brunhes–Matuyama and Matuyama–Gauss boundaries and two each for the three known events in the Matuyama. Counting the percentage of the core in which any ash is observed (Fig. 2) provides the probability (Pa) of ash occurring in any part of the core. Combined with the number (N) of observed coincidences of ash and polarity change (Fig. 2 and Table 1) we can then express the probability (Pa,g) of the observed occurrences of simultaneous polarity change and ash deposition being a simple coincidence of two independent variables. Other statistical approaches are possible.

In cores E27–03 and E32–47 there is only a 0·02 probability that the ash and polarity change coincidences occur by chance only (Table 1). In cores E27–04 and E33–15 the probability is less than 0·16; but this rises to 0·30 in E32–02, where the ash content is very high. In cores E27–02 and E33–16, fewer coincidences and high ash content, respectively, reduce Pa,g to almost even chance. Only in core 27–21 is Pa,g greater than 0·50, meaning that in this core there is a better than even chance that the three observed coincidences of polarity change and ash are random events.

Fig. 2 (right) contains the results from one other core. E32–51 contains a disconformity, which has resulted in only a few centimetres of Brunhes sediment being above the oldest Matuyama, and what appears to be a complete Gauss section. Micropalaeontological evidence confirms these correlation lines. Both the Kaena and Mammoth events are clearly displayed: no additional events in the Gauss are indicated. Volcanic materials are at a maxima very close to these events and limits of the Gauss epoch.

Palaeomagnetic and potassium–argon analyses of lavas on Nunivak Island[28] (Fig. 3a) show that volcanic maxima on the island are most marked close to the geomagnetic polarity reversals defined. McDougall and Chamalaun's[29] results from the Indian Ocean islands of Mauritius, Réunion, and Rodriguez (Fig. 3b) show frequent volcanism during the Brunhes epoch and there may be a correlation between volcanic activity and polarity change during the preceding 3 m.y. A coincidence of ash layers and geomagnetic polarity change is included in the results of Ninkovitch[25] from study of a series of cores taken from east of New Zealand (Fig. 3c). Here five megascopically distinct tephra are reported. One is closely associated with the polarity change at the end of the Jaramillo event, two straddle the Brunhes–Matuyama polarity change, and two occur within the Brunhes. Similarly, Ninkovitch et al.[30] report megascopic ash layers in the North Pacific (core V20–119) which coincide with the Jaramillo event (Fig. 3d) and Genter et al.[31] report an ash layer in the north-east Indian Ocean (core V19–153) which coincides with the Brunhes–Matuyama boundary. While we find these observed correlations of volcanism and polarity change difficult to accept as purely coincidental we stress our awareness of the fact that many more data are required to prove that the implied relationship is meaningful.

We now briefly examine the relevance of our data to problems of climatic changes and microfaunal extinctions and appearances. If significant widespread increases in volcanism have occurred during restricted intervals of geological time, then climatic changes are very likely, particularly at high latitudes, because volcanic ash, which remains in the atmosphere for prolonged intervals of time and moves poleward at high altitudes, inhibits solar radiation at the surface[32,33]. In this context it is therefore possibly significant that a distinct world-wide cooling

occurred near the Brunhes–Matuyama boundary[13,34,36], and that other climatic coolings may be associated with the Gilsa event and Matuyama–Gauss boundary[37]. Increases in volcanism within the Brunhes epoch as indicated by us in the Southern Oceans occur in New Zealand (personal communication from H. W. Wellman). This coincides with, and may be related to, the much greater climatic fluctuations of the past 0·7 m.y., compared with the preceding million years[34,36,38,39].

Of eight observed microfaunal extinctions and appearances in the Southern Ocean, six occur either during or very close to geomagnetic polarity changes[40]. We suggest that the climatic changes which can result from volcanic maxima are much more plausibly the cause of such extinctions and appearances than increased radiation at the water surface during any dipole collapse accompanying a polarity change, although other explanations must be considered.

We conclude that our results provide sufficient evidence to justify serious consideration of Heirtzler's[20] speculations of a connexion between geomagnetic polarity change and upper mantle activity, which was also proposed by Hide[41]. Volcanic maxima during polarity changes may be expected to have an influence on climatic conditions, particularly at very high latitudes, and may therefore also be the indirect cause of those microfaunal extinctions which have occurred during geomagnetic polarity changes.

The work is supported by US National Science Foundation grants and by a grant from the Florida State University who also provided computer time. This research was carried out at Florida State University.

[1] Cox, A., *Science*, **163**, 237 (1969).
[2] Heirtzler, J. R., Dickson, G. O., Herron, E. M., Pitman, W. C., and Le Pichon, X., *J. Geophys. Res.*, **73**, 2119 (1968).
[3] Vine, F. J., *Science*, **154**, 1405 (1966).
[4] Pitman, W. C., and Heirtzler, J. R., *Science*, **154**, 1164 (1966).
[5] Holmes, A., *Trans. Geol. Soc., Glasgow*, **18**, 599 (1928).
[6] Hess, H. H., *Geol. Soc. Amer. Buddington*, **599** (1962).
[7] Dietz, R. S., *Amer. Geophys. Un., Geophys. Mono.*, **6**, 11 (1962).
[8] Dietz, R. S., *Amer. J. Sci.*, **264**, 177 (1966).
[9] Opdyke, N. D., Glass, B., Hays, J. D., and Foster, J., *Science*, **154**, 349 (1966).
[10] Watkins, N. D., and Goodell, H. G., *Science*, **156**, 1083 (1967).
[11] Summary by: Hays, J. D., Saito, T., Opdyke, N. D., and Burkle, L. H., *Bull. Geol. Soc. Amer.*, **80**, 1481 (1969).
[12] Harrison, C. G. A., and Funnell, B. M., *Nature*, **204**, 556 (1964).
[13] Hays, J. D., and Opdyke, N. D., *Science*, **158**, 1001 (1967).
[14] Cox, A., and Dalrymple, G. B., *J. Geophys. Res.*, **72**, 2603 (1967).
[15] Harrison, C. G. A., *Nature*, **217**, 46 (1968).
[16] McCormac, B. M., and Evans, J. E., *Nature*, **223**, 1255 (1969).
[17] Uffen, R. J., *Nature*, **198**, 143 (1963).
[18] Black, D. I., *Earth. Planet. Sci. Lett.*, **3**, 225 (1967).
[19] Waddington, C. J., *Science*, **158**, 913 (1967).
[20] Heirtzler, J. R., *Sci. Amer.*, **219**, 60 (1968).
[21] Manshina, L., and Smylie, D. E., *Science*, **161**, 1127 (1968).
[22] Watkins, N. D., and Goodell, H. G., *Earth. Planet. Sci. Lett.*, **2**, 123 (1967).
[23] Doell, R. R., and Cox, A., *Methods in Palaeomagnetism*, 241 (Elsevier, Amsterdam, 1967).
[24] Foster, J. H., *Earth Planet. Sci. Lett.*, **1**, 463 (1966).
[25] Ninkovitch, D., *Earth Planet. Sci. Lett.*, **4**, 89 (1968).
[26] Ninkovitch, D., Heezen, B. C., Connolly, J. R., and Burckle, L. H., *Deep Sea Res.*, **11**, 605 (1964).
[27] Fisher, R. V., *J. Geophys. Res.*, **69**, 341 (1964).
[28] Hoare, J. M., Condon, W. H., Cox, A., and Dalrymple, G. B., *Geol. Soc. Amer. Mem.*, **116**, 249 (1968).
[29] McDougall, I., and Chamalaun, F. H., *Bull. Geol. Soc. Amer.*, **80**, 1419 (1969).
[30] Ninkovitch, D., Opdyke, N. D., Heezen, B. C., and Foster, J. H., *Earth Planet. Sci. Lett.*, **1**, 476 (1966).
[31] Genter, W., Glass, B. P., Storzer, D., and Wagner, G. A., *Science*, **168**, 359 (1970).
[32] Lamb, H. H., *Rev. Geog. Phys. Geol. Dynamique*, **11**, 3 (1969).
[33] Mitchell, J. M., *Proc. AAAS Symposium on Global Effects of Environmental Pollution* (edit. by Singer, F. S.) (in the press).
[34] McIntyre, A., *Program, Annual Meeting Geol. Soc. Amer.*, Atlantic City, 146 (1969) (abst.).
[35] Bandy, O. L., Casey, R. E., and Wright, B. C., *Proc. Eighth Int. Quaternary Assoc. Congress*, Paris, 62 (1969) (abst.).
[36] Ruddiman, W. F., *Program, Annual Meeting Geol. Soc. Amer.*, Atlantic City, 193 (1969) (abst.).
[37] Beard, J. H., *Trans. Gulf Coast Assoc. Geol. Soc.*, **19**, 535 (1969).
[38] Kennett, J. P., *Deep Sea Res.* (in the press).
[39] Herman, Y., and Vergnaud Gruzzini, C., *Proc. Eighth Int. Quaternary Assoc. Congress*, Paris, 68 (1969) (abst.).
[40] Hays, J. D., *Program, Annual Meeting Geol. Soc. Amer.*, Atlantic City, 92 (1969) (abst.).
[41] Hide, R., *Science*, **157**, 55 (1967).
[42] McDougall, I., and Chamalaun, F. H., *Nature*, **212**, 1415 (1966).

34

Reprinted from Earth and Planetary Sci. Letters **26**:370–376 (1975)

ON POSSIBLE RELATIONSHIPS BETWEEN UPPER QUATERNARY GLACIATIONS, GEOMAGNETISM, AND VULCANISM

J. CHAPPEL

Department of Geography, S.G.S., Australian National University, Canberra, A.C.T. (Australia)

Received January 28, 1975
Revised version received May 2, 1975

Sea-level changes, representing ice-volume changes during the last 0.25 m.y., now have been well identified from raised coral reefs dated by $^{230}Th/^{234}U$. This detailed record allows hypotheses relating glaciation to magnetism or vulcanism to be tested more accurately than previously. There is strong temporal correlation between 7 transgression-regression cycles and those variations of seasonal radiation which arise from the effects of orbital eccentricity and precession. Similar close correlation does not exist with known records of magnetic intensity and dip, or with vulcanism, identified in deep-sea cores. It is argued that orbital perturbations theoretically are unlikely to affect magnetism or vulcanism, and these also are unlikely to be affected by global isostatic changes associated with changing ice and water loads. However, certain testing is needed, by finer-scale analysis of deep-sea core volcanic records, and by eliminating chronologic vagueness between deep-sea and coral-reef records.

1. Introduction

Several different cause and effect relationships have been proposed between Quaternary glaciations, vulcanism, and reversals or excursions of the magnetic field. Changes of magnetic intensity have been suggested to alter mean global temperatures and possibly to regulate glaciation, on the grounds that the flux of cosmic rays into the upper atmosphere increases as magnetic field strength decreases [1]. Additionally, excursions of the principal dipole may affect climate, because the field itself may exert a steering effect on the positions of troughs in the circum-polar circulation [2]. Substantiation has been claimed, from correlation in deep-sea cores between paleoclimatic indicators and variations of magnetic intensity and dip [1]. Alternatively, increased vulcanism is seen by others as a cause of glaciation [2,3], and has been further linked to magnetic reversals, on the grounds that heightened seismic activity might induce earth wobble sufficient to perturb the magnetic field [4]. These relationships also are claimed to be substantiated in deep-sea cores [5]. Yet another suggested relationship is that glaciation and deglaciation affect the water loading on great ocean basins, to such a degree that

tectonic processes are affected, and hence may vary the rate of vulcanism [6].

These hypotheses are re-evaluated here, in the light of certain recent Quaternary studies. Firstly, the record of sea-level changes for the last 0.25 m.y., known from accurately dated flights of raised coral reefs, shows that glacial fluctuations strongly have been regulated by those variations of seasonal radiation which stem from orbital precession [7]. Secondly, analyses of global glacio- and hydro-isostasy [8,9] provide a basis for assessing the consequences of changing ocean loads, and thus for discussing possible effects of such isostatic processes on vulcanism. Consideration of the various hypotheses, mentioned above, commences with examination of orbital variations, ice volumes and magnetic variations, over the last 0.25 m.y.

2. Ice volumes, orbital perturbation, and magnetic variations

Fig. 1 shows variations of orbital parameters, ice volumes, and selected deep-sea core magnetic data for the last 0.25 m.y. Curve A shows obliquity changes,

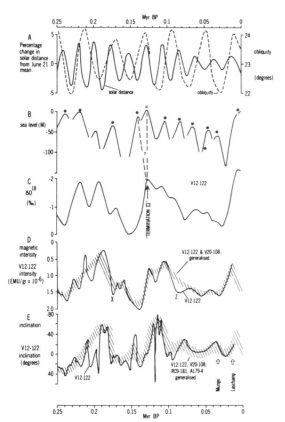

Fig. 1. A. Variations of solar distance for June 21 and orbital obliquity, for last 0.25 m.y. (from [10]). B. Sea-level changes (from [7]). Points marked * dated by multiple ^{230}Th/^{234}U determinations (see text). C. δ^{18}O variations in Caribbean core V12-122 [16]. D. Magnetic intensity variations in V12-122 (solid line), and generalised variations for V12-122 and V20-108 [1]. E. Magnetic dip variations in V12-122 (solid line) and generalised variations for V12-122, V20-108, RC9-181, and A179-4 (data from [1]).

and variations of solar distance on June 21 (i.e., combined precession and orbit eccentricity), from Vernekar's [10] recent revision of the astronomical calculations. The time scale here rests essentially on precise observations of planetary motions; the time scales for the other curves of Fig. 1 are subject to some uncertainties, as briefly will be discussed.

Curve B shows sea-level changes [7], identified from flights of raised coral reefs in New Guinea and Barbados, well dated by ^{230}Th/^{234}U. Sea-level trans-

gressions and regressions relative to the reef flights were identified stratigraphically by various authors, who then subtracted tectonic movements, estimated on a separate basis [11–15]: results from Barbados and New Guinea are in good agreement, and are supported by similar results from raised reefs in the Ryukyu Islands [16,17]. The sea-level maxima at about 0.06, 0.084, 0.105, and 0.12 m.y. are well dated in both Barbados [11] and New Guinea [15]. The statistical error of a single good ^{230}Th/^{234}U date in this range is around 5%, but actual uncertainty may be greater owing to minor diagenesis — a problem difficult to eliminate absolutely in the case of potentially changeable coral skeletons. However, because of the very good consistency between multiple samples from both New Guinea and Barbados, the age error for these particular maxima is put at ±4%. Concerning the other peaks in curve B, that at 0.04 m.y. is well dated in both New Guinea [15] and Ryukyus [16,17]. The 0.028 m.y. peak is recognised only in New Guinea, where it occurs as a separate reef so far dated only by ^{14}C [14,15]. The transgression peak at 0.135–0.14 m.y. preceding the 0.12 m.y. maximum, is recognised in New Guinea where it is dated from 2 samples [12,15], and is supported by recent results from Jamaica [18]. The transgression terminating around 0.18 m.y. is stratigraphically well defined in New Guinea [14], but dated only by 2 determinations from a single sample: the error here is about ±10%. The 0.22- and 0.24-m.y. peaks appear to be represented in Barbados [11,19], but have been stratigraphically separated only in New Guinea, where they are dated from several samples [14], having group errors around 7%.

Curves C, D, E respectively show variations of ^{18}O [20], magnetic intensity [1], and magnetic dip [1], in Caribbean core V12-122. The hatchured band overlaid on D generalises the intensity curves of cores V12-122 and V20-108 (compiled from [1]); the hatchured band over E is a generalisation of dip records in cores V12-122 (Caribbean), V20-108 (N. Pacific), RC9-181 (Mediterranean), and A179-4 (Caribbean) (compiled from [1]). The Late Quaternary Laschamp and Mungo excursions, identified elsewhere [21–23], are arrowed against curve E. Time scales for these and other deep-sea cores have been debated, as none of the principal dating methods have yielded unequivocal results. Earlier U-series datings of cores indicated an age around 0.10 m.y. for the widely recognized peak labelled "Termination II" in curve C [24–26]. However, Broecker and co-workers argue for revision [27,28], placing "Termination II" close to 0.13 m.y. on the basis of their own and all previous U-series core analyses. Support comes from the tying of certain cores to the Brunhes/Matuyama magnetic reversal: placement by Ericson and Wollin [29] of the U/V paleontologic boundary at about 0.4 m.y. agrees well with Broecker's expanded time scale [28]. Further corroboration comes from Shackleton and Opdyke's magnetically-tied chronology of core V28-238, which indicates Termination II close to 0.13 m.y. [30]. Another method of estimating core chronology, by correlating against dated terrestrial sequences, is less sound because debatable assumptions about global ice-age phenomena are entailed. For example, the $\delta^{18}O$ record and ^{230}Th chronology obtained by Duplessy et al., [31,32] from French speleotherms is used by Emiliani [26] to support an age close to 0.1 m.y. for "Termination II". His correlation is uncertain for 2 reasons: firstly, Emiliani [26] inverts Duplessy's paleotemperature interpretation of the $\delta^{18}O$ record to support his argument, and secondly, there is no strong reason for paleotemperatures in southern France to have exactly paralleled northern ice advances during earliest phases of the last glaciation (see [33]). Consequently, the expanded time scale of Broecker, Shackleton, and others is used for the core records in Fig. 1.

That uncertainty remains about time-scale relationships between sea-level curve B and the deep-sea core records is suggested by the apparent age discrepancy, of about 0.01 m.y., between "Termination II" in ^{18}O curve C, and the major marine transgression which is dated in New Guinea [12,15] and Jamaica [18] to have terminated around 0.135–0.14 m.y. The dominating effect of ice volume on oceanic ^{18}O [21] argues that these "terminations" are the same event.

Despite age uncertainties, however, two clear relationships appear in Fig. 1:

(1) Ice volume maxima and minima correlate with maxima and minima of solar distance for June 21, at those times when orbital eccentricity is relatively large (this relationship has been pointed out previously [7,11], and climatologic explanations recently have been advanced [33,34]). Obliquity changes show no correlation with ice volumes.

(2) Wollen et al. [1] point out that times of higher magnetic intensity are generally associated with glaciation, and that times of lower intensity are relatively ice-free. Although this may merely reflect glacial-interglacial variations of core lithologies, the effect is in the theoretically expected sense [1]. This relationship is clear between curves B and D, at a general level, and is improved if time scales are adjusted to bring "Termination II" and the 0.135 m.y. transgression peak into line.

However, hard to perceive are other relationships, such as might support the idea that magnetic field changes directly modulate glaciation. Ice volumes shows no detailed correspondence with the deep-sea core magnetic records, either of intensity or dip, although the many "wriggles" in the latter are suggestive of a finer structure, perhaps similar to the approximately 0.025-m.y. period in the sea-level curve. Lack of correspondence may be due to statistical variability within the core records themselves. To illustrate, between the two nearby Caribbean cores V12-122 and A179-4 are substantial differences in their dip records (data in [1]). Further, considerable differences in detail exist amongst the eight late Quaternary cores from the Gulf of Mexico reported by [22].

Relationships are yet harder to perceive between glaciation and volcanic ash records reported from deepsea cores. Analysis by Kennett and Watkins [5] for nine southern Pacific cores is at too coarse a level for any meaningful comparison with the detailed records shown in Fig. 1. For any testing the idea that vulcanism causes glaciation, the volcanic record in deep-sea cores not only must be analysed at the finest level of detail, but also must be determined and correlated at very many stations around the globe. For volcanic dust veil to be estimated — the critical parameter in the vulcanism/glaciation hypothesis [2,35, 36] — statistical density of explosive vulcanism must be established. The problem of reading this from deep-sea cores is illustrated by the fact that, while several very massive explosive events occurred in New Zealand in Latest Quaternary times [37], none of these have been traced to ash layers identified by Ninkovitch [38] in cores from off the New Zealand coast.

Summarizing thus far, the empirical side of questions about relationships between glaciation and magnetism, and/or vulcanism, appears quite open at present. From the records given in Fig. 1, the strongest relationship apparent so far is between cycles of icecap growth on one hand, and the orbital precession cycle, which dominates the seasonal solar distance curve, on the other. However, the open question of the other connections with glaciation is an interesting one, and is discussed more theoretically in the remainder of this paper. Changes of magnetism and/or vulcanism, if at all synchronous with glaciation, must be so related in one of two ways. In the first case, magnetism and/ or vulcanism might cause the glacial cycles, in which case these phenomena themselves must be modulated by orbital perturbations. In the second case, changing ice and water loads might induce magnetic and/or volcanic changes. These possibilities are discussed in turn.

3. Possible effects of orbital perturbations

Three forces within the earth stem from its motion: the usual centrifugal and Coriolis forces, plus the precessional force:

$$(\widetilde{\omega} \times \widetilde{\Omega}) \times \widetilde{r} \tag{1}$$

where $\widetilde{r}$ is position vector, from center of the earth, and $\widetilde{\omega}$ and $\widetilde{\Omega}$ are rotation and precession vectors, respectively (named the *Poincare force* by Malkus [39]). Effects stemming from interaction of these forces within the mantle are negligible [40]. Within the molten core, on the other hand, interplay of these forces is likely to be important [39,41], and Malkus [39] suggests that the Poincare force might sustain turbulent hydromagnetic flow, and hence might drive the magnetic field. It is still debated however, whether flow within the core is turbulent, or whether large-scale smooth flow occurs, of a relatively complicated type [42]. If the precessional force is a determinant of the magnetic field, then orbital perturbation conceivably might correlate with magnetic changes.

To a close approximation, precessional angular velocity is:

$$\Omega = \frac{3G}{2\omega} \frac{C - A}{C} \left(\frac{M_m}{r^3} + \frac{M_s}{R^3} \right) \cos \epsilon \tag{2}$$

where C, A are the earth's moments of inertia about equatorial and polar axes, ϵ is angle of obliquity, M_m

and M_s are masses of the moon and sun with r and R being their respective distances. Between conditions of minimum and maximum obliquity the strength of the precessional force varies by about 8% (from eqs. 1 and 2) — about an order of magnitude smaller than the observed variations of magnetic intensity. No correlation appears in Fig. 1, between obliquity and the magnetic intensity curve. It can be noted that Malkus [39] suggests that magnetoturbulence in the core might switch between higher and lower turbulent states, and that a jump to the higher state becomes more probable as precessional force increases. However, nothing suggestive of such "triggering" appears in Fig. 1, even at time of greatest orbital eccentricity, when the precessional force will seasonally be a few percent stronger.

Concerning orbital perturbations and vulcanism, elsewhere [35] I have argued that the former cannot cause variations of the latter. Direct effects of the perturbations on flow within the mantle are negligible, as stated above, as are any possible indirect effects stemming from coupling of core motions into the mantle.

Direct modulations of terrestrial magnetism and vulcanism by orbital perturbations, in Upper Quaternary times, thus appears unlikely. The possibility of indirect influence by oscillatory isostatic processes, themselves modulated by the orbital perturbations *via* glacial cycles, is now considered.

4. Possible effects of oscillatory isostasy

Global and regional results of changing ice and water loads, during Quaternary glaciations, include instantaneous elastic effects and visco-elastic relaxation effects. Recent calculations of both factors [8,9,43] match up with various evidences of Late Quaternary isostasy, including Fennoscandian and Laurentide deformed strandlines and gravity anomalies, non-tidal acceleration of the earth's rotation, global variations in the depth of strandlines formed during the last glacial low sea level, variations between continental coasts and oceanic islands of Holocene strandlines, and geodesic relevelling results showing current uplift of inland United States [9].

Major consequences of global isostatic adjustments to changing ice and water loads are:

(1) Changes occur in moment of inertia, as deflection of the great ocean floors is reflected in the 2nd degree harmonic, with a relaxation time around 0.01 m.y. [9,43].

(2) There is transfer of mass within the mantle from beneath the continents to beneath the oceans, or vice versa, according to whether icecaps are increasing and sea levels falling, or the reverse. The flows are of large magnitude: at maximum, such as at termination of the last deglaciation ($\sim$7000 years B.P.), the gross mass transfer from beneath oceans to beneath continents is around 8×10^{17} cm^3/year [9]. Although this is 3 times greater than the total mass transfer by subduction at plate boundaries, viscous dissipation is very much less because shearing rates associated with the isostatic flows are up to 2 orders of magnitude smaller [9]. Assuming that about half the total flux occurs within the lower-viscosity part of the upper mantle, from lithosphere base to the 400-km phase transition, the mean isostatic flow velocity beneath continental margins could reach 1 cm/year, within the upper mantle.

The possibility that isostatic effects might interfere with volcanic processes now can be considered. Beneath ocean regions, isostatic mantle flows will be small, compared with continental margins. The simplest possible mechanisms for influencing vulcanism involve only the direct pressure effects of glacio-eustatism, amounting to about 15 bars variation. Viewed simply, magma genesis beneath mid-ocean ridges seems more sensitive to pressure variation than for other oceanic areas according to phase-boundary models by Wyllie [44]. The glacio-eustatic pressure variation is <0.05% of the pressure at the solidus boundary in these models, which seems unlikely substantially to modulate magma genesis at mid-ocean ridges. Similarly, direct effects on melting are unlikely at ocean-margin volcanic arcs. Although the isostatic upper mantle flows might intersect subduction slabs, at the rate of 1 cm/year, only 100 m or so of mass translocation would occur before the flow reversed. Compared with the tens of kilometers rise necessary for magma genesis by diapiric processes in the subduction zone [44,45], such oscillating translocations are small.

More difficult to estimate are the effects of isostatic stresses at the upper boundary of a magma "pool" beneath a lithosphere plate. Weertman [46]

suggests that self-propagating magma-filled cracks nucleate whenever tensile stresses greater than zero occur above a magma pool. Any such process might be sensitive to periodic changes in horizontal stress gradients. Stress gradients resulting from glacio-eustatic changes are very small across ocean basins, but vary up to at least 10^{-3} bars/m (10^2 N/m^3) at ocean margins, changing at rates of up to 3×10^{-8} bars/sec. Whether such low temporal rates of stress change might substantially affect the stress condition near lithoscope base, in places where it verges on tension, is beyond the scope of this paper. Comparison is noted, however, with stress gradients associated with earth tides, which significantly affect eruption timing [47]. Earth-tide spatial stress gradients are 10^5 times smaller than maximal hydro-isostatic gradients, but change 10^3 times more quickly.

In short, the probability seems small that eustatic changes might modulate vulcanism; however, the question could be empirically resolved by very detailed analysis of the volcanic record in deep-sea cores, along the lines described by Watkins and co-workers [48,49]. Potentially most useful here would be cores from locations "downwind" of those volcanic arcs closest to ocean margins (such as Peru–Chile) where hydro-isostatic flexures and flows are strongest.

Finally, concerning the question of oscillatory isostasy affecting magnetism, the possibility appears even smaller than for vulcanism. Viewed simply, any such connection might come from hydro-isostatic changes of the earth's figure. The precessional force increases as the equatorial moment of inertia (C) increases (eqs. 1 and 2). Further, magnetic field is likely to become stronger as ellipticity of the core increases [39,50]. During glacial advance, rebound of ocean basins increases the equatorial bulge and hence C; the effect of the water uptake by near-polar icecaps is to reduce C. The net result of the last deglaciation is that C is at present decreasing. The resulting rotational acceleration is small, however (smaller than the tidal deceleration [51,52]). Variations of this type influence precessional torque by <0.1%, which seems much too small to affect the precessional role in the geodynamo, as envisaged by Malkus [39]. The direct effect of changing ellipticity is also too small. This eccentricity variation amounts to about 0.3% (reckoned from global elastic effects, and relaxation effects of ice-volume curve B from Fig. 1). According to Suess [50],

394

the strength of westward drift in the core — itself probably proportional to field strength — is about proportional to eccentricity of the core. The upper limit value of 0.3% variation thus is much too small to explain the intensity variations shown in Fig. 1D.

5. Conclusions

Questions concerning possible major climatic effects of changing geomagnetism or vulcanism are important to answer, if for no object other than that there is widespread and growing concern about climatic deterioration. The geological records of ice volumes and magnetism, summarised in Fig. 1, show that no simple relationship exists. The same holds for vulcanism and glaciation, as far as is known. On the other hand, Upper Quaternary glaciation has strongly been influenced by changes of the seasonal distribution of solar radiation, resulting from combined orbital precession and eccentricity effects. A connection between these latter factors and magnetism or vulcanism appears theoretically very unlikely. It seems almost as unlikely that global isostatic effects of changing ice and water loads, during successive glaciation, could affect either magnetism or vulcanism. However, whether any such relationships exist should be tested by further detailed analyses of deep-sea cores.

One particular relationship in Fig. 1, difficult to explain, is the general association of higher magnetic intensities with glaciation. It seems unlikely that substantially higher intensities should result from slight increases of global oblateness, during glaciation. The possibility remains that a stronger field induces global cooling to such an extent that icecaps appear, to become subject to the radiation changes resulting from orbital perturbations. This would mean that glacial initiation was determined by magnetic changes, occurring for quite independent reasons. For this to be the case, the magnetic maxima marked as X and Z on curve D would have to be around 0.12 and 0.22 m.y. — i.e., about 0.03 m.y. earlier than shown. This, and other issues about relationships between the curves of Fig. 1, can be answered only by carefully re-examining the chronologies.

References

1 G. Wollin, D.B. Ericson, W.B.J. Ryan and J.H. Foster, Magnetism of the earth and climatic changes, Earth Planet. Sci. Lett. 12 (1971) 175.
2 H.H. Lamb, Climate, Present, Past, and Future, vol. 1 (Methuen, London, 1972).
3 M.I. Budyko, Climate change, Soviet Geog. 10 (1969) 429.
4 J.R. Heirtzler, Sea floor spreading, Sci. Am. 219 (1968) 60.
5 J.P. Kennett and N.D. Watkins, Geomagnetic polarity change, volcanic maxima, and faunal extinction in the South Pacific, Nature 227 (1970) 930.
6 R.K. Matthews, Tectonic implications of glacio-eustatic sea-level fluctuations, Earth Planet. Sci. Lett. 5 (1969) 459.
7 J. Chappell, Relationships between orbital perturbations, $\delta^{18}O$ variations and sea changes, Nature 252 (1974) 199.
8 R.J. Walcott, Past sea levels, eustasy, and deformation of the earth, Quaternary Res. 2 (1972) 1.
9 J. Chappell, Late Quaternary glacio- and hydro-isostasy on a layered earth, Quaternary Res. 4 (1974) 405.
10 A.D. Vernekar, Long period global variations of incoming solar radiation, Met. Monogr. 12 (1972) 6.
11 W.S. Broecker, D.L. Thurber, J. Goddard, T-L Ku, R.K. Matthews and K.J. Mesolella, Milankovitch hypothesis supported by precise dating of coral reef and deep sea sediments, Science 159 (1968) 297.
12 H.H. Veeh and J. Chappell, Astronomic theory of climatic change: support from New Guinea, Science 167 (1970) 862.
13 R.P. Steinen, R.S. Harrison and R.K. Matthews, Eustatic low stand of sea level between 125,000 and 105,000 B.P.: evidence from the sub-surface of Barbados, West Indies, Geol. Soc. Am. Bull. 84 (1973) 63.
14 J. Chappell, Geology of coral terraces, Huon Peninsula, New Guinea: a study of Quaternary tectonic movements and sea level changes, Geol. Soc. Am. Bull. 85 (1974) 553.
15 A.L. Bloom, W.S. Broecker, J. Chappell, R.K. Matthews and K.J. Mesolella, Quaternary sea level fluctuations on a tectonic coast: new $^{230}Th/^{234}U$ dates from the Huon Peninsula, New Guinea, Quaternary Res. 4 (1974) 185.
16 K. Konishi, S.O. Schlanger and A. Omura, Neotectonic rates in the central Ryukyu Is. derived from Th^{230} coral ages, Marine Geol. 9 (1970) 225.
17 K. Konishi, A. Omura and O. Naksmichi, Radiometric coral ages and sea level records from the Late Quaternary reef complexes of the Ryukyu Is., Proc. 2nd Int. Coral Reef Symp., 2 (1974) 595.
18 W.S. Moore and B.L.K. Somayajula, Age determinations of fossil corals using $^{230}Th/^{227}Th$, J. Geophys. Res. 79 (1974).
19 K.J. Mesolella, R.K. Matthews, W.S. Broecker and D.L. Thurber, The astronomical theory of climatic change: Barbados data, J. Geol. 77 (1969) 250.

20 W.S. Broecker and J. van Donk, Insolation changes, ice volumes, and the ^{18}O record in deep sea cores, Rev. Geophys. Space Phys. 8 (1970) 169.

21 N. Bonhommet and J. Zahringer, Paleomagnetism and potassium argon age determinations of the Laschamp geomagnetic polarity event, Earth Planet. Sci. Lett. 6 (1969) 43.

22 H.C. Clark and J.P. Kennett, Paleomagnetic excursion recorded in latest Pleistocene deep-sea sediment, Gulf of Mexico, Earth Planet. Sci. Lett. 19 (1973) 267.

23 M. Barbetti, Evidence for a geomagnetic excursion at 30,000 B.P., from Aboriginal fireplaces in Australia, Nature 239 (1972) 327.

24 E. Rona and C. Emiliani, Absolute dating of Caribbean cores P6304-8 and P6304-9, Science 163 (1969) 66.

25 C. Emiliani and E. Rona, Caribbean cores P6304-8 and P6304-9: new analysis of absolute chronology – A reply, Science 166 (1969) 1551.

26 C. Emiliani, The Last Interglacial: paleotemperature and chronology, Science 171 (1971) 571.

27 W.S. Broecker and T.-L. Ku, Caribbean cores P6304-8 and P6304-9: new analysis of absolute chronology, Science 166 (1969) 404.

28 W.S. Broecker and J. van Donk, Insolation changes, ice volumes and the ^{18}O record in deep-sea cores, Rev. Geophys. Space Phys. 8 (1970) 169.

29 D.B. Ericson and G. Wollin, Pleistocene climates and chronology in deep sea sediments, Science 162 (1968) 1227.

30 N.J. Shackleton and D.D. Opdyke, Oxygen isotope and paleomagnetic stratigraphy of equatorial Pacific core V28-238, Quaternary Res. 3 (1973) 39.

31 J.C. Duplessy, C. Lalou and A.C. Vinot, Differential isotopic fractionation in benthonic foraminifera and paleotemperatures reassessed, Science 168 (1970) 250.

32 J.C. Duplessy, J. Labeyrie, C. Lalou and H.V. Nguyen, La mesure des variations climatiques continentales application à la periode comprise entre 130,000 et 90,000 ans B.P., Quaternary Res. 1 (1971) 162.

33 J. Chappell, Causes of the last ice age, Nature (1975) in press.

34 G. Kukla, Milankovitch and climate: the missing link?, Nature (1975) in press.

35 J. Chappel, Astronomical theory of climatic change: status and problem, Quaternary Res. 3 (1973) 221.

36 D. Deirmendjian, on Volcanic and other particular turbidity anomalies, Advances in Geophys. 16 (1973) 267.

37 C.G. Vucetich and W.A. Pullar, Stratigraphy and chronology of Late Pleistocene ash beds in central North Island, New Zealand, New Zealand J. Geol. Geophys. 12 (1969) 784.

38 D. Ninkovitch, Pleistocene volcanic eruptions in New Zealand recorded in deep-sea sediments, Earth Planet. Sci. Lett. 4 (1968) 89.

39 W.V.R. Malkus, Precession of the earth as a cause of geomagnetism, Science 160 (1968) 259.

40 D.C. Tozer, Heat transfer and convection currents, Phil. Trans. R. Soc. London 1088 (1965) 252.

41 R.J. Gans, on Hydromagnetic precession in a cylinder, J. Fluid Mech 45 (1970) 111.

42 F.E.M. Lilley, Geomagnetic reversals and the position of the north magnetic pole, Nature 227 (1970) 1336.

43 R.J. O'Connell, Pleistocene glaciation and the viscosity of the lower mantle, Geophys. J. R. Astron. Soc. 23 (1971) 299.

44 P.J. Wyllie, Role of water in magma generation and initiation of diapiric uprise in the mantle, J. Geophys. Res. 76 (1971) 1328.

45 T.H. Green and A.H. Ringwood, Genesis of the calc-alkaline igneous rock suite, Contrib. Mineral. Petrol. 18 (1968) 105.

46 J. Weertman, Theory of water-filled crevasses in glaciers applied to vertical magma transport beneath ocean ridges, J. Geophys. Res. 76 (1971) 1171.

47 F.J. Mauk and M.J.S. Johnston, On the triggering of volcanic eruptions by earth tides, J. Geophys. Res. 78 (1973) 3356.

48 T.C. Huang, N.D. Watkins, D.M. Shaw and J.P. Kennett, Atmospherically transported volcanic dust in South Pacific deep-sea sedimentary cores at distances over 3000 km from the eruptive source, Earth Planet Sci. Lett.

49 T.C. Huang and N.D. Watkins, Atmospherically transported volcanic glass in deep-sea sediments: development of a separation and counting technique, Deep Sea Res. 21 (1974).

50 S.T. Suess, Some effects of gravitational tides on a model earth's core, J. Geophys. Res. 75 (1970) 6650.

51 W.H. Munk and G.J.F. Macdonald, The Rotation of the Earth (University Press, Cambridge, 1960).

52 R. Newton, Secular accelerations of the earth and moon, Science 166 (1969) 825.

396

Part VIII

THE NOMENCLATURE OF MAGNETOSTRATIGRAPHY

Editor's Comments
on Papers 35 and 36

35 THE POLARITY TIME SCALE SUBCOMMISSION
Magnetic Polarity Time Scale

36 WATKINS
Polarity Subcommission Sets Up Some Guidelines

Magnetostratigraphy, as a new branch of stratigraphy, organizes strata according to their magnetic properties acquired at the time of deposition. As stressed previously, because the polarity of the earth's magnetic field reversed repeatedly in the geologic past, especially in the Cenozoic and late Cretaceous, and because polarity transitions, while lasting only a few centuries or millenia, are synchronous over the entire globe, their record in marine or land-based sediments provides isochrons applicable to worldwide correlation. Magnetostratigraphic approaches differ from the chronologic techniques that provided the widely used polarity time scale (Cox, Paper 1). The latter was constructed by organizing radiometrically dated samples of known polarity according to the isotopic age, often in the absence of knowledge on stratigraphic relations between individual samples and particularly between groups of samples. Magnetostratigraphy, on the other hand, is normally employed in the complete absence of any direct radiometric data on the sedimentary sections being examined. Because of the repetitive nature of reversals, the identification of particular polarity events within incomplete sequences is only possible by comparisons with other stratigraphic or radiometric data. Such independent criteria normally employed are well-established biostratigraphic schemes.

Magnetostratigraphy, like any branch of stratigraphy, requires a clear set of definitions of the stratigraphic terms employed to reduce potential ambiguities that normally arise from studies carried out by numerous investigators. As summarized by Watkins (Paper 35), there now exists a well-established nomenclatural convention in magnetic polarity scale studies. Periods of one polarity of the order of 1 m.y. are termed *epochs* and those lasting about 50,000 to 100,000 years are termed *events*. The four epochs recognized for the Pleistocene and Pliocene (Brunhes, Matuyama, Gauss, and Gilbert) are named after past workers in terrestrial magnetism, while paleomagnetic events have been named after the type locations where they were first recognized. Older paleomagnetic epochs have been numbered. This is similar to the numerical system developed by the Lamont group during the 1960s for sea-floor magnetic anomalies, but the numbers that identify the different polarity events do not coincide between the two systems. On the whole, the stratigraphic nomenclatural and subdivision scheme that has been employed in magnetostratigraphy has worked well, although conflicts have inevitably arisen in relation to nomenclature and terminology. Any system used should be flexible enough to allow for polarity intervals discovered at a later date to be conveniently and unambiguously inserted into the existing system. In the classification of scientific data, there has also been a strong tendency in the past to create large numbers of names. To some extent, this has resulted from the desire of scientists to be attributed to any name that is later accepted and utilized. A potential also exists in magnetostratigraphy to follow this trend and name all the polarity epochs and events as they are discovered and documented. This was anticipated by Irving (1972), who correctly pointed out that the procedure of naming each polarity interval, if extended, would create a vast nomenclatural system requiring much memorization, thus decreasing the practicality and utility of magnetostratigraphy. The success of the numerical classification of ocean-floor magnetic anomalies was used by Irving to demonstrate the relative success of such an approach. Another illustration of the success of numerical nomenclature in stratigraphic subdivision is the Cenozoic tropical planktonic foraminiferal biostratigraphy of Blow (1969). In large part, this system has been widely used because of its ease of use by a wide spectrum of geologists and geophysicists who may not be familiar with the microfossils themselves.

By the early 1970s it had become apparent that inevitable inconsistencies and nomenclatural conflicts in magnetic stratigraphy required some resolution. In response to this need, Norman Watkins in 1972 recommended that an international subcommission on the geomagnetic polarity time scale be formed to assist in the definition of stratigraphic nomenclature and problems related to definition of the polarity history (Papers 35 and 36). This proposition directly resulted in the establishment in 1972 of the Subcommission on a Magnetic Polarity Time Scale (SMPTS) as part of the International Commission on Stratigraphy, which in turn is a part of the International Union of Geological Sciences. In a series of meetings, this commission produced a set of recommendations in an attempt to establish an unambiguous nomenclature in magnetostratigraphy, which is as consistent as possible with conventional stratigraphic terminology. Recommendations by the subcommission, which were rather general, are summarized in Papers 35 and 36. Other than the various recommendations made, the greatest importance of the subcommission is that it provided a platform for discussion of magnetostratigraphic nomenclature; it assisted in integrating magnetostratigraphic nomenclature within stratigraphic nomenclature as a whole; and it assisted in drawing attention to the various options available for magnetostratigraphic subdivision. The contributions of this committee, however, failed to fully satisfy the International Subcommission on Stratigraphic Classification (ISSC) with respect to proposed terminology or with respect to the relations of proposed magnetostratigraphic units to stratigraphic classification as a whole. The Subcommission on Stratigraphic Classification under the chairmanship of Dr. Hollis Hedberg has thus produced a document on magnetostratigraphic polarity units (August 1978), which was also reviewed by members of the Subcommission on a Magnetic Polarity Time Scale (more recently under the chairmanship of Dr. N. Opdyke). The resulting report will be published shortly as a supplementary chapter of the *International Stratigraphic Guide* and is the prepared statement of both the IUGS International Subcommission on Stratigraphic Classification (ISSC) and the IUGS Subcommission on a Magnetic Polarity Time Scale (SMPTS). The most significant changes in terminology adopted in this document is the replacement of the word "epoch," as in Brunhes Epoch, with the word *"chron"* as in *Brunhes Chron.* Furthermore, the word "event," as in Jaramillo Event, is replaced by the word *"subchron,"* as in *Jaramillo Subchron.* These are chronological (geochronological) units. The chronostratigraphic equivalents for these terms as adopted are "chronozone" for "chron" and "subchronozone" for "subchron." The detailed justification of these and other changes are outlined in the document to be published (Hedberg, et al., in press). Recommended terminology for magnetostratigraphic polarity units and their geochronologic and chronostratigraphic equivalents is summarized in the following table:

Magnetostratigraphic polarity units	Geochronologic equivalent	Chronostratigraphic equivalent
Polarity superzone	Chron (or superchron)	Chronozone (or superchronozone)
Polarity zone	Chron	Chronozone
Polarity subzone	Chron (or subchron)	Chronozone (or subchronozone)

REFERENCES

Blow, W. H. 1969. Late Middle Eocene to Recent planktonic foraminiferal biostratigraphy. *Plank. Microfossils*, 1st International Conference, Geneva, 199 pp.

Hedberg, H. D., and members of ISSC and SMPTS. Magnetostratigraphic polarity units—A supplementary chapter to the ISSC International Stratigraphic Guide (in press).

Irving, E. 1972. Paleomagnetic stratigraphy—Names or numbers? *Comments Earth Sci.: Geophys.* **2**:125–130.

ADDITIONAL READINGS

Editorial Comment. 1973. Magnetostratigraphic nomenclature. *Nature Phys. Sci.* **242**:65.

On the second meeting of the IUGS Subcommission on the Magnetic Polarity Time Scale. 1975. *Jour. Geol. Soc. Japan* **81**:211–212.

Subcommission on the Magnetic Polarity Time Scale. 1973. *Jour. Geol. Soc. Japan* **79**:319–322.

Watkins, N. D. 1975. Correlating stratigraphic zones and magnetic polarities. *Geotimes* **20**:26–27.

Watkins, N. D., and the Subcommission on the Magnetic Polarity Time Scale. 1973. Magnetostratigraphic nomenclature I—Terminology. *Comments Earth Sci.: Geophys.* **3**:55–58.

35

Reprinted from Geotimes 18:21–22 (1973)

magnetic polarity time scale

The Polarity Time Scale Subcommission

Although some paleomagnetic evidence for past reversals of the Earth's magnetic field was first published over half a century ago, it was not until 1963 that systematic attempts were initiated to define the geomagnetic-polarity history. This history has assumed great importance in the geological sciences. Polarity reversals are globally synchronous and recorded by all rock types, thus providing horizons of potentially unparalleled resolution for use by the stratigrapher and other geologists. The most spectacular use of the geomagnetic polarity time scale has followed its recognition in the form of linear magnetic anomalies resulting from sea-floor spreading.

The polarity time scale, which is well defined for the past 5 million years and known in some (but far from complete) detail for selected earlier periods, is now being applied to stratigraphic correlation and dating problems in sediments of various types. These studies, in addition to volcanological problems, demand a nomenclature system that does not conflict with present stratigraphic terminology and usages. A method of numbering successive periods of normal and reversed polarity (which was proposed during the late 1950s) was succeeded by the 'epoch' and 'event' nomenclatural system (Alan Cox, R.R. Doell, G.B. Dalrymple, 1964, Science, v. 144, p. 1,537) for the last 4 million years or so: 'epochs' are periods of dominantly one polarity lasting of the order of 1 million years, whereas 'events' are understood to be periods of one polarity lasting of the order of 0.01 to 0.1 million years. This convention is now widely accepted but for the period prior to 5 million years ago, substantial conflicts exist (for example: Kenneth M. Creer, 1971, Nature, v. 233, p. 545; A.H. Johnson, A.E.M. Nairn & D.N. Peterson, 1972, Nature, v. 237, p. 9; Michael W. McElhinny & P.J. Burek, 1971, Nature, v. 232, p. 98).

A sub-commission of the International Commission on Stratigraphy (within the International Union of Geological Sciences) has been established to facilitate an unambiguous development of the polarity time scale and its associated nomenclature. These countries are represented on the subcommission: Australia, Canada, Czechoslovakia, Germany, Iceland, India, Japan, New Zealand, the United Kingdom, the U.S.A., and the U.S.S.R. The first series of meetings was held Aug. 21, 23, and 24, 1972, during the 24th International Geological Congress, in Montreal. This was preceded by written communications between all members. The group restricted its activities to consideration of a preferred unambiguous nomenclatural system, involving polarity changes only: no consideration was given to nomenclature for such other known geomagnetic characteristics as deviations or excursions of the geomagnetic field (which are suspected to last of the order of 10^2 to 10^3 years), and apparent long-term polar wondering. Data-acceptability criteria, and endorsement of those several polarity scales so far proposed in the literature for the period prior to 5 million years

Sub-commission members: Robert N. Carter (New Zealand); H. Basil S. Cooke (Canada); Kenneth M. Creer (U.K.); Trausti Einnarsson (Iceland); James D. Hays (U.S.A.); Charles E. Helsley (U.S.A.); Edward Irving (Canada); A. Khramov (U.S.S.R.); Jiri Kukla (Czechoslovakia); Ian McDougall (Australia); Michael W. McElhinny (Australia); Hisao Nakagawa (Japan); Neil Opdyke (U.S.A.); D.M. Perchesky (U.S.S.R.); Norman D. Watkins (U.S.A.), convenor. Correspondents: Orville Bandy (U.S.A.); Peter Burek (Germany); Alan Cox (U.S.A.); R.K. Verma and P.C. Paul (India); Walter C. Pitman (U.S.A.); Alexandre Roche (France); R. Henry Spall (U.S.A.).

Summary of recommended unit-terms and hierarchies in paleomagnetic stratigraphy

magnetostratigraphic units	approximate duration in years	chronostratigraphic units	chronologic units
polarity subzone	10^4—10^5	polarity subinterval	polarity event
polarity zone	10^5—10^6	polarity interval	polarity epoch
polarity superzone	10^6—10^7	polarity superinterval	polarity period
polarity hyperzone	10^7—10^8	polarity hyperinterval	polarity era

ago, were similarly left for consideration at future meetings.

The following 11 points were agreed upon (Edward Irving dissenting) by the subcommission:

1. A new category of stratigraphic classification should be recognized as *magnetostratigraphy*, defined as that element of stratigraphy which is concerned with the organization of strata into units based on the polarity of their remanent magnetism.

2. The purpose of magnetostratigraphic classification is to organize rock strata of the Earth into designed units

The geomagnetic polarity scale for the period t = 0 to 4.5 million years, according to Alan Cox (1969, *Science*, v. 163, p. 237). The diagram includes constituent data points, inferred periods of normal (black) and reversed (open) polarity, and established nomenclature of the polarity events and epochs. Although this polarity scale is currently the most commonly used version, some doubt exists concerning the reality of the Laschamp event as a global polarity change, and the event nomenclature within the lower Matuyama is subject to debate.

based on the recording of former geomagnetic polarities by the rocks.

3. A magnetostratigraphic unit may be defined as a local body of rock strata which is unified by its magnetic polarity and thus differentiated from adjacent strata.

4. The basic local magnetostratigraphic unit is the *magnetozone:* but the term 'zone' may be used without the prefix when the usage is clear.

5. A magnetozone has its upper and lower limits defined by the positions of boundaries, which are termed *transitions*, marking a change of polarity between opposite senses. Each transition must be clearly defined in terms of the recognized lithologic and/or paleontologic landmarks in the stratigraphic section within the geographic area concerned. (For example: 'Magnetozone R7 has its lower transition 4.0 m above the top of the Twin Rivers Member of the Jones Formation in the Chalk Bluff Group, and its upper transition 1.5 m below the base of the Queens Valley Formation in the Red Fork Group.')

6. Each local magnetozone should be designated in such a way that it can be referred to readily in discussion. (This could be by letter, or by letter and number within a group or formation, or by name.) At this time, the use of new stratigraphic names should be discouraged to avoid proliferation of terms.

7. Although each magnetozone would normally be characterized by its constant magnetic polarity, it is realized that later work, particularly that involving geographic extension of associated mapping, may reveal the existence of shorter periods of polarity change within the magnetozone. Thus *subzones* would be recognized. Similarly it may become necessary to regroup a magnetozone into a larger unit, and thus superzones can be expected to become defined.

The natural geomagnetic polarity reversal spectrum is believed to extend from the order of 10^4 to 10^7 years per reversal (B.E. McMahon & D.W. Strangway, *Science*, v. 155, p. 1012). Shorter-period reversals are suspected to have occurred, but are not yet accepted as definite phenomena. Periods of recognizable polarity reversal character can be expected: for example, long periods of mixed polarity, or long periods of polarity changes irregularly spaced in time. This natural polarity reversal frequency or signal can be conveniently incorporated into a no-

menclatural system, as shown in the table. This use of absolute durations necessitates attempts to estimate the approximate duration of any magnetostratigraphic unit. As in conventional stratigraphic systems, eventual elevation or other reassignment of initially proposed terms can be expected in some cases, in response to additional data.

8. In order to permit designation of units of unknown duration, the term *episode* is suggested, this term to be usable both for magnetostratigraphic and for chronostratigraphic purposes.

9. It is proposed that the magnetostratigraphic units be used essentially In the conventional lithostratigraphic context, as is the case with biostratigraphic units. Partly to accommodate usage that has already become well established, it is recommended that special chronostratigraphic and chronologic terms be recognized (see table).

10. It is strongly recommended that the magnetostratigraphic units must be initially local ('provincial') or informal units until wide, preferably global, correlations and therefore polarity history (formal units) can be recognized. The time-representative component inherent in chronostratigraphic terms will permit their use in a time sense pending the adoption of internationally approved chronologic terms.

11. No specific polarity scales are recommended for the period prior to 5 million years ago, and no recommendations are made for preferred prefix or suffix systems at this time. It is recognized, however, that the use of Brunhes, Matuyama, Gauss, and Gilbert for the polarity epochs of the past 5 million years are now part of established terminology, and this special case (the period t = 0 to 5.0 million years) is therefore endorsed. The terms now applied to the associated polarity events are similarly recognized as firmly established, although final resolution of nomenclature priorities for events within the Matuyama paleomagnetic epoch are still not yet resolved (N.D. Opdyke, 1972, *Reviews of geophysics*, v. 10, p. 213. Norman D. Watkins, Geological Society of America *Bulletin*, v. 83, p. 551.)

Since the sub-commission is now officially recognized by the International Commission on Stratigraphy it is anticipated that the several factors (such as specific polarity scales and terms) not taken up by the sub-commission at its first meeting will receive early consideration in future agendas.

403

36

Reprinted from *Geotimes* **21**:18–20 (1976)

POLARITY SUBCOMMISSION SETS UP SOME GUIDELINES

N. D. Watkins

Last summer in France, the Polarity Time Scale Subcommission set up rules for evaluating data in magneto-stratigraphic studies and defined terms used to describe rapid large movement in the geomagnetic field. Both actions seem sure to affect future research.

Data-acceptability criteria. Because of rapid growth in application of the paleomagnetic method to stratigraphic problems, it was proposed that, in addition to establishing a uniform magnetostratigraphy terminology, the Subcommission should consider recommending a series of minimum data-acceptability criteria in order to discourage use of possibly misleading results in magnetostratigraphic studies.

Igneous rocks. Early work on the polarity time scale included proposals that any paleomagnetic data and K:Ar ages involved should be based on samples from the same body, and that more than one geographically orient-

ed sample should be used for the paleomagnetic determinations. In addition, laboratory methods (such as alternating magnetic field demagnetizing treatment) should be used to minimize unstable components. The Subcommission recommended that the same philosophy should govern all magnetostratigraphic work in igneous provinces: at least 2 separate geographically oriented samples should be taken from each separate body, and unstable components should be minimized by appropriate conventional laboratory procedures. Although the Subcommission recognizes that measurement of polarity at the outcrop, using a compass or a magnetometer, can provide valuable stratigraphic information, it believes that the possibility of later overprinting of the original polarity (especially during the present period of normal polarity) has been demonstrated to be sufficiently high to require unstable component examination for any definitive magnetostratigraphic data.

Sediments. It is widely known that ancient geomagnetic field directions may be imperfectly recorded in sediments because of dynamic distortion during original deposition. Moreover, post-depositional processes (particularly those of a chemical nature) can lead to overprinting of the original detrital remanent magnetism. The Subcommission therefore recommends that no definitive magnetostratigraphy be formally proposed unless multiple sampling of horizons has been made, and laboratory measurements have included methods appropriate to the removal of unstable components or overprinting. For various reasons, formal magnetostratigraphy based on sediments should be derived from sequences involving different sedimentation rates, lithologies, and depositional and post-depositional environments.

Nomenclature for rapid large movement of the geomagnetic field: the Subcommission has hitherto consid-

This chart is a revised and more legible version of that published in Geotimes in June 1975. It is now divided into 3 parts: Neogene, Paleogene, and part of the Cretaceous. These parts show correlations between established stratigraphic zonations, and magnetic polarities derived from (left to right): magnetic-anomaly analyses; K:Ar and paleomagnetic studies of igneous rocks (polarity-epoch names); deep-sea sedimentary core analyses (polarity-epoch numbers). Polarities (black= normal; clear=reversed) are derived largely from K:Ar and paleomagnetic studies of igneous rocks for the youngest 5 million years of the scale, and from sea-floor spreading analyses of marine magnetic anomalies for the rest of the time represented. This chart is to encourage use of magnetostratigraphic methods and terms, and is not intended to represent a final unambiguous correlation scheme. (Diagram by William A. Berggren, Neil D. Opdyke & Norman D. Watkins.)

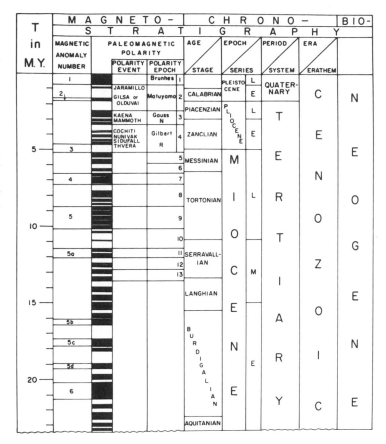

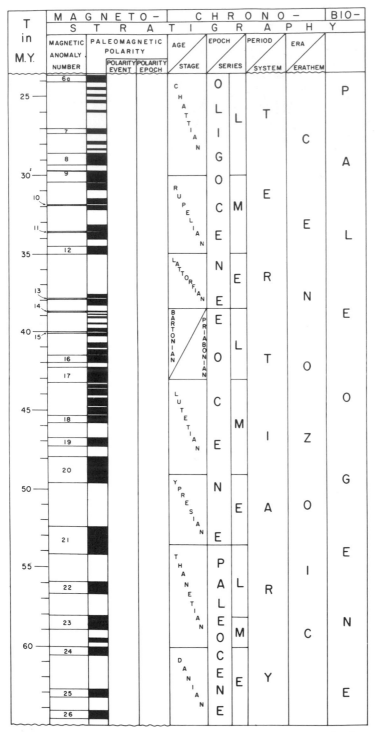

MAGNETO-			CHRONO-				BIO-
S	T	R	A	T	I	G	R A P H Y

Polarity Subcommission —a history

The Polarity Time Scale Subcommission is part of the Internal Commission on Stratigraphy, which is in turn part of the International Union of Geological Sciences. The Subcommission—formed in 1972, and comprising representatives from 8 countries—has met twice, first in August 1972 in Montreal during the 25th International Geological Congress, and second in November 1974 in Miami Beach during the annual meeting of the Geological Society of America.

A series of recommendations for establishing an unambiguous nomenclature in magnetic-polarity stratigraphy, which is as consistent as possible with conventional stratigraphic terminology, was produced during those first meetings, and several publications have resulted (see the list below). Also, the Subcommission has been recognized by the International Union of Geodesy & Geophysics.

The Subcommission met for a third time last Sept. 2-3 in Grenoble, France, at the 16th Congress of the IUGG immediately after sessions of the International Association of Geomagnetism & Aeronomy. More than 20 members of IAGA from 14 countries attended and took part in the discussion.

Publications resulting from the first 2 meetings:
1 *Magnetic polarity time scale. Geotimes* (May 1973), p. 21-22.
2 *Subcommission on magnetic polarity time scale. Journal of Geological Society of Japan*, v. 79 (1973), p. 319-322.
3 *Editorial comment. Nature physical science*, v. 242 (1973), p. 65.
4 *Magnetostratigraphic nomenclature I—terminology. Comments on the Earth sciences: Geophysics*, v. 3 (1973), p. 55-58.
5 *Correlating stratigraphic zones and magnetic polarities. Geotimes* (June 1975), p. 26-27.
6 *On the second meeting of the IUGS Subcommission on the Magnetic Polarity Time Scale. Journal of the Geological Society of Japan*, v. 81 (1975), p. 211-212.

ered nomenclatural problems involving only polarity changes. A polarity event is presently understood to be a chronologic unit characterized by a single geomagnetic polarity that lasted 10^4 to 10^5 years. Polarity epochs, periods, and eras are chronologic units that last much longer. Recently it has been suspected that an even shorter geomagnetic behavior, involving migration of the virtual geomagnetic pole (VGP) to the opposite hemisphere for a period of perhaps only 10^2 to 10^3 years, may have occurred. It is yet to be demonstrated that such behavior is global, and thus due to reversal of the dipole field, but the phenomenon has been recorded in sediments and some basalts, and has been given in the literature names such as 'short event', 'excursion', 'departure', and even 'flip'. The Subcommission recommends that the term

MAGNETO-			CHRONO-				BIO-
T in M.Y.	S T R A T I	G R A P H Y					Y
	MAGNETIC ANOMALY NUMBER	PALEOMAGNETIC POLARITY (POLARITY EVENT / POLARITY EPOCH)	AGE / STAGE	EPOCH / SERIES	PERIOD / SYSTEM	ERA / ERATHEM	
	27		MAESTRICHTIAN	SENONIAN	CRETACEOUS	MESOZOIC	
	28						
	29						
70	30		CAMPANIAN				
	31						
75	32						
	33		SANTONIAN				
80							

'polarity excursion' be employed for the phenomenon, and that all other terms be dropped. The logic for this recommendation is contained in the Subcommission's recommended definition of a polarity excursion—a sequence of virtual geomagnetic poles which may reach intermediate latitudes and which may extend beyond 135° of latitude from the pole, for a short interval of time, before returning to the original polarity. The polarity excursion can, as with other magnetostratigraphic terms, be locally identified using an appropriate geographic prefix. The preferred definition leaves open the questions of the limits of the duration of the excursion (although our definition of a polarity event requires excursion to be less than 10^4 to 10^5 years), and whether it is due to dipole or non-dipole field activity. We are assuming that the associated paleomagnetic data reflect the ancient geomagnetic field direction and the term cannot therefore be applied to data which may be due to other effects.

As the final part of this report, the Subcommission presents a more legible and corrected version of the chart that appeared last June in *Geotimes*. Without intending any recommendation that the chart be formally adopted, it was presented to demonstrate the Subcommission's vision of the ultimate aim of magnetostratigraphic studies, which is the combining of the magnetic-anomaly numbering system (based on marine magnetic-anomaly analysis), the named polarity epoch and event system (based on K:Ar and paleomagnetic studies of outcropping Pliocene and Pleistocene igneous rocks), and the polarity-epoch numbering system (based on micropaleontological and paleomagnetic analysis of piston cores).

BIBLIOGRAPHY OF
NORMAN D. WATKINS

Deutsch, E. R., and N. D. Watkins. 1961. Direction of the geomagnetic field during the Triassic period in Siberia. *Nature* **189**:543–545.

Watkins, N. D. 1961. The relative contributions of remanent and induced magnetism to the observed magnetic field in northeastern Alberta. *Geophys. Prosp.* **9**:421–426.

Watkins, N. D. 1963. The behaviour of the geomagnetic field during the Miocene period of southeastern Oregon. *Nature* **197**:126–128.

Watkins, N. D. 1964. Structural implications of palaeomagnetism in Miocene lavas of northeastern Oregon, southeastern Washington, and west central Idaho. *Nature* **203**:830–832.

Watkins, N. D. 1965. Paleomagnetism of the Columbia Plateaus. *Jour. Geophys. Research* **70**:1379–1406.

Watkins, N. D. 1965. Frequency of extrusion of some lavas in Oregon during an apparent transition of the polarity of the Miocene geomagnetic field. *Nature* **206**:801–803.

Watkins, N. D. 1965. A palaeomagnetic observation of Miocene geomagnetic secular variation in Oregon. *Nature* **206**:879–882.

Watkins, N. D., and S. E. Haggerty. 1965. Some magnetic properties and the possible petrogenic significance of oxidized zones in an Icelandic olivine basalt. *Nature* **206**:747.

Watkins, N. D., and E. E. Larson. 1966. Combined paleomagnetic and petrological delineation of faulting in southeastern Oregon. *Geol. Mag.* **103**:166–178.

Watkins, N. D., A. Richardson, and R. G. Mason. 1966. Palaeomagnetism of the Macaronesian Insular Region: Maderia. *Earth and Planetary Sci. Letters* **1**:471–475.

Watkins, N. D., A. Richardson, and R. G. Mason. 1966. Palaeomagnetism of the Macaronesian Insular Region: The Canary Islands. *Earth and Planetary Sci. Letters* **1**:225–231.

Watkins, N. D., C. W. Holmes, and S. E. Haggerty. 1967. Primary oxidation variation and the distribution of uranium and thorium in a lava flow. *Science* **155**:579–581.

Wilson, R. L., and N. D. Watkins. 1967. Correlation of magnetic polarity and petrological properties in Columbia Plateau basalts. *Geophys. Jour.* **12**:405–424.

Watkins, N. D. 1967. Comments on a paper by W. H. Taubeneck: "An evaluation of tectonic rotation in the Pacific Northwest." *Jour. Geophys. Research* **72**:1411–1414.

Watkins, N. D., and S. E. Haggerty. 1967. Primary oxidation variation and petrogenesis in a single lava. *Contr. Mineralogy and Petrology* **15**: 251–271.

Watkins, N. D. 1967. Unstable components and paleomagnetic evidence for a geomagnetic polarity transition. *Jour. Geomagnetism and Geoelectricity* **19**:63–76.

Watkins, N. D., and H. G. Goodell. 1967. Confirmation of the reality of the Gilsa geomagnetic polarity event. *Earth and Planetary Sci. Letters* **2**: 123–129.

Watkins, N. D., and H. G. Goodell. 1967. Geomagnetic polarity change and faunal extinction in the Southern Ocean. *Science* **156**:1083–1087.

Dagley, P., R. L. Wilson, J. M. Ade-Hall, G. P. L. Walker, S. E. Haggerty, T. Sigurgeirsson, N. D. Watkins, P. J. Smith, J. Edwards, and R. L. Grasty. 1967. Geomagnetic polarity zones for a sequence of Icelandic lavas. *Nature* **216**:25–29.

Richardson, A., and N. D. Watkins. 1967. Atlantic island paleomagnetism: Fernando Noronha. *Nature* **215**:1470–1473.

Baski, A., D. York, and N. D. Watkins. 1967. The age of the Steens Mountain geomagnetic polarity transition. *Jour. Geophys. Research* **72**:6299–6308.

Goodell, H. G., and N. D. Watkins. 1968. Paleomagnetic stratigraphy of the Southern Ocean: 20° W to 160° E longitude. *Deep-Sea Research* **15**:89–112.

Watkins, N. D., and A. Richardson. 1968. Paleomagnetism of the Lisbon Volcanics. *Geophys. Jour.* **15**:287–304.

Watkins, N. D., and S. E. Haggerty. 1968. Oxidation and polarity variation in Icelandic lavas and dikes. *Geophys. Jour.* **15**:305–316.

Dymond, J., N. D. Watkins, and Y. R. Nayudu. 1968. Age of the Cobb Seamount. *Jour. Geophys. Research* **73**:3977–3979.

Wilson, R. L., and N. D. Watkins. 1968. Reply to comments on 1967 paper by E. Larson and D. W. Strangway. *Geophys. Jour.* **15**:443–447.

Goodell, H. G., N. D. Watkins, T. T. Mather, and S. Koster. 1968. Antarctic glacial history in the sediments of the Southern Ocean. *Palaeogeography, Palaeoclimatology, Palaeoecology* **5**:41–62.

Watkins, N. D. 1968. Comments on the interpretation of linear magnetic anomalies. *Pure and Appl. Geophysics* **69**:179–192.

Wilson, R. L., S. E. Haggerty, and N. D. Watkins. 1968. Variation of paleomagnetic stability and other parameters in a vertical transverse of a single Icelandic lava. *Geophys. Jour.* **16**:179–192.

Watkins, N. D., and A. Richardson. 1968. Comments on the relationship between magnetic anomalies, crustal spreading and continental drift. *Earth and Planetary Sci. Letters* **4**:257–264.

Watkins, N. D., A. Richardson, and R. G. Mason. 1968. Palaeomagnetism of the Macronesian insular region: The Cape Verde Islands. *Geophys. Jour.* **16**:119–140.

Watkins, N. D., and A. Richardson. 1968. Reply to comment on 1968 paper with H. G. Goodell by R. Van derVoo. *Geophys. Jour.* **16**:549–551.

Watkins, N. D. 1968. Short-period geomagnetic polarity events in deep-sea sedimentary cores. *Earth and Planetary Sci. Letters* **4**:341–349.

Gunn, B., and N. D. Watkins. 1969. The petrochemical effect of the simultaneous cooling of adjoining basaltic and rhyolitic magmas. *Geochim. et Cosmochim. Acta* **33**:341–356.

Watkins, N. D. 1969. Non-dipole behavior during a Miocene geomagnetic polarity transition in Oregon. *Geophys. Jour.* **17**:121–149.

Prevot, M., and N. D. Watkins. 1969. Essais de détermination de l'intensité du champ magnétique terrestre au cours d'un reversement de polarité. *Annales Géophysique* **25**:351–369.

Watkins, N. D., and B. Gunn. 1969. Magmatic pulsing and changes in titanium concentration during Miocene volcanism in Oregon. *Nature* **224**:360–362.

Watkins, N. D., B. Gunn, and R. Coy-Yll. 1970. Major trace element variations during the initial cooling of an Icelandic lava. *Am. Jour. Sci.* **268**:24–49.

Krause, D., and N. D. Watkins. 1970. North Atlantic crustal genesis in the vicinity of the Azores. *Geophys. Jour.* **19**:261–283.

Gunn, B., and N. D. Watkins. 1970. Geochemistry of the Steens Mountain basalts. *Geol. Soc. America Bull.* **81**:1497–1516.

Watkins, N. D., and A. Richardson. 1970. Rotation of the Iberian Peninsula. *Science* **167**:209.

Watkins, N. D., T. Paster, and J. Ade-Hall. 1970. Variation of magnetic properties in a single deep-sea pillow basalt. *Earth and Planetary Sci. Letters* **8**:322–328.

Ade-Hall, J., and N. D. Watkins. 1970. Absence of correlation between opaque petrology and natural remanence polarity in Canary Island lavas. *Geophys. Jour.* **19**:351–360.

Gunn, B. M., R. Coy-Yll, N. D. Watkins, C. E. Abranson, and J. Nougier. 1970. Geochemistry of an oceanite-ankaramite-basalt suite from East Island, Crozet Archipelago. *Contr. Mineralogy and Petrology* **28**:319–338.

Kennett, J. P., and N. D. Watkins. 1970. Geomagnetic polarity change, volcanic maxima and faunal extinction in the Southern Ocean. *Nature* **227**:930–934.

Watkins, N. D., and R. Self. 1970. An examination of the "Eltanin" dredged rocks from the Scotia Sea. *Biology of Antarctic Seas IV, Antarctic Research Series 15.* Washington, D.C.: American Geophysical Union, pp. 327–343.

Watkins, N. D., and T. Paster. 1970. Magnetic properties of submarine igneous rocks. *Royal Soc. (London) Proc.* **268**:507–550.

Gunn, B. M., and N. D. Watkins. 1970. Magnetic properties and geochemistry of some rocks dredged from the Macquarie Ridge. *New Zealand Jour. Geology and Geophysics* **14**:153–168.

Watkins, N. D., and A. Richardson. 1970. Intrusives, extrusives and linear magnetic anomalies. *Geophys. Jour.* **23**:1–13.

Watkins, N. D., and A. Abdel-Monem. 1970. Detection of the Gilsa geomagnetic polarity event on the island of Maderia. *Geol. Soc. America Bull.* **82**:191–198.

Watkins, N. D., and F. W. Cambray. 1970. Paleomagnetism of Cretaceous dikes from Jamaica. *Geophys. Jour.* **22**:163–179.

Watkins, N. D., and J. P. Kennett. 1971. Paleomagnetic and associated studies of Eltanin deep-sea sedimentary cores. *Antarctic Jour. U.S.* **5**:183.

Watkins, N. D., B. M. Gunn, A. K. Baksi, D. York, and J. Ade-Hall. 1971. Paleomagnetism, geochemistry, and potassium-argon ages of the Rio Grande de Santiago volcanics, Central Mexico. *Geol. Soc. America Bull.* **82**:1955–1968.

Hart, S. R., B. Gunn, and N. D. Watkins. 1971. Intralava variation of alkali elements in Icelandic basalt. *Am. Jour. Sci.* **270**:315–318.

Kennett, J. P., N. D. Watkins, and P. Vella. 1971. Paleomagnetic chronology of Pliocene-Early Pleistocene climates and the Plio-Pleistocene boundary in New Zealand. *Science* **171**:276–279.

Watkins, N. D., and J. P. Kennett. 1971. Antarctic bottom water: Major change in velocity during the Late Cenozoic between Australia and Antarctica. *Science* **173**:813–818.

Gunn, B. M., C. E. Abranson, J. Nougier, N. D. Watkins, and A. Hajash. 1971. Amsterdam Island: An isolated volcano in the southern Indian Ocean. *Contr. Mineralogy and Petrology* **32**:79–92.

Watkins, N. D., and A. Richardson. 1971. Reply to "Comments on paleomagnetism of the Lisbon volcanics by N. D. Watkins and A. Richardson" by K. Storetvedt. *Geophys. Jour.* **22**:446–448.

Abdel-Monem, A., N. D. Watkins, and P. W. Gast. 1971. Geochronology and paleomagnetism of volcanism on Lanzarote, Fuerteventura, Gran Canaria, and LaGomera (Canary Islands). *Am. Jour. Sci.* **271**:490–521.

Watkins, N. D. 1971. Definition of short geomagnetic polarity events, and the problem of the "reinforcement syndrome." *Comments Earth Sci.: Geophys.* **2**:44–51.

Watkins, N. D., and R. Self. 1972. Examination of the "Eltanin" dredged rocks from high latitudes of the South Pacific Ocean. *Symposium on Antarctic Geology and Solid Earth Geophysics, Oslo, 1970*, edited by R. J. Adie. Oslo: Universitetsforlaget, pp. 61–70.

Gunn, B. M., E. C. Abranson, N. D. Watkins, and J. Nougier. 1972. Petrology and geochemistry of Iles Crozet—A summary. *Symposium on Antarctic Geology and Solid Earth Geophysics, Oslo, 1970*, edited by R. J. Adie. Oslo: Universitetsforlaget, pp. 825–829.

Watkins, N. D., and J. P. Kennett. 1972. Regional sedimentary disconformities and Upper Cenozoic changes in bottom water velocities between Australasia and Antarctica. *Antarctic Oceanology II, Antarctric Research Series 19.* Washington, D.C.: American Geophysical Union, pp. 273–294.

Watkins, N. D. 1972. Hemispherical contrasts in support for the offset dipole hypothesis: The case for an equal coaxial dipole pair as the geomagnetic field source. *Geophys. Jour.* **28**:193–212.

Watkins, N. D. 1972. A review of the development of the geomagnetic polarity time scale and a discussion of prospects for its finer definition. *Geol. Soc. America Bull.* **83**:551–574.

Watkins, N. D., and A. Richardson. 1972. Reply to "Comments on a paper by N. D. Watkins and A. Richardson 'Intrusives, extrusives, and linear magnetic anomalies' by C. G. A. Harrison." *Geophys. Jour.* **28**:191–192.

Abdel-Monem, A., N. D. Watkins, and P. W. Gast. 1972. Potassium-argon ages, volcanic stratigraphy and geomagnetic polarity history of the Canary Islands: Tenerife, La Palma, and Hierro. *Am. Jour. Sci.* **272**:805–825.

Watkins, N. D., A. Hajash, and C. E. Abranson. 1972. Geomagnetic secular variation during the Brunhes Epoch in the Indian and Atlantic Ocean regions. *Geophys. Jour.* **28**:1–25.

Wilson, R. L., N. D. Watkins, T. Einarsson, T. Sigurgeirsson, S. E. Haggerty, P. J. Smith, P. Dagley, and A. G. McCormack. 1972. Paleomagnetism of ten lava sequences from south-western Iceland. *Geophys. Jour.* **29**:459–472.

Watkins, N. D., and J. P. Kennett. 1973. Response of deep-sea sediments to changes in physical oceanography resulting from separation of Australasia and Antarctica. In *Continental Drift, Sea Floor Spreading and Plate Tectonics: Implications for the Earth Sciences,* Vol. 2, edited by D. Tarling and S. K. Runcorn. New York: Academic Press, pp. 787–798.

Watkins, N. D. 1973. Paleomagnetism of the Canary Islands and Madeira. *Geophys. Jour.* **32**:249–267.

Baksi, A. K., and N. D. Watkins. 1973. Volcanic production rates: Comparison of oceanic ridges, islands, and the Columbia Plateaus. *Science* **180**:493–497.

Huang, T.-C., N. D. Watkins, D. M. Shaw, and J. P. Kennett. 1973. Atmospherically transported volcanic dust in South Pacific deep-sea sedimentary cores at distances over 3000 km from the eruptive source. *Earth and Planetary Sci. Letters* **20**:119–124.

Watkins, N. D., and J. Nougier. 1973. Excursions and secular variation of the Brunhes Epoch geomagnetic field in the Indian Ocean region. *Jour. Geophys. Research* **78**:6060–6068.

Watkins, N. D. 1973. Brunhes Epoch geomagnetic secular variation on Reunion Island. *Jour. Geophys. Research* **78**:7763–7768.

Watkins, N. D., J. P. Kennett, and P. Vella. 1973. Paleomagnetism and the *Globorotalia truncatulinoides* datum in the Tasman Sea and Southern Ocean. *Nature Phys. Sci.* **244**:45–47.

McDougall, I., and N. D. Watkins. 1973. Age and duration of the Reunion geomagnetic polarity event. *Earth and Planetary Sci. Letters* **19**:443–452.

Ellwood, B., N. D. Watkins, C. Amerigian, and R. Self. 1973. Brunhes Epoch geomagnetic secular variation on Terceira Island, Central North Atlantic. *Jour. Geophys. Research* **78**:8699–8710.

Watkins, N. D. 1973. Marine geology (review of ten year's research from USNS *Eltanin*). *Antarctic Jour. U.S.* **8**:69–78.

Prevot, M., and N. D. Watkins. 1973. Sue la variation de l'intensité du champ magnétique terrestre pendant une période de transition de la transition de la polarité. *Acad. Sci. Comptes Rendus,* Ser. D, **276**:2431–2433.

Watkins, N. D., S. K. Runcorn, and P. J. Smith. 1973. A comparison between Earth Science University systems in Britain and the United States. *Comments Earth Sci.: Geophys.* **3**:77–81.

Ridley, W., N. D. Watkins, and D. J. MacFarlane. 1974. The oceanic islands: The Azores. In *The Ocean Basins and Margins,* edited by F. Stehli and A. Nairn. New York: Plenum Press, pp. 445–484.

Huang, T.-C., R. H. Fillon, N. D. Watkins, and D. M. Shaw. 1974. Volcanism and siliceous microfaunal diversity in the southwest Pacific during the Pleistocene period. *Deep-Sea Research* **21**:377–384.

Watkins, N. D., and A. K. Baksi. 1974. Magnetostratigraphy and oroclinal folding of the Columbia River, Steens, and Owyhee basalts in Oregon, Washington and Idaho. *Am. Jour. Sci.* **274**:148–189.

Watkins, N. D., B. Gunn. J. Nougier, and A. Baksi. 1974. Kerguelen—Continental fragment or oceanic island? *Geol. Soc. America Bull.* **85**:201–212.

Hedge, C. E., N. D. Watkins, R. A. Hildreth, and W. P. Doering. 1974. Sr[87]/Sr[86] ratios in basalts from islands in the Indian Ocean. *Earth and Planetary Sci. Letters* **21**:29–34.

Kennett, J. P., and N. D. Watkins. 1974. Late Miocene-Early Pliocene paleomagnetic stratigraphy, paleoclimatology, and biostratigraphy in New Zealand. *Geol. Soc. America Bull.* **85**:1385–1398.

Shaw, D. M., N. D. Watkins, and T.-C. Huang. 1974. Atmospherically transported volcanic glass in deep-sea sediments: Theoretical considerations. *Jour. Geophys. Research* **79**:3087–3094.

Watkins, N. D., D. R. Kester, and J. P. Kennett. 1974. Paleomagnetism of the type Pliocene-Pleistocene boundary section at Santa Maria de Catanzaro, Italy, and the problems of post-depositional precipitation of magnetic minerals. *Earth and Planetary Sci. Letters* **24**:113–119.

Watkins, N. D., J. Keany, M. T. Ledbetter, and T.-C. Huang. 1974. Antarctic glacial history derivation by analysis of ice-rafted deposits in marine sediments—A new model and the initial testing. *Science* **186**:533–536.

Watkins, N. D. 1974. Magnetic telechemistry is elegant but nature is complex. *Nature* **25**:497–498.

Huang, T.-C., N. D. Watkins, and D. M. Shaw. 1975. Atmospherically transported volcanic glass in deep-sea sediments: Development of a separation and counting technique. *Deep-Sea Research* **21**:377–383.

Amerigian, C., N. D. Watkins, and B. B. Ellwood. 1975. Brunhes Epoch geomagnetic secular variation on Marion Island: Contribution to evidence for a long-term regional geomagnetic secular variation maxima. *Jour. Geomagnetism and Geoelectricity* **26**:429–442.

Huang, T.-C., N. D. Watkins, and D. M. Shaw. 1975. Atmospherically transported volcanic glass in deep-sea sediments: Volcanism in sub-Antarctic latitudes of the South Pacific during the Upper Pliocene and Pleistocene. *Geol. Soc. America Bull.* **86**:1305–1315.

Gunn, B. M., N. D. Watkins, W. E. Trzcienski, Jr., and J. Nougier. 1975. The Amsterdam-St. Paul volcanic province and the formation of low Al Tholeiitic andesites. *Lithos* **8**:137–149.

Watkins, N. D., I. McDougall, and J. Nougier. 1975. Paleomagnetism and potassium-argon ages of St. Paul Island, southeastern Indian Ocean: Contrasts in geomagnetic secular variation during the Brunhes Epoch. *Earth and Planetary Sci. Letters* **24**:377–384.

Freed, W. K., and N. D. Watkins. 1975. Contribution of volcanic eruptions to magnetism in deep-sea sediments downwind from the Azores. *Science* **188**:1203–1205.

Kennett, J. P., and N. D. Watkins. 1975. Deep-sea erosion and manganese nodule development in the SE Indian Ocean. *Science* **188**:1011–1013.

Watkins, N. D., and A. Richardson. 1975. Analysis of Brunhes Epoch paleomagnetic date in terms of geocentric and offset axial dipole fields: Long-term flattening of the dipole field? *Geophys. Jour.* **43**:501–516. Abridged version also published in *Proceedings of Takesi Nagata Conference*, edited by R. Fisher et al. New York: Goddard Space Flight Center, pp. 145–171.

Watkins, N. D., I. McDougall, and L. Kristjansson. 1975. A detailed paleomagnetic survey of the type location for the Gilsa geomagnetic polarity event. *Earth and Planetary Sci. Letters* **27**:436–444.

Vella, P., B. B. Ellwood, and N. D. Watkins. 1975. Surface water temperature changes in the Southern Ocean southwest of Australia during the last one million years. In *Quaternary Studies: A Selection of Papers Presented at the IX INQUA Congress*, edited by R. P. Suggate and M. M. Cresswell. Wellington, New Zealand: Royal Society of New Zealand, pp. 297–309.

Kennett, J. P., and N. D. Watkins. 1976. Regional deep-sea dynamic processes recorded by Late Cenozoic sediments of the southeastern Indian Ocean. *Geol. Soc. America Bull.* **87**:321–339.

Ellwood, B. B., and N. D. Watkins. 1976. Diagnosis of emplacement mode of basalts in Holes 319A and 321. In *Initial Reports of the Deep Sea Drilling Project*, Vol. XXXIV, edited by R. S. Yeats et al. Washington, D.C.: U.S. Government Printing Office, pp. 495–499.

McDougall, I., N. D. Watkins, G. P. L. Walker, and L. Kristjansson. 1976. Potassium-argon and paleomagnetic analysis of Icelandic lava flows: Limits on the age of "Anomaly 5." *Jour. Geophys. Research* **81**:1505–1512.

Keany, J., M. T. Ledbetter, N. D. Watkins, and T.-C. Huang. 1976. Diachronous deposition of ice-rafted debris in sub-Antarctic deep-sea sediments. *Geol. Soc. America Bull.* **87**:873–882.

Gunn, B. M., and N. D. Watkins. 1976. Geochemistry of the Cape Verde Islands and Fernando de Noronha. *Geol. Soc. America Bull.* **87**:1089–1100.

Ellwood, B. B., and N. D. Watkins. 1976. Comparison of observed intrusive to extrusive ratios in Iceland and the Troodos Massif with results of experimental emplacement mode analysis of DSDP igneous rocks. *Jour. Geophys. Research* **81**:4152–4156.

Huang, T.-C., and N. D. Watkins. 1976. Volcanic dust in deep-sea sediments—Relationship of microfeatures to explosivity estimates. *Science* **193**:576–579.

McDougall, I., N. D. Watkins, and L. Kristjansson. 1976. Geochronology and paleomagnetism of a Miocene-Pliocene lava sequence at Bessastadaa, eastern Iceland. *Am. Jour. Sci.* **276**:1078–1095.

Ellwood, B. B., and N. D. Watkins. 1977. Some magnetic properties of specimens from Holes 332B, 334 and 335, and corresponding analysis in terms of emplacement mode. In *Initial Reports of the Deep Sea Drilling Project*, Vol. XXXVII, edited by F. Aumento et al. Washington, D.C.: U.S. Government Printing Office, pp. 511–514.

Huang, T.-C., and N. D. Watkins. 1977. Contrasts between the Brunhes and Matuyama sedimentary records of bottom water activity in the South Pacific. *Marine Geology* **23**:113–132.

McDougall, I., K. Saemundsson, H. Johannesson, N. D. Watkins, and L. Kristjansson. 1977. Extension of the geomagnetic polarity time scale to 6.5 m.y.: K/Ar dating, geological, and paleomagnetic study of a 3500 meters lava succession in western Iceland. *Geol. Soc. America Bull.* **88**:1–15.

Watkins, N. D., and J. P. Kennett. 1977. Erosion of deep sea sediments in the Southern Ocean between longitudes 70°E and 190°E and contrasts in manganese nodule development. *Marine Geology* **23**:103–111.

Watkins, N. D., I. McDougall, and L. Kristjansson. 1977. Upper Miocene and Pliocene geomagnetic secular variation in the Borgarfjordar area of western Iceland. *Geophys. Jour.* **49**:609–632.

Watkins, N. D., and G. P. L. Walker. 1977. Magnetostratigraphy of eastern Iceland. *Am. Jour. Sci.* **277**:513–584.

Watkins, N. D., and T.-C. Huang. 1977. Tephras in abyssal sediments east of the North Island, New Zealand: Chronology, paleowind velocity, and paleoexplosivity. *New Zealand Jour. Geology and Geophysics* **20**:179–198.

Ellwood, B. B., and N. D. Watkins. 1977. Experimental emplacement mode determination of basalt in Hole 396B. In *Initial Reports of the Deep Sea Drilling Project*, Vol. XLVI, edited by L. Dmitriev et al. Washington, D.C.: U.S. Government Printing Office, pp. 363–367.

Harrison, C. G. A., and N. D. Watkins. 1977. Shallow inclinations of remnant magnetism in Deep Sea Drilling Project igneous cores: Geomagnetic field behavior or post-emplacement effects? *Jour. Geophys. Research* **82**:4869–4877.

Kristjansson, L., and N. D. Watkins. 1977. Magnetic studies of basalt fragments recovered by deep drilling in Iceland, and the "magnetic layer" concept. *Earth and Planetary Sci. Letters* **34**:365–374.

Watkins, N. D., R. S. J. Sparks, H. Sigurdsson, T.-C. Huang, A. Federman, S. Carey, and D. Ninkovich. 1978. Volume and extent of the Minoan tephra from Santorini: New evidence from deep-sea sediment cores. *Nature* **271**:122–126.

Sparks, R. S. J., H. Sigurdsson, and N. D. Watkins. 1978. The Thera eruption and Late Minoan-IB destruction on Crete. *Nature* **271**:122.

Wilson, L., R. S. J. Sparks, T.-C. Huang, and N. D. Watkins. 1978. The control of volcanic column heights by eruption energetics and dynamics. *Jour. Geophys. Research* **83**:1829–1836.

Ledbetter, M. T., and N. D. Watkins. 1978. Separation of primary ice-rafted debris from lag deposits utilizing manganese micro-nodule accumulation rates in abyssal sediments of the Southern Ocean. *Geol. Soc. America Bull.* **89**:1619–1629.

Hooper, P. R., C. R. Knowles, and N. D. Watkins. 1979. Magnetostratigraphy of the Imnaha and Grand Ronde basalts in the southeast part of the Columbia Plateau. *Am. Jour. Sci.* **279**:737–754.

Huang, T.-C., N. D. Watkins, and L. Wilson. 1979. Deep-sea tephra: Azores volcanism during the past 300,000 years, Part I. *Geol. Soc. America Bull.* **90**:131–133; Deep-sea tephra: Azores volcanism during the past 300,000 years, Part II. *Geol. Soc. America Bull.* **90**:235–288.

Federman, A., N. D. Watkins, and H. Sigurdsson. 1979. Volcanic history of the Scotia Arc inferred from studies of volcanic dust in abyssal piston cores downwind from the islands. *Proceedings of the 3rd Antarctic Symposium on Geology and Geophysics*, in press.

Huang, T.-C., M. T. Ledbetter, and N. D. Watkins. 1979. Contrasts in Antarctic bottom water between the Brunhes and Matuyama Epochs in the South Pacific. *Proceedings of the 3rd Antarctic Symposium on Geology and Geophysics*, in press.

Watkins, N. D., M. T. Ledbetter, and T.-C. Huang. 1979. Antarctic glacial history using spatial and temporal variations of ice-rafted debris in abyssal sediments of the Southern Ocean. *Proceedings of the 3rd Antarctic Symposium on Geology and Geophysics*, in press.

Harrison, C. G. A., and N. D. Watkins. 1979. Comparison of the offset dipole and zonal non-dipole geomagnetic field models using Icelandic paleomagnetic data. *Jour. Geophys. Research* **84**:627–635.

Harrison, C. G. A., I. McDougall, and N. D. Watkins. 1979. A geomagnetic field reversal time scale back to 13.0 million years before present. *Earth and Planetary Sci. Letters* **42**:143–152.

AUTHOR CITATION INDEX

Palmer, J. D., 383
Parker, F. L., 113, 122, 167, 168
Parker, R. L., 60, 306
Parry, L. G., 43, 225, 235
Passerini, P., 214
Paster, T. P., 45, 411
Payne, R. R., 347
Pecherski, D. M., 44
Perch-Nielsen, K., 348
Perry, L. G., 383
Pessagno, E. A., Jr., 214, 215
Peter, G., 44
Peterson, D. N., 12, 89, 226, 263, 402
Peterson, M. N. A., 122, 248
Petrushevskaya, M. G., 269
Pevzner, M. A., 44
Phillipi, E., 269
Phillips, J. D., 21, 41, 44, 72, 73, 110, 112, 165, 187, 193, 206, 248, 329
Picard, M. D., 226, 235
Piper, J. D. A., 60
Pirini Radrizzani, C., 215
Pisias, N., 312, 348
Pitman, W. C., III, 11, 12, 20, 43, 44, 60, 111, 112, 167, 177, 189, 214, 215, 226, 360, 383, 389
Pittman, U. J., 383
Polach, H. A., 60
Porthault, B., 215
Postuma, J. A., 215
Powers, J. W., 322
Preikstas, R., 167
Premoli Silva, I., 187, 188, 214, 215, 360
Prevot, M., 411, 413
Price, P. B., 317
Prospero, J. M., 356
Pullaiah, G., 12, 189, 226
Pullar, W. A., 396
Punning, J., 286

Quilty, P. G., 348

Rad, U. von, 75
Radhakrishnamurty, C., 112
Radinov, V. P., 383
Raff, A. D., 215
Raukas, A., 286
Rees, A. I., 75
Reeside, J. B., Jr., 214
Reinhold, T. H., 177
Renard, A. F., 269
Renz, O., 215
Repenning, C. A., 249
Revelle, R. R., 167
Reyment, R. A., 165
Richardson, A., 45, 409, 411, 412, 414

Richmond, G. M., 41, 286
Ridley, W., 413
Riedel, W. R., 113, 122, 168
Rikitake, T., 20, 122, 364
Rimbert, F., 122
Ringwood, A. H., 396
Roberts, W. O., 356
Robertson, J. H., 313
Robinson, E., 189
Roche, A., 44, 60
Rodionov, V. P., 21
Roggenthen, W. M., 188, 214, 360
Rona, E., 396
Rona, P. A., 348
Rosholt, J. N., 168
Rosselot, F. L., 348
Rubin, M., 168
Ruddiman, W. F., 44, 168, 269, 286, 301, 323, 348, 389
Runcorn, S. K., 44, 60, 226, 235, 413
Russell, J. R., III, 235
Rutford, R. H., 134
Rutten, M. G., 44, 134, 168, 193
Ryan, W. B. F., 45, 73, 188, 206, 207, 262, 357, 383, 395

Sackett, W. M., 168
Saemundsson, K., 60, 215, 415
Sagan, C., 21
Sagri, M., 214
Saito, T., 43, 73, 74, 75, 111, 112, 166, 168, 177, 188, 193, 206, 207, 262, 264, 285, 286, 301, 302, 306, 383, 389
Sakai, T., 188, 262
Sancetta, C., 312, 348
Sanfilippo, A., 168
Savage, D. E., 42, 113, 248, 249
Savin, S. M., 6, 301
Schink, D. R., 214
Schlanger, S. O., 395
Schlich, R., 73, 215
Schott, W., 262
Schreiber, L. C., 312, 317
Schneider, E. D., 21
Sciarrillo, J. R., 306
Scientific Staff, DSDP Leg 37, 215
Scientific Staff, DSDP Leg 39, 215
Sclater, J. G., 12, 188, 215, 226
Scott, G. H., 44
Scott, G. S., 44
Seidemann, D. E., 6
Self, R., 411, 412, 413
Selli, R., 166, 168, 193
Senftle, F. E., 317
Serebryanny, L., 286
Sestini, G., 214

SUBJECT INDEX

About the Editor

JAMES P. KENNETT is professor of oceanography at the University of Rhode Island. He has held this position since 1970. From 1965 to 1966 he worked as a scientist at the New Zealand Oceanographic Institute before emigrating to the United States where he was a post-doctoral fellow at the University of Southern California, Los Angeles, until 1968. He held his first teaching position from 1968 to 1970 at Florida State University, Tallahassee, where he began a long collaboration with N. D. Watkins on paleomagnetic stratigraphy of marine sediments, which continued after both workers moved to the University of Rhode Island in 1970.

Dr. Kennett's research interests are largely directed toward a better understanding of the paleoceanographic and biotic evolution of the world's oceans. He has participated in two cruises of the Deep Sea Drilling Project (Legs 21 and 29) and in numerous other cruises. He received his geological education at Victoria University of Wellington, New Zealand. His Ph.D. thesis, completed in 1965, dealt with Late Miocene biostratigraphy and paleoceanography of New Zealand for which he received the McKay Hammer Award (in 1968) from the New Zealand Geological Society.

Dr. Kennett has been a member of the JOIDES Antarctic Advisory Committee from 1970 to 1975, of the IPOD Planning Committee from 1975 to 1978, of the IPOD Paleoenvironmental Panel from 1975 to the present, and of several other international and national earth science commissions. He is also on the editorial board of the journal *Marine Micropaleontology*.

He and his wife, Diana, have two children (Douglas and Mary) born in the United States. In 1978 he became a naturalized U.S. citizen. He is a dedicated long-distance runner having completed several marathons including the Boston Marathon.